BASIC SCIENCE AND THE HUMAN BODY

ANATOMY AND PHYSIOLOGY

BASIC SCIENCE
AND THE HUMAN BODY
ANATOMY AND PHYSIOLOGY

Stewart M. Brooks

Chairman of Science Curriculum,
Newton-Wellesley Hospital School of Nursing,
Newton Lower Falls, Massachusetts

with 386 illustrations

The C. V. Mosby Company

Saint Louis   1975

Copyright © 1975 by The C. V. Mosby Company

All rights reserved. No part of this book may be reproduced in any manner without written permission of the publisher.

Printed in the United States of America

Distributed in Great Britain by Henry Kimpton, London

Library of Congress Cataloging in Publication Data

Brooks, Stewart M
    Basic science and the human body.

    1. Human physiology. 2. Anatomy, Human.
I. Title. [DNLM: 1. Anatomy. 2. Physiology.
QS4 B873b]
QP34.5.B72   612   74-10679
ISBN 0-8016-0809-0

GW/VH/VH   9 8 7 6 5 4 3 2 1

TO
DOREEN AND PATTY

PREFACE

I have tried very hard to make this textbook relevant to the usual anatomy and physiology course offered in the allied health field. What is more, I have tried to accentuate this relevancy within the framework of supportive physics and chemistry, both of which are clearly fundamental to a solid understanding of the body's structure and function. Thus, the first three chapters are intended to strengthen all those that follow, and hopefully in a way more palatable than the practice of taking time out to explain specific gravity, osmosis, oxidation-reduction, and so on.

Further, I have supplied the student with the highlights of appropriate and important diseases to enhance and perfect an understanding of the normal via the abnormal. Teachers of anatomy and physiology have embellished their lectures in this way since time immemorial, so why not put it in print? At the very least a little pathology seems fundamental to the type of course under consideration here. Disease, of course, encompasses infection, and fundamental to infection is microbiology—the subject of Chapter 4. By no means should instructors feel necessarily constrained to formalize (cover in class) the first four chapters. These can be handled as collateral or outside reading. In my own teaching experience, I have done it both ways, depending on circumstances.

Finally, with regard to the questions at the end of each chapter, they are, relative to competing works, numerous and searching. Above all, they are not afterthoughts. Many center on tie-in knowledge and some are intended to stimulate outside reading.

People everywhere have been more than kind in assisting me. In addition to the consulting authorities, I wish to thank all the many scientists, authors, publishers, and manufacturers for the use of illustrative materials. I am especially indebted to my illustrator, Marie E. Litterer, and my wife, Natalie.

Stewart M. Brooks

ACKNOWLEDGEMENT OF CONSULTING AUTHORITIES

The authorities listed below critically reviewed the chapters (in manuscript form) in their respective areas of expertise. Through their involvement, I hope to have eliminated any inaccuracies and to have brought the presentation of all information up to contemporary standards. I gained much from this experience and wish here to express acknowledgement and deep appreciation for their cooperation and valuable criticisms and suggestions. It should be noted that the assistance of these individuals does not necessarily constitute their endorsement of the chapters. Full responsibility for the finished product and for any errors existing therein is mine alone.

PRAPHULLA K. BAJPAI
Department of Biology
University of Dayton

JAMES T. BARRETT
Department of Microbiology
School of Medicine
University of Missouri, Columbia

DAVID S. BRUCE
Department of Biology
Seattle Pacific College

ROBERT H. CATLETT
Department of Zoology
California State University, San Diego

ROGER D. FARLEY
Department of Biology
University of California, Riverside

DONALD E. GROGAN
Department of Biology
University of Missouri, St. Louis

DAVID N. MENTON
Department of Anatomy
Washington University
School of Medicine

ROBERT A. ROUSE
Department of Chemistry
University of Missouri, St. Louis

ANITA M. SANDRETTO
Nutrition Program
School of Public Health
University of Michigan

TULLY H. TURNEY
Department of Biology
Hampden-Sydney College

RONALD L. WILEY
Department of Zoology
Miami University

CONTENTS

BASIC SCIENCE AND THE HUMAN BODY

ANATOMY AND PHYSIOLOGY

CHAPTER 1

MEASUREMENT

The need to measure is a fundamental fact of life. We measure everything. How long is it? How wide is it? How much does it weigh? How much does it hold? Indeed, science and measurement are inseparable. But measurements are of little value if their accuracy and precision are not known. A given measurement is no better than the technique and instrument we use to make it.

The major systems of measurement are the British imperial system, the United States customary system, and the metric (or international) system.* The latter is used throughout most of the civilized world, and almost exclusively in scientific work. It is the only system ever to receive specific legislative sanction by the United States Congress (long ago in 1866!).

The metric system (Table 1-1) is a decimal system, that is, one in which all so-called "derived units" are multiples of ten. The appropriate prefixes, in combination with the basic unit names, provide the multiples and submultiples. For example, kilo- plus the base meter produces the kilometer—or 1,000 meters; milli- plus meter produces the millimeter—or 0.001 meter. One

thousand millimeters equals a meter, 1,000 meters a kilometer, and so forth, just like counting change. Ten pennies to the dime, ten dimes to the dollar, captures the whole idea of the metric system.

LENGTH

The fundamental unit of length in the metric system is the meter (m), equal to 39.37 inches. Originally defined in 1790 as one ten-millionth of the earth's quadrant passing through Paris, in 1960 it was redefined as the length equal to 1,650,763.73 wavelengths of the orange-red radiation of krypton-86 in a vacuum. In scientific work the most commonly encountered derived units of the meter are the millimeter (mm), the centimeter (cm), the micrometer (μm), and the nanometer (nm). Cells, for example, are measured in micrometers, viruses and electromagnetic waves in nanometers. In the realm of atomic physics we encounter the angstrom (Å), a unit of length equal to one tenth of a nanometer.

MASS AND WEIGHT

Mass is quantity of matter. Weight is the product of mass times the acceleration due to gravity. Mass is a constant, whereas weight varies with changes in gravitational force. The

---

*The United States customary system and the British imperial system are commonly lumped together as the "English system."

TABLE 1-1

Metric system

Length	
1 nanometer (nm)	= 0.000000001 meter
1 micrometer (μm)*	= 0.000001 meter
1 millimeter (mm)	= 0.001 meter
1 centimeter (cm)	= 0.01 meter
1 decimeter (dm)	= 0.1 meter
Meter (m)	= 1.0 meter
1 dekameter (dkm)	= 10 meters
1 hectometer (hm)	= 100 meters
1 kilometer (km)	= 1,000 meters

Weight	
1 microgram (μg)	= 0.000001 gram
1 milligram (mg)	= 0.001 gram
1 centigram (cg)	= 0.01 gram
1 decigram (dg)	= 0.1 gram
Gram (g)	= 1.0 gram
1 dekagram (dkg)	= 10 grams
1 hectogram (hg)	= 100 grams
1 kilogram (kg)	= 1,000 grams

Volume	
1 microliter (μl)	= 0.000001 liter
1 milliliter (ml)	= 0.001 liter
1 centiliter (cl)	= 0.01 liter
1 deciliter (dl)	= 0.1 liter
Liter (l)	= 1.0 liter
1 dekaliter (dkl)	= 10 liters
1 hectoliter (hl)	= 100 liters
1 kiloliter (kl)	= 1,000 liters

Metric-English equivalents	
1 meter (m)	= 39.37 inches
1 centimeter (cm)	= 0.3937 inch
1 kilometer (km)	= 0.62 mile
1 kilogram (kg)	= 2.204 pounds
1 liter (l)	= 1.057 quarts (liquid)
1 yard (yd)	= 0.914 meter
1 foot (ft)	= 30.48 centimeters
1 inch (in)	= 2.54 centimeters
1 mile (mi)	= 1.61 kilometers
1 ounce (oz) (avoir.)	= 28.35 grams
1 pound (lb)	= 453.6 grams
1 quart (qt) (liquid)	= 0.956 liter
1 quart (dry)	= 1.101 liters

*Formerly, micron (μ).

astronaut weighs less on the moon than on earth, and nothing at all in space! In common practice, however, mass and weight can be taken to be the same because equal masses have equal weights under identical conditions (such as on earth). In this work we shall generally use the familiar term "weight."

The fundamental metric unit of weight is the kilogram (kg), which is equal to 2.2046 pounds. A cylinder made of platinum and iridium (the international prototype kilogram) serves as the standard kilogram and is housed at the International Bureau of Weights and Measures in Sèvres, France. In scientific work the most commonly encountered derived units of the kilogram are the microgram (μg), the milligram (mg), and the gram (g).*

VOLUME

The metric unit of volume is the *liter* (l), defined to be the volume occupied by 1 kg of water at its maximum density (4° C). Although the milliliter (ml) should equal the cubic centimeter (cc), 1 kg of water (which is equal to 1,000 ml) occupies 1,000.28 cc instead of the expected and intended 1,000 cc, making the milliliter slightly larger than the cubic centimeter. However, for usual measurements the difference may be ignored and the terms used interchangeably.

ENERGY AND WORK

Because no one knows precisely what energy is, we must define it in terms of what it can do. What energy can do is work—and work can be measured. We may therefore define energy as the ability to do work.

Work is done when a force moves through a distance, or arithmetically, work equals force times distance. In engineering practice, units of

---

*Although it is certainly convenient to look upon the gram as the fundamental unit of weight, such is not the case. Perhaps the reader may wish to inquire into the matter.

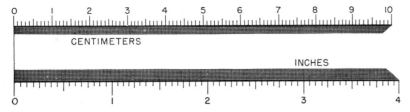

FIG. 1-1
Comparison of inch and centimeter.

work are based upon units of weight, in pounds, and units of distance, in feet. A unit of work, the foot-pound, is the force of an object weighing 1 pound moving through a distance of 1 foot. For example, we do 60 foot-pounds of work in lifting a 20-pound object 3 feet. In the metric system the unit of energy is the joule (J), which is the amount of energy equal to the work done when the point of application of a force of 1 newton (N)* moves through a distance of 1 m in the direction of the force. The unit used for measuring very small amounts of energy is the erg, which equals the work done (or energy expended) when a force of 1 dyne† moves through a distance of 1 cm. This is approximately the amount of muscular energy expended in winking the eye. By way of comparison, 1 joule equals 10 million ergs. It is necessary to have standard units like the joule and the erg in order to relate the various forms of energy and work to one another. For example, 1 foot-pound of mechanical energy or work equals 1.3558 joules and 1 calorie (cal) of heat equals 4.186 joules. When both quantities are expressed in joules, a meaningful comparison can be made. Further, power, which is the rate of doing work or ex-

pending energy, is defined in the international system directly in terms of the joule. One watt, the basic unit of power, is equal to a rate of 1 joule per second.

TEMPERATURE AND CALORIES

Temperature is the intensity of hotness or coldness, and we measure it in degrees on the thermometer. Quantity of heat, on the other hand, must be measured and expressed in a different way. Here we encounter the British thermal unit (BTU) used in engineering and commercial work, the large Calorie of nutrition, and the small calorie of science. We shall talk more about temperature and calories in the next chapter.

QUESTIONS

1. What is your height in centimeters?
2. What is your weight in kilograms?
3. A red cell has a diameter of about 8 μm. How many, then, would it take to span an inch?
4. How many milligrams are there to the ounce?
5. Which contains more, a quart or a liter of blood?
6. The cold sore virus is about 100 nm in diameter. How much larger than this is a red cell? (See question 3.)
7. How many nanometers are there to the angstrom?
8. The adult body averages about 60% water by weight. What, then, is your water content in kilograms and, more appropriately, in liters?
9. At each heart beat about 70 ml of blood is pumped into the circulation. With a heart rate of 70 beats

---

*In physics, force is that which causes a moving body to accelerate. The newton (after Sir Isaac Newton) is a unit of force of such size that under its influence a body whose mass is 1 kg would experience an acceleration of 1 m/sec/sec.
†The dyne is a unit of force of such size that under its influence a body whose mass is 1 g would experience an acceleration of 1 cm/sec/sec.

per minute, how many liters of blood does the heart pump in a minute? How many quarts?

10. A cylindrical vessel is 10 inches high and has a diameter of 1 inch. What is its volume in cubic centimeters? In milliliters?

11. How much work (in foot-pounds) is performed in lifting a 50 kg object 2 feet?

12. In the above question, what is the energy expenditure in joules? In calories?

13. An ounce of breakfast cereal supplies 101 Cal.* How many joules is this?

14. Mass and weight are used interchangeably, as are the milliliter and the cubic centimeter. Discuss.

15. Compare energy, work, force, and power.

————

*One Cal (large calorie) equals 1,000 cal (small calories).

CHAPTER 2

MATTER AND ENERGY

Living things consist of matter and energy. Matter is anything that occupies space, has mass (weight), and under ordinary conditions is neither created nor destroyed. Mass, or the quantity of matter, may be measured in several ways, the most familiar being gravitational attraction or weight. Energy is the ability to do work, and under ordinary conditions is neither created nor destroyed. The various forms of energy, however, such as heat, light, and electricity, may be freely transformed into one another. The toaster, for instance, converts electrical energy into heat energy. Also, energy may be either *kinetic* or *potential,* the former referring to the energy possessed by a moving body, the latter to stored energy or energy of position. Chemical energy is an important type of potential energy. Heat and motion are just two of the forms in which the chemical energy of food is released in the body.

LAWS OF CONSERVATION

As noted, under ordinary conditions matter is neither created nor destroyed (the *law of conservation of mass*). The same applies to energy (the *law of conservation of energy*). Under extraordinary conditions, however, such as an atomic blast, mass can be converted into energy, and thereby destroyed; and under extraordinary conditions, such as encountered in atomic ma-

chines, energy can be converted into mass. These phenomena are in accordance with Einstein's equation $E = mc^2$ (energy equals mass times the speed of light squared). Thus in a strict sense there is but a single law of conservation: The total amount of mass and energy in the universe is constant.

THE MOLECULE

Robert Brown was the first to note that pollen grains and other minute particles suspended in water randomly dance about (Fig. 2-1). He made this observation in 1829, but the explanation did not come until the 1870's. Brownian movement, as the random motion of the particles soon became known, results from the rapid movement of the water molecules and their bombardment of the suspended particles. Brownian movement is encountered in any situation where minute particles of any substance are suspended in a liquid or gas. Perhaps the most familiar example is afforded by the dust particles that can be seen dancing about in the air when a ray of light streams through a darkened room.

Robert Brown's discovery led to the vital concept that all matter is composed of molecules. Although we know today that the molecule is composed of atoms, and the atom of certain

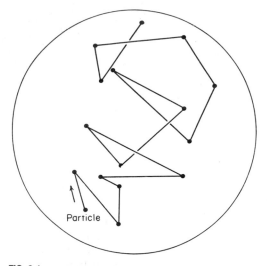

FIG. 2-1
Brownian movement.

subatomic particles, the general physical behavior of matter is usually best understood in a molecular framework. This is more than just convenience, because the fundamental physical characteristics of molecules have been shown to be largely unrelated to their composition. For example, oxygen and nitrogen respond in basically the same way to changes in pressure and temperature.

Molecules vary tremendously in size; the largest have masses millions of times greater than those of the smallest. The largest, the so-called macromolecules, can actually be seen with an electron microscope, whereas the size of the smallest molecules defies easy comprehension. By way of example, an ounce of air contains roughly 100 billion molecules!

STATES OF MATTER

Because molecules are always in motion, they possess kinetic energy. Least motion, or least kinetic energy, exists in the solid state; most motion, or most kinetic energy, exists in the gaseous state. The third state matter can take,

the liquid state, may be thought of as an intermediate state relative to molecular motion. All this is well visualized in heating ice to steam: the sluggish ice molecules pick up speed to become water molecules, and the latter then pick up even more speed to become steam molecules. Thus, the three states are most fundamentally accounted for on a basis of molecular motion and energy content.

GAS LAWS

Provided the temperature of a gas remains constant, the volume it occupies varies inversely with the pressure exerted upon it. This is *Boyle's law*. Provided the pressure remains constant, the volume of a gas varies directly with the absolute temperature.* This is *Charles's law*. Both laws can be demonstrated with an inflated tire. By releasing the pressure in the tire, the air escapes to occupy a larger volume. Similarly, the tire expands when it is hot and contracts when it is cold.

Two other important gas laws are *Dalton's law* and *Henry's law*. Provided no chemical reaction takes place, each gas in a mixture of gases behaves independently of the other gases present. According to Dalton's law the pressure of such a mixture is the sum of the *partial pressures* of the constituent gases. (The partial pressure of a gas is defined as the pressure the gas would exert if it occupied the same volume alone at the same temperature.) The air we breathe follows Dalton's law because it is a mixture of gases—nitrogen, oxygen, carbon dioxide, and traces of a dozen others.

Henry's law deals with the pressure of a gas and its solubility in a liquid, and states that the greater the partial pressure of a gas above a liquid, the greater its solubility in that liquid. Every time we open a bottle of soda pop we

---

*The absolute temperature, given in degrees Kelvin (K°), is equal to the centigrade temperature added algebraically to 273°. For example, −20° C is equal to 253° K.

demonstrate this law. When the cap is removed, the pressure is released, and a lively outrush of carbon dioxide gas ensues. A lively outrush of gas (air) also occurs from the blood of persons who after working in areas of high pressure are suddenly subjected to decompression, a pathological event not uncommon among divers and aviators. The symptoms of decompression sickness mostly arise from the bubbles that plug up key blood vessels.

The relationship of temperature to the solubility of a gas is just the opposite of Henry's law. An increase in temperature decreases the solubility of a gas, and a decrease in temperature increases the solubility. A common example is a glass of cold water that releases bubbles when left in a warm room.

DENSITY AND SPECIFIC GRAVITY

Density and specific gravity are arithmetical expressions always included in the physical characterization of substance, whether it be a solid, liquid, or gas. Density means weight per unit volume and is usually expressed in the metric system in grams per cubic centimeter (g/cc). If 25 cc of a certain liquid weighs 30 g, then it has a density of 30 over 25, or 1.2 g/cc. Because volume varies with the temperature, the latter must always be indicated in precise work. For example, water has a density of 1 g/cc at 4° C.

Specific gravity is the weight of a substance divided by the weight of an equal volume of water, and here it is convenient to remember that 1 cc of water weighs 1 g. In the above example of density, the liquid would have a specific gravity of 30 over 25, or 1.2 (note that no units are involved). This means that this particular liquid is 1.2 times heavier than water. A substance with a specific gravity of 3 is three times heavier than water; a substance with a specific gravity of 0.5 is half as heavy as water; and so on. The specific gravity of liquids is con-

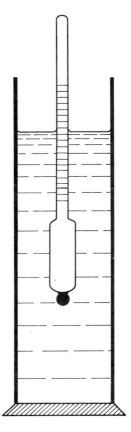

FIG 2-2

Hydrometer. Instrument operates on principle that object floats "high" in liquid of high specific gravity and "low" in liquid of low specific gravity. Specific gravity is read where surface of liquid crosses graduated scale.

veniently measured by means of an instrument called a hydrometer (Fig. 2-2).

SURFACE TENSION AND CAPILLARITY

At all surfaces liquids have a tendency to contract to the smallest possible area. This is shown, for example, by the energy it takes to expand a soap film into a soap bubble, which has a much greater surface area than the flat film. This surface phenomenon, aptly referred to as surface tension, can be explained at the molecular level in simple terms. Whereas the molecules in the interior of a liquid are attracted equally on all sides by the surrounding molecules, those at the surface are pulled inward, thereby tightening and contracting the surface. Agents that have the ability to lower surface tension (sometimes called surfactants) are extensively used as emulsifiers, detergents, and germicides. Surface-active germicides owe a good part of their effectiveness to their ability to spread over the infected area and penetrate bacterial protoplasm. The mucus that lines the passageways of the lungs contains a potent surfactant.

Capillarity is the action by which a liquid rises along a surface that it wets and is depressed by a surface that it does not wet. For example, water rises in a glass tube, but mercury is depressed. The extent of the rise of liquid in a tube varies inversely with the density of the liquid and the caliber of the tube and directly with the surface tension.

SOLUTIONS

A solution is a homogeneous mixture of two or more substances that are present as molecules or atoms. The component that is present in excess, or that undergoes no change of state, is called the solvent; the component present in lesser amounts is the solute. Tincture of iodine is prepared by dissolving iodine, the solute, in alcohol, the solvent. Water, of course, is the "universal solvent." When the solute and solvent are both liquids and mix in all proportions, they are said to be miscible (for example, water and alcohol). Liquids that do not mix (for example, vegetable oil and water) are said to be immiscible.

Solutions have certain properties that are related solely to the concentration of the solute rather than to the chemical nature of the solute. These properties, freezing point, boiling point, and osmotic pressure, are known as colligative properties. Though radically different chemically, when dissolved in water both salt and sugar

lower the freezing point and raise the boiling point.

The ways to express the concentration or strength of a solution are almost without number. The chemist, the physicist, the pharmacist, the physiologist, and the physician each have a preferred system. In medicine we often encounter the expression "milligrams percent" (mg%), or milligrams of solute per 100 ml of solution. The classic means of expressing concentration, and the one everyone should know, is percent composition, which unless otherwise specified means the weight of the solute in grams divided by the volume of the solution in milliliters, times 100. For example, if we dissolve 3 g of salt in enough water to make 15 ml of solution, the concentration is 20%. In the case of liquid solutes, the volume of the solution is often divided into the volume of the solute. But this must be specified. For example, if the label on a bottle of rubbing alcohol reads "70% ethyl alcohol by volume," this means that each 100 ml of this solution contains 70 ml of ethyl alcohol.

DIFFUSION AND OSMOSIS

When a lovely perfumed lady enters the room, our sense of smell realizes it in a matter of seconds. The perfume molecules spread out among the air molecules, eventually permeating the whole room. This process is called diffusion, and common sense tells us that molecules diffuse from an area of greater concentration to an area of lesser concentration. Diffusion is at the heart of molecular life and helps explain a great many biological systems and processes, perhaps most especially osmosis.

When a 5% sugar solution and a 10% sugar solution are separated by a semipermeable membrane (one that is freely permeable to water molecules but not to sugar molecules), more water will diffuse into the 10% sugar solution than into the 5% sugar solution because more water molecules are on the 5% side of the mem-

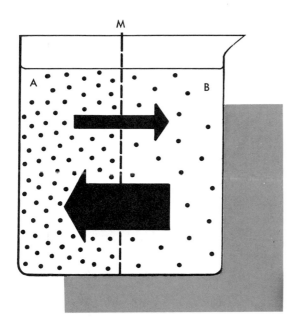

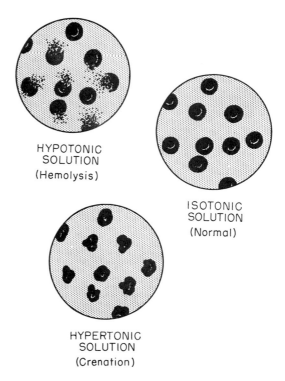

FIG. 2-3

Osmosis. Water diffuses freely through semipermeable membrane (*M*), but chief flow is from less concentrated to more concentrated solution. Black spherules represent solute particles. (From Brooks, S. M.: Basic facts of body water and ions, ed. 3, New York, 1973, Springer Publishing Co., Inc.)

HYPOTONIC SOLUTION (Hemolysis)

ISOTONIC SOLUTION (Normal)

HYPERTONIC SOLUTION (Crenation)

FIG. 2-4

Effect of hypotonic, isotonic, and hypertonic solutions on red cell.

brane than on the 10% side. Indeed, this state of affairs will remain until there are the same number of water molecules on both sides, at which time the rate of diffusion is the same in both directions—a true state of dynamic equilibrium. This process is called osmosis, and may be defined as the passage of water (or other solvent) through a semipermeable membrane separating two solutions, the principal flow being from the less concentrated to the more concentrated (Fig. 2-3). By concentration we mean here the number of particles of solute, in this case sugar, dissolved in a given volume of solution.

Let us now turn to a few quantitative aspects of osmosis. When solvent molecules diffuse into the more concentrated solution, they do so at a par-

ticular pressure, a phenomenon underscored by the fact that pressure must be applied to the more concentrated solution to prevent osmosis. Thus, we speak of a solution's osmotic pressure, or the pressure that must be applied to a solution to prevent osmosis across a semipermeable membrane separating this solution from the pure solvent. A 10% sugar solution has an osmotic pressure twice that of a 5% sugar solution. The central idea is this: the greater the solute particle concentration, the greater the osmotic pressure.

Solutions of the same osmotic pressure are said to be *isotonic* or *isosmotic*. The terms *hypotonic* and *hypertonic* are used to characterize solutions of lesser and greater osmotic pressures, respectively. When red blood cells are added to distilled water, they swell and burst, but when added to a 5% sodium chloride solution, they shrivel and shrink (Fig. 2-4). Thus, relative to red cell protoplasm, water is hypotonic and 5% sodium chloride solution is hypertonic. The only sodium chloride solution in which the red cell feels really at home is one with a 0.9% concentration (isotonic sodium chloride). The concentration of an isotonic dextrose solution is 5%, however, which indicates quite clearly that concentration expressed in percent composition is of limited value in comparing the osmotic pressures of solutions of different solutes. It so happens that 0.9% sodium chloride solution and 5% dextrose solution are isotonic for the fundamental reason that, for a given volume, both contain the same number of solute particles. On a weight-to-weight basis, sodium chloride yields more solute particles than dextrose.

COLLOIDS

A little sugar stirred in a glass of water produces a clear *solution*. However, when sand is added to water it settles to the bottom. A mixture such as sand and water is a *suspension*.

Between these two extremes is a state neither clear nor settling. Vigorously shaking a mixture of water and ultrafine clay produces a cloudy mixture that remains cloudy for a long time. Such a mixture is called a colloid. In this particular case the colloid consists of a solid (clay) dispersed in a liquid (water). In addition to this type of colloidal system, there are seven others—a solid dispersed in a solid (e.g., colored glass), a solid dispersed in a gas (smoke), a liquid dispersed in a solid (cheese), a liquid dispersed in a liquid (milk), a liquid dispersed in a gas (fog), a gas dispersed in a solid (pumice), and a gas dispersed in a liquid (whipped cream). Solids dispersed in liquids are collectively referred to as *sols*, and liquids dispersed in liquids are collectively referred to as *emulsions*. Colloid is

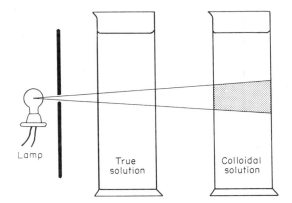

FIG. 2-5
Path of light is visible in colloidal solution (Tyndall effect).

derived from the Greek word *kolla,* meaning glue, and glue is indeed a typical sol.

The colloidal state depends on the size of the dispersed particles, which must be too small to settle out and too large to be unseen (Fig. 2-5). Whereas the particles of a solution are either ions or molecules of fantastically small dimensions, colloidal particles are either macromolecules or aggregates of molecules, and huge in comparision. Still, they are small enough to be bumped around by the molecules of the dispersing medium, which accounts for their pronounced brownian movement and their resulting tendency to remain dispersed. Because they characteristically carry an electric charge (all have the same charge in a given colloidal system), colloidal particles repel one another, an effect that further inhibits settling.

In sum, a colloid is a solid, liquid, or gas dispersed—or subdivided—to such a degree that the size of the particles becomes an important factor in determining its properties. The behavior of a substance depends to a significant degree on its state of subdivision. A small cube of charcoal has little power to adsorb odors, but the same cube ground to colloidal dimensions proves an excellent deodorant. In nature, colloids

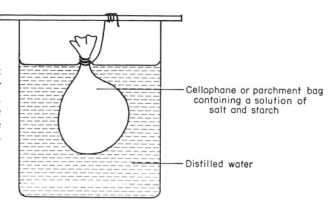

FIG. 2-6

Demonstration of dialysis. Mixture of boiled starch and salt is placed in cellophane bag, which is then immersed in jar of water. Salt, a crystalloid, diffuses through bag, but starch, a colloid, does not. This can be proved by adding a little $AgNO_3$ (to test for salt) and iodine (to test for starch) to the water. (From Brooks, S. M.: Basic facts of general chemistry, Philadelphia, 1956, W. B. Saunders Co.)

Cellophane or parchment bag containing a solution of salt and starch

Distilled water

are the rule, not the exception. We can find no better example of this than protoplasm, which is best described as a polyphasic colloid.

Dialysis

Substances that form true solutions, such as sugar and salt, are called *crystalloids*. The reason they do form true solutions, as noted above, relates to the smallness of their ions and molecules. Their particles are small enough to pass easily through certain selective membranes through which colloidal particles cannot pass because of their size. Thus, a mixture of a crystalloid and a colloid can be easily separated by means of such a membrane (Fig. 2-6), a process called dialysis.

Without question, the most spectacular application of dialysis is the artificial kidney. This device allows the patient's blood to pass through an extensive coil of cellophane tubing wound around a metal cylinder. The entire assembly is immersed in a solution of substances vital to the body (glucose, salt, etc.). Wastes, which are crystalloids, diffuse through the cellophane into the solution, while blood cells, proteins, and other vital colloids remain in the blood. Vital crystalloids also escape into the solution, but this loss is balanced by the diffusion of the same

substances back into the blood. The solution surrounding the cellophane is regulated carefully to ensure that the blood is cleansed safely.

HEAT

Heat is a form of energy and a cardinal manifestation of life. The human body is a heat engine in almost every sense of the word, and even has a thermostat. Whether on an iceberg or on a sand dune, body temperature in a healthy individual hovers incredibly close to 98.6° F.

Heat must be thought of in terms of quality and quantity. Quality is characterized by the *temperature,* which may be defined as the intensity of heat. The temperature of a body is proportional to the average kinetic energy of its constituent molecules; that is, a hot body has a higher temperature than a cold body because the molecules of the former are moving faster and have more kinetic energy than the molecules of the latter. As a cold body is heated up to higher and higher temperatures, the sluggish molecules gain more and more speed.

Temperature is registered on a thermometer in degrees of a particular scale. On the familiar Fahrenheit scale, water freezes at 32° and boils at 212°; on the Celsius, or centigrade, scale, water freezes at 0° and boils at 100°. The Celsius scale is used throughout the world in scientific work, and in many countries for all purposes. In this country the Farhenheit scale is still the most popular at the commercial, household, and clinical level. To convert degrees Celsius to degrees Fahrenheit, simply multiply by 1.8 and add 32 (or $F = 1.8 \times C + 32$). Conversely, to go from Fahrenheit to Celsius, subtract 32 and divide by 1.8 (or $C = [F-32]/1.8$).

We know from experience that it takes longer to boil a quart of water than a cup of water, a good indication that a quart of water holds more heat then a cup of water, even though both register 100° C when boiling. Similarly, a cup of hot water is hotter than a tub of warm water, but the latter holds more heat than the former. Temperature, then, does not tell us the quantity, or amount, of heat. The unit used to measure quantity of heat is the calorie, of which there are two kinds. The small calorie (cal) is the amount of heat required to raise the temperature of 1 g of water 1° C. The large Calorie (Cal), equal to 1,000 calories, and thus sometimes called a kilocalorie, is the amount of heat required to raise the temperature of 1 kg of water 1° C.

Food stores the potential energy necessary for life. We measure this potential with an instrument called the bomb calorimeter (Fig. 2-7). An accurately weighed sample of food is burned in the "bomb," and the rise in temperature of the surrounding water is noted on a centigrade thermometer. The complete combustion of 1 g of a certain breakfast cereal, for example, raises the temperature of 1 kg of water 3.56° C, indicating an energy content of 3.56 Cal/g. Therefore, one cup (1 ounce, or 28.35 g) supplies 101 Cal.

LIGHT

Light is a form of energy that makes its presence known by its effect on the eye. It is an electromagnetic wave that travels at fantastic

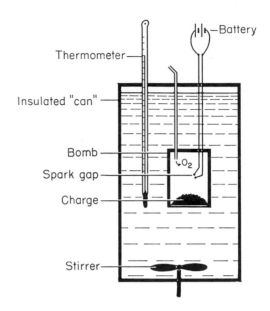

FIG. 2-7
Bomb calorimeter.

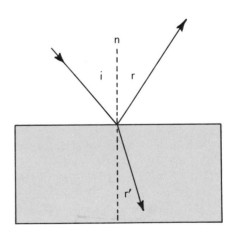

FIG. 2-8
Reflection and refraction. When ray of light strikes smooth surface of glass plate, it is reflected and refracted. Note that angles of incidence, i, and reflection, r, are equal and that angle of refraction, r', is less than angle of incidence. All three angles are measured from a perpendicular to the surface, called the normal, n.

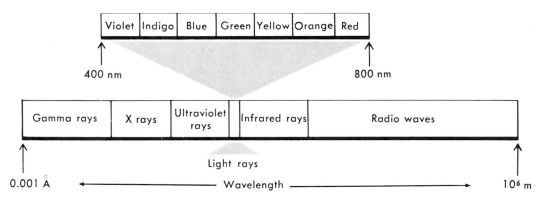

FIG. 2-9

Electromagnetic spectrum. Å, angstrom; m, meter; nm, nanometers.

speeds, passing through a vacuum at 186,000 miles per second, through the atmosphere slightly more slowly, and through denser media at substantially reduced speeds. Through the diamond, light travels at a mere 75,000 miles per second.

When a ray of light strikes a reflecting surface obliquely, it is reflected obliquely in a precise geometric way. The angle of reflection is always equal to the angle of incidence (Fig. 2-8). This relationship is commonly referred to as the *law of reflection*. The opposite of reflection is refraction, the bending of light in passing from one medium into another of different density. When a ray of light passes obliquely from air into glass, it is bent toward the normal (Fig. 2-8); conversely, a ray of light is bent away from the normal in passing from glass into air. In such a situation, the only time light is not bent, or refracted, is when the light ray is perpendicular to the surface of contact between the two media. The cause of refraction relates to the slower speed of light in the denser medium (in the above instance, glass). The rule is that light is bent toward the normal when entering a denser medium, and away from the normal when entering a less dense medium. Were it not for the

refractive power of the lens of the eye, light rays striking the reader's eye would not fall on the retina, with blurred vision the result.

Color can be explained in terms of wavelength. Waves on water, as we well know, have crests and troughs. A wavelength is the distance from one crest to the next crest, or the distance from one trough to the next trough. Every train of such waves has a certain wavelength, and so it is with electromagnetic waves. The visible electromagnetic spectrum (Fig. 2-9) embraces red, with a wavelength of 800 nm, orange (650 nm), yellow (600 nm), green (550 nm), blue (450 nm), and violet (400 nm). An electromagnetic wave train with a wavelength of 800 nm causes certain retinal cells to transmit a certain signal to the brain, creating a sensation arbitrarily labeled red. A wavelength of 650 nm is interpreted as orange, and so on. Black is the absence of light, white the presence of all the colors of the spectrum. When a ray of sunlight passes through a glass prism, it becomes refracted into the entire visible spectrum—red, orange, yellow, green, blue, and violet, in that order.

The color of nonluminous objects is easily explained if we stick to the idea of wavelength. Hemoglobin is red because it absorbs all visual

FIG. 2-10

Originally there were nine 14-month-old mice in each group. One group, **A,** received large but nonlethal dose of radiation as young adults; the other, **B,** did not. Untreated group are healthy; only three irradiated mice survive, and they are senile and gray. (Courtesy Howard J. Curtis, Brookhaven National Laboratory, Upton, N. Y.)

wavelengths except 800 nm (red), which is reflected to the observer. Similarly, red glass is red because it absorbs light of all visual wavelengths except those in the region of 800 nm (red), which pass through the glass and on to the observer. White is white because it reflects all colors. Black is black because it absorbs all colors.

RADIATION BEYOND THE VISUAL

Beyond the visible spectrum in both directions (Fig. 2-9) are electromagnetic waves of vast importance to biology in general and to man in particular. Beyond 800 nm we encounter infrared rays and radio waves; beyond 400 nm, ultraviolet rays, X rays, and gamma rays. Biologically and medically, the most useful (and most damaging!) of these radiations are X rays and gamma rays. As these rays pass through protoplasm, they cause all kinds of chemical havoc, especially in areas of rapidly multiplying cells. This is why radiation is beneficial in the treatment of cancer but detrimetal to the devel-

oping fetus. No less interesting, radiation speeds the aging process (Fig. 2-10).

Not all radiation is electromagnetic, or nonparticulate, in nature. Alpha (α) rays consist of helium nuclei, beta (β) rays of high-speed electrons, and cosmic rays of protons and other kinds of subatomic particles. Cosmic rays are the most powerful and penetrating rays known and represent one of the most important sources of energy in the universe. Most originate somewhere in outer space.

Radiation can be measured and expressed in a number of ways. The principal units are the roentgen (R), the curie (Ci), the rad, and the rem. Simultaneously exposing the whole body to 700 R is fatal; a dose of 100 R is not immediately fatal, but may cause leukemia or cancer. For the general public, the Federal Radiation Council has recommended that the yearly wholebody exposure to radiation not exceed 0.17 R, and that the cumulative 30-year exposure be less than 5 R, exclusive of natural background radiation.

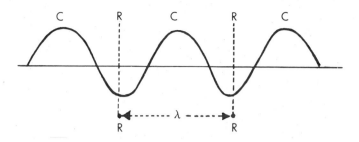

FIG. 2-11

Sound wave. *C*, Crest of wave, representing compression of air molecules; *R*, trough of wave, representing rarefaction of air molecules. Distance between corresponding points is called the wavelength (λ).

SOUND

Sound is a molecular disturbance produced by a substance in vibration. As a metal band vibrates back and forth, the air molecules adjacent to the sides of it are alternately compressed and rarefied. These molecules in turn compress and rarify the air molecules adjacent to them, thereby propagating the disturbance in all directions. Upon reaching the eardrum, the disturbance causes the eardrum to vibrate. This vibration is passed on to the inner ear, where special receptors are stimulated to produce a nerve impulse that the brain interprets as sound.

For convenience, this molecular disturbance may be likened to a train of waves, as in the case of light. A wavelength is taken to be the distance from crest to crest or from trough to trough (Fig. 2-11). Wavelength bears a simple relationship to speed and frequency: wavelength equals the speed of sound divided by the frequency (the number of vibrations, or cycles, per second). If one metal band vibrates 100 times per second and another 200 times per second, the former produces a sound wave twice the length of the latter. Sound travels in air at 1,087 feet per second (at 0° C), and in water at 5,000 feet per second. The difference in speed is readily understandable if we bear in mind that water is much denser than air (i.e., the molecules are much closer together).

A sound or tone is characterized by pitch, loudness, and quality. By *pitch* is meant the response of the brain to frequency. If the tones of two tuning forks of different frequencies are compared, all persons with normal ears will agree that the fork with the higher frequency emits the tone with higher pitch. The human ear is insensitive to frequencies of less than 16, or more than 20,000 cycles per second, but cats and dogs pick up sound far above the latter value. *Loudness* is somewhat self-explanatory in that the harder the drummer beats the drum, the louder is the sound. Quality, or *timbre,* is a characteristic of a tone that enables the ear to distinguish a particular tone from others of the same intensity and pitch. A musical string not only vibrates as a whole, producing a fundamental tone, but also in parts, each of which emits waves that embellish the fundamental. The tones produced by the vibrating parts are called overtones, and it is they that make the distinction between tones of like intensity and pitch possible. Indeed, overtones are one of the important features of what we call music.

ELECTRICITY

Electricity is a form of energy that is said to be either static or dynamic. Static electricity relates to stationary electrical charges, of which there are two kinds, positive and negative. When a glass rod is rubbed with silk and placed in a freely rotating stirrup, it is repelled by another silk-rubbed glass rod. If the experiment is repeated using two ebonite rods that have been rubbed with fur, we note the same result. However, a charged glass rod and a charged ebonite rod attract each other. Benjamin Franklin, who thought that electricity was a "tenuous, invisible, weightless fluid," and that a charged body had either too much or too little of such fluid, dubbed the glass rod positive (because of excess fluid) and the rubber rod negative (because of insufficient fluid). Although we still retain Franklin's designations, today's explanation is that the ebonite rod, or any negatively charged body, has an excess of negative particles (called electrons), whereas the glass rod, or any positively charged body, has a deficiency of electrons.

Dynamic electricity deals with electrons in motion, or electric currents, of which there are two kinds. In a direct current (DC) the electrons flow in one direction; in an alternating current (AC) the electrons flow first one way in the conductor and then the other way. Both DC and AC are produced by special generators, but only DC is produced by batteries and nerve fibers. The most fundamental rule relating to electric currents is *Ohm's law,* which states that the current in a given circuit varies directly with the *electromotive force* (the force pushing the electrons along), and inversely with the *resistance* (to the flow of electrons) encountered in the conductor. The unit of current is called the ampere; the unit of electromotive force, the volt; and the unit of resistance, the ohm. Arithmetically rendered, Ohm's law becomes amperes equal volts divided by ohms.

Highly characteristic of living tissue is the electrical charge, or potential, across the cell membrane. The outside of the membrane is positive and the inside negative, just like the poles of a battery. Stimulating the membrane by mechanical pressure, electric shock, or other means

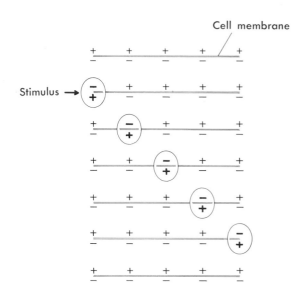

FIG. 2-12

Cell membrane and action potential. Stimulus causes depolarization that moves over membrane as electrical wave, or action potential.

momentarily reverses the potential (depolarization) and thereby sets up an electrical current or wave that spreads out from the point of stimulation. This wave is called the *action potential* (Fig. 2-12). Immediately afterward, the cell membrane returns to its normal resting potential. Muscle and nerve tissues function by this electrical process, and in fact the nerve impulse is an action potential. The heart and brain produce electrical currents of sufficient magnitude to be measured and recorded at the surface of the body by the electrocardiograph and electroencephalograph.

MACHINES AND THE LEVER

A machine is a system, usually of rigid bodies, formed and connected to alter, transmit, and direct applied forces in such a manner as to perform useful work. Simple machines include the inclined plane, the pulley, and the lever.

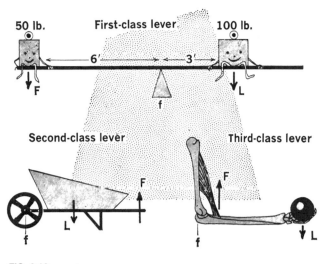

FIG. 2-13
Classes of levers.

Of special interest to us is the lever, which is a rigid bar arranged so as to permit rotation about an axis called the fulcrum. In the operation of this machine a force (F) is applied at one point in order to move an object or load (L) located at some other point (Fig. 2-13). The distance between the force and the fulcrum is called the force arm (Fa); the distance between the load and the fulcrum, the load arm (La). It turns out that the applied force times the force arm equals the load times the load arm. For example, on the seesaw (a first-class lever) a 100-pound boy sitting 3 feet from the fulcrum can be raised and balanced by a 50-pound boy sitting 6 feet from the fulcrum: $100 \times 3 = 50 \times 6$. The lever permits a small force to move a stronger opposing force; it performs useful work.

The seesaw is called a first class lever because the fulcrum is between F and L. In second-class levers, L is between the fulcrum and F, and in third-class levers F is between the fulcrum and L. The skeletal system is a mass of levers, with the joints serving as fulcrums and the muscles supplying the force. In raising the head, for example, the first vertebra, the atlas, is the fulcrum, the muscles on the back of the neck provide the force, and the head is the load.

QUESTIONS

1. $E = mc^2$ disproves the law of conservation of mass. Explain.
2. Robert Brown, in one sense, discovered the molecule. Explain.
3. Some proteins are described as macromolecules. Why?
4. Describe the molecular events that occur when water is transformed from steam to ice.
5. When water freezes, heat is given off; when ice melts, heat is absorbed. Explain.
6. Express Boyle's law as a proportion, letting v equal the old volume, V the new volume, p the old pressure, and P the new pressure.
7. Express Charles's law as a proportion, letting v equal the old volume, V the new volume, t the old temperature, and T the new temperature.
8. Dalton's law may be expressed a $P = p_1 + p_2 + p_3$. . . . What does this mean?

9. During periods of extreme heat, fish die by the thousands in certain lakes and ponds. Why?

10. Osmosis does not occur across a freely permeable membrane. Why not?

11. Intravenous solutions should be as nearly isotonic to blood cells as possible. Why?

12. Why do prunes swell when placed in water?

13. Relative to osmosis, what is meant by dynamic equilibrium?

14. A drop of alcohol spreads out more than a drop of water. Which has the greater surface tension?

15. What physical factors influence the rise of water in the stem of a plant?

16. Ice floats in water. What does this tell us about the density and specific gravity of ice?

17. Urine always has a specific gravity greater than 1. Explain.

18. Glycerine has a density of 1.25 g/cc. What does this mean, precisely?

19. What is the specific gravity of glycerine in the above question?

20. A substance weighs less in water than in air because it is buoyed up by a force equal to the weight of the displaced water. If a certain substance weighs 10 g in air and 8.5 g in water, what is its specific gravity?

21. Do you think the human body has a specific gravity above or below 1? Offer proof for your answer.

22. For every 1,000 feet of altitude, the barometric pressure drops 1 inch. What, therefore, is the pressure at 15,000 feet in inches, centimeters, and millimeters of mercury?

23. Show by arithmetic that 32° F and 212° F are equal to 0° C and 100° C, respectively.

24. What is the normal body temperature on the Celsius scale?

25. How many Calories of heat are required to raise the temperature of 3 kg of water from 16° C to 18° C. How many calories would this be?

26. The complete combustion of 1 g of sugar yields 4 Cal. To what temperature would the complete combustion of 2 g of sugar raise 2 kg of water at 20° C?

27. Is the speed of light greater through air or through water? Why?

28. When you are looking at an object below the surface of a body of water, why does it appear to be where it is not?

29. What is meant by the expression "visual electromagnetic spectrum"?

30. How many nanometers are there in an inch?

31. Why is white paper white?

32. What is the color of a yellow sweater viewed in red light?

33. Why do gamma rays have a greater biological effect than infrared rays?

34. What rays and waves compose the nonvisual electromagnetic spectrum?

35. Some persons cannot distinguish between red and green. Offer an idea or two about the cause of their difficulty.

36. Some of the most severe cases of sunburn at the beach occur on cloudy days. Explain.

37. By consulting another source, explain what *thermography* is.

38. Why is cancerous tissue more vulnerable to radiation than normal tissue?

39. An X ray is essentially a shadow photo. Explain.

40. What are the implications of radiation to heredity?

41. Look up the definitions for rad and rem.

42. What are the origins of the terms roentgen and curie?

43. Does sound pass through a vacuum?

44. By means of a drawing, demonstrate that a sound with a frequency of 4 cycles per second has a wavelength one half that of a sound of 2 cycles per second.

45. Other things being equal, in which medium— air or water—is the wavelength of sound greater?

46. One tuning fork is marked "256," and a second, "260." Which one emits the higher pitch?

47. What does it mean for a body to have a positive charge?

48. Explain how a body with a known positive charge can be used to tell the charge of another body.

49. What happens to the current in a circuit when the resistance increases?

50. Look up "electrocardiography" and "electro-encephalography" and write a brief paragraph about each.

51. When light passes through a colloid, it makes a visible path (the Tyndall effect). Why is this not true of solutions?

52. Proteins and starches do not form true solutions. Why not?

53. What is meant by saying that protoplasm is a "polyphasic colloid"?

54. Explain how you would separate a mixture of potassium chloride and egg white.

55. Soap is an emulsifying agent. How does this explain its cleansing action?

56. The concentration of glucose in the blood is given as "90 mg%." What does this mean?

57. How much sodium chloride is needed to prepare 1 liter of "normal saline" (0.9% solute)?

58. The biological preservative formalin is an aqueous solution of 37% formaldehyde. How would you prepare 1 liter of 10% formaldehyde using this solution?

59. A solution contains 700 ml of methyl alcohol and 1,300 ml of distilled water. What is the concentration in percent alcohol by volume?

60. What force must be applied to balance a weight of 5 pounds, when the force arm is 20 inches and the load arm is 8 inches?

CHAPTER 3

COMPOUNDS AND REACTIONS

Simply put, there are three kinds of matter—elements, compounds, and mixtures. An element (Table 3-1) is a substance that cannot be broken down into simpler material by ordinary chemical means. The earth and everything that inhabits it are constituted by 92 elements in an enormous number of combinations. In a general way, elements are commonly categorized as metals and nonmetals. Metals are marked by opacity, luster, and the ability to conduct heat and electricity. Such elements as iron, tin, lead, copper, silver, and gold are familiar to us as metals. Although many nonmetals (oxygen, nitrogen, iodine, phosphorus, and sulfur, to name a few) are common, we usually do not think of them as a group because of their highly variable natures. Some elements look like metals but behave chemically as nonmetals, and vice versa. Each element is represented by a symbol formed from one or two letters of its English or Latin name: for example, the symbol for nitrogen is N; that for sodium, Na (from *natrium*).

A compound is a substance consisting of two or more elements united chemically, that is, united in such a manner that each loses its identity. The elements of a compound charac-teristically combine in a definite proportion by weight. Thus, when sodium and chlorine come together in Istanbul, they produce the same compound—sodium chloride (table salt)—as when the synthesis occurs in New York. This striking feature of matter is known as the *law of definite proportions*.

Most substances in nature are neither elements nor compounds, but mixtures. A mixture may be defined as two or more elements, or compounds, or elements and compounds not chemically combined and present in any proportion. Any substance that is neither a pure element nor a pure compound is a mixture. Candy, blood, air, soil, milk, and toothpaste are all mixtures.

ATOMIC THEORY

Elements, compounds, and mixtures describe matter in a gross sense. But the structure and behavior of matter at the most fundamental levels are also of importance. If we were to take a piece of matter, say a banana, and by some means break it up into an infinite number of smaller and smaller pieces, what would the ultimate look like? That is, is there a basic unit of matter? Although the Greeks answered the

TABLE 3-1

Common elements

Element	Symbol	Atomic number	Atomic weight
Aluminum	Al	13	26.981
Barium	Ba	56	137.34
Boron	B	5	10.811
Bromine	Br	35	79.909
Calcium	Ca	20	40.08
Carbon	C	6	12
Chlorine	Cl	17	35.5
Chromium	Cr	24	51.996
Cobalt	Co	27	58.933
Copper	Cu	29	63.54
Fluorine	F	9	18.998
Gold	Au	79	196.967
Helium	He	2	4.003
Hydrogen	H	1	1.008
Iodine	I	53	126.9
Iron	Fe	26	55.904
Lead	Pb	82	207.19
Magnesium	Mg	12	24.312
Manganese	Mn	25	54.938
Mercury	Hg	80	200.59
Nickel	Ni	28	58.71
Nitrogen	N	7	14.007
Oxygen	O	8	15.999
Phosphorus	P	15	30.974
Platinum	Pt	78	195.09
Potassium	K	19	39.102
Silicon	Si	14	28.086
Silver	Ag	47	107.870
Sodium	Na	11	22.989
Sulfur	S	16	32.064
Tin	Sn	50	118.69
Uranium	U	92	238.03
Zinc	Zn	30	65.37

1. All matter is composed of very small particles called atoms.
2. All atoms of the same element are alike in size, shape, and weight and differ from the atoms of every other element.
3. Atoms of most elements are able to unite with atoms of other elements.
4. In all chemical changes atoms do not break up but act as individual units.

As we shall see, the atom is not exactly the ultimate particle, but it does serve as a practical unit of matter. An element is made up of like atoms; a compound, of two or more different kinds of atoms in a definite proportion.

MODERN ATOMIC THEORY

After the discovery of the *electron* at the turn of the century, it became clear that Dalton's small particles were composed of still smaller particles. Gradually, the picture developed of electrons (negative particles) coursing about a *nucleus* composed of positive particles called *protons* and neutral particles called *neutrons* (Fig. 3-1). In fact, electrons travel around the nucleus at discrete distances (*energy levels*) from the nucleus. Electrons that are farther from the nucleus have greater potential energy, in much the same sense that an object orbiting around the earth has greater potential energy the farther it is from the earth. Some of this potential energy would be released in one form or another if a given electron were to jump to a lower energy level. An *electron shell* is the average distance from the nucleus to the electrons at a particular energy level. Each electron shell may contain specific regions, or *orbitals*, in which the various electrons travel at that level.

The maximum number of electrons in the first shell is 2, in the second 8, in the third 18, in the fourth 32, in the fifth 32, in the sixth 18, and in the seventh and last, 8. The third, fourth, fifth, and sixth shells cannot have the maximum number of electrons indicated if there are no

question in the affirmative, calling the unit the atom, the question remained outside the realm of science and in the domain of philosophy until 1803 when the English chemist John Dalton proposed a suitable theory. Bearing in mind the law of conservation of mass and the law of definite proportions, Dalton made the following four assumptions:

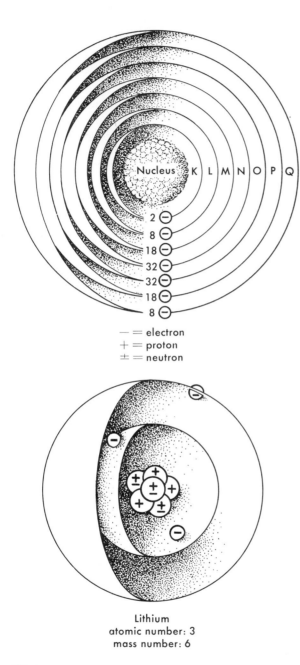

$-$ = electron
$+$ = proton
$\pm$ = neutron

Lithium
atomic number: 3
mass number: 6

FIG. 3-1
Upper model shows maximum number of electrons per shell; lower model depicts atom of lithium.

other shells outside them. In the outermost shell the maximum number of electrons that can be present is eight. In the neutral atom the number of electrons always equals the number of protons.

We now encounter three key terms: *atomic number, mass number,* and *atomic weight.* The atomic number is defined as the number of protons. The mass number is the sum of the number of neutrons and protons. The atomic weight of an element is the weight of its atoms compared to the weight of a carbon atom, which has been designated as 12 atomic units. Since for a single atom the atomic weight is equal to the mass number, we would expect all the elements to have integral atomic weights. However, a glance at Table 3-1 will disclose that most elements do not. This is because an element is usually composed of two or three varieties of atoms with differing mass numbers. Thus, the atomic weight (35.5) of the gas chlorine represents the statistical average of its two kinds of atoms. One kind, with 17 protons and 18 neutrons, and thus a mass number of 35, constitutes 75% of the gas; the other, with 17 protons and 20 neutrons, yielding a mass number of 37, constitutes 25% of the gas. Atoms that are alike except in atomic weight are called *isotopes.* Chlorine has two isotopes: chlorine-35 (^{35}Cl) and chlorine-37 (^{37}Cl).

VALENCE AND FORMULAS

The key to an element's ordinary chemical behavior resides in its atomic number. Atoms strive for maximum stability, which occurs when the outer shell has the maximum number of electrons. Atoms react in order to arrive at this state. When sodium, with one electron in its outer shell, reacts with chlorine, with seven electrons in its outer shell, to form the compound sodium chloride, each atom of sodium donates one electron to each atom of chlorine, thereby yielding an outer shell of eight electrons for both (Fig. 3-2). When magnesium, with two electrons

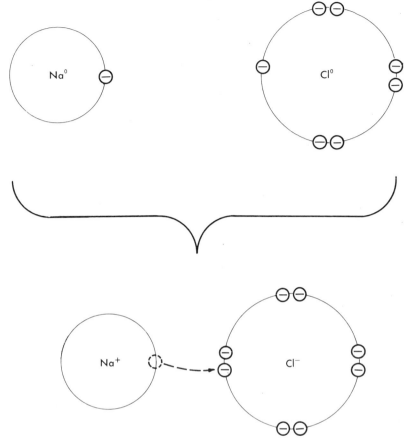

FIG. 3-2

Union of chlorine and sodium to form NaCl. Only outer shell is shown. The symbols Na^0 and Cl^0 indicate the neutral or uncharged atoms; Na^+ and Cl^- are ions held together by their opposite charges.

in its outer shell, combines with chlorine to form magnesium chloride, each atom of magnesium donates two electrons and reacts with two atoms of chlorine, each of which accepts an electron. The simplest possible particle of sodium chloride consists of two atoms, one of sodium and one of chlorine; the simplest possible particle of magnesium chloride consists of three atoms, one of magnesium and two of chlorine. These simplest possible particles are called molecules, the composition of which is expressed by a *chemical formula*. The formula for magnesium chloride is $MgCl_2$, meaning that each molecule of this compound consists of one atom of magnesium* and two atoms of chlorine.

The number of electrons an atom can gain or lose is called its *valence*. Sodium and chlorine have valences of 1, magnesium a valence of 2 (Table 3-2). In writing the formula for a com-

---

*A symbol without a subscript (in this case Mg) is understood to represent one atom.

TABLE 3-2

Valences of common elements and radicals

Electrovalent		
METALS		
Aluminum	3	(Al^{+++})
Barium	2	(Ba^{++})
Bismuth	3,5	$(Bi^{+++}), (Bi^{+++++})$
Calcium	2	(Ca^{++})
Copper	1,2	$(Cu^{+}), (Cu^{++})$
Hydrogen	1	(H^{+})
Iron	2,3	$(Fe^{++}), (Fe^{+++})$
Lead	2	(Pb^{++})
Magnesium	2	(Mg^{++})
Mercury	1,2	$(Hg^{+}), (Hg^{++})$
Nickel	2	(Ni^{++})
Potassium	1	(K^{+})
Silver	1	(Ag^{+})
Sodium	1	(Na^{+})
Tin	2,4	$(Sn^{++}), (Sn^{++++})$
Zinc	2	(Zn^{++})
NONMETALS		
Bromine	1	(Br^{-})
Chlorine	1	(Cl^{-})
Fluorine	1	(F^{-})
Iodine	1	(I^{-})
Oxygen	2	$(O^{=})$
Sulfur	2	$(S^{=})$
RADICALS		
Ammonium	1	(NH_4^{+})
Bicarbonate	1	(HCO_3^{-})
Carbonate	2	$(CO_3^{=})$
Chlorate	1	(ClO_3^{-})
Hydroxyl	1	(OH^{-})
Nitrate	1	(NO_3^{-})
Nitrite	1	(NO_2^{-})
Phosphate	3	$(PO_4^{\equiv})$
Sulfate	2	$(SO_4^{=})$
Sulfite	2	$(SO_3^{=})$
Covalent		
Carbon	4	
Hydrogen	1	
Nitrogen	1,2,3,4,5	
Oxygen	2	
Phosphorus	3,5	
Sulfur	4,6	

pound, one must always balance the valences. The formula for sodium chloride is NaCl because both elements have a valence of 1 (1 = 1); the formula for magnesium chloride is $MgCl_2$ because two chlorines are needed to balance the valence of magnesium ($2 = 2 \times 1$). Sometimes we must use two subscripts. The only way to balance aluminum and oxygen in aluminum oxide is Al_2O_3 (2×3 [valence of Al] $= 3 \times 2$ [valence of O]). The smallest possible subscripts are always used.

IONS AND ELECTROVALENT COMPOUNDS

Instead of saying that sodium has a valence of 1, more specifically we say it has a valence of +1 ("plus one"), and we write the symbol as Na^{+} to show this. In the same way, magnesium has a valence of +2 (Mg^{++}). The "+" stands for positive, and the reason sodium and magnesium become positive is that the loss of electrons leaves the atoms with an excess of protons. In the same fashion, when chlorine accepts an electron, it becomes negative (Cl^{-}) because it has an excess electron per atom. Na^{+}, Mg^{++}, and Cl^{-} are charged atoms called *ions*.

Electrovalent compounds are compounds that are composed of ions. Passing an electric current through an aqueous solution of such a compound demonstrates its ionic composition (Fig.

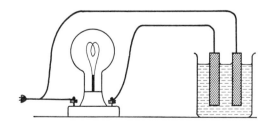

FIG. 3-3

Apparatus used to test ability of solution to conduct electric current. (From Brooks, S. M.: Basic facts of general chemistry, Philadelphia, 1956, W. B. Saunders Co.)

3-3). An electric current cannot pass through pure water, but it readily passes through a salt solution. Clearly, electrical charges must be present to carry the current. These charges are the ions set free when the compound dissolves, a process referred to as *dissociation,* or, less correctly, ionization. Electrovalent compounds are called *electrolytes*.

We can stick our finger into a salt solution, or any other electrolyte solution, without getting a shock because the sum total of the electrical charges carried by the positive ions (called *cations*) is equal to the sum total of the electrical charges carried by the negative ions *(anions)*. The solution is neutral. One trillion molecules of NaCl yield 1 trillion Na^+ ions and 1 trillion Cl^- ions. One trillion molecules of $MgCl_2$ yield 1 trillion Mg^{++} ions (2 trillion positive charges), and 2 trillion Cl^- ions (2 trillion negative charges).

A group of atoms may act like a single atom or ion. Such a group is called a *radical*. Sodium bicarbonate (baking soda), which in solution yields the cation Na^+ and the anion HCO_3^-,

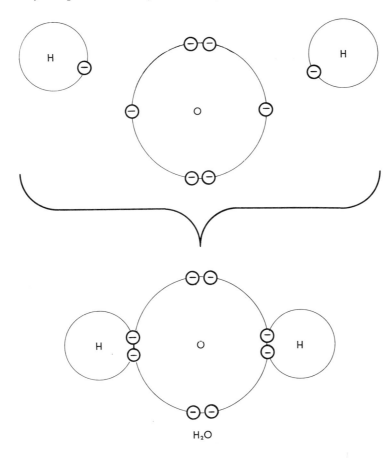

FIG. 3-4

Combining of oxygen and hydrogen to form water. Only outer shell of oxygen is shown. Note that oxygen atom shares two pairs of electrons, giving it a covalence of 2.

has the formula $NaHCO_3$. One Na^+ ion balances one HCO_3^- radical. The formula of magnesium bicarbonate is $Mg(HCO_3)_2$. One Mg^{++} ion balances two HCO_3^- radicals. Note the radical is within parentheses when the formula calls for subscripts.

COVALENCE

The gain and loss of electrons is not the only way atoms obtain the maximum number of electrons in the outer shell. Many compounds exist in which the atoms are bonded together by sharing of electrons. This is called covalent bonding. In water two pairs of electrons are shared (Fig. 3-4). Each oxygen atom shares two pairs, while each hydrogen atom shares one pair. By definition this means that oxygen has a covalence of 2, and hydrogen a covalence of 1, giving water the formula H_2O. Thus, we write the formula for a covalent compound in exactly the same manner as we write the formula for an electrovalent compound. The actual writing of a formula hinges on what might be called the valence number, that is, the number of electrons gained or lost, or the number of pairs of electrons shared.

The major difference between electrovalent and covalent compounds has already been indicated. Electrovalent compounds are electrolytes, whereas covalent compounds, such as water, are nonelectrolytes, because in the typical covalent compound there are no ions. Electrovalent compounds are composed of *ions*, whereas covalent compounds are composed of neutral atoms.

MULTIPLE VALENCE

As shown in Table 3-2, many elements have more than one valence. Copper can have a valence of +1 (Cu^+) or +2 (Cu^{++}), depending on the circumstances. Copper reacts with chlorine under one set of conditions to give $CuCl$, and under another set of conditions to give $CuCl_2$.

The expression "copper chloride" is ambiguous because it does not indicate the valence involved. To eliminate the ambiguity we call $CuCl$ *cuprous* chloride and $CuCl_2$ *cupric* chloride. The general rule is: Use the suffix *-ous* for the lower valence and suffix *-ic* for the higher valence. *Ferrous* sulfate is $FeSO_4$; *ferric* sulfate is $Fe_2(SO_4)_3$.

PERIODIC TABLE

In the latter half of the nineteenth century, the Russian chemist Mendeleeff discovered that when the elements (omitting hydrogen) were arranged in order of their increasing atomic weights, similar chemical properties periodically recurred. He found that if these "periods" were placed under one another to form a table, similar elements fell into the same column or group. Mendeleeff's table, however, proved to have certain discrepancies, which prompted Moseley in England to arrange the elements according to their atomic numbers.

The secret of the table (Fig. 3-5) resides in the number of electrons in an element's outer shell. Elements in the same group are similar in that they have the same valence. Lithium, sodium, and potassium from Group IA have one electron in the outer shell, and therefore a valence of +1. Similarly, elements from Group VIIA have seven electrons in the outer shell and a consequent strong affinity for an extra electron to fill the shell to maximum capacity, giving them a valence of −1. The elements of Group 0 have full outer shells and hence no valence, which explains why these elements are chemically inert under ordinary circumstances. There is a gradual transition from metallic to nonmetallic properties as we pass from the elements on the left hand side of the table to those on the right.

OXIDES, ACIDS, BASES, AND SALTS

The inorganic world is composed of some 50,000 compounds. Fortunately, and most con-

Period	IA	IIA	IIIB	IVB	VB	VIB	VIIB	VIIIB			IB	IIB	IIIA	IVA	VA	VIA	VIIA	0
1	1 H 1.008																	2 He 4.003
2	3 Li 6.939	4 Be 9.012											5 B 10.811	6 C 12.01	7 N 14.007	8 O 15.999	9 F 18.998	10 Ne 20.183
3	11 Na 22.990	12 Mg 24.312											13 Al 26.982	14 Si 28.086	15 P 30.974	16 S 32.064	17 Cl 35.453	18 Ar 39.948
4	19 K 39.102	20 Ca 40.08	21 Sc 44.956	22 Ti 47.90	23 V 50.942	24 Cr 51.996	25 Mn 54.938	26 Fe 55.847	27 Co 58.933	28 Ni 58.71	29 Cu 63.546	30 Zn 65.37	31 Ga 69.72	32 Ge 72.59	33 As 74.922	34 Se 78.96	35 Br 79.904	36 Kr 83.80
5	37 Rb 85.47	38 Sr 87.62	39 Y 88.905	40 Zr 91.22	41 Nb 92.906	42 Mo 95.94	43 Tc (99)	44 Ru 101.07	45 Rh 102.905	46 Pd 106.4	47 Ag 107.868	48 Cd 112.40	49 In 114.82	50 Sn 118.69	51 Sb 121.75	52 Te 127.60	53 I 126.904	54 Xe 131.30
6	55 Cs 132.905	56 Ba 137.34	57 to 71	72 Hf 178.49	73 Ta 180.948	74 W 183.85	75 Re 186.2	76 Os 190.2	77 Ir 192.2	78 Pt 195.09	79 Au 196.967	80 Hg 200.59	81 Tl 204.37	82 Pb 207.19	83 Bi 208.980	84 Po (210)	85 At (210)	86 Rn (222)
7	87 Fr (223)	88 Ra (226)	89 to 103	104 Ku* (257)	105 Ha*													

*104 synthesized by the Russians and named Kurchatovium (Ku);
104 also synthesized at Berkeley and tentatively named rutherfordium (Rf).
105 tentatively named hahnium (Ha).

57 La 138.91	58 Ce 140.12	59 Pr 140.907	60 Nd 144.24	61 Pm (147)	62 Sm 150.35	63 Eu 151.96	64 Gd 157.25	65 Tb 158.924	66 Dy 162.50	67 Ho 164.930	68 Er 167.26	69 Tm 168.934	70 Yb 173.04	71 Lu 174.97
89 Ac (227)	90 Th 232.038	91 Pa (231)	92 U 238.03	93 Np (237)	94 Pu (242)	95 Am (243)	96 Cm (247)	97 Bk (249)	98 Cf (249)	99 Es (254)	100 Fm (253)	101 Md (256)	102 No (254)	103 Lw (257)

FIG. 3-5
Periodic table of elements.

veniently, these fall into four major categories—oxides, acids, bases, and salts.

Oxides

An oxide is a compound of oxygen and one other element. The most abundant compound on earth is the oxide water (H_2O)—hydrogen monoxide. Oxides result from oxidation.* Iron (Fe) becomes oxidized in air to ferric oxide (Fe_2O_3), commonly called rust. This is a slow process, as we know. Rapid oxidation, in contrast, amounts to what we call burning, or combustion. The burning or explosion of hydrogen yields water.

---

*The term oxidation is not always used to mean simply the addition of oxygen, as will be discussed later.

Acids

Acids have the following attributes: they taste sour; they turn litmus from blue to red; and they react with bases to form salts. Acids have these properties in common because they furnish *hydrogen* ions.* The chemical strength of an acid depends on the degree to which it dissociates in solution. Weak acids, such as carbonic (H_2CO_3), furnish few hydrogen ions, while strong acids, such as hydrochloric (HCl), dissociate almost completely, thus furnishing many hydrogen ions. We can indicate this on paper by using arrows of unequal length to show the extent of dissociation:

---

*According to the Brönsted theory, acids are compounds that donate protons, which are, of course, hydrogen ions.

$$HCl \rightleftharpoons H^+ + Cl^-$$
$$H_2CO_3 \rightleftharpoons H^+ + HCO_3^-$$

Binary acids—those composed of hydrogen and one other element—are named by combining the stem of the other element with the prefix *hydro-* and suffix *-ic*. Ternary acids—those composed of hydrogen, oxygen, and a third element—are named, for the most part, by adding the endings *-ous* and *-ic* to the stem of the third element, with the suffix *-ic* indicating the acid with the most oxygen. HBr, HNO_2, and HNO_3, for example, are named hydrobromic acid, nitrous acid, and nitric acid, respectively.

Bases

Bases have the following characteristics: they taste bitter; they turn litmus from red to blue; and they react with acids to produce salts. Bases have these properties in common because they typically contain the OH (hydroxyl) group combined with some metal and in solution furnish OH^- ions.* Weak bases are partially dissociated, while strong bases (or alkalies) are highly dissociated. For example:

$$NaOH \rightleftharpoons Na^+ + OH^-$$
$$NH_4OH \rightleftharpoons NH_4^+ + OH^-$$

The name of a base derives from the metal plus the word hydroxide: e.g., sodium hydroxide, potassium hydroxide, aluminum hydroxide.

Salts

Salts are formed in the reaction between acids and bases, a reaction aptly called neutralization. An acid reacts with a base to yield a salt plus water. Sodium hydroxide reacts with hydrochloric acid to produce the salt sodium chloride plus water:

---

*According to the Brönsted theory, a base is defined as a proton (H^+) acceptor, which the OH^- ion certainly is ($H^+ + OH^- \rightarrow H_2O$). Our definition is restricted to water systems.

$$NaOH + HCl \longrightarrow NaCl + H_2O$$

Note that in this reaction the metal of the base combines with the nonmetal of the acid to form the salt and that the hydrogen of the acid combines with the OH group of the base to form water. A salt, then, may be defined as a compound of a metal and a nonmetal.

Salts are named after the metal and the corresponding acid. Those derived from binary acids are named simply by affixing the suffix *-ide* to the stem; those derived from a ternary acid carry the suffix *-ate* for acids ending in *-ic,* and *-ite* for acids ending in *-ous.* The sodium salts of hydrochloric acid, nitrous acid, and nitric acid are sodium chloride, sodium nitrite, and sodium nitrate, respectively. Acid salts are those in which not all the hydrogen in the acid has been replaced by the metal of the base. A reaction between sodium hydroxide (NaOH) and carbonic acid (H_2CO_3) may yield Na_2CO_3 or $NaHCO_3$, depending on the conditions. Na_2CO_3 is called sodium carbonate, and $NaHCO_3$ sodium bicarbonate, the prefix *bi-* indicating the presence of the unreplaced hydrogen.

pH

The hydrogen ion concentration of a solution is expressed as pH and is a measurement of the solution's acidity or alkalinity. The pH scale (Fig. 3-6) runs from 0 to 14, 0 representing strong acid and 14 representing strong base (or alkali). Chemically pure water has a pH of 7—the midpoint of the scale—because the hydrogen ion concentration equals the hydroxyl ion concentration. Water is neutral, and pH 7 indicates neutrality. A solution with a pH of 6.6 is slightly more acidic than a solution with a pH of 6.8. On the other hand, a solution with a pH of 7.4 is slightly more basic, or alkaline, than a solution with a pH of 7.3. Living things are ultrasensitive to variations in pH; even a minute change can mean trouble, sometimes death.

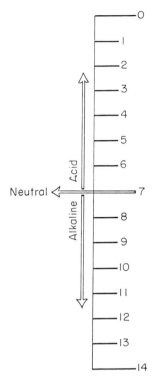

FIG. 3-6

pH, a measure of hydrogen ion concentration, is the negative logarithm of the amount of hydrogen ions per liter of solution. A solution with pH of 7 contains 10^{-7}g of hydrogen ions per liter; a pH of 8 signifies a concentration of 10^{-8} g per liter; and so on.

The most accurate way to measure pH is with a pH meter, an instrument that measures electrically the so-called "*potential of hydrogen ion*" (hence, the expression pH). Much less accurate, but very useful, is "pH paper." This test paper changes color when dipped into a solution. The resultant color is matched against an accompanying precoded color chart to determine the pH of the solution.

ORGANIC CHEMISTRY

Originally, organic chemistry was defined as the chemistry of living things and their products.

In 1828 the German chemist Wöhler startled the scientific world with the discovery that heat converted the inorganic compound ammonium cyanate into urea, one of the chief waste products in urine. This had a tremendous impact on chemistry and biology, for it essentially removed the barrier between the two sciences. Today, we define organic chemistry as the study of carbon compounds. The old definition outlines the realm of biochemistry.

The chemistry of carbon is special because of the ability of the carbon atom to combine with other carbon atoms in theoretically endless chains and groups. Consequently, there are millions of carbon compounds. Whereas an inorganic molecule of a dozen atoms is considered fairly large, an organic molecule commonly consists of hundreds of thousands of atoms. Several different organic compounds may have the same empirical formula, just as several structurally different brick houses could be built from the same kind and number of bricks. Methyl ether and ethyl alcohol, for example, both have the formula C_2H_6O. Such a relationship is called *isomerism,* and the compounds themselves are called *isomers.* An empirical formula is of little value to the organic chemist because his central concern is the way the atoms of a particular molecule join together. For such a purpose he uses a *structural formula.*

Methyl ether and ethyl alcohol have the following structures:

$$
\begin{array}{cc}
\begin{array}{ccccc}
 & H & & H & \\
 & | & & | & \\
H - & C & - O - & C & - H \\
 & | & & | & \\
 & H & & H & \\
\end{array}
&
\begin{array}{ccccc}
 & H & H & & \\
 & | & | & & \\
H - & C & - C & - O - H \\
 & | & | & & \\
 & H & H & & \\
\end{array}
\\
\text{Methyl ether} & \text{Ethyl alcohol}
\end{array}
$$

In such formulas the lines between atoms are called bonds and represent one pair of shared electrons. The number of bonds around an atom must equal its valence. In the above structures, note that each atom of hydrogen, which has a valence of 1, has one bond; each atom of oxygen, which has a valence of 2, has two bonds; and each atom of carbon, which has a valence of 4, has four bonds. With these rules, plus the evidence turned up in the laboratory, the organic chemist can arrive at the structure of a given compound. The "—O—" of methyl ether gives the compound certain characteristic properties, just as the "—O—H" group of ethyl alcohol does to that compound. Finding out such things can prove difficult. It took a dozen years for Frederick Sanger and his associates to piece together the 777 atoms of insulin, a relatively small protein molecule.

Classification of organic compounds

When dealing with a multitude of things, be they organic compounds, plants, or animals, there must be some system of classification lest chaos ensue. Theoretically, all organic compounds may be considered to have derived from hydrocarbons, a class of compounds composed of only two elements, carbon and hydrogen. Hydrocarbons range in complexity from methane (CH_4), the simplest organic compound, to compounds with hundreds of atoms per molecule. Many naturally occur in petroleum and coal tar, and all fall into one of two major categories— those in which the carbon atoms are joined in open chains (*aliphatic* hydrocarbons), and those in which the carbon atoms are joined in closed chains (*cyclic* hydrocarbons). Propane (C_3H_8), an aliphatic hydrocarbon, has the following structure:

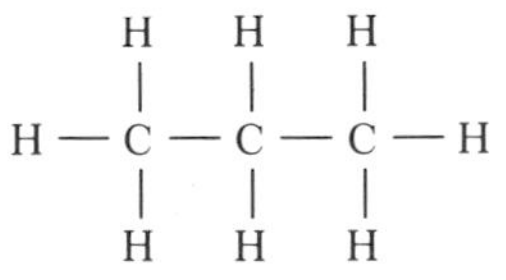

Cyclopropane (C_3H_6), a cyclic hydrocarbon, has the following structure:

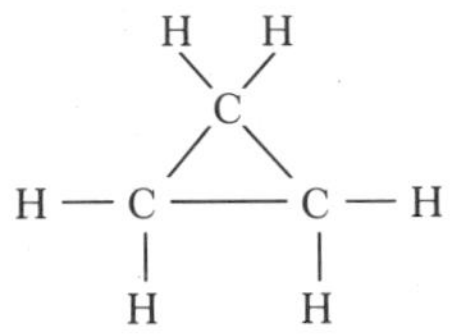

The difference of two atoms of hydrogen per molecule causes the compounds to differ considerably in chemical and physical properties. Whereas propane is physiologically insignificant, cyclopropane is one of the most potent anesthetics known.

One of the more important hydrocarbons is benzene, a cyclic compound with the empirical formula C_6H_6 and the structure:

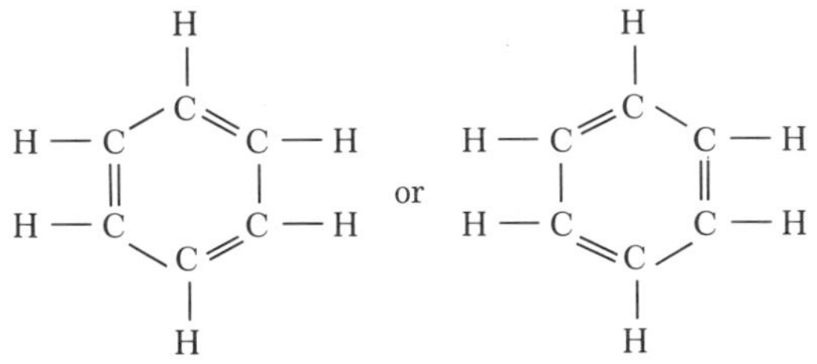

Abbreviated these structures are

A   B   C

The existence of *double bonds* satisfies the valence of carbon (4). Their presence in a hydrocarbon, or any other organic compound, usually gives the compound certain signal features. Indeed, a double bond may spell the difference between a valuable and worthless substance. Compounds with double bonds are said to be *unsaturated* because they are able to take on more hydrogen. Compounds without double bonds are said to be *saturated*. The unsaturated hydrocarbon ethylene (C_2H_4) reacts with hy-

---

*This is a so-called resonance structure. The dotted circle indicates the two possible forms of benzene (i.e., A and B).

drogen to form the saturated hydrocarbon ethane (C_2H_6):

$$
\begin{array}{c}
\text{H}\quad\text{H} \\
|\quad\quad| \\
\text{C}=\text{C}+\text{H}_2 \longrightarrow \text{H}-\text{C}-\text{C}-\text{H} \\
|\quad\quad| \\
\text{H}\quad\text{H}
\end{array}
$$

The various classes of organic compounds may be thought of in terms of a generic formula *RX*, where *R* stands for some hydrocarbon *residue*, and *X* for some characteristic group. The generic formula for alcohols is *R*OH, because the OH, or hydroxyl group, is common to all compounds of this class. Methyl alcohol (CH_3OH) is derived from the hydrocarbon methane (CH_4), ethyl alcohol (C_2H_5OH) from the hydrocarbon ethane (C_2H_6), and so on. Other fundamental organic classes include aldehydes (*R*CHO), acids (*R*COOH), ethers (*R*O*R'*), ketones (*R*CO*R'*), and esters (*R*COO*R'*). Here the designations *R* and *R'* indicate that the hydrocarbon residues need not be the same. Just as there are relationships among acids, bases, and salts, so there are relationships among organic classes (for example, the partial oxidation of an alcohol yields an aldehyde; the oxidation of an aldehyde yields an acid; the reaction of an acid with an alcohol yields an ester).

Carbohydrates

Carbohydrates are compounds of carbon, hydrogen, and oxygen in which the ratio of hydrogen to oxygen is characteristically two to one. Specifically, they are polyhydroxy aldehydes or polyhydroxy ketones. Carbohydrates include starches, sugars, cellulose, and closely related substances.

Sugars are crystalline, soluble in water, and sweet. Simple sugars contain five (pentoses) or six (hexoses) carbon atoms per molecule, double sugars 12. Simple and double sugars are referred to formally as *mono*saccharides and *disac*-charides, respectively. Of biological significance

are the monosaccharides glucose, fructose, and galactose (isomers with the empirical formula $C_6H_{12}O_6$) and the disaccharides sucrose, maltose, and lactose (isomers with the empirical formula $C_{12}H_{22}O_{11}$). Heated water to which a drop or two of acid has been added will split disaccharides into monosaccharides. One molecule of sucrose, for example, yields one molecule of glucose and one molecule of fructose:

$$C_{12}H_{22}O_{11} + H_2O \longrightarrow C_6H_{12}O_6 + C_6H_{12}O_6$$

The disaccharide molecule may be considered two monosaccharide molecules joined together by the splitting out of water.

Starches are *poly*saccharides. Their molecules are made up of hundreds of glucose units joined together as follows:

Their generic formula is $(C_6H_{10}O_5)_x$, where *x* stands for the number of glucose units (usually about 1,000) present in a particular starch. Polysaccharides of exceptional importance include dextrins, pectins, glycogen ("animal starch"), and cellulose (the chief constituent of the plant cell wall). All but the pectins, which are distributed widely in fruits and berries, are polymers of glucose with the generic formula given above.

Lipids

Lipids are fats and fat-like substances. They are characterized by insolubility in water and solubility in ether, chloroform, benzene, and certain other organic solvents. For the most part, they fall into three categories: simple lipids, compound lipids, and sterols.

Simple lipids are esters of *fatty acids* and constitute what we commonly call fats, oils, and waxes. A typical fat is an ester of a saturated fatty acid and the alcohol glycerol; a typical oil is an ester of an unsaturated fatty acid and glycerol. A few fats are fatty acid esters of the alcohol cholesterol. One of the chief fats in human flesh is glyceryl tripalmitate, formed from glycerol and palmitic acid according to the following equation:

$$C_3H_5(OH)_3 + 3C_{15}H_{31}COOH \rightarrow C_3H_5(C_{15}H_{31}CO_2)_3 + 3H_2O$$

Glycerol   Palmitic acid   Glyceryl   Water
tripalmitate

The equation reads: one molecule of glycerol plus three molecules of palmitic acid yields one molecule of glyceryl tripalmitate ·plus three molecules of water. Glycerol and palmitic acid are coupled by the splitting out of water, three

H's coming from the fatty acid and three OH's from the glycerol ($3H + 3OH \rightarrow 3H_2O$).

Compound lipids are substances that can be degraded into some products besides alcohols and fatty acids. Those of particular biological importance include phospholipids, which are compounds of an alcohol, a fatty acid, phosphoric acid, and a nitrogenous base (such as choline), and glycolipids, which are compounds of sphingosine (an amino alcohol), a sugar (either glucose or galactose), and a fatty acid. Glycolipids are present in most tissues of the body; phospholipids are present in all tissues and most particularly those of the nervous system.

Sterols are high molecular weight monohydroxy alcohols derived from a cyclic hydrocarbon called cyclopentanoperhydrophenanthrene (pronounced cyclo-pentano-perhydro-phenanthrene). The formula is written as follows, with the rings numbered to aid in naming derivatives:

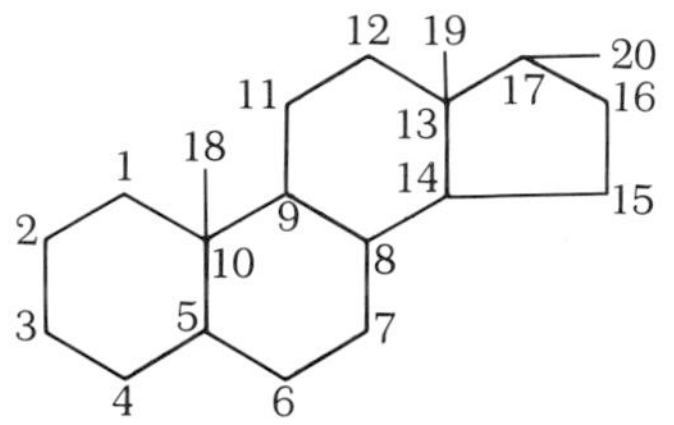

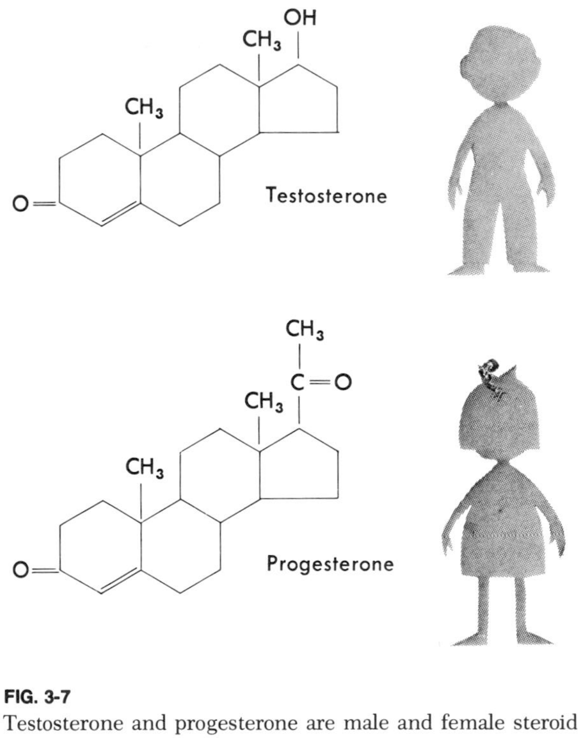

FIG. 3-7
Testosterone and progesterone are male and female steroid hormones, respectively.

The most famous sterol is cholesterol, a compound commonly known today as the alleged culprit in atherosclerosis and coronary heart disease. Another sterol, 7-dehydrocholesterol, is found in the skin and forms vitamin D when exposed to ultraviolet light. Sex hormones (Fig. 3-7), adrenocortical hormones, and bile salts closely resemble sterols in basic molecular structure and are therefore commonly called *steroids*.

Proteins

Proteins, the most complex substances known, form the organic backbone of protoplasm. For the most part, they are huge macromolecules, some of which have molecular weights* in excess of 1 billion! They are giant *polymers* formed from *amino acids,* that is, amino acids joined together in all kinds of twisted chains. Although there are only about 20 amino acids, the number of possible combinations of them, and consequently the number of proteins, is enormous.

Consider the building of a hypothetical protein from the amino acid glycine. Like all amino acids, glycine has an amino group (NH_2) and an acid, or carboxyl, group (COOH). Under appropriate conditions, the OH from the COOH group of one molecule combines with an H from the NH_2 group of another molecule to form water. A bond then forms between the C of the COOH group and the N of the NH_2 group from which the OH and H have split off, yielding glycylglycine:

$$NH_2 \qquad\qquad H - \!\!\! - N - H$$
$$|\qquad\qquad\qquad\qquad |$$
$$CH_2 - CO\,|OH\ +\quad|\quad CH_2 - COOH \longrightarrow$$

$$NH_2 \quad O \quad H$$
$$|\qquad \parallel \quad |$$
$$CH_2 - C - N - CH_2 - COOH + H_2O$$
Glycylglycine

Glycylglycine combines with another molecule of glycine to form diglycylglycine, which in turn combines with another molecule of glycine to form triglycylglycine, and on and on the chain grows to give us a protein.

The angles of the bonds between the amino acid residues in such a chain cause the chain to coil in a spiral shape known as a helix. This helical conformation is stabilized by the existence of *hydrogen bonds* between the residues.

---

*The *molecular weight* of a substance is the sum of the atomic weights of its constituent atoms as indicated by the molecular formula; for example, the molecular weight of sulfuric acid (H_2SO_4) is 98.

Hydrogen bonds result from the electronegativity of oxygen and nitrogen. When an oxygen or a nitrogen atom forms a bond with another atom, the shared electrons are closer to the oxygen or nitrogen atom, resulting in a partial polarization of the bond. A partial negative charge (δ^-) arises near the oxygen or nitrogen atom, and a partial positive charge (δ^+), near the other atom. When a partial negative charge and a partial positive charge come near each other, there is a weak attraction, called a hydrogen bond. Although individual hydrogen bonds are weak, the effect of a large number of them can be significant, as the protein molecule demonstrates. The partial polarization of the N—H bond in one residue and of the C=O bond in another will result in a hydrogen bond between the residues if they are close together:

$$\overset{\delta^+}{\underset{|}{C}}=\overset{\delta^-}{O}\ \cdots\cdots\ \overset{\delta^+}{H} - \overset{\delta^-}{\underset{|}{N}} -$$

Each residue in the chain is linked to two others in such a manner, thereby stabilizing the helical shape of the molecule.

The protein helix in turn is folded into a specific three-dimensional configuration, which brings certain amino acid residues close together and buries others, thus giving an individual protein a unique chemical behavior. This accounts for the specificity of the proteins known as enzymes.

A protein may be simple or conjugated. Simple proteins are those which, upon chemical degradation, yield only amino acids. Albumins, globulins, histones, and protamines are all simple proteins. Conjugated proteins upon degradation yield some substances besides amino acids. Important examples are nucleoproteins, glycoproteins, phosphoproteins, lipoproteins, and chromoproteins. The chromoprotein hemoglobin, for instance, yields an iron-containing substance called heme.

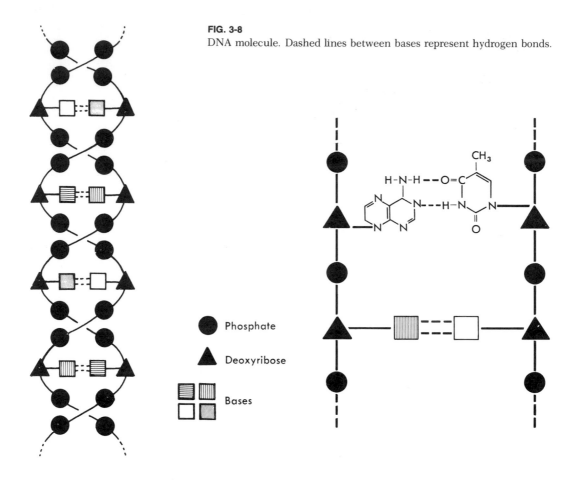

FIG. 3-8

DNA molecule. Dashed lines between bases represent hydrogen bonds.

Phosphate

Deoxyribose

Bases

Nucleic acids

Nucleoproteins, so called because of their close ties with the nucleus of the cell, yield amino acids and nucleic acids upon chemical degradation. There are two kinds of nucleic acid, deoxyribonucleic acid (DNA) and ribonucleic acid (RNA). The building blocks of DNA are phosphoric acid, deoxyribose, adenine, guanine, cytosine, and thymine; the building blocks of RNA are phosphoric acid, ribose, adenine, guanine, cytosine, and uracil. Ribose and deoxyribose are pentose sugars; adenine and guanine are purine bases (so called because they are basic and derived from purine); and cytosine, thymine, and uracil are pyrimidine bases (so called because they are basic and derived from pyrimidine). When ribose or deoxyribose is linked to one of these bases, the resulting molecule is called a nucleo*side;* when a nucleoside is linked to phosphoric acid, the resulting molecule is called a nucleo*tide.* Thus, adenine plus ribose yields the nucleoside adenosine, which in turn may react with phosphoric acid to give the nucleotide adenylic acid (also called adenosine monophosphate, or AMP). The nucleotide adenosine triphosphate (ATP) serves as the key chemical in the storage of cellular energy.

The DNA molecule is made up of two *polynucleotide* chains twisted upon each other to form the famed double helix (Fig. 3-8). These chains are held together by hydrogen bonds acting between the purine and pyrimidine bases, adenine always bonding with thymine, and guanine always bonding with cytosine. This particular bonding, together with the sequence of the bases, explains genetic action and much about the mystery of life (Chapter 6).

RADIOACTIVITY

Certain materials are said to be radioactive. This radioactivity originates from atoms whose nuclei are unstable and spontaneously decay. This decay process involves the emission of radiation and a change in the number of protons or neutrons or both in the unstable nuclei. The result of radioactive decay is a new atom that will ordinarily belong to a different element; it will contain a different number of protons than the original unstable nucleus. (Remember that each element has its own atomic number, that is, number of protons in the nucleus.) The decay process proceeds at a constant and well-defined rate for each element. This rate, called the half-life of the radioactive nucleus, is the time required for one half of a sample of radioactive atoms to spontaneously decay. It varies greatly among the elements. Polonium has a half-life of less than a millionth of a second. Carbon-14 has a half-life of about 5,600 years, and one of the isotopes of uranium has a half-life of almost 5 billion years.

The radiations given off in the decay process are of three kinds (Fig. 3-9)—alpha rays (α), beta rays (β), and gamma rays (γ). Alpha and beta rays are particulate. An alpha particle consists of two protons and two neutrons bound together. Beta rays are streams of electrons that result from the breakdown of neutrons (a proton

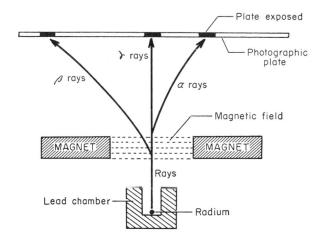

FIG. 3-9

Effect of magnetic field upon rays emitted by radioactive substance. Magnetic field is perpendicular to and directed into plane of paper.

and electron bound together). Gamma rays are electromagnetic in nature (p. 14) and among the most powerful and penetrating rays known.

Radioactive isotopes

Whereas uranium and radium are naturally occurring elements, most radioactive materials in use today are radioactive isotopes derived artifically from a stable element via atomic reactors and associated devices. Bombarding regular sodium chloride (NaCl) with neutrons, for instance, converts stable sodium with a mass number of 23 into its radioactive isotope with a mass number of 24.

Radioactive isotopes and compounds formed from them are employed in many ways. In medicine, they are used in the treatment of cancer and in the diagnosis of various diseases; in biological research, they are used to tag compounds in order to trace metabolic pathways within the tissues of plants and animals via the emission of tell-tale rays. Carbon dioxide prepared from carbon-14, for example, enables the researcher to follow the chemistry of photosynthesis. The tracer technique has opened many doors heretofore closed to conventional chemistry.

THE REACTION

We say a chemical reaction occurs when a substance undergoes decomposition or, more particularly, when two or more substances react with one another to produce one or more new substances. The souring of milk, the rusting of a nail, the burning of a piece of paper, and the digestion of a hamburger are all chemical reactions. Some reactions generate light (e.g., the firefly), some generate electricity (the battery), and most generate heat. Such energy may be released slowly, as in the oxidation of sugar in the body, or rapidly, as in the explosion of gunpowder. Thermodynamically, reactions are either feasible (*exothermic*) or unfeasible (*endothermic*); kinetically, they are either slow or fast. In a hypothetical reaction of A and B to form B and C, thermodynamics tells us whether we will end up with more A and B or more C and D, whereas kinetics relates to the nature of and the ease by which A and B are converted to C and D, or vice versa.

The equation

What happens in the test tube is described on paper by an equation. To write an equation, we simply place the formulas for the *reactants* to the left of an arrow and the formulas for the *products* to the right. A formula is preceded by an appropriate number (*coefficient*), if needed, in order to have the same number of atoms of each element on both sides of the arrow. The equation must balance.

Sodium chloride and silver nitrate react to form sodium nitrate and silver chloride according to the following equation:

$$NaCl + AgNO_3 \longrightarrow NaNO_3 + AgCl$$

Here, one molecule of sodium chloride plus one molecule of silver nitrate yields one molecule of sodium nitrate plus one molecule of silver chloride. This equation requires no tampering with because it is already balanced.

Let us consider the reaction between calcium chloride and silver nitrate to yield calcium nitrate and silver chloride. Before being balanced, the equation is as follows:

$$CaCl_2 + AgNO_3 \longrightarrow Ca(NO_3)_2 + AgCl$$

Here the chlorides and nitrates do not balance. We must find the lowest possible coefficients to correct the situation:

$$CaCl_2 + 2AgNO_3 \longrightarrow Ca(NO_3)_2 + 2AgCl$$

As a check, we have one calcium on the left and one on the right; two chlorides on the left and two on the right; two silvers on the left and two on the right; and two nitrates on the left, and two on the right. The balanced equation reads: one molecule of calcium chloride plus two molecules of silver nitrate yields one molecule of calcium nitrate plus two molecules of silver chloride.

Types of reactions

All reactions fall into two basic categories, oxidation-reduction and Lewis acid-base. Oxidation-reduction reactions involve a loss (*oxidation*) and gain (*reduction*) of electrons. There can be no gain without a corresponding loss because electrons do not float around free—an electron gained must come from someplace and vice versa. For example, in the reaction between zinc and copper sulfate to form zinc sulfate and copper, the electrons lost by the copper are gained by the zinc. The equation is:

$$Zn + CuSO_4 \longrightarrow ZnSO_4 + Cu$$
$$\text{or}$$
$$Zn + Cu^{++} + SO_4^{=} \longrightarrow Zn^{++} + SO_4^{=} + Cu$$

Note that zinc is oxidized and copper reduced.

Lewis acid-base reactions involve the formation of a covalent bond. A Lewis acid is any species that accepts a pair of electrons in the formation of the bond, a Lewis base any species that donates the electrons. In the reaction between NaCl and $AgNO_3$, a covalent bond is formed between silver and chloride, producing the precipitate AgCl. The equation is:

$$AgNO_3 + NaCl \longrightarrow NaNO_3 + AgCl$$
$$\text{or}$$
$$Ag^+ + NO_3^- + Na^+ + Cl^- \longrightarrow Na^+ + NO_3^- + Ag\!:\!\ddot{\underset{..}{C}l}\!:\ ^*$$

Reversible reactions

In many instances, it is likely that the product or products of a reaction will turn around and yield the original reactants; that is, the reaction is reversible. To express such a reaction, we use double arrows, as shown in the following equation for the reaction between phosphorus trichloride and chlorine:

$$PCl_3 + Cl_2 \rightleftarrows PCl_5$$

The equation actually describes two reactions going on simultaneously and in opposite directions—a situation aptly labeled *"dynamic equilibrium."* Many reactions that take place within the cell are reversible. The liver, for example, converts glucose to glycogen and vice versa.

Rate of reaction

Reactions vary in rate, or speed. The rusting of a nail and the explosion of gunpowder give an indication of the extremes. A given rate is usually altered by changing the temperature or the concentration or by adding a catalyst. A rise of 10° C just about doubles the reaction rate. Consider the increase in heart rate during a fever, which indicates a speed-up of the chemistry in that organ. An increase in concentration increases the rate of reaction. If A reacts with B to form C, we can speed up the reaction by

---

* Dots indicate electrons in the outer shell; the two between Ag and Cl serve as the covalent bond.

increasing the concentration of either A or B, or by removing C. This is called the *law of mass action*. A *catalytic* agent alters the speed of a reaction without itself undergoing a change. The production of oxygen from potassium chlorate ($KClO_3$) is an excellent example. Without a little manganese dioxide (MnO_2), extreme heat is needed for the reaction to occur; with MnO_2 present, oxygen is released rapidly at moderate temperatures:

$$2KClO_3 \xrightarrow[\Delta]{MnO_2} 2KCl + 3O_2 \uparrow$$

The delta symbol (Δ) indicates heat (energy). The writing of the catalyst above the arrow is standard procedure. Note that oxygen is O_2, not O, because it, like many gases, exists in the diatomic condition.

ENZYMES

The hundreds of reactions that take place within living cells involve a multitude of protein catalysts called enzymes. The application of hydrogen peroxide (H_2O_2) to an open wound creates a lively effervescence, a simple reminder of the omnipresence of enzymes in living tissue. In this case, the enzyme catalase speeds up the decomposition of H_2O_2 into water and oxygen, normally a sluggish reaction. The role of enzymes in the cell is the catalytic stimulation of chemical reactions. Enzymes lower the *activation energy* of a reaction; that is, they reduce the amount of energy needed to bring about the reaction. They allow reactions that otherwise would occur much too slowly to sustain life to take place at the modest temperatures of living cells. In the example given, catalase reduces the activation energy by a factor of approximately ten.

Hypothetically, an enzyme (E) first combines with the substance acted upon, called the substrate (S), to form a complex (ES). The complex then dissociates into the product(s) (P) and the

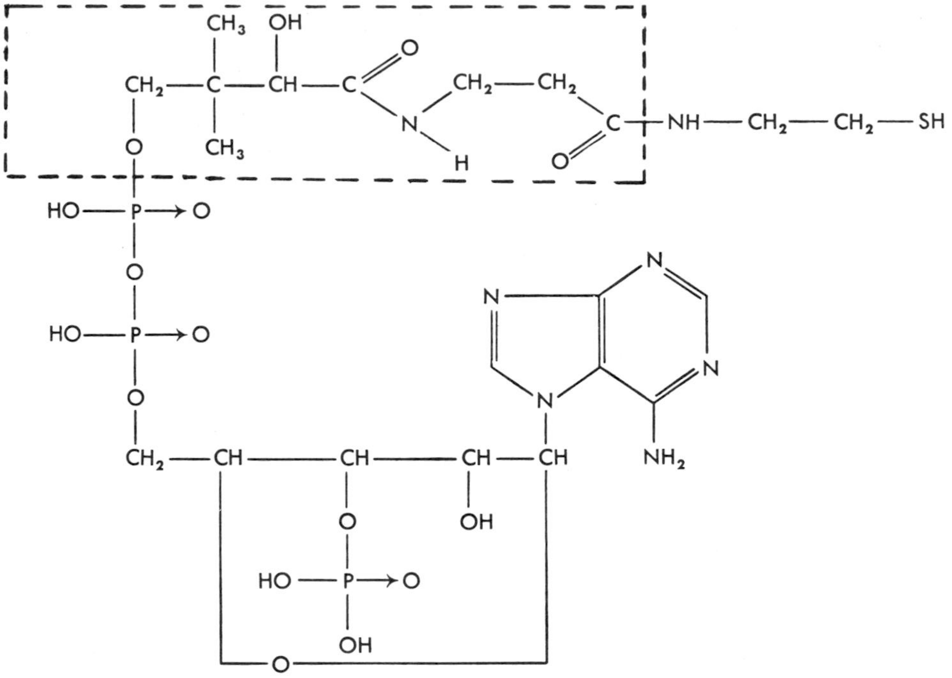

FIG. 3-10
Coenzyme A, essential to normal metabolism. Portion of molecule outlined by dotted lines is contributed by pantothenic acid, a B complex vitamin.

free enzyme. This reaction is typically a reversible one:

$$E + S \rightleftharpoons ES \rightleftharpoons P + E$$

Because E is released from ES, enzymes are true catalysts; they are not used up in the reaction.

Enzymes are characterized by their phenomenal potency and specificity. In 1 second, one molecule of catalase will decompose 44,000 molecules of hydrogen peroxide! Catalase acts only on hydrogen peroxide, a feature called absolute specificity. Some enzymes, however, attack two or more chemically related compounds. Such action is called relative specificity.

Enzyme action is critically related to temperature and pH. At low temperatures, enzymes are

impotent; at high temperatures (70° C and up), most are destroyed. Their optimum effective temperature may differ from plant to plant and animal to animal. The optimum temperature for the enzymes of human tissue is the normal body temperature, 37° C or 98.6° F. Different enzymes function most efficiently at different pH's. The enzymes pepsin, urease, and trypsin do best at pH 2, 7, and 9, respectively. For an enzyme to perform with maximum efficiency, the temperature and pH must be just right. Such is the case in normal cells.

Coenzymes

Some enzymes are entirely protein. Others are composed of a protein (called the *apoenzyme*) and a nonprotein component in tandem. The nonprotein component is referred to as a *prosthetic* group if it is firmly attached to the protein moiety. Loosely attached organic components called coenzymes are essential to the functioning of most enzyme systems (Fig. 3-10). Vitamins are usually the chief component of coenzymes.

Proenzymes and activators

Many enzymes are produced in an inactive form, called a proenzyme or *zymogen,* which must then be activated by an activator. Pepsin, the digestive enzyme in gastric juice, is secreted as the proenzyme pepsinogen and activated (to pepsin) by the hydrogen ion of hydrochloric acid. Other ions that serve as activators include Cl^-, Fe^{++}, Co^{++}, Mn^{++}, Zn^{++}, and Mg^{++}.

Inhibitors

Inhibitors slow down or stop enzyme action. Vast numbers of poisons and drugs depend upon inhibition for their effects. Cyanide inhibits respiratory enzymes, thereby depriving the tissues of oxygen; penicillin kills bacteria by inhibiting enzyme systems involved in building cell walls; certain psychotropic drugs stimulate mental activity by inhibiting monoamine oxidase, an enzyme of brain tissue.

Classification

Enzymes are classified in a number of ways, the only common ground of which seems to be the use of the suffix -*ase*. Even this is not universal, however. Enzyme names such as ptyal*in*, peps*in*, and renn*in* are highly revered antiques and will probably be with us for a long time. The International Union of Biochemistry set up the following six major categories of enzymes:

1. *Synthetases,* catalyzing the joining together of two or more molecules.
2. *Isomerases,* catalyzing the conversion of one isomer to another.
3. *Transferases,* catalyzing the transfer of a chemical group from one molecule to another.
4. *Lyases,* catalyzing the addition or removal of some chemical group without hydrolysis, oxidation, or reduction.
5. *Hydrolases,* catalyzing reactions involving water.
6. *Oxido-reductases,* catalyzing reactions of oxidation and reduction.

To give more meaning to the classification, let us consider the enzymes sucrase and pancreatic lipase. Sucrase is the digestive enzyme of the intestinal juice that catalyzes the reaction between the sugar sucrose and water. The reaction yields glucose and fructose, as follows:

$$C_{12}H_{22}O_{11} + H_2O \xrightarrow{\text{Sucrase}} C_6H_{12}O_6 + C_6H_{12}O_6$$
$$\text{Sucrose} \qquad\qquad \text{Glucose} \quad \text{Fructose}$$

Thus, sucrase is a hydrolase and, more particularly, a *carbohydrase* (any hydrolase that acts on a carbohydrate). Pancreatic lipase (also called steapsin) catalyzes the reaction between the fat stearin (glyceryl tristearate) and water. The reaction yields glycerol and stearic acid, as follows:

$$
\begin{array}{c}
\phantom{C_{17}H_{35}COO -}\, 3H \;\vdots\; (3\,OH) \\
C_{17}H_{35}COO -\!\!\mid- CH_2 \qquad\qquad C_{17}H_{35}COOH \quad HO-CH_2 \\
\mid \\
C_{17}H_{35}COO -\!\!\mid- CH + 3H_2O \xrightarrow{\text{Steapsin}} C_{17}H_{35}COOH + HO-CH \\
\mid \\
C_{17}H_{35}COO -\!\!\mid- CH_2 \qquad\qquad C_{17}H_{35}COOH \quad HO-CH_2 \\
\text{Stearin} \qquad\qquad\qquad \text{Stearic acid} \quad \text{Glycerol}
\end{array}
$$

Pancreatic lipase is a hydrolase, but unlike sucrase, it belongs to the subcategory esterases because it acts on fats (which are esters of fatty acids and glycerol).

QUESTIONS

1. Give evidence to show that air is a mixture and not a compound.
2. An atom is electrically neutral. Explain.

3. The atomic number of an atom is equal to the number of its electrons or the number of its protons. Does this hold true for an ion?

4. A certain element has an atomic number of 11 and an atomic weight of 23. Draw a picture of this atom.

5. Distinguish between atomic weight and mass number.

6. The key to an element's chemical behavior resides in its atomic number. Why is this so?

7. Compare metals and nonmetals, relative to the number of electrons in the outer shell.

8. Write the chemical formula for potassium chloride, magnesium bicarbonate, calcium phosphate, ferrous sulfate, cupric oxide, sodium nitrate, and ammonium bromide.

9. How can you tell whether an unknown compound is electrovalent or covalent?

10. When DC electrodes are placed in a solution of an electrolyte, the cations go to the cathode (negative terminal) and the anions go to the anode (positive terminal). Explain.

11. When sodium chloride dissolves in water, we commonly say that it "ionizes" into Na^+ and Cl^-. Some chemists strongly object to this term, preferring to say "dissociate." Strictly speaking, which is correct and why?

12. If an element belongs to Group VIIA of the periodic table, what do we know about its electronic configuration and its valence?

13. Why are the elements of Group 0 inert?

14. What is meant by oxidation?

15. What makes an acid an acid?

16. Nitric acid is very strong acid. Indicate this, using double arrows.

17. Name the following acids: HI, H_2SO_3, H_2SO_4, H_3PO_4.

18. What makes a base a base?

19. What is the formula for aluminum hydroxide?

20. Using double arrows, indicate that calcium hydroxide is a weak base.

21. Write the equation for the reaction between potassium hydroxide and nitric acid.

22. How would you define a salt?

23. What do we call the salts of HCl, HBr, H_2SO_3, H_2SO_4, HNO_3, and H_3PO_4?

24. A neutral solution has a hydrogen ion concentration of 10^{-7} g of hydrogen ion per liter, and we say it has a pH of 7; a solution with a hydrogen ion concentration of 10^{-3} has a pH of 3; and so on. What, then, is the mathematical definition of pH?

25. Demonstrate that your definition in the above question explains why the pH decreases as the acidity increases.

26. There are two butane isomers (C_4H_{10}). Write their structural formulas.

27. Cyclohexane, a cyclic hydrocarbon with the empirical formula C_6H_{12}, takes the form of a six-membered ring. Write its structural formula.

28. Propylene is an unsaturated aliphatic hydrocarbon with the formula C_3H_6. Draw its structural formula.

29. Write the structural formula for glycerol (glycerine), which has the empirical formula $C_3H_5(OH)_3$. (Note: put OH's on separate carbons.)

30. The chemical name for nitroglycerine is glyceryl trinitrate. What class of organic compound is this, and from what is it made?

31. In the process of digestion, disaccharides are broken down into monosaccharides. Indicate this action by means of a chemical equation. (Do not forget that water is involved.)

32. What is the basic molecular difference between potato starch and corn starch?

33. A polysaccharide is a polymer of a monosaccharide. What does this mean?

34. Elaborate your answer to the above question to explain the buildup of starch in the roots of plants and the buildup of glycogen in the liver.

35. What is the basic physical difference between a true fat and a true oil? What is the basic chemical difference?

36. Most shortenings are prepared by the hydrogenation of vegetable oils. Explain precisely what happens here.

37. Write the chemical equation for the formation of glyceryl tristearate (a common fat of living tissue) from glycerol and stearic acid ($C_{17}H_{35}COOH$).

38. Laymen and popular writings often refer to or think of cholesterol as a fat. Is this chemically correct?

39. When a laboratory does a blood analysis, it re-

ports both free cholesterol and cholesterol esters. Why?

40. In the process of digestion, what do you think proteins are broken down into?

41. Starches and proteins are polymers. Explain.

42. How can you account for two proteins being made up of the same number of amino acids and still being different?

43. How do you explain two proteins being composed of the same kinds of amino acids and still being different?

44. All proteins contain what four elements?

45. Account for the derivation of the expression "amino acid."

46. Chlorophyll is a plant protein containing the metal magnesium. To which of the two classes of proteins does chlorophyll belong?

47. Ribose and deoxyribose and pentoses. What does this mean?

48. DNA and RNA differ in regard to "the sugar" and one base. Explain.

49. Distinguish between a nucleoside and a nucleotide.

50. What do starches, proteins, and polynucleotides have in common?

51. The ability of DNA to make exact copies of itself resides in its bases. Explain.

52. Radioactive isotopes for internal use should have a short half-life. Why?

53. How can radioactivity be detected?

54. Iodine-127 is stable and iodine-131 is radioactive. What does this mean?

55. By consulting other sources, find and briefly discuss two ways in which radioactive isotopes are used in medical diagnosis.

56. The pharmacologist may use sulfur-tagged sulfadiazine in studying that drug. What can he learn from this procedure?

57. In a sense, a chemical equation is a statement of the law of conservation of mass. Explain.

58. Write equations for the burning of magnesium to form magnesium oxide, the reaction between aluminum hydroxide and hydrochloric acid to form aluminum chloride and water, the reaction between iron and cupric chloride to produce ferrous chloride and copper, and the decomposition of sodium bicarbonate in-to sodium carbonate, carbon dioxide, and water.

59. Indicate the types of reactions illustrated in the above question.

60. When a direct current is passed through a solution of cupric chloride, the cations travel to the cathode to become copper (Cu) and the anions travel to the anode to become chlorine (Cl_2). Thus, reduction occurs at the cathode and oxidation at the anode. Explain.

61. In bringing about the oxidation of another element or compound, an oxidizing agent becomes reduced. By the same token, a reducing agent becomes oxidized. Explain.

62. A popular test for the presence of glucose in urine depends on the formation of Cu_2O (a reddish compound) from oxygen and Cu^{++} (blue in color). Thus, glucose is said to be a reducing agent. Explain.

63. Acetic acid (CH_3COOH) is a weak acid:

$$CH_3COOH \rightleftharpoons H^+ + CH_3COO^-$$

On the basis of the law of mass action, explain why the addition of sodium acetate or any other soluble acetate to a solution of acetic acid increases the pH.

64. To enable pancreatic lipase to split fats into fatty acids and glycerol effectively enough for the process to be of digestive value, the body uses bile to speed up the reaction. Technically, we say that the bile emulsifies fats and thereby increases their effective concentrations. What does this mean?

65. When most people think of catalysis, they do so in a "positive" sense. Is there such a thing as negative catalysis?

66. When a piece of fresh liver is added to hydrogen peroxide, a lively effervescence ensues. Such is not the case with *cooked* liver. Explain.

67. A drop or two of fresh blood added to hydrogen peroxide produces a lively effervescence. Such is not the case if the blood has been treated with a pinch of cyanide. Explain.

68. The coenzyme cocarboxylase has the chemical name "thiamine pyrophosphate." What is the nutritional significance of this name?

69. $KClO_3$ must be heated to about 400° C to release

oxygen. The addition of a little MnO_2 lowers the temperature at which oxygen release occurs to about 200° C. What does this tell us about the mechanism of action of enzymes?

70. Trypsinogen does not digest protein in the absence of an activator called enterokinase. What does the enterokinase do?

71. Proteinases and amylases (enzymes that act on proteins and starches, respectively) are classified as hydrolases. Why?

CHAPTER 4

MICROBES AND INFECTION

Organisms too small to be seen or studied without the aid of a microscope are appropriately called microorganisms or, less formally, microbes. Contrary to possible first impressions, most microbes are beneficial; through the process of decay and putrefaction they make our planet inhabitable. They aid in the manufacture of many things, from sour cream to streptomycin. One day microbial food may well usurp the shelves at the corner grocery. Our concern here, however, centers on maleficent microbes and the infections they cause.

CLASSIFICATION

There is no one system for classifying plants and animals (*taxonomy*) that is acceptable to all biologists. Textbooks vary widely, even in regard to spelling; it seems only wise, therefore, to eschew taxonomic dogmatism. For our purpose, let us say the various kinds of microbes compose the kingdom Protista, a category first proposed in 1886 by the German naturalist Ernst Haekel to embrace lowly forms of life that are neither true plants nor true animals.

Protista, like all kingdoms, is divided into phyla, classes, orders, families, genera, and species, in that order. A certain number of species makes up a given genus, a certain number of genera, a given family, and so on. By inter-national agreement, the naming of any organism, including a microorganism, follows the binomial system of nomenclature. The name consists of two parts, the genus and the species to which the organism belongs. Both parts of the name are italicized (or underlined), but only the genus is capitalized. The causative organism of lockjaw is *Clostridium tetani*, just as we are *Homo sapiens*. For convenience in treating modifications within a species, variety (var.) names are sometimes used. The head louse and body louse are named, respectively, *Pediculus humanus* var. *capitis* and *Pediculus humanus* var. *corporis*.

There is little or no disagreement about the kinds of microorganisms. There are seven kinds: bacteria, rickettsiae, fungi, algae, slime molds, protozoa, and viruses. We shall discuss each of these except algae and slime molds, which do not cause disease, and for the sake of completeness, we shall say something about parasitic worms and ectoparasites. Although these last organisms are not microbial in the usual sense of the word, they cannot be studied without the aid of the microscope.

BACTERIA

Bacteria are unicellular organisms, the vast majority of which do not contain chlorophyll.

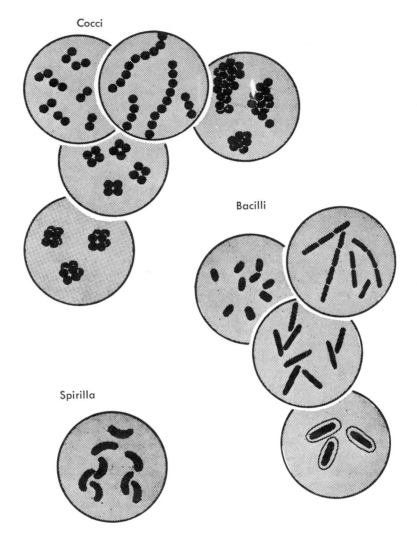

Cocci

Bacilli

Spirilla

FIG. 4-1
Morphology of bacteria. Shown here are spirilla and various types of cocci and bacilli. Note bacilli with capsules.

The approximately 2,000 species fall into a number of rather well-defined orders, including Eubacteriales, Actinomycetales, Spirochaetales, Pseudomonadales, and Mycoplasmatales (see below).

Bacterial orders

The members of the order Eubacteriales, sometimes referred to as "true bacteria," cause the bulk of all infectious diseases. These organisms are of three basic morphological types: rodlike, spiral, and spherical (Fig. 4-1). The rodlike forms are called bacilli (sing., bacillus); the rigid, spiral forms, spirilla (sing., spirillum); and the spherical forms, cocci (sing., coccus). The spherical forms are characterized by the groupings they assume after cell division: diplococci (in pairs), streptococci (in chains), staphylococci (in grapelike bunches), sarcinae (in cuboidal packets of eight cells), and gaffkyae (in cuboidal packets of four cells). The principal pathogenic species are *Salmonella typhosa* (typhoid fever), *Salmonella paratyphi* (paratyphoid fever), *Yersinia pestis* (plague), *Francisella tularensis* (tularemia), *Haemophilus influenzae* (respiratory infections, meningitis), *Bordetella pertussis* (whooping cough), *Neisseria gonorrhoeae* (gonorrhea), *Neisseria meningitidis* (epidemic meningitis), *Staphylococcus aureus* ("staph infections"), *Diplococcus pneumoniae* (pneumococcal pneumonia), *Streptococcus pyogenes* (scarlet fever), *Corynebacterium diphtheriae* (diphtheria), *Bacillus anthracis* (anthrax), *Clostridium botulinum* (botulism), *Clostridium* spp.* (gas gangrene), and *Shigella* spp. (bacillary dysentery).

The order Actinomycetales is made up of moldlike bacteria with elongated, frequently filamentous, cells which have a tendency to branch. The chief pathogenic actinomycetes are *Mycobacterium tuberculosis* (tuberculosis) and *Mycobacterium leprae* (leprosy). A dozen or more species of the genus *Streptomyces* yield valuable antibiotics. *Streptomyces griseus*, for example, produces streptomycin.

The order Spirochaetales is composed of slender, flexuous, spiral organisms that closely resemble protozoa. Most are highly motile. They differ from one another morphologically in length and in number of spirals. The chief pathogenic spirochetes are *Treponema pallidum*

---

*The expression "spp." indicates that two or more species of the genus may cause the disease.

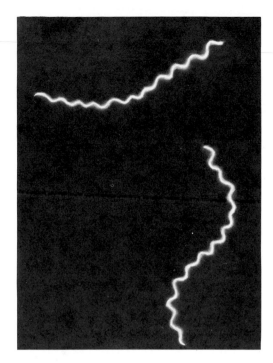

FIG. 4-2

Spirochete *Treponema pallidum*, the cause of syphilis. (× 6,000.) (From Kraus, S. J.: J.A.M.A. **211:**2140, 1970.)

(syphilis, Fig. 4-2), *Treponema pertenue* (yaws), and *Leptospira icterohaemorrhagiae* (Weil's disease).

The order Pseudomonadales is made up of a number of species of bacilli and spirilla distributed widely throughout nature. Many are short, nonsporing rods that move by means of polar flagella. The pathogens of chief concern are *Vibrio cholerae* (Asiatic cholera) and *Pseudomonas aeruginosa*, a common invader of wounds, burns, the urinary tract, and the middle ear.

The order Mycoplasmatales is composed of the smallest free-living organisms known. Delicate, nonmotile cells without a cell wall, they display a variety of sizes and shapes. Because the first species discovered causes pleuropneumonia in cattle, these organisms commonly are referred

Spore types

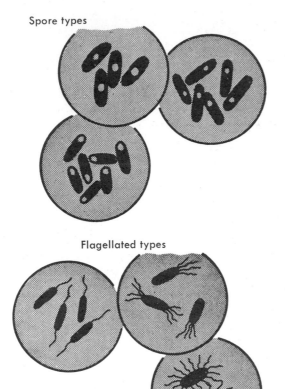

Flagellated types

FIG. 4-3
Spore types and flagellated types of bacteria.

to as pleuropneumonia-like organisms (PPLO). The species *Mycoplasma pneumoniae* is a common cause of pneumonia in man.

The bacterial cell

Bacterial cells and the cells of blue-green algae are the most primitive known. Called *procaryotic* cells, they differ substantially from all other living cells which are called *eucaryotic* cells. The transition from the procaryotic to the eucaryotic level was a cataclysmic evolutionary

event and accounts for the practice of some authors to relegate the procaryata to the kingdom Monera.

The hallmark of the bacterial cell, and of all procaryotic cells, is the absence of a nuclear membrane. They have neither mitochondria nor chloroplasts, and reproduce by simple division, or binary fission, rather than mitosis. The bacterial cell ranges in size from 0.5 to 15 μm in its greatest dimension and in some instances is protected by a gelatinous outer wall called the *capsule* (Fig. 4-1). Although rare among the cocci, whiplike locomotor extensions called *flagella* (Fig. 4-3) are found in most bacilli and spirilla. The manner in which flagella are distributed about the surface of a cell helps to identify the cell.

Some bacilli form round or oval resistant structures of condensed protoplasm, called *spores* (Fig. 4-3), for protection against high temperatures, desiccation, and other adverse environmental conditions. As a result, all methods of sterilization must be aimed at the spores as well as at the vegetative cell. Spores aid in the classification and identification of a bacterial species, on the basis of not only their presence or absence, but also their size, shape, and location within the cell. *Clostridium tetani,* the cause of tetanus, is identified by its large, round, terminal spores.

Growth requirements

Bacteria display a variety of growth requirements that must be met if the organisms are to be cultured successfully in the laboratory. *Autotrophic* bacteria do well on a simple diet of carbon dioxide, inorganic salts, and water; *heterotrophic* bacteria (the saprophytes and parasites) demand organic nutrients; *obligate* parasites cannot be cultured except in living tissue. Other important considerations relate to pH, temperature, and oxygen. Bacteria, like all microbes, are quite sensitive to changes in pH;

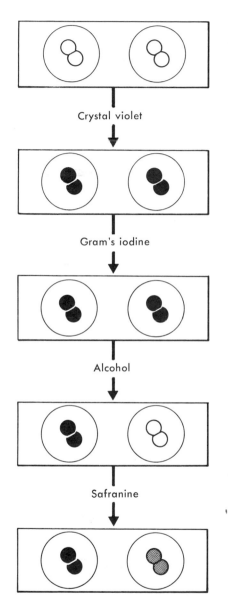

FIG. 4-4

Gram staining. Species of diplococci to left is gram positive; that to right is gram negative.

the majority culture best at a pH of about 7.5. Some bacteria thrive at high temperatures, some at low, and some are killed at both extremes. Saprophytes generally do best at temperatures between 25° and 30° C, and parasites within a range roughly 10° higher. Certain autotrophs have optimum temperature ranges far below or far above these figures. Some bacteria cannot live without oxygen (obligate aerobes), some cannot live with it (obligate anaerobes), and others (the facultative anaerobes) can adapt to either condition

Staining

A special feature of bacteriology is the use of stains for the purpose of visualization and identification. In the *Gram technique,* the most widely used staining procedure, bacterial cells are smeared on a glass slide and stained with crystal violet (Fig 4-4). The smear is treated with Gram's iodine, then decolorized with alcohol, and finally counterstained with safranin. (After each application the slide is washed with water.) Provided all goes well, some bacteria take the first stain (gram positive), and others take the counterstain (gram negative). Although the precise mechanism of Gram staining is not understood, most studies indicate that the cell wall is of central importance. The alcohol treatment is thought to dissolve much of the lipid from the wall of gram-negative bacteria, thereby permitting the escape of the crystal violet-iodine complex. In contrast, dehydration by alcohol of the gram-positive cell reduces the size of the pores in the cell wall, making decolorization much more difficult. As the Gram reaction indicates, there is a profound difference between gram-positive and gram-negative bacteria. Penicillin is most effective in the treatment of infections caused by gram-positive bacteria, whereas streptomycin is most effective against gram-negative organisms.

Another important staining procedure is "acid-

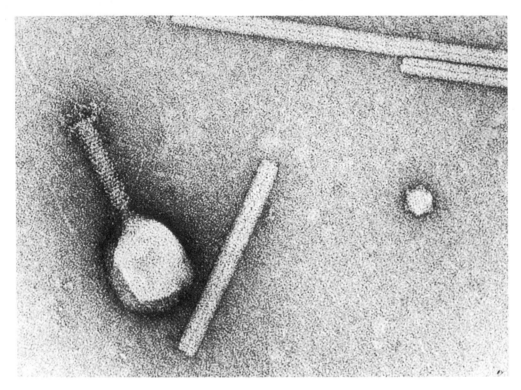

FIG. 4-5
Electron micrograph showing differences in size and shape among viruses. Flask-shaped virion is T4 bacteriophage; rod-shaped virions are tobacco mosaic virus; spherical virion is φX174 bacteriophage. (× 45,000.) (Courtesy F. A. Eiserling, Department of Bacteriology, University of California at Los Angeles.)

fast staining." Acid-fast bacteria retain a stain even when washed with acid-alcohol. This highly characteristic property is thought to result from the presence of certain fatty substances in the cellular membrane. In the Ziehl-Neelsen version of acid-fast staining, the prepared smear is covered with carbolfuchsin and heated gently for about 5 minutes. The smear is then decolorized with acid-alcohol and counterstained with methylene blue. When this procedure is carried out properly, acid-fast organisms are stained red, and all other organisms are stained blue.

VIRUSES

Because they are not cellular and do not carry on metabolism, viruses can be removed from the area of biological taxonomy. They are not bona fide microbes and at very best occupy some sort of twilight zone of ambiguous life. Only for convenience have we here labeled them Protists, and only for convenience does Bergey* relegate them to the order Virales (class Microtatobiotes, phylum Protophyta, kingdom Plantae).

*Breed, R. S., Murray, E. G. D., and Smith, N. R.: Bergey's manual of determinative bacteriology, ed. 7, Baltimore, 1957, The Williams & Wilkins Co.

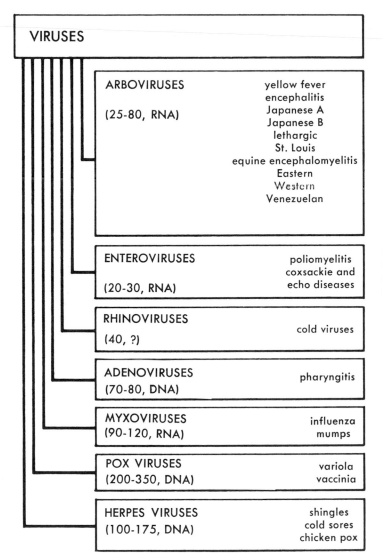

VIRUSES		
ARBOVIRUSES (25-80, RNA)	yellow fever encephalitis Japanese A Japanese B lethargic St. Louis equine encephalomyelitis Eastern Western Venezuelan	
ENTEROVIRUSES (20-30, RNA)	poliomyelitis coxsackie and echo diseases	
RHINOVIRUSES (40, ?)	cold viruses	
ADENOVIRUSES (70-80, DNA)	pharyngitis	
MYXOVIRUSES (90-120, RNA)	influenza mumps	
POX VIRUSES (200-350, DNA)	variola vaccinia	
HERPES VIRUSES (100-175, DNA)	shingles cold sores chicken pox	

FIG. 4-6
Classification of viruses on basis of particle size (in nanometers), nucleic acid content (DNA or RNA), and diseases caused.

Classification (nonbiological)

Virus particles, or virions, range in size from about 10 to 350 nm. All that are smaller than 200 nm are beyond the reach of the light microscope, and hence are called "ultramicroscopic."

Some are spherical, some cuboidal, some rod shaped, and some look like surrealistic tadpoles (Fig. 4-5). The virions of any given virus are composed of either RNA or DNA tightly folded and packed within a protein coat called a *capsid*.

FIG. 4-7

Bottle cultures of monkey kidney cells, on which enteroviruses can grow, were used in detecting mixtures of viruses. Clear spaces (plaques) are areas in which cells have been infected by virus and have degenerated. Bottle at left contains two kinds of plaques: large ones characteristic of poliovirus and small, irregular ones typical of echovirus. In second bottle, polio antiserum inhibited growth of poliovirus, and only echovirus plaques appeared. In third bottle, echovirus antiserum suppressed growth of echovirus and permitted isolation of poliovirus. In bottle at far right, antisera to both poliovirus and echovirus were present, and no plaques developed. (Courtesy M. Benyesh-Melnick and J. L. Melnick, Baylor University College of Medicine, Waco, Texas.)

Capsids are organized into a characteristic number of subunits called *capsomeres*. Plant viruses contain RNA, animal viruses contain RNA or DNA, and most bacterial viruses contain DNA. Animal viruses fall into seven categories on the basis of particle size and nucleic acid content (Fig. 4-6). The myxoviruses, for example, measure between 90 and 120 nm and contain RNA.

Tissue cultures

At one time the embryonated hen's egg served as the virologist's chief test tube, but today to propagate viruses he generally employs the tissue, or cell, culture. Some of the more commonly used cell lines are chick embryo cells, monkey kidney cells, and human cancer cells. When a bottle or Petri dish is used, these cells, bathed in the appropriate culture fluid, attach to the bottom and grow as a thin sheet called a monolayer or lawn. When the virus destroys the cells of the lawn *(cytopathic effect),* colorless areas called plaques form, each of which represents the progeny of a single virion (Fig. 4-7).

Tissue cultures are used in three ways: to determine the virus content of a specimen on the basis of the number of plaques; to produce a virus in bulk amounts for the preparation of vaccines; and to identify viruses by the type of plaque formed or the cytopathic inhibition caused by a type-specific immune serum.

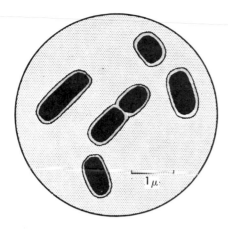

FIG. 4-8

Rickettsia rickettsii, the causative agent of Rocky Mountain spotted fever. (From Brooks, S. M.: Basic facts of medical microbiology, ed. 2, Philadelphia, 1962, W. B. Saunders Co.)

Cancer

Viruses cause certain malignant tumors (cancer) in animals, including, among others, Rous sarcoma (chickens), myxomatosis (rabbits), leukemias (mice, chickens), and polyomas (rabbits, mice, hamsters). In man, the common wart, a benign tumor, is caused by a virus, but as yet no virus has been isolated from any type of human cancer. The prospect of finding such a virus, however, appears to be just a matter of time, and already there have been a number of reported isolations.

RICKETTSIAE

Rickettsiae are unicellular organisms situated between bacteria and viruses in size and complexity. They are pleomorphic, gram-negative, bacilloid structures (Fig. 4-8) that culture only in living tissue. Like viruses, they are intracellular parasites, but otherwise more closely resemble bacteria. They attack man and animals, with rodents and ticks serving as the chief reservoirs of infection. Mites, lice, and fleas serve as vectors, transmitting rickettsiae from animal to animal, from animal to man, and from man to man. The principal species of medical concern (Fig. 4-9) are *Rickettsia prowazekii* (epidemic typhus), *Rickettsia mooseri* (endemic typhus), *Rickettsia tsutsugamushi* (scrub typhus), and *Rickettsia rickettsii* (Rocky Mountain spotted fever).

FUNGI

Classically, the term "fungi" referred to chlorophyll-free "plant protists" and included bacteria, slime molds, and "true fungi." Today, however, the term is restricted to the 100,000 or so true fungi: morels, truffles, cup fungi, mildews, mushrooms, puff balls, smuts, rusts, and, most especially, molds and yeasts. A mold is a fuzzy growth of interlacing filaments called *hyphae* that reproduce by means of *spores* (Fig. 4-10). A tuft of hyphae is referred to as a *mycelium*. In some species the hyphae are divided by partial septa and appear to be multicellular; in others, the hyphae are nonseptate and clearly unicellular. It is still common practice to define yeasts as unicellular organisms that reproduce by *budding* (Fig. 4-10), but this definition is actually somewhat tenuous. Some yeasts reproduce by fission, and many, under certain conditions, produce mycelia. It seems more realistic to consider yeasts as organisms that usually are single-celled and that usually reproduce by budding. For the most part, molds and yeasts do well at room temperatures and prefer slightly acidic media. Sabouraud's dextrose agar usually produces good growths.

Most fungal infections (*mycoses*) are caused by yeasts and molds belonging to the class Fungi Imperfecti. Mycoses of the skin and hair (dermatomycoses) involve dozens of species, notably those of the genera *Trichophyton* and *Epidermophyton* in the case of athlete's foot. Systemic, or deep, mycoses are dangerous and sometimes fatal. These include, among others, candidiasis (*Candida albicans*), histoplasmosis (*Histoplas-*

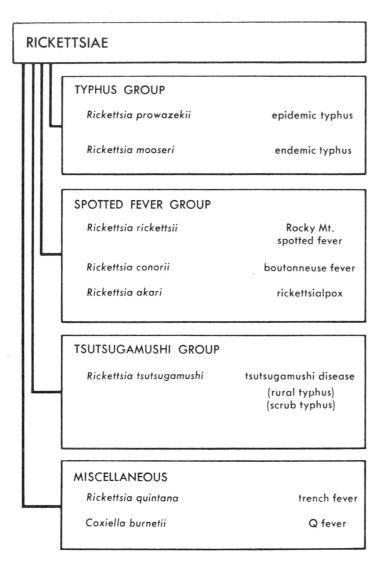

FIG. 4-9
Clinical classification of rickettsiae.

ma capsulatum), coccidioidomycosis (*Coccidioides immitis*), blastomycosis (*Blastomyces dermatitidis*), and cryptococcosis (*Cryptococcus neoformans*).

PROTOZOA

Protozoa are unicellular animal or animal-like organisms composing phylum Protozoa (king-dom Protista, or subkingdom Protozoa, king-dom Animalia). Some of the estimated 100,000 species are holophytic and free living; others are parasitic and pathogenic. The euglenoids and dinoflagellates are capable of autotropic nutrition and, therefore, perch on the hazy fence between botany and zoology. As a group, pro-tozoa are much larger than the largest bacterium

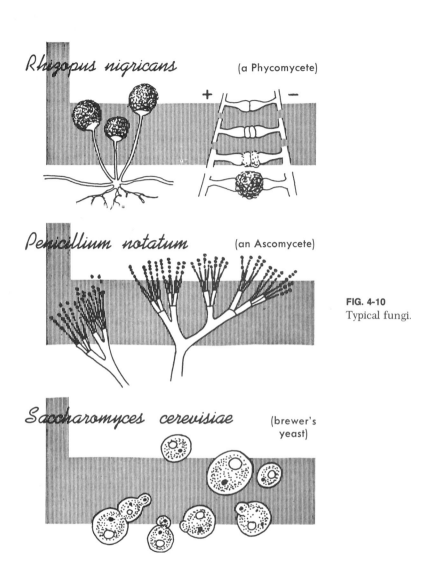

Rhizopus nigricans (a Phycomycete)

+ −

Penicillium notatum (an Ascomycete)

FIG. 4-10
Typical fungi.

Saccharomyces cerevisiae (brewer's yeast)

and possess well-defined nuclei and vacuoles. They exhibit a rather complex internal structure and appear to be the most advanced form of the one-celled state. Some species may exist in either an active (the *trophozoite* stage) or an inactive form (the *cyst* stage), the latter serving as a means of weathering adverse conditions. Protozoa naturally divide into five classes (Fig. 4-11) according to their mode of locomotion: Sarcodina or Rhizopoda (movement by means of pseudopodia), Flagellata or Mastigophora (movement by means of flagella), Ciliata or Infusoria (movement by means of cilia), Suctoria (cilia only in early stages of development), and Sporozoa (no movement in the adult form).

Most protozoa are harmless or beneficial, but

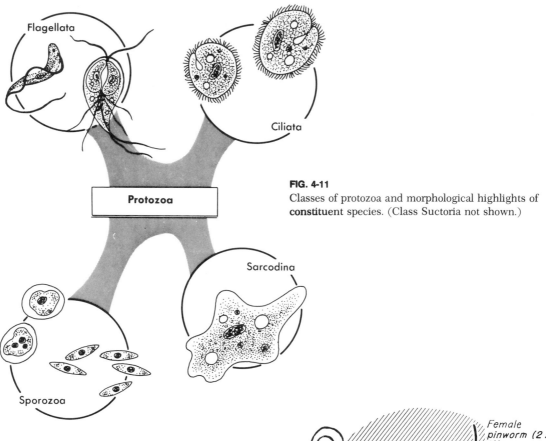

FIG. 4-11
Classes of protozoa and morphological highlights of **constituent** species. (Class Suctoria not shown.)

a number of species do cause serious infections in man and animals. Some of the more common and better known are *Plasmodium* spp. (the cause of malaria), *Trypanosoma* spp. (African sleeping sickness), *Entamoeba histolytica* (amebiasis), and *Trichomonas vaginalis* (trichomonas vaginitis).

PARASITIC WORMS

Worm infestation may very well be the most common ailment of man in areas and countries with poor sanitation. Such parasites fall into three categories: nematodes, cestodes, and trematodes (Fig. 4-12).

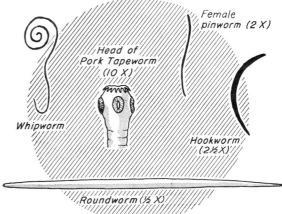

FIG. 4-12
Parasitic worms. (From Brooks, S. M.: Basic facts of pharmacology, ed. 2, Philadelphia, 1963, W. B. Saunders Co.)

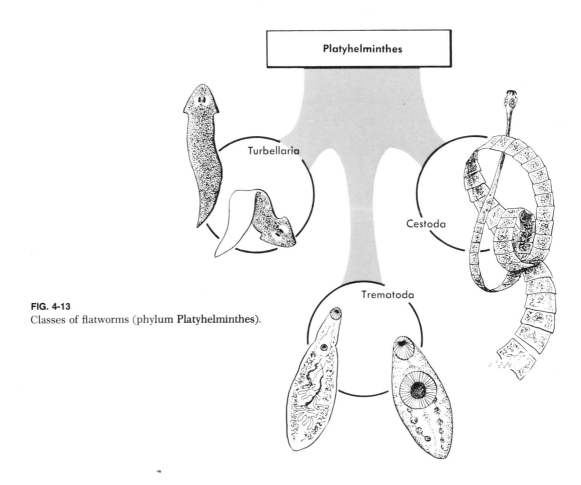

FIG. 4-13
Classes of flatworms (phylum **Platyhelminthes**).

Nematodes

These worms compose a class of the animal phylum Aschelminthes. They have an elongated cylindrical body, usually pointed at both ends, without segmentation. Nematodes illustrate the evolutionary step of "tube-in-tube" body structure, which consists of an outer tube (the body wall) that encloses a cavity (the pseudocoel) within which is another tube (the digestive tract) open at both ends. Reproduction is sexual, and the sexes are separate. Some nematodes are free living and harmless, but others cause serious infestations in man and animals. The species

of chief medical concern include hookworm, whipworm, pinworm, roundworm, *Trichinella spiralis* (causes trichinosis), and *Wuchereria bancrofti* (filariasis). The roundworm (*Ascaris lumbricoides*), a typical nematode, demonstrates the characteristics of the class as a whole. The adult, which averages about 10 inches in length, inhabits the intestine of man, the definitive host.* After being fertilized by the male, the female lays ova, which escape into the feces

---

*A *definitive host* is the host in which the sexual reproduction of a parasite takes place.

and gain entrance into the intestinal tract of a new host via contaminated food or water. Here the ova liberate larvae, which enter the blood and escape into the lungs. From the lungs the larvae travel up the trachea, down the esophagus, through the stomach, and back into the intestine, where they develop into adult worms. In a severe case, dozens of worms may be present.

Cestodes

Cestodes, or tapeworms, compose a class of the animal phylum Platyhelminthes (Fig. 4-13). They are ribbonlike intestinal parasites ranging from 20 to 50 feet in length. The head (scolex) has suckers, or hooklets, or both, and the body is divided off into hundreds of segments (proglottids), each of which is morphologically and functionally autonomous. The beef tapeworm (*Taenia saginata*), the most common cause of tapeworm infestation (cestodiasis) in man, averages about 25 feet in length and has an extremely small head. In the intestine the hermaphroditic adult worm releases fertile eggs that appear in the feces and thereby contaminate soil and water. When ingested by cattle (the intermediate host), the ova develop into embryos (oncospheres), which in turn penetrate the tissues and infiltrate the muscles. Once in the muscles, the oncospheres undergo further development to form cysticerci, which are encapsulated scolices. When contaminated beef is eaten, the scolices are set free and hook onto the intestinal lining, where they produce the adult generation.

Trematodes

Trematodes are unsegmented flat worms, commonly called flukes, equipped with suckers or hooks for anchoring themselves to the host (Fig. 4-13). They have complex life cycles with sexual and asexual generations, and generally two or more hosts are involved. Fluke infestations are common in man and animals. Human

schistosomiasis, caused by blood flukes belonging to the genus *Schistosoma*, now ranks next to malaria as a world health problem. In the typical fluke life cycle, the adult passes eggs, which leave the body of the host in the feces or the urine. In time, an egg develops into a ciliated stage, called a miracidium, which in the presence of water swims about looking for a certain species of snail. Once within the tissues of such an intermediate host, the miracidia form cysts, within which wormlike forms called rediae develop. Rediae, in turn, produce tadpolelike organisms called cercariae. (In the genus *Schistosoma* the cercariae are produced without a rediae stage.) Some cercariae escape into the water, where they await an obliging definitive host in whose tissues they will develop into adult flukes. Other cercariae become encysted metacercariae in a plant or animal and become adult when this plant or animal is eaten.

ECTOPARASITES

Ectoparasites live on the exterior of another organism, which in the case of man means the skin. The principal species include the itch mite (*Sarcoptes scabiei*), the chigger (*Trombicula irritans*), the head louse (*Pediculus humanus* var. *capitis*), the body louse (*Pediculus humanus* var. *corporis*), the crab louse (*Phthirus pubis*), and the human flea (*Pulex irritans*).

INFECTION

Essentially, infection is a tug-of-war between the pathogen and the host. To identify a given species as the cause of a particular disease, certain requirements (*Koch's postulates*) must be met: The organism must be present in every case of the disease; it must be isolated and grown in pure culture; in pure culture it must cause the disease when inoculated into susceptible animals; and it must be recovered from the inoculated animals and shown to be the same as the original. With relatively few exceptions, the

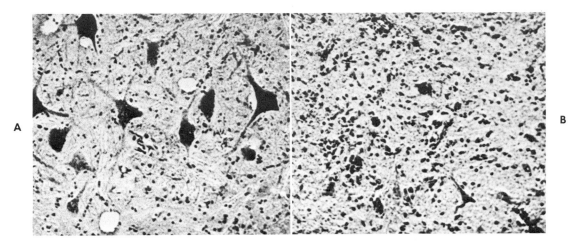

FIG. 4-14

Poliovirus attacking nervous tissue. Normal spidery nerve cells of monkey spinal cord, **A,** are all but obliterated, **B,** in the wake of the attack. (From Burrows, W.: Textbook of microbiology, ed. 17, Philadelphia, 1959, W. B. Saunders Co.)

acknowledged pathogens of man meet these requirements.

Source and transmission of pathogens

Man and animals serve as the permanent and primary reservoirs of infection. The human reservoir includes patients and carriers. The patient harbors a pathogen only during the period of infection; the carrier harbors a pathogen for months, years, or sometimes a lifetime. Carriers threaten the health of a community because unless identified and treated, they can infect thousands of others, as did an itinerant cook known to history as Typhoid Mary. The animal world harbors a number of vicious pathogens, including those responsible for anthrax, encephalitis, brucellosis, and rabies.

The manner in which pathogens travel from source to victim is clearly of paramount importance, and many great steps in the control of disease have involved this area. Generally, it may be said that transmission is effected either directly by actual body contact and droplet infection (sneezing, coughing) or indirectly via air,

water, soil, food, contaminated objects (fomites), and vectors. A *vector* is a carrier (usually an insect) that transmits a pathogen from one host to another in a biological or a mechanical fashion. In biological transmission the vector is essential to the survival of the pathogen and the infection because the pathogen passes through part of its life cycle in the vector. Thus, if we eradicate the vector, we eradicate the disease, a principle underscored dramatically in the instance of the mosquito and yellow fever. A mechanical vector, by contrast, is not essential to the life cycle of the pathogen; it merely carries the pathogen on its body surface. The common housefly is a good example of a mechanical vector.

Pathogenesis

An infection develops from cellular destruction, or intoxicating mechanisms, or both (Fig. 4-14). Exotoxins, the most deadly biological poisons known, are responsible for a number of diseases including tetanus, diphtheria, gas gangrene, anthrax, plague, and botulism. Much less deadly but still capable of causing violent responses are endotoxins, the products released by certain gram-negative bacteria. Exotoxins are heat-labile proteins readily released from the bacterial cell: endotoxins are heat-stable lipopolysaccharide-protein complexes not readily released. Other key differences relate to antigenicity and pyrogenicity. Exotoxins are highly antigenic, stimulating the output of neutralizing antibodies called antitoxins; endotoxins are weakly antigenic. Endotoxins are highly pyrogenic (fever-inducing), whereas exotoxins are not.

In addition to exotoxins and endotoxins, microbes manufacture a variety of other toxic products including hemolysins (destroy red cells), leukocidins (destroy white cells), coagulase (facilitates clotting of plasma), hyaluronidase (dissolves intercellular cement), and kinases. Kinases dissolve clots, or inhibit their formation, by an activating mechanism. Streptokinase, produced by certain streptococci, converts inactive plasminogen of the plasma into active plasmin, or fibrinolysin, which dissolves blood clots. Another streptococcal agent, streptodornase, works with streptokinase in the depolymerization of DNA.

Immunity

In addition to an inherited general resistance to infection, we acquire immunity through the process of living. Acquired immunity may be

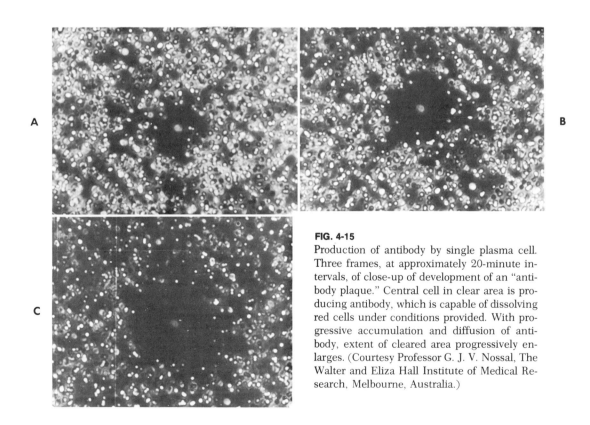

FIG. 4-15
Production of antibody by single plasma cell. Three frames, at approximately 20-minute intervals, of close-up of development of an "antibody plaque." Central cell in clear area is producing antibody, which is capable of dissolving red cells under conditions provided. With progressive accumulation and diffusion of antibody, extent of cleared area progressively enlarges. (Courtesy Professor G. J. V. Nossal, The Walter and Eliza Hall Institute of Medical Research, Melbourne, Australia.)

active or passive. In the active variety the body builds up its resistance to infection according to a scheme whereby an antigen (the pathogen or its toxic products) stimulates the plasma cells of the reticuloendothelial system (p. 97) to produce antibodies (Fig. 4-15), the specific immune substances that appear in the gamma globulin fraction of blood serum. Antibodies destroy or neutralize the pathogen or its toxic products, thereby protecting the host. In naturally acquired active immunity, antibody is provoked incident to the actual microbial invasion. Measles, for instance, attacks only once because it leaves in its wake sensitized plasma cells all set to pour forth protective concentrations of antibody in the event of subsequent encounters with the virus. Ideally, acquired active immunity is provoked artificially via vaccines and toxoids, antigenic preparations made from pathogens and exo-

toxins, respectively. Measles vaccine is made from the attenuated virus and provokes a potent resistance against measles; tetanus toxoid, made from the neutralized exotoxin of *Clostridium tetani,* provokes a potent resistance against tetanus.

When antibodies from one person are injected into another, the immunity acquired is characterized as passive because the plasma cells of the recipient play no role. Although such immunity is short lived, the procedure is used to protect dangerously exposed susceptibles and to treat those already ill. In actual practice, antisera are used, the most important of which are those used against diphtheria (Fig. 4-16), tetanus, gas gangrene, rabies, and botulism. These antisera are generally made from the blood of vaccinated horses, but human antisera now are available for the management of tetanus and rabies in

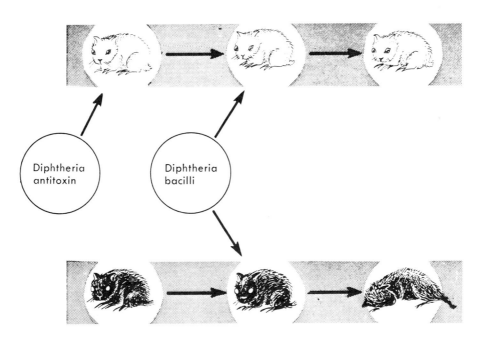

FIG. 4-16
Use of guinea pig to demonstrate neutralizing action of diphtheria antitoxin. (From Brooks, S. M.: Basic facts of medical microbiology, ed. 2, Philadelphia, 1962, W. B. Saunders Co.)

those allergic to horse serum. Human gamma globulin also has proved valuable in the treatment of measles, hepatitis, and poliomyelitis.

While the administration of immune products is an artificial measure, naturally acquired passive immunity is encountered in the newborn as a consequence of the diffusion of maternal antibodies across the placenta into the fetal circulation. For the first several weeks of life, the baby is protected against many of the same infections as the mother.

Antigen-antibody reactions

The reaction between an antigen and its homologous antibody is chemical in nature, and a variety of serologic tests have been devised that are based on such reactions. An unknown organism can be identified by employing known antibody; unknown antibody can be identified by employing a known organism. When blood serum containing antibody against the bacterium *Salmonella typhosa* (the cause of typhoid fever) is added to a suspension of that organism, clumping, or *agglutination,* of the bacterial cells occurs. Thus, if a suspension of *Salmonella typhosa* is agglutinated by the serum under study, it shows that antityphoid antibody is present in the serum and that the individual from whom the serum was taken has encountered *Salmonella typhosa,* and may in fact have typhoid fever. By the same token, if the bacterium under study is strongly agglutinated by known antityphoid serum, it pinpoints the organism as *Salmonella typhosa.*

By convention, antibodies are classified according to the serologic tests used to demonstrate their presence. The principal kinds of antibody include agglutinins, which agglutinate microbial cells such as in the above example; precipitins, which produce a precipitate when added to a solution of the homologous antigen; *lysins,* which cause dissolution of bacterial or other cells; *antitoxins,* which neutralize toxins;

opsonins, which enhance phagocytosis; and *neutralizing antibodies,* which deactivate viruses.

Although antigen-antibody reactions are characterized by specificity, there are a number of instances in which unrelated microbes react with the same antibody. This occurs when microbes have common antigens. A classic example concerns certain strains of the bacterium *Proteus vulgaris,* an organism that is agglutinated by antibodies provoked by several species of rickettsiae. This is the basis of the *Weil-Felix* test used in the differential diagnosis of rickettsial diseases.

Complement-fixation test. One of the most commonly used serologic procedures for the diagnosis of infectious diseases and the identification of microorganisms is the complement-fixation test. This test is based on the fact that a blood serum protein group called *complement* enters into the reaction between antigen and antibody. Antigen (A) plus antibody (B) plus

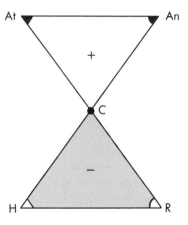

FIG. 4-17

Complement fixation at a glance. If patient's serum contains antibody, *At,* against the test antigen, *An,* complement, *C,* becomes fixed and hemolysin, *H,* is unable to lyse red cells (positive test, upper triangle). Conversely, if patient's serum does not contain antibody against test antigen, complement is free to interact with *H* and *R* (red cells) to effect hemolysis (negative test, lower triangle).

complement (C) yields a chemical complex (A-B-C) in which complement is said to be "fixed." If complement becomes fixed when it and known antigen are added to a specimen under investigation, the presence of the homologous antibody is demonstrated. If complement does not become fixed, the homologous antibody is not present. By the same token, complement and known antibody can be used to signal the presence or absence of the homologous antigen. To detect the fixation, sheep red cells and rabbit se-

rum containing antisheep hemolysins are used. In the presence of complement, hemolysins cause red cells to *lyse* (disintegrate), thereby producing a red color as a result of the release of hemoglobin. As in all such antigen-antibody reactions, complement becomes fixed.

In practice (Fig. 4-17), the test is carried out in two steps. In step one, known antigen and unknown antibody—or vice versa—and a measured amount of complement are added to a tube and incubated. In step two, sheep red cells and rab-

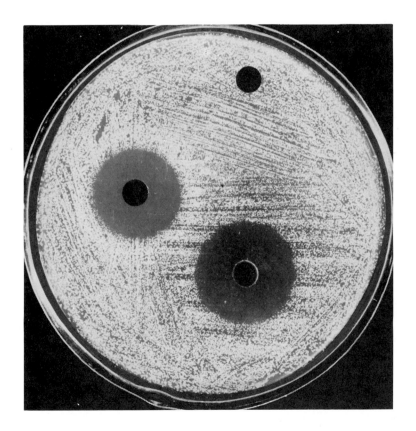

FIG. 4-18
Determination of sensitivity of bacteria to different antibiotics. Petri dish of agar is heavily inoculated with test organism. Disks of filter paper impregnated with different antibiotics are dropped on freshly inoculated surface. Bacterial growth is white. Note zone of inhibition of bacterial growth around two of the three disks. Zone of inhibition indicates growth of organism would probably be limited in patient receiving that antibiotic. (From Smith, A. L.: Principles of microbiology, ed. 6, St. Louis, 1969, The C. V. Mosby Co.)

bit antiserum are added to this mixture. If hemolysis occurs, complement was not fixed in step one, meaning that the homologous antibody or antigen in question is not present. On the other hand, if hemolysis does not occur, the homologous antibody or antigen is present.

Hemagglutination inhibition test. This serologic test is used to identify certain viruses and viral antibodies. It is based on the fact that some viruses attach themselves to the surface of red blood cells and in so doing cause the red cells to agglutinate. If such a virus has been mixed with its specific antibody, however, it cannot attach to red cells and cause agglutination. Thus, if a serum under study prevents a known agglutinating virus from producing agglutination, it proves that the individual from whom the serum was taken has come into contact with this virus. By the same token, a virus under study that fails to agglutinate red cells after having been mixed with known antibody is thereby identified, provided, of course, that the virus is known to produce agglutination in the absence of antibody.

Diagnosis of infection

The diagnosis of infection is made on the basis of signs and symptoms, skin tests, and laboratory findings. The signs and symptoms of some infections are so characteristic that a diagnosis can be made based solely on them. The mother who has reared two or three children, for instance, can often spot a case of measles with no more difficulty than the young doctor.

Skin tests are diagnostic procedures that employ antigen or antibody. In the Schultz-Charlton test, if an injection of scarlet fever antitoxin into an area of rash causes the rash to blanch, the person with the rash has scarlet fever.

Any and all procedures carried out in the laboratory for the purpose of aiding the physician in his diagnosis are referred to as laboratory findings. Included are microbiologic investigations of suitable specimens (blood, spinal fluid, feces, urine, and swabbings) and a variety of serologic tests, including those referred to above.

Treatment and prevention

Specific treatment of an infection refers to the use of antiinfective drugs such as antibiotics (Fig. 4-18), sulfonamides, and antisera. The patient, however, must not be forgotten. Things must be done to ameliorate his discomfort. Sometimes symptomatic treatment is all that can be given. All of us realize, for example, that the doctor can do little more for a cold than treat the symptoms and wait for the virus to leave the body.

Obviously, our ultimate aim is to discover ways and means to protect the body against its microbial enemies. This is a difficult task, for it entails immunization, isolation, identification of carriers, eradication of vectors, and good sanitation. For best results against most infections, a variety of such measures must be used together.

QUESTIONS

1. What are some of the mechanisms by which disease-producing microbes attack and injure the body?
2. In the Gram method of staining, why are the microbes that take the counterstain referred to as "negative"?
3. About how many influenza virions would it take to span an inch?
4. What is the fundamental difference between viruses and rickettsiae?
5. The great majority of viruses cannot be seen with

the light microscope. Explain this fact on a basis of wavelength.

6. Penicillin inhibits or destroys bacteria by interfering with the synthesis of the cell wall. This is its so-called mechanism of action. Investigate the mechanism of action of the sulfonamides ("sulfa drugs").

7. Distinguish among diplococci, streptococci, and staphylococci.

8. Describe a virus and the way in which it attacks living tissue.

9. Describe the different types of protozoa.

10. Is a bacterial spore a "mode" of reproduction? Explain.

11. What is an antibiotic?

12. In the complement-fixation test, hemolysis signals a negative result. Explain.

13. An object that has been boiled in water may or may not be sterile. Explain.

14. Pasteurized milk is not sterile. Explain.

15. Name an infection involving a mechanical vector and an infection involving a biological vector.

16. A patient with an infectious disease may be clinically cured, but not bacteriologically cured. What does this mean?

17. There are two great categories of disease, infectious and noninfectious. What is the basic difference?

18. All contagious diseases are infectious, but not all infectious diseases are contagious. Explain and illustrate.

19. Distinguish between an exotoxin and an endotoxin.

20. Hyaluronidase has been referred to as a "spreading factor." Why?

21. Gastric juice does much to protect us against infection. Can you explain how?

22. What is the relationship between antigens and antibodies?

23. In previously immunized persons, a tetanus booster (tetanus toxoid) is considered sufficient protection for most injuries. This is not true for the nonimmunized. Why?

24. The Dick and the Schick tests are important skin tests not specifically mentioned in this chapter. For what are they used and what are their major similarities and differences.

25. According to many authorities, there are three kingdoms of living things. Explain.

26. Describe a typical bacterial cell.

27. Distinguish between spirilla and spirochetes.

28. Discuss the microbes belonging to the orders Eubacteriales and Actinomycetales.

29. In the microscopic study of a bacterial culture, why is it important to know the age of the culture?

30. Compare mitosis and binary fission.

31. All parasitic bacteria are heterotrophic. Explain.

32. Distinguish between an obligate aerobe and a facultative anaerobe.

33. Most bacteria do not grow in heavy syrups. Explain.

34. Why must all laboratory media be sterilized before use?

35. Sabouraud's dextrose agar has a pH adjusted to about 5. Why?

36. Explain why, when stained according to the Gram technique, gram-positive bacteria stain blue, and gram-negative bacteria stain red.

37. Describe the technique for doing an acid-fast stain according to the Ziehl-Neelsen method.

38. What is meant by "the normal flora"?

39. Compare the composition and use of vaccines, antisera, and toxoids.

40. Distinguish among the terms fungus, mold, and yeast.

41. How do molds and yeasts usually reproduce?

42. Compare the terms mycosis and dermatomycosis.

43. Parasitic worms fall into three categories. What are they?

44. Why are helminthiases (worm infestions) so common?

45. Consulting outside sources, describe in detail the treatment for beef tapeworm.

46. Name five protozoiases.

47. Why are lice and fleas called ectoparasites?

CHAPTER 5

OVERVIEW OF THE HUMAN BODY

This chapter takes a general look at the structure and function of the human body.

PLANES AND POSITIONS

For descriptive purposes, the body, or any of its parts, is divided by three planes (or sections): *sagittal, coronal,* and *transverse.* A sagittal plane runs vertically from front to back and divides the body into left and right portions; a midsagittal plane is one passed through the midline, dividing the body into left and right halves. A coronal, or frontal, plane is a vertical plane that divides the body into front and back. A transverse, or horizontal, plane divides the body into upper and lower parts.

There are many anatomical terms used to indicate direction or position: anterior or ventral (in front of), posterior or dorsal (in back of), superior (above), inferior (below), cranial (nearest or toward the head), caudal (away from the head), medial (nearer the midsagittal plane), lateral (farthest from the midsagittal plane), proximal (nearest, or closer to the point of origin of a part), and distal (remote, or farther from the point of origin of a part).

PARTS AND REGIONS

Precise terms are used in referring to a particular part or region of the body. These are familiar names for the most part, but we must be certain that we use them with their true anatomical meaning (Fig. 5-1). Especially useful for locating internal organs are the nine imaginary regions formed by the intersection of the horizontal and vertical lines drawn in the manner shown in Fig. 5-2.

BODY CAVITIES

The human body is a bilaterally symmetric structure with a backbone and an array of organs housed in a dorsal and a ventral cavity (Fig. 5-3). The ventral, or anterior, cavity lies before the spinal column and contains the bulk of the body's internal organs or *viscera.* A musculo-membranous partition called the diaphragm divides it into the thoracic cavity above and the abdominopelvic cavity below. The thoracic cavity, in turn, is subdivided into three other parts: the two pleural cavities containing the lungs; and the mediastinal cavity (mediastinum) containing the heart (enclosed in its pericardi-

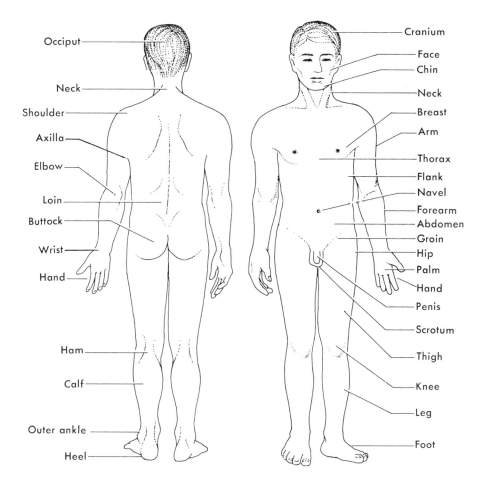

FIG. 5-1
Parts and regions of the body.

um), the thymus gland, the trachea, the bronchi, the esophagus, and certain muscles, nerves, veins, arteries, and lymphatic vessels. The abdominopelvic cavity is subdivided into an upper portion, the abdominal cavity, which contains the liver, the gallbladder, the stomach, the spleen, the kidneys, the pancreas, the ureters, and the intestines, and a lower portion, the pelvic cavity, which contains the bladder, rectum, sigmoid colon, and reproductive organs.

The dorsal, or posterior, cavity is a continuous space enclosed by the bones of the cranium and the vertebrae and lined by the dura mater (the innermost of the three membranes, or meninges, which envelop the brain and spinal cord). It is divided into the cranial cavity, which houses the brain, and the spinal cavity, which houses the spinal cord.

ORGANIZATION AND SYSTEMS

The cell is generally regarded as the basic structural and functional unit in living things.

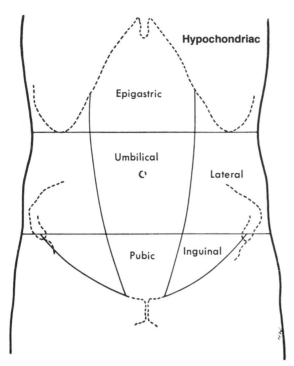

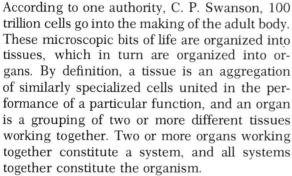

FIG. 5-2

Diagram of abdominal region. (From Francis, C. C: Introduction to human anatomy, ed. 6, St. Louis, 1973, The C. V. Mosby Co.)

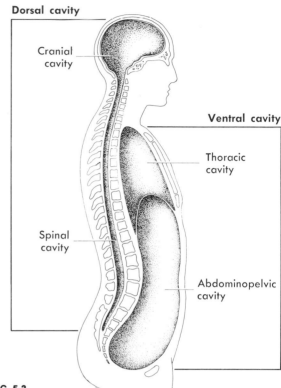

FIG. 5-3

Body cavities.

According to one authority, C. P. Swanson, 100 trillion cells go into the making of the adult body. These microscopic bits of life are organized into tissues, which in turn are organized into organs. By definition, a tissue is an aggregation of similarly specialized cells united in the performance of a particular function, and an organ is a grouping of two or more different tissues working together. Two or more organs working together constitute a system, and all systems together constitute the organism.

The classic approach to the study of the human body centers on the various systems. We shall take this approach, always keeping in mind that the systems must be considered in the framework of the body as a whole.

Nervous system

The nervous system is composed of the brain, the spinal cord, and the peripheral nerves. It has two functions: to permit the body to react to and act upon its environment; and to regulate the *milieu interne,* or internal environment. To respond to our environment, we first must be informed about it via sight, hearing, smell, taste, touch, pressure, pain, vibration, position of the body, and tension in the muscles. The nerves that carry such stimuli and the parts of the brain that analyze them constitute the *sensory* portion of the nervous system. Receiving continual appraisal of what is transpiring, the *motor* portions of the brain effect a response by transmitting impulses over the motor nerves that supply the skeletal muscles. So it is that we run when we see a snake!

Autonomic nervous system. The autonomic system, so named because it operates without conscious effort, refers to those peripheral nerves that supply smooth muscles, glands, and the heart. In the typical situation, an organ or gland receives two physiologically antagonistic autonomic fibers. In the case of muscle, impulses over one cause contraction, while impulses over the other permit relaxation. In the case of a gland, one fiber will stimulate secretion, and the other will inhibit it. The physiologist divides the autonomic system into two divisions, the *sympathetic* and *parasympathetic*. The sympathetic division acts, as Cannon said, to prepare the body for "fight or flight." For example, stimulation of the sympathetic fibers to the heart causes the heart to increase in rate and strength of contraction; impulses transmitted over the parasympathetic fibers slow the heart.

Skeletal system

The skeleton is the basic framework of the body. It is composed of 206 separate bones that articulate with one another at joints. The location and design of a particular joint determines the type and extent of movement it allows. In addition to supporting the body, the skeleton plays a key role in blood cell formation and calcium metabolism.

Muscular system

Properly speaking, the muscular system refers to all muscles of the body considered together, but usually the expression is used to refer to skeletal muscles only. Skeletal muscles are attached to bones by tendons; when they contract in response to nerve impulses, tension is exerted on the tendon, and movement is thereby effected. Other types of muscle include cardiac muscle, of which the heart is composed, and smooth muscle, which helps make up the walls of various other internal structures.

Circulatory system

The circulatory system is composed of the blood, the heart, and the vessels. The "right side" of the heart pumps blood to the lungs, where it picks up oxygen and releases carbon dioxide; the "left side" pumps oxygenated blood to all other parts of the body. We apply the term *pulmonary circulation* to the flow of blood between the heart and lungs, and the term *systemic circulation* to the flow between the heart and the body as a whole. A special division of the systemic circulation brings blood from the intestines and spleen through the liver and is called the *portal* circulatory system. The complex chemical machinery of the liver acts on the nutrients that have been absorbed from the intestine and detoxifies certain noxious agents before they have an opportunity to insult the body at large.

Blood flows from the heart via the arteries and returns via the veins. Between these major vessels, microscopic channels called *capillaries* permit the diffusion of oxygen and nutrients from the blood into the tissues and the diffusion of wastes from the tissues into the blood. In addition to oxygen, nutrients, and wastes, the blood transports an endless variety of substances, such as hormones, antibodies, and enzymes. Certain blood cells destroy microbes and other foreign agents via engulfment or *phagocytosis*. Blood plays a vital role in regulating the temperature and pH of the body.

Lymphatic system

The lymphatic system transports lymph, a watery fluid derived from interstitial fluid, from the tissues to the blood. As interstitial fluid accumulates in the tissues, it forces its way into the lymphatic capillaries and from there into valved lymphatic vessels. Two major vessels lead the lymph to veins in the neck. Thus, the interstitial fluid, which comes from the blood, is returned to the blood via the lymphatic system. Unlike the circulatory system, the lymphatic system does not require a pump. Instead, the movement of the body in general and the contraction of the muscles in particular are of sufficient force to move the lymph through the system.

Located along the lymphatic vessels are lymph nodes, through which the lymph must pass on its journey to the blood. Here, all large particles (old debris of dead tissue, bacteria, and the like) are filtered out and destroyed by phagocytic cells. Moreover, it is principally here that plasma cells (large lymphocytes) manufacture antibodies. The lymphatic system is one of the chief defense centers of the body.

Respiratory system

In a narrow sense, the respiratory system concerns the exchange between the air and the blood of the gases oxygen, carbon dioxide, and water vapor. In a broader and more realistic sense, however, respiration includes not only this exchange in the lungs, but also the exchange of gases between the blood and the tissues, and the transport of gases by the blood between the lungs and the tissues. The fundamental portions of the system include the lungs, the air passageways (nose, pharynx, larynx, trachea, and bronchi), and the respiratory muscles (the intercostal muscles and the diaphragm).

Digestive system

That the digestive organs constitute the bulk of the abdominal viscera is an indication of not only man's desire for food but also the importance of the processing of that food for body needs. The digestive apparatus includes the gastrointestinal tract (mouth, stomach, and intestines), the salivary glands, the liver, and the pancreas. The purpose of the system is to physically and chemically extract nutrients from the diet and permit the absorption of them into the blood through the intestinal mucosa.

Digestion, or the breakdown of food, is accomplished through digestive juices and enzymes secreted into the gut by the stomach, the intestine, the salivary glands, the pancreas, and the liver. These agents split carbohydrates, fats, and proteins into molecules small enough for the blood to absorb. The major nutrients include amino acids (from protein), glucose (from carbohydrates), glycerol and fatty acids (from fat), water, minerals, and vitamins. Once in the blood, these substances are transported throughout the body and ultimately enter the complex machinery of the cell, where they are burned for energy or forged into the materials of life.

Metabolism. In its broadest sense, metabolism simply means the sum total of life's chemical reactions. Stated otherwise, metabolism is what the body does with products of digestion. It includes *anabolism* (synthesis) and *catabolism* (degradation). The basic anabolic process is assimilation, the formation of new proteins and nucleic acids; the basic catabolic processes are digestion, both intracellular and extracellular, and cellular respiration.

Urinary system

As the blood courses through the organs and tissues, it collects the refuse of metabolism. Unwanted and noxious substances such as urea, uric acid, creatinine, and phenols must be taken out of the blood. It is the kidneys' job to do this. Situated as they are in the circulatory path, the kidneys seem to know what to take out and what to leave in. In the process of this picking and choosing, they produce urine, a concentrated aqueous solution of the waste products mentioned above. As urine emerges, drop by drop, from the microscopic collecting tubules of the kidneys, it works its way down the ureters to the bladder, where it waits to be discharged.

Sometimes substances appear in the urine that do not belong there. The presence of sugar, blood, albumin, or pus is a strong indication that all is not well along the genitourinary tract or in some other area of the body. Thus, urine serves as a valuable aid in the diagnosis of disease.

Endocrine system

Scattered throughout the body are islands of glandular tissue (ductless or endocrine glands) that release potent chemical agents called *hormones* directly into the blood. In concert with the autonomic nervous system, hormones regulate the metabolic profile and the physiologic processes of the body. The importance of the system immediately becomes apparent when we realize that infinitesimal changes in the output of these tiny glands can mean the difference between health and disease, and sometimes between life and death.

Reproductive system

Whereas the other systems are concerned exclusively with running the body, the reproductive systems function for posterity.

The ovum and the sperm cell, released by the ovary and the testis, respectively, contain 23 chromosomes each. In the act of fertilization, the two *gametes* fuse into a 46 chromosome cell called a *zygote*. The zygote divides again and again, and the *morula* stage attaches itself to the rich uterine lining, where it undergoes rapid development and growth. By the time the *embryo* stage has been reached, a full-fledged exchange of nutrients and wastes has been established via the umbilical cord and the placenta. Nine months following fertilization the development of the *fetus* is complete, and birth ensues.

Homeostasis

The cells throughout the body are aquatic, being surrounded for the most part by an aqueous medium known as intercellular fluid, interstitial fluid, or simply tissue fluid. This fluid has well-defined concentrations of mineral matter, nutrients, wastes, and the like. When the concentration of one of these solutes departs significantly from what it should be, the cells are in for trouble, and the body as a whole is headed away from the normal toward the abnormal, that is, toward disease. In the long run, the myriad physiological functions of the body are directly or indirectly concerned with maintaining a steady state in this internal environment (*milieu interne*), a tendency called homeostasis. The various body systems work together to achieve this state. The circulatory system, for instance, delivers oxygen and nutrients to the tissues and removes the wastes; the urinary system, in turn, rids the body of these wastes.

QUESTIONS

1. Order the following terms according to complexity, starting with the simplest: cells, atoms, organisms, systems, molecules, organs, tissues.
2. Supply synonyms for the following terms: posterior, anterior, vertebral canal, internal environment, steady state, frontal plane, horizontal plane, interstitial fluid.
3. Supply the appropriate antonym for inferior, dorsal, medial, distal.
4. What regions are lateral to the epigastric?
5. What region is inferior to the umbilical?
6. What positional relation do the knuckles and nails bear to the wrist?
7. What positional relation do the ankles bear to the knees?
8. What positional relation do the hams bear to the knees?
9. What is the positional relation of the nose to the chin?
10. The thigh lies between what two reference points?

11. Anterior and ventral are the same for man, but not for the dog. Explain.
12. The blood takes nutrients and oxygen to the tissues and wastes from the tissues. Name three other systems called into play incident to this function.
13. Damage to the bone marrow can prove fatal. Why?
14. If ductless glands are called endocrine, what do you think those with a duct are called?
15. Would you say all systems are vital to homeostasis? Discuss.
16. Skeletal muscle is often referred to as voluntary muscle, and smooth muscle, as involuntary muscle. Why?
17. Which autonomic system prevails in time of physical and emotional relaxation?
18. In speaking of chromosomes, we encounter the terms "haploid" and "diploid." What do they mean?
19. Commonly, an infection is marked by "swollen glands" (lymph nodes). Why?
20. The veins are part of which system?
21. Vital proteins escape into the tissue fluid as the blood travels through the capillaries. Are these proteins lost for good?
22. Why does oxygen diffuse from the blood into the tissues and not the other way around?
23. Why, precisely, must food molecules be broken down?
24. Assuming that the taste buds are normal, what do you think could have gone wrong in a person who has lost his sense of taste? (There are two possibilities.)
25. Distinguish among the terms: anatomy, histology, and cytology.

CHAPTER 6

THE CELL

According to the *American College Dictionary,* a cell, among other things, is "any small compartment." Robert Hooke apparently had this in mind when he applied the term in 1665 to the empty microscopic compartments he saw in cork. Hooke was referring to the nonliving cellulose walls, the shells of life. This empty-space idea held sway until 1835 when Dujardin emphasized the contents of the cell rather than the walls. He called the living stuff "sarcode," a term replaced 11 years later by Hugo von Mohl's *protoplasm.* Today we think of a cell as a circumscribed mass of cytoplasm with a nucleus.

The idea that the cell is the fundamental unit of living things—the so-called "cell theory"—was expressed in one way or another by several biologists, including Hugo von Mohl, but most authors usually credit the revelation to the German botanist Matthias Schleiden and the German zoologist Theodor Schwann. Schleiden put forward the theory in terms of plant tissues in 1838, and a year later Schwann applied the theory to all living things. In Schwann's words, "all organized bodies are composed of essentially similar parts, namely cells."

BASIC DESIGN

Cells have various shapes and sizes, each kind designed and equipped to do a particular task.

Surface cells protect; muscle cells contract; nerve cells relay electrical impulses; and so on. Cells vary in size from as small as 0.1 μm to as large as the largest bird's egg. Human cells have an average diameter somewhere in the vicinity of 10 μm.

A cell, as noted, is a circumscribed mass of *cytoplasm* with a nucleus (Fig. 6-1). All cells are circumscribed by a cell membrane, and plant cells are further circumscribed by an outer, nonliving cell wall. The fundamental parts of the cell are the nucleus, the cell membrane, and the cytosome or body of the cell. Most cells contain a single nucleus, but there are notable exceptions. Some organisms are without a distinct nucleus at all, and others are multinucleate. The classic idea of cell individuality has many exceptions. In some tissues the cell membranes (the boundaries between the cells) dissolve away, leaving one huge cytosome with many nuclei throughout its substance. Such a situation is referred to as a *syncytium.*

Nucleus

The nucleus is a roundish mass of specialized protoplasm usually situated near the center of the cell. It consists of a nuclear membrane and *nucleosome* (all the materials and structures enclosed by the membrane). The continuous sub-

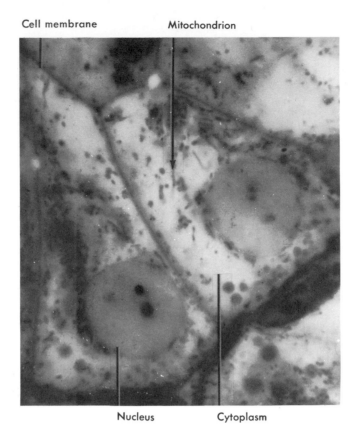

Cell membrane   Mitochondrion

Nucleus   Cytoplasm

FIG 6-1

Liver cells of turtle. Note nucleoli within nucleus. (× 1,000.) (From Bevelander, G.: Essentials of histology, ed. 6, St. Louis, 1970, The C. V. Mosby Co.)

stance of the nucleosome is *nucleoplasm,* suspended in which are spherical bodies called *nucleoli* and a filamentous network called *chromatin.* Chromatin is chiefly a DNA nucleoprotein; the nucleoli are composed chiefly of RNA nucleoprotein. DNA serves as the replicating master plan, which in conjunction with RNA directs the building and workings of the cell.

The nuclear membrane is a nucleoplasmic envelope with a high content of protein and lipid substances. As shown in Fig. 6-2, the nuclear membrane is double and continuous with the all-important endoplasmic reticulum

(see below). It is selectively permeable, like all animal membranes, and functions so as to regulate the exchange of ions and molecules between the nucleosome and the cytosome.

Cytosome

The cytosome, the part of the cell lying outside the nucleus, is composed of a continuous basic substance called the cytoplasm, in which is suspended an array of specialized structures and bodies called *organelles.* Typically, the outer portion of the cytoplasm (the *ectoplasm*) is clear, whereas the inner portion (the *endoplasm*) is

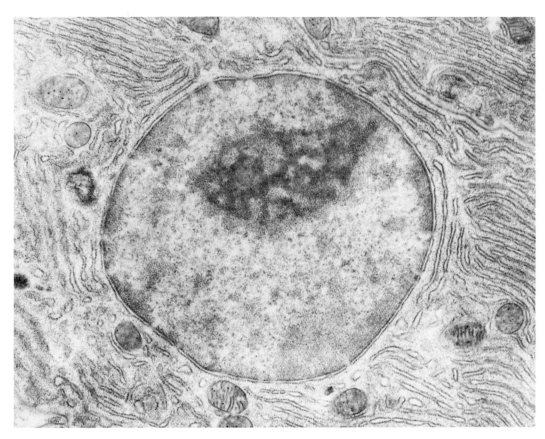

FIG. 6-2
Electron micrograph of nucleus of bat pancreas cell. Note pores in nuclear membrane through which nucleus possibly communicates with cytoplasm. In cytoplasm are round mitochondria and long, thin endoplasmic reticulum studded with ribosomes. ($\times$ 28,400.) (Courtesy Don W. Fawcett, M. D., Harvard Medical School, Boston.)

granular. The consistency of the cytoplasmic mass ranges from a rather freely moving sol to a rather firm gel.

A network of fine membranes, called the *endoplasmic reticulum* (Fig. 6-2), runs through the cytoplasm and serves to interconnect the nuclear membrane, the cell membrane, and all the organelles. Although the streaming and shifting about of the cytoplasm and the organelles strikes one as a mobile, hit-or-miss form of engineering,

the actual setup is extremely stable. Because of the endoplasmic reticulum, when things move they move together.

The organelles take all shapes and sizes and have diverse functions. Organelles recognized by most authors include centrosomes, fibrils, Golgi bodies, lysosomes, mitochondria, plastids, ribosomes, and vacuoles. The *centrosome*, or central body, is present in lower plants and in practically all animal cells. Each centrosome contains one or one pair of granular structures, called centrioles, which participate in cellular reproduction. *Fibrils* are filamentous structures that play a role in contraction and conduction. *Golgi bodies* are composed of one or more stacks of tiny flattened sacs called saccules. They are present in all cells, most conspicuously in glandular tissue, and serve as the primary site for the synthesis of carbohydrates and the packaging of secretions released by the cell. *Lysosomes* are membranous sacs of digestive enzymes produced by Golgi bodies. Substances to be digested are incorporated into these structures, and the resulting breakdown products are released to the cytoplasm. *Mitochondria* are roundish or rod-shaped organelles present in all but the most primitive cells. They are charged with respiratory enzymes and function as the chief factories in the output of energy-rich ATP. *Plastids* are spherical, egg-shaped, or disk-shaped organelles that play a key role in the synthetic chemistry of plant cells. The chloroplast, the best known plastid, contains all the machinery needed to carry on photosynthesis. *Ribosomes* are granular bodies extending throughout the endoplasmic reticulum. Composed of RNA and an assortment of enzymes, they synthesize proteins from amino acids. *Vacuoles* are fluid-filled membranous sacs. Sap vacuoles occur in plant cells; contractile vacuoles occur in unicellular organisms and function in osmoregulation; food vacuoles, which occur in both unicellular and multicellular organisms, serve as temporary digestive sites.

CELL MEMBRANE (TRANSPORT AND OSMOSIS)

The cell, or plasma, membrane is the outside living boundary of all cells. Composed of a thin layer of phospholipids sandwiched between two layers of protein, it is selectively permeable ("semipermeable") to the ions and molecules dissolved in the surrounding intercellular fluid. This selectivity is a key of life. When a cell dies, the membrane becomes freely permeable to all substances.

Probably in all but a few cases, a cell must expend energy to allow an ion or molecule to pass through its membrane. The task is especially difficult in those instances where the membrane has to chemically "pump" a substance from a lower to a higher concentration. The red cell has a much higher potassium content than the surrounding blood plasma and has to work hard to maintain it.

Two spectacular mechanisms employed by certain cells for moving substances through cell membranes in an active fashion, but only in one direction, inward, are *phagocytosis* and *pinocytosis*. In phagocytosis a segment of the cell membrane forms a pocket around a bit of solid material, and in pinocytosis a segment of cell membrane forms around a bit of liquid material. Such segments eventually pinch off and become cytoplasmic food vacuoles.

Also of importance is the outside influence of certain hormones, a lack of which can actually starve a cell. For example, glucose cannot enter most cells in significant amounts unless insulin is present. The mechanism of action here relates to some sort of stimulation of the membrane's transport system.

The cell membrane's selective permeability permits the osmotic movement of water. The most extensive shifts occur in solutions where the solute is entirely excluded from the cell. In *hypotonic* solutions, cells gain water and swell, in the case of animal cells sometimes to the point of bursting (a situation called *plasmoptysis* or, in red cells, *hemolysis*); in *hypertonic* solutions, cells lose water and shrink (*plasmolysis*). Only in an *isotonic* environment where the inflow and outflow are in balance, do cells retain their exact shape. Because of the restraining effect of the cell wall, plant cells are not nearly so sensitive to osmotic activity as animal cells.

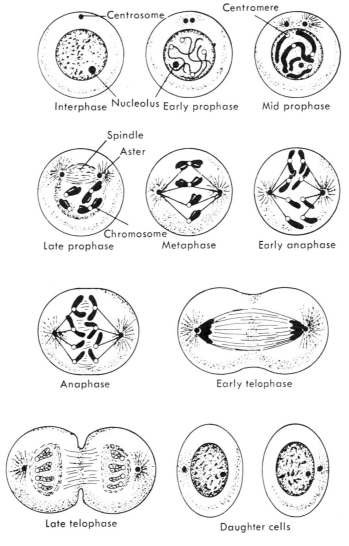

Labels in figure:
Centrosome · Centromere
Interphase · Nucleolus · Early prophase · Mid prophase
Spindle · Aster
Late prophase · Metaphase · Early anaphase
Chromosome
Anaphase · Early telophase
Late telophase · Daughter cells

FIG. 6-3

Diagram of mitosis in animal cell with four chromosomes. (From Levine, L.: Biology of the gene, ed. 2, St. Louis, 1973, The C. V. Mosby Co.)

MITOSIS

The ability of protoplasm to duplicate itself is the essence of life. At the molecular level, this is a matter of DNA replication; at the cellular level, it is a matter of division. Cellular division is a necessary consequence of DNA replication. Typically, higher cells divide via mitosis, a process that involves division of the cytoplasm (*cytokinesis*) and a complicated indirect division of the nucleus (*karyokinesis*). In successful mitosis the offspring (daughter cells) end up with all the chemical information possessed by the parent cell. Although the process is continuous, for instructional purposes it is best viewed in terms of four stages: prophase, metaphase, anaphase, and telophase (Fig. 6-3). The period between divisions is known as the interphase or resting stage. These terms are used also for meiosis (see below). Mitosis may take anywhere from a few minutes to several hours, depending on the type of cell, the species, and the chemical and physical environment.

Prophase

During prophase the chromatin material condenses into rodlike structures called *chromosomes,* each of which is composed of two spiral filaments (*chromatids*) held together by a tiny body called the *centromere*. The number of chromosomes is usually constant for each species—man has 46. Highlights of the prophase stage include the disappearance of the nucleoli and the nuclear membrane and, in the case of animal cells, the migration of the centrioles to opposite sides of the nucleus before its membrane is entirely broken down.

Metaphase

Metaphase begins with the formation of the *spindle,* an array of fibers radiating between the poles of the cell. In animal cells the poles are marked by the centrioles, which by this point in the mitotic process are surrounded by a radiating structure called the *aster*. Once the spindle is established, the centromeres of each chromosome become attached to the spindle fibers and arrange themselves in a plane (the equator) midway between the poles.

Anaphase

Each centromere divides, and the two resulting centromeres migrate along the spindle fibers to the opposite poles of the cell, each dragging along a chromatid in the process. Thus, the two chromatids of each chromosome part company and become new chromosomes. A chromatid is a chromatid until the division of the centromere, after which it changes its name.

Telophase

In telophase, the final stage, the chromosomes again become the thin threads of the chromatin network. The nucleoli and the nuclear membrane then emerge to complete the nucleus. In animal cells, this is accompanied by a division of the centriole and a pinching in two of the cytoplasm. By contrast, cytokinesis in plant cells is characterized by the formation, at the equator of the *cell plate,* a structure that gives rise to a new cell wall and thereby to two new cells.

MEIOSIS

The cells of an organism except its sex cells constitute the *somatoplasm* (or somaplasm); the sex cells are said to constitute the *germ plasm*. The cardinal difference between these two protoplasmic categories is the number of chromosomes, the cells of the somatoplasm having twice as many as those of the germ plasm. The sexual process involves the "addition" of chromosomes; if the chromosome number were not cut in half prior to fertilization, the result would be progressive doubling. Man, for instance, maintains his chromosome number of 46 (the *diploid* number) via sex cells with a complement of 23 chromosomes (the *haploid* or *monoploid* number). That is, sperm + egg = 46.

The chromosomes of diploid cells are homologous; they can be arranged in pairs of look-alikes. One chromosome of each such pair is of paternal origin, and the other of maternal origin (Fig. 6-4). By contrast, in haploid cells the chromosomes are nonidentical, the members of the various homologous pairs having parted company in the special process of cell division called meiosis. Whereas diploid cells are the

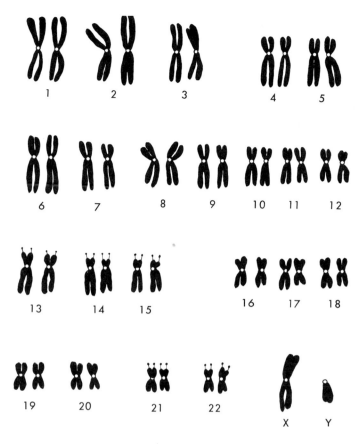

FIG. 6-4

Diagram (ideogram) of chromosomal constitution (karyotype) of human male at meiotic metaphase. Note arrangement and numbering of homologous pairs. (From McKusick, V. A.: Human genetics, Englewood Cliffs, N. J., 1964, Prentice-Hall, Inc.)

result of mitosis, haploid cells are the result of meiosis ("reduction division").

Meiosis consists of two consecutive cell divisions (Fig. 6-5), each involving the same four stages as mitosis. The first division is a reduction division because the number of chromosomes is reduced by half. For some yet unknown reason, in the prophase of the first meiotic division the homologous chromosomes are attracted to each other and pair up in a process called *synapsis*. Thus, instead of the 46 independently acting chromosomes present for mitosis, meiosis begins with 23 pairs. Just as in mitosis, each chromosome consists of two chromatids at this stage; the physical pairing of two homologous chromosomes is referred to as a *tetrad,* because it consists of four chromatids. During metaphase, the centromeres become attached to the spindle fibers, and with the onset of anaphase each tetrad is pulled apart, with the two chromosomes (each consisting of two chromatids) going to their respective poles. At the end of the telophase of the first meiotic division, two daughter cells emerge, each containing a member of each ho-

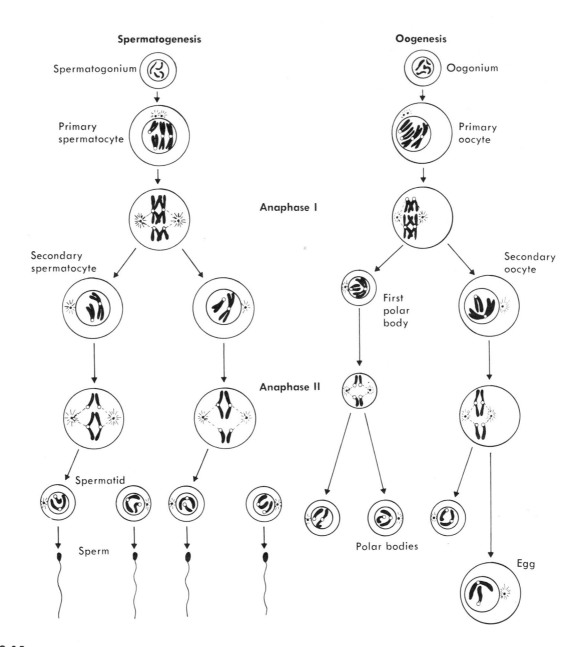

FIG. 6-5

Meiosis (gametogenesis) in male (spermatogenesis) and female (oogenesis). Note that starting cells have four chromosomes, and sex cells have two. Polar bodies are nonfunctional. (From Levine, L.: Biology of the gene, ed. 2, St. Louis, 1973, The C. V. Mosby Co.)

NH₂ ... (adenosine triphosphate structural diagram)

$$NH_2$$

Adenosine triphosphate ($\sim$ is high energy bond)

mologous pair of chromosomes that was originally present in the parent cell.

During the second division, which varies considerably in time of onset, the two daughter cells divide in essentially a mitotic fashion. At metaphase, the centromeres align themselves along the equatorial plane and duplicate; at anaphase the daughter centromeres, with the assistance of the spindle fibers, move to the respective poles, each dragging along a chromatid (now called a chromosome). Four haploid cells result, each containing an assortment of nonidentical chromosomes.

CHEMISTRY AND METABOLISM

Physical chemistry would describe protoplasm as a polyphasic colloid ranging in consistency from a sticky solution to a jellylike semisolid. There is nothing unique about its constituent elements or, for that matter, many of its compounds. The chemical key to the mystery of life resides in the way atoms and molecules are arranged and put together.

Usually 75% of protoplasm is water. The other inorganic constituents may be characterized as salts, the most important being chlorides, phosphates, and sulfates of sodium, calcium, and potassium. The organic constituents may be categorized as carbohydrates, lipids, proteins, and nucleic acids. Carbohydrates serve as fuel for energy; lipids serve as reserve fuel and to a lesser degree as structural elements; proteins serve as enzymes and collectively as the basic fabric, the "molecular backbone"; and nucleic acids act as "chemical information."

The sum total of the chemical processes of protoplasm constitutes metabolism. Metabolic reactions involving the conversion of simple substances into complex substances constitute anabolism; those involving the breakdown of complex substances into simple substances constitute catabolism. Anabolism is "constructive metabolism," and catabolism is "destructive metabolism." The key anabolic event is the DNA-directed synthesis of proteins, for with new protein comes new protoplasm. Like all anabolic reactions, this requires energy in the form of adenosine triphosphate (ATP), the "high power" molecule produced in the catabolic reactions of respiration. Indeed, most cellular activity is fueled by ATP (shown above). The energy of this molecule resides in its phosphate bonds ($\sim P$). Once they are broken, their pentup chemical energy becomes available immediately to the cell. Food energy is not immediately available; that is, the chemical bonds of food molecules are not energetic enough to be useful, as such, and it is therefore necessary—via respiration—to concentrate and package their potential energy in the form of fast-acting ATP.

CELLULAR RESPIRATION

As already indicated, the orderly release of energy sequestered in the chemical bonds of food molecules constitutes the process of respiration. Whereas food burns outside the cell in a simple,

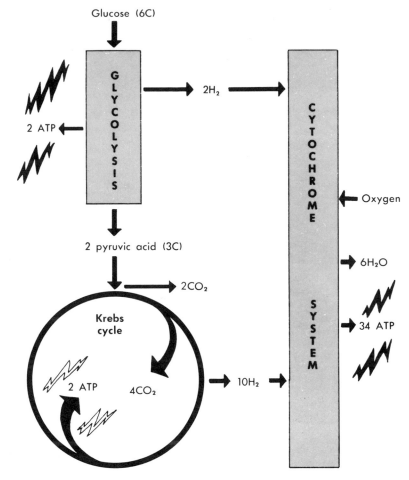

FIG. 6-6

Overview of cellular respiration. One molecule of glucose has six atoms of carbon and yields a net total of 38 molecules of ATP, two in glycolysis, two in Krebs cycle, and 34 in cytochrome system.

one-step manner, within the cell the process involves a galaxy of cunningly integrated enzymatic maneuvers.

Overview

The complexity of the process notwithstanding, the basic mechanism of cellular respiration can be expressed in two words, oxidation and phosphorylation. Oxidation involves the removal of hydrogen or electrons or both, and phosphorylation refers to reactions involving phosphate, particularly the formation of ATP from adenosine diphosphate (ADP) and inorganic phosphate (Pi). Oxidation stimulates phosphorylation and is responsible for the bulk of the ATP produced in the respiratory process. Oxidation and phosphorylation play out their roles within the framework of three integrated stages: glycolysis, the citric acid (Krebs) cycle, and the cytochrome system (Fig. 6-6).

Glucose (six carbons), the cell's chief fuel, commences glycolysis and emerges as pyruvic acid (three carbons), generating ATP in the process. Under anaerobic conditions, one molecule of glucose yields two molecules of pyruvic acid, four atoms of hydrogen, and a net synthesis of two molecules of ATP. The four atoms of hydrogen reduce the pyruvic acid to lactic acid (*acid fermentation*) in the case of muscle tissue and certain bacterial cells, or to ethyl alcohol and carbon dioxide (*alcoholic fermentation*) in the case of yeast cells. Under aerobic conditions, the four atoms of hydrogen enter the cytochrome system and there produce six molecules of ATP.

Under aerobic conditions, pyruvic acid enters the citric acid cycle and is broken down to carbon dioxide, ATP, and hydrogen, the latter entering the cytochrome system to turn out more ATP. Specifically, one molecule of pyruvic acid yields three molecules of carbon dioxide and 15 molecules of ATP. Since one molecule of glucose yields two molecules of pyruvic acid, the citric acid cycle, in conjunction with the cytochrome system, generates a total of 30 molecules of ATP. This plus the eight molecules of ATP generated in the glycolysis-cytochrome system amounts to a grand total of 38 molecules. The waste products in this whole process are the six molecules of carbon dioxide generated in the citric acid cycle and the six molecules of water generated in the cytochrome system. The overall aerobic respiratory process, then, can be expressed as follows:

$$C_6H_{12}O_6 + 6O_2 \longrightarrow 6CO_2 + 6H_2O + 38 \text{ molecules ATP}$$

With this overview firmly in mind, let us take a closer look at each of the three stages, starting at the beginning with glycolysis.

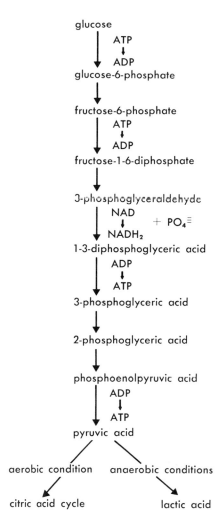

FIG. 6-7
Biochemical highlights of glycolysis via Embden-Meyerhof pathway. The many and various enzymes involved are not shown.

Glycolysis

The first step in glycolysis (Fig. 6-7) is the phosphorylation of glucose ($C_6H_{12}O_6$) to glucose phosphate. As in all cellular reactions, an enzyme is needed, in this case hexokinase. The phosphate is supplied by ATP. Glucose phosphate is isomerized and phosphorylated to fructose diphosphate, which in turn is split into phosphoglyceraldehyde; specifically, one molecule of six-carbon fructose diphosphate yields two molecules of three-carbon phosphoglyceraldehyde. The latter compound then undergoes oxidation and phosphorylation to become diphosphoglyceric acid.* Next, diphosphoglyceric acid reacts with ADP to produce ATP and phosphoglyceric acid, which then dehydrates to become phosphoenolpyruvic acid; in turn, phosphoenolpyruvic acid reacts with ADP to form ATP and the metabolite pyruvic acid. Under anaerobic conditions, $NADH_2$ reduces pyruvic acid either to lactic acid (acid fermentation) or to ethyl alcohol and carbon dioxide (alcoholic fermentation)†; under aerobic conditions, pyruvic acid enters the Krebs cycle, where it is degraded to carbon dioxide and hydrogen, and $NADH_2$ enters the cytochrome system.

The ATP's generated in glycolysis proper number four: one molecule of glucose yields two molecules of phosphoglyceraldehyde, each of which in turn yields two ATP's. However, since it takes two molecules of ATP to convert one molecule of glucose to fructose diphosphate, there is a net gain of only two molecules. Under aerobic conditions, the four atoms of hydrogen transferred to the cytochrome system generate six molecules of ATP. Thus, under anaerobic conditions glycolysis generates two molecules of ATP, whereas under aerobic conditions glycolysis, in conjunction with the cytochrome system, generates a net total of eight molecules of ATP.

Citric acid (Krebs) cycle

Under aerobic conditions three-carbon pyruvic acid diffuses into the mitochondria and reacts

---

*Oxidation is effected by the removal of hydrogen; that is, nicotinamide adenine dinucleotide (NAD) plus hydrogen yields $NADH_2$.
†In the plant world.

with coenzyme A and NAD to produce $NADH_2$, carbon dioxide, and two-carbon acetyl-coenzyme A. This is the beginning of the citric acid cycle, the highlights of which are shown graphically in Fig. 6-8. This pathway is called a cycle because the four-carbon oxaloacetic acid produced in the last step reacts with acetyl-coenzyme A in the first step.

Dehydrogenation occurs at five points in the cycle. At four, NAD is involved, and at one, between succinic acid and fumaric acid, hydrogen enters the cytochrome system without the transfer assistance of NAD. One molecule of pyruvic acid supplies five hydrogen atoms to the cytochrome system, resulting in the generation of 14 molecules of ATP. Additionally, one molecule of ATP is generated in the citric acid cycle proper in the conversion of succinyl-coenzyme A to succinic acid. In all, then, the citric acid cycle, in conjunction with the cytochrome system, grinds out 15 molecules of ATP from one molecule of pyruvic acid.

Three molecules of the waste carbon dioxide are produced, one at the start of the cycle, one between oxalosuccinic acid and α-ketoglutaric acid, and one between the latter and succinyl-coenzyme A. This accounts for all the carbon in pyruvic acid.

Cytochrome system

The cytochrome, or electron transport, system is a mitochrondrial complex of enzymes and coenzymes especially designed for the transfer of electrons and the generation of ATP. Electrons (derived from hydrogen) are passed down a line of coenzymes, resulting in a series of oxidation-reduction reactions and the release of energy to synthesize ATP from ADP and inorganic phosphate. The major components include NAD, FAD (flavin adenine dinucleotide), cytochrome c, cytochrome b, cytochrome a, and cytochrome a_3 (cytochrome oxidase). The cytochromes contain an atom of iron per molecule, and this atom

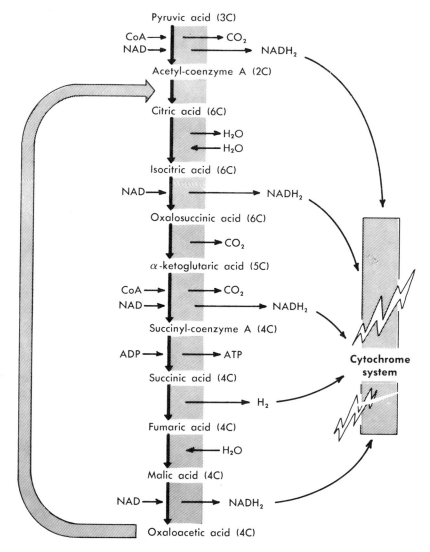

FIG. 6-8

Citric acid cycle. Read as follows: Pyruvic acid (three carbons per molecule) reacts with CoA (coenzyme A) and NAD (nicotinamide adenine dinucleotide) to produce CO_2, $NADH_2$, and acetyl-coenzyme A; acetyl-coenzyme A then reacts with oxaloacetic acid to produce citric acid; and so on.

is responsible for the reduced (Fe^{++}) and oxidized (Fe^{+++}) state of these agents. We may represent oxidized and reduced cytochrome c as Cc^{+++} and Cc^{++}, respectively.

As shown in Fig. 6-9, $NADH_2$ initiates the system by passing on its hydrogen to FAD. Next, $FADH_2$ donates one electron to each of two molecules of oxidized cytochrome b (Cb^{+++}) and two

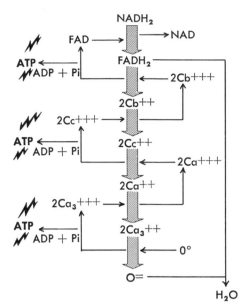

FIG. 6-9

Cytochrome system. Read as follows: FAD reacts with NADH$_2$ to produce NAD and FADH$_2$; FADH$_2$ reacts with Cb^{+++} to produce FAD, Cb^{++} and 2H$^+$; Cb^{++} reacts with Cc^{+++} to produce Cb^{+++} and Cc^{++}; and so on. Electrons pass through the system, releasing ATP at three points and ultimately reacting with 2H$^+$ and oxygen to form water. Pi, inorganic phosphate.

protons (hydrogen nuclei) to the mitochondrial fluid. Reduced cytochrome b (Cb^{++}) passes on one electron to oxidized cytochrome c (Cc^{+++}), and, in turn, reduced cytochrome c (Cc^{++}) passes on one electron to oxidized cytochrome a (Ca^{+++}), reducing it to Ca^{++}. Oxidized cytochrome a$_3$ (Ca$_3{}^{+++}$) is then reduced to Ca$_3{}^{++}$ by Ca^{++}, and Ca$_3{}^{++}$ passes on one electron to neutral oxygen; more precisely, two molecules of Ca$_3{}^{++}$ donate one electron each to one atom of oxygen (O), reducing it to O$^=$. In the final step, O$^=$ couples with the two protons (2H$^+$) to form a molecule of H$_2$O.

For each pair of electrons passed down the cytochrome chain, three molecules of ATP are produced. In the dehydrogenation of succinic acid in the Krebs cycle, however, the "NAD step" is not involved, and only two molecules of ATP are produced. In sum, for each molecule of glucose, the cytochrome system produces six molecules of ATP from the two pairs of hydrogen atoms derived from glycolysis, and 28 molecules (14 for each molecule of pyruvic acid) from the ten pairs of hydrogen atoms derived from the Krebs cycle. This total of 34 molecules of ATP plus the four molecules produced outside the cytochrome system amount to a grand total of 38 molecules of ATP per molecule of glucose.

Respiratory efficiency

From the standpoint of physics, the cell holds its own in the conversion of energy. From 1 mole (180 g) of glucose, with an energy content of 690,000 calories, the respiratory process produces 38 moles of ATP, with an energy content of 380,000 calories, and 310,000 calories of "lost" heat. This amounts to an efficiency of 55% in the conversion of glucose chemical energy into ATP chemical energy. By comparison, an internal combustion engine has an efficiency of about 25%. Moreover, respiratory heat is not actually "lost," for heat, besides keeping us warm, is essential to the cell's myriad enzymatic reactions.

METABOLIC INTEGRATION

The respiratory process serves as the common denominator for almost all the catabolic chemical reactions going on within the cell. Several intermediates produced in the catabolism of other foods link in with the metabolism of glucose. As shown in Fig. 6-10, glycerol, a product of fat digestion, enters glycolysis as phosphoglyceraldehyde; fatty acids, the other products of fat digestion, enter the Krebs cycle as acetylcoenzyme A; and amino acids, the products of protein digestion, enter the Krebs cycle as acetylcoenzyme A, oxaloacetic acid, and α-ketoglu-

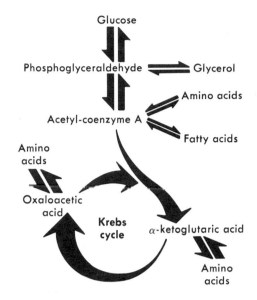

FIG. 6-10

Metabolic integration, indicating how protein, as amino acids, and fat, as glycerol and fatty acids, flow into same molecular pathways as glucose.

taric acid. Equally important is the feature of convertibility: Glucose, in excess, is converted to fat; amino acids and fats can be converted to glucose; amino acids can be converted to fat; and so on. In short, to understand cellular respiration is to understand metabolism.

GENETIC CODE

The chromosome is composed of nucleoprotein, a rather loose sort of compound consisting of protein, RNA, and most especially DNA. DNA is a single molecule (p. 34) made up of two polynucleotide chains wound around each other in a helical fashion (the famed Watson-Crick double helix). The chains are joined at their bases in a special way—adenine is bonded to thymine, and cytosine, to guanine. During mitotic prophase when the chromosomes duplicate, these bonds are broken, and the double helix becomes unzipped and unwound. Each chain

is now free to string together a new partner chain exactly like the old one. This is called replication.

The information coded on DNA is read through the mechanisms of transcription and translation, in that order. In transcription, the unwound DNA chain effects the synthesis of a chain of messenger RNA (mRNA), which thereupon leaves the nucleus via the endoplasmic reticulum and concentrates in the ribosomes.* Here, in translation, transfer RNA's (tRNA), each coupled with a particular amino acid, are aligned along the mRNA chain according to their matching bases, and the respective amino acids are then enzymatically clamped into a polypeptide chain or protein.

Replication versus transcription

In replication, DNA's two polynucleotide helices unwind and separate along the line of the weak hydrogen bonds between the bases (p. 34). This occurs in an environment rich in triphosphate nucleotides, which are attracted to the unsatisfied bonds, deoxyadenylic acid coupling with deoxythymidylic acid, and deoxycytidylic acid coupling with deoxyguanylic acid.† Finally, the nucleotides are joined together by DNA polymerase. Thus, the parent double helix duplicates itself, each daughter double helix possessing one of the original polynucleotide chains.

In transcription, the double helix also unwinds and separates, but this time one of the two polynucleotide helices (apparently only one is involved) strings together a polynucleotide chain of mRNA (Fig. 6-11). The mechanism is basi-

---

* Recent work has turned up a particular kind of DNA called informational DNA (iDNA), which, in at least some cases, is the actual molecule that effects the synthesis of mRNA.
† For example, deoxyadenosine triphosphate bonds to thymine as a consequence of the splitting away of two high energy phosphate groups, leaving the monophosphate (deoxyadenylic acid) in position.

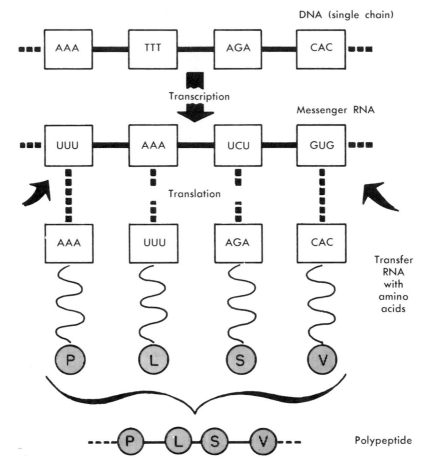

DNA (single chain)

| AAA | TTT | AGA | CAC |

Transcription

Messenger RNA

| UUU | AAA | UCU | GUG |

Translation

| AAA | UUU | AGA | CAC |

Transfer RNA with amino acids

P L S V

P L S V   Polypeptide

FIG. 6-11

Overall look at transcription and translation in making of a polypeptide. *A*, adenine; *T*, thymine; *G*, guanine; *C*, cytosine; *U*, uracil; *P*, phenylalanine; *L*, lysine; *S*, serine; *V*, valine.

cally the same as that in replication, except for the use of uridylic acid instead of deoxythymidylic acid. Uridylic acid, via its base uracil, is attracted to the unsatisfied bonds of adenine (of deoxyadenylic acid). The final stringing together is accomplished by RNA polymerase. Once formed, the single chain of mRNA peels from the DNA template and makes its way to the ribosomes of the cytoplasm.

The code

By transcription, then, DNA passes along to mRNA the genetic code, which is eventually translated into polypeptides and proteins. The code resides in the four bases—the four "letters" —in different combinations or "words." Two-letter words are not enough because they provide a language of only 16 words, and at least 20 words are needed, one for each amino acid.

Three-letter combinations are more than enough, providing a language of 64 words. Much brilliant research has shown that the genetic language is indeed a three-letter word affair. Although these three-letter words are actually nucleotide triplets called *codons,* it is customary to think merely in terms of the bases. In speaking of adenine (A), for example, the nucleotides adenylic acid (in RNA) and deoxyadenylic acid (in DNA) are understood.*

One gene, one polypeptide

The better known polypeptides average about 150 amino acid units per molecule, suggesting an mRNA template of 150 codons, and in turn, a DNA template of the same length. Such a DNA segment turns out to be what has been referred to somewhat loosely as a "gene." Thus, the genetic code boils down to one gene, one polypeptide, or according to some authors, "one gene, one protein," or "one gene, one enzyme."

Translation

The translation of the message carried by mRNA into a polypeptide or protein involves ribosomes and tRNA. Each amino acid is picked up and transfered to the appropriate position along the mRNA polynucleotide chain by a particular tRNA. Specifically, tRNA picks up the amino acid at its attachment site at one end of the molecule by means of ATP and a certain enzyme and then positions itself and the amino acid by means of its recognition site at the other end. The recognition site is a nucleotide triplet that bonds with the complementary mRNA codon according to the "base-pairing rule" (i.e., U-A and C-G). In this manner, the requisite amino acids are assembled, positioned, and bonded to produce the polypeptide (Fig. 6-11). These fantastic events occur in the ribosome, a

*Likewise, C, G, T, and U stand for the bases cytosine, guanine, thymine, and uracil, respectively, and their corresponding nucleotides.

cytoplasmic organelle composed of another kind of RNA. The ribosome clearly takes an active part in the whole process.

Errors

Nature makes mistakes, to be sure, and on occasion the genetic code amounts to a message of misery or even death. Errors can occur in the process of either replication or transcription. The wrong nucleotide, an extra nucleotide, or a missing nucleotide are three possibilities.

The classic genetic error is sickle cell anemia, classic because it was the first disease to yield its genetic secrets in such unbelievable detail. The disease gets its name from the distorted red cells, some of which appear sickle shaped (Fig. 6-12). Such cells are exceptionally fragile and readily destroyed, hence the anemia. Also, they clump together, clogging the blood vessels and interfering with the circulation of the blood. The victim of sickle cell anemia seldom lives beyond his teens.

The faulty red cells stem from faulty hemoglobin. A molecule of hemoglobin consists of four polypeptides (two alpha and two beta) interwoven among four iron-containing heme groups. The only difference between normal hemoglobin (HbA) and sickle cell hemoglobin (HbS) relates to a single amino acid in the beta chains. Whereas the HbA molecule has glutamic acid at a certain position, HbS has valine. Theoretically, the codon GAG (which spells glutamic acid) may have been altered to the codon GUG (which spells valine) by a DNA change of CTC to CAC. In short, a single nucleotide, indeed, a single base, can make the difference between life and death!

Regulation

A master plan must include built-in schemes and devices for self-regulation. No one pretends to know the regulatory secrets of the genetic code, life's master plan, but thanks to François

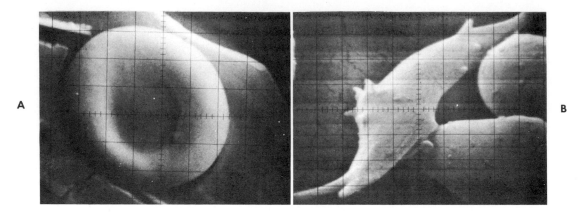

FIG. 6-12
A, Normal red blood cells. **B,** Those of sickle cell anemia. (× 10,000.) (Courtesy Patricia Farnsworth, Ph.D.; micrograph taken by Irene Piscopo of Philips Electronic Instruments, Mt. Vernon, N. Y., on a Philips EM 300 electron microscope with scanning attachment.)

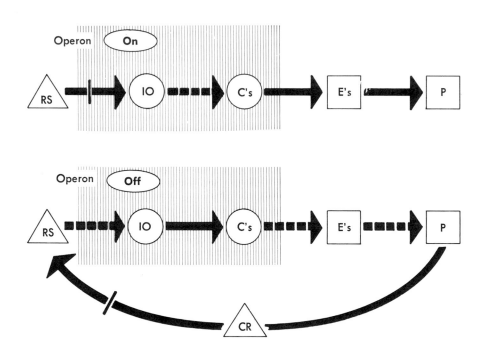

FIG. 6-13
One version of "on again-off again" mechanism suggested by Jacob-Monod genetic model. Operon on: Repressor substance (*RS*) blocks inhibitory operator (*IO*), thus allowing (dashed arrow) cistrons (*C's*) to produce enzymes (*E's*), which in turn produce product (*P*). Operon off: *P*, acting as corepressor (*CR*), blocks *RS*, thereby allowing (dashed arrow) *IO* to inhibit (solid arrow) *C's* and curtail (dashed arrows) output of *E's* and *P*.

Jacob and Jacques Monod we have some idea of what such secrets possibly may be. According to the Jacob-Monod model, the 450-nucleotide segment of DNA that we have been calling the gene is more appropriately called a structural gene, or *cistron*. A group of cooperating cistrons linked to a so-called operator gene, or *operator*, constitutes a DNA segment called an *operon*. In a way, the operon is a big gene whose function it is to turn out all the mRNA's needed to do a certain job. The "lac operon" of the bacterium *Escherichia coli,* for instance, turns out all the enzymes needed to metabolize the sugar lactose.

To hypothesize further, the operator, depending on the particular operon, either stimulates ("turns on") or inhibits ("turns off") the constituent cistrons and is itself stimulated or inhibited by a nearby gene, called a *regulator,* that is not part of the operon. A given regulator may regulate a number of operons. Presumably the regulator is on all the time, producing a repressor substance that turns the operator off. At times an agent called a *corepressor* may neutralize the repressor substance and thus in high enough concentration turn the operator on. Any number of things can act as a corepressor—light, heat, hormones, growth factors, or the operon's products, to suggest a few. One version of the Jacob-Monod model works like this (Fig. 6-13): If a certain operon with its inhibitory operator (IO) off codes the necessary enzymes (E) needed for the manufacture of a certain product (P), the latter at a certain concentration turns the inhibitory operator on—and thereby the operon cistrons (C) off—by acting as a corepressor against the regulator's repressive substance that normally turns the inhibitory operator off and the operon on. Conversely, when the concentration of this certain product begins to drop, the repressor substance is free to act, turning the inhibitory operator off and the operon cistrons on. This negative feedback situation or something like it must exist, since otherwise the cell could not keep its balance, producing things when needed and curtailing their production when not needed.

Differentiation

The Jacob-Monod model also affords some insight into one of the great mysteries of life—cell differentiation. The sperm fertilizes the egg, and the resulting zygote goes on to become a ball of cells, each of which in time transforms itself into a "cooperating individual." Some cells become blood cells, some become nerve cells, some become muscle cells, some manufacture insulin, and so on. The cells differentiate even though they carry the same genetic information.

How do we explain all this? How do we explain, for instance, that all cells carry the instructions for making insulin but only certain cells of the pancreas actually produce the hormone? The answer may be that many cistrons do not function until turned on, after which they stay on. Something turns on the insulin cistrons in the cells of the pancreas, and something keeps them off elsewhere. Quite conceivably, the insulin-producing cistrons have their inhibitory operator in the on position in all cells except those certain few in the pancreas, where the operator might be repressed by some repressor substance. Inasmuch as the body needs insulin to burn sugar, sugar is an ideal candidate for the role of repressor. Such speculation, of course, always brings forth its own brand of questions, and in this case one well may ask why sugar should act as a repressor in some cells and not in others. The extent of our scientific knowledge increases through the propagation and answering of such questions.

QUESTIONS

1. The cells lining the inside of the cheek average some 50 μm in width. How many of these cells would it take to span an inch?

2. Distinguish between the terms cytosome and cytoplasm.
3. Compare the cell membrane and cell wall with regard to structure and function.
4. What is sol? a gel?
5. Do lysosomes manufacture enzymes?
6. Compare the function of plastids and mitochondria.
7. In a sense, the cell membrane is continuous with the inside of the cell. Explain.
8. What is the role of RNA?
9. What are vacuoles?
10. The cell membrane is said to be semipermeable. Discuss.
11. What is lignin? (Consult outside sources.)
12. Distinguish between plasmoptysis and plasmolysis.
13. Normal saline solution (0.9% sodium chloride) is said to be isotonic. What does this mean?
14. Distinguish between chromosome and chromatid.
15. Distinguish between centromere and centrosome.
16. Are chromosomes visible during the interphase?
17. Meiosis is often referred to as "reduction division." Why?
18. In meiosis, why is a pair of homologous chromosomes called a tetrad?
19. Does synapsis occur in mitosis?
20. Sex cells contain nonidentical chromosomes. Why?
21. Give the formula for potassium chloride, sodium sulfate, and calcium phosphate.
22. Distinguish between lipids and fats.
23. What is the source of energy used in the synthesis of ATP?
24. Illustrate the difference between active and passive transport.
25. Give the derivation of syncytium, cytosome, chromosome, ectoplasm, isotonic, karyokinesis, somatoplasm, synapsis, plasmolysis, homologous.
26. What is the plural of helix?
27. What is the difference between a helix and a spiral?
28. What is the difference between ribose and deoxyribose?
29. What is the difference between a nucleoside and a nucleotide?
30. What is the difference between a DNA polynucleotide and an RNA polynucleotide?
31. The bases of the DNA molecule are joined by hydrogen bonds. What does this mean?
32. Compare messenger RNA and transfer RNA. Are there other kinds of RNA?
33. Using the analogy of a coded message, define replication, transcription, and translation.
34. Many authors use the terms peptides, polypeptides, and proteins interchangeably. Discuss.
35. In replication, each daughter double helix ends up with an intact, original polynucleotide chain. What does this mean?
36. "Once formed, the single chain of mRNA peels from the DNA template. . . . " What is the meaning of "template"?
37. A four-letter alphabet and two-letter combinations afford a language of only 16 words. Account for this fact.
38. In your own words, state the central idea of "One gene, one polypeptide."
39. Distinguish between adenylic acid and deoxyadenylic acid.
40. Reference is made to the "base-pairing rule." What is this rule?
41. Explain how amino acids are linked together (polymerized) into a polypeptide.
42. On an overall basis, an error of replication has far greater implications than an error of transcription. Why?
43. Consulting outside sources, discover what basic points the following diseases have in common: porphyria, phenylketonuria, galactosemia, and Tay-Sachs disease.
44. Errors in genetic coding can and do occur by

chance. For instance, by chance the right nucleotide does not always go where it is supposed to go. Suggest possible causes other than chance.

45. François Jacob and Jacques Monod received the Nobel Prize. Give the year, area of the award (chemistry or medicine), and the nature of their work.

46. In your own words, define the following kinds of genes: operator, regulator, cistron, and operon.

47. In the negative feedback control system outlined in the text the cistrons were on when the inhibitory operator was off. In the framework of the Jacob-Monod model, could negative feedback be explained if the situation were reversed (i.e., cistrons off when inhibitory operator was on)? (Do not forget the regulator.)

48. Aside from regulator genes (per Jacob-Monod), extrinsic regulators must also be considered. Do you think that such things as metallic ions, oxygen, or hormones could affect or alter genetic action? Support your answer.

49. It is now well established that a developing cell is under varying degrees of influence from its neighbors. Propose some ideas as to how this might play a key role in embryogenesis and differentiation.

50. Respiration involves the conversion of food chemical energy into ATP chemical energy. Why doesn't the cell tap food energy directly?

51. ATP is said to have high-energy bonds. What does this mean?

52. Give three definitions of "oxidation."

53. Where there is oxidation, there is reduction, and vice versa. Explain.

54. What is a reducing agent? An oxidizing agent?

55. What is the literal meaning of glycolysis?

56. During glycolysis proper, one molecule of glucose yields four molecules of ATP, but this amounts to a *net gain* of only two molecules. Explain.

57. Under aerobic conditions, what happens to the $NADH_2$ produced in glycolysis?

58. What happens to the $NADH_2$ from glycolysis under anaerobic conditions?

59. What is the total ATP yield from two molecules of glucose under anaerobic conditions?

60. What is the total ATP yield from two molecules of glucose under aerobic conditions?

61. What happens to the phosphate group in the conversion of diphosphoglyceric acid to phosphoglyceric acid?

62. For yeast to ferment glucose, air must be excluded. Why?

63. What is the carbon dioxide yield in the citric acid cycle from one molecule of pyruvic acid?

64. What is the total number of hydrogen atoms passed on to the cytochrome system in the catabolism of one molecule of glucose?

65. The cytochrome system turns out a total of 34 molecules of ATP per molecule of glucose. Account for this figure.

66. The citric acid cycle, in conjunction with the cytochrome system, turns out a total of 30 molecules of ATP per molecule of glucose. Account for this figure.

67. A severe deficiency of nicotinamide, a B complex vitamin, causes a disorder called pellagra. What is the underlying pathologic mechanism?

68. Pantothenic acid is an integral part of coenzyme A. What is pantothenic acid?

69. $Cb^{+++} \rightarrow Cb^{++}$ is an example of reduction. Explain.

70. The last step in the cytochrome system involves the reduction of oxygen. Explain.

71. On the basis of Fig. 6-9 explain how glucose is converted into fat.

72. For amino acids to enter the citric acid cycle, they must first be deaminated. What does this mean?

73. Glycogen is a polymer of glucose. What does this mean?

74. In severe exercise there is an accumulation of lactic acid in muscle tissue and a concomitant "oxygen debt." Explain.

CHAPTER 7

TISSUES, GLANDS, AND MEMBRANES

The body's trillions of cells are organized into four basic kinds of tissue—epithelial, connective, muscular, and nervous. These differ in regard to size, architecture, and arrangement of the constituent cells, type and quantity of intercellular substance, location, and function. For the most part, it is not difficult to tell one kind from another. The basic tissues occur in several types and varieties, each designed for a special purpose.

EPITHELIAL TISSUE

Epithelial tissue, or epithelium, consists of cells tightly joined together by small amounts of intercellular cement (Fig. 7-1). The scarcity of intercellular space is highly characteristic. Since this tissue is devoid of a blood supply, it must derive its sustenance from the underlying tissue fluid. Epithelial tissue generally is fashioned into a membrane covering external and internal surfaces, including vessels and other small channels and cavities, and serves the functions of protection, secretion, excretion, adsorption, and sensory reception. The chief types of this tissue are discussed below.

Simple squamous

Simple squamous epithelium consists of a single layer of platelike cells arranged in an attractive and highly characteristic mosaic. This type of tissue lines the alveoli, the serous cavities, the heart, the vessels, the crystalline lens of the eye, and the labyrinth of the inner ear. The terms *endothelium* and *mesothelium* are applied to the simple squamous epithelium that lines the circulatory organs and the serous cavities, respectively.

Stratified squamous

Stratified squamous epithelium consists of several layers of cells that, as a rule, range in shape from cylindrical in the deepest layer to simple squamous at the surface (Fig. 7-1). Stratified squamous epithelium forms the outer layer of skin and lines the nose, the mouth, and the anus.

The cells in the deeper layers are continually multiplying and migrating upward. In so doing they are flattened, dehydrated, and hardened. The simple squamous cells at the top (the old cells) are continually being rubbed off to make

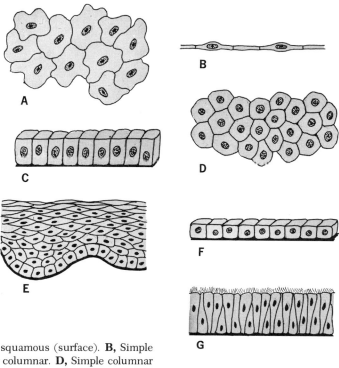

FIG. 7-1

Epithelial tissue. **A,** Single squamous (surface). **B,** Simple squamous (side). **C,** Simple columnar. **D,** Simple columnar (surface). **E,** Stratified squamous. **F,** Cuboidal. **G,** Pseudostratified ciliated columnar. **H,** Transitional.

room for a new surface. This remarkable sequence of events makes stratified squamous epithelium the most logical covering to protect body surfaces, particularly for those areas under constant environmental attack. This unique type of epithelium not only wards off outside forces, but also prevents the loss of body fluids. Furthermore, it contains certain microscopic structures (*receptors*) responsive to environmental stimuli.

Transitional

Transitional epithelium is similar to stratified squamous epithelium in that the cells are ar-

ranged in layers, but it differs in that there are fewer layers and especially in that the surface cells are not flattened. Transitional epithelium lines most of the urinary tract.

Simple columnar

The simple columnar type of epithelial tissue is composed of a single layer of tall, upright, cylindrical, or prismatic cells fitted together in a very orderly fashion (Fig. 7-1). Highly characteristic is the presence of *goblet* cells, unicellular glands that secrete mucus. This accounts for the expression *mucous membranes* (or mucosa).

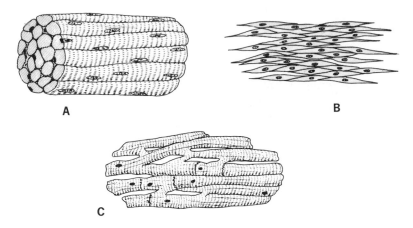

FIG. 7-2
Types of muscle. **A,** Striated. **B,** Smooth. **C,** Cardiac.

In its truest form, simple columnar epithelium lines the gastrointestinal tract from the lower esophagus to the anal opening. Throughout this extensive area it secretes digestive juices and absorbs fluids and digested food.

Simple ciliated columnar

Simple ciliated columnar epithelium differs from the simple columnar type in possessing at its free or exposed surface protoplasmic projections called *cilia*. These whiplike microscopic processes impel secretions and microscopic particles along the surface. This is accomplished not by the cilia lashing madly back and forth *en masse* but rather by a calculated succession of movements in one direction. Because of this and because of their astronomical number, the cilia generate a current of fantastic power. Simple ciliated columnar epithelium lines the nose, the uterine tubes, and the upper part of the uterus.

Pseudostratified ciliated columnar

Pseudostratified ciliated columnar epithelium is actually simple ciliated columnar epithelium modified by pressures occasioned by its location. The cells are squeezed in such a fashion that the displaced nuclei present a picture of stratification—hence, the expression pseudostratified (Fig. 7-1). This tissue lines the respiratory passageways.

MUSCLE TISSUE

Muscle tissue is characterized by its ability to contract. There are three types: skeletal, smooth, and cardiac.

Skeletal muscle

Skeletal muscle is composed of fibers marked by transverse bands (Fig. 7-2), and for this reason is often referred to as striated or striped muscle. Because it can be controlled by the individual, the expression voluntary muscle is also used. Skeletal muscle constitutes about two thirds of the weight of the body.

Smooth muscle

Whereas skeletal muscle is remarkably uniform in appearance and functional characteristics, smooth muscle is extremely variable. In general, however, smooth muscle is composed of elongated, cigar-shaped cells without striations (Fig. 7-2)—hence, the expression "smooth."

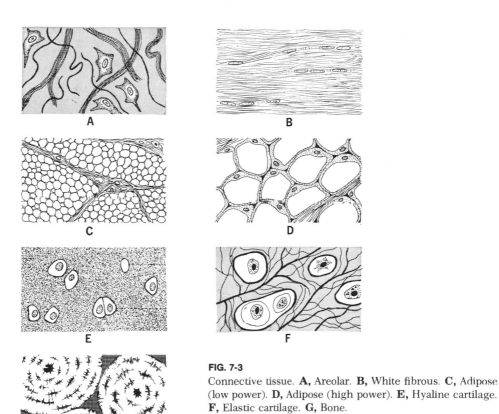

FIG. 7-3

Connective tissue. **A,** Areolar. **B,** White fibrous. **C,** Adipose (low power). **D,** Adipose (high power). **E,** Hyaline cartilage. **F,** Elastic cartilage. **G,** Bone.

Since it is the muscle of the viscera and largely under involuntary control, the terms visceral and involuntary are often used to describe this particular tissue. In contrast to the powerful and energetic movements of skeletal muscle, smooth muscle generally contracts in a less forceful manner. The waves of motion through the intestine (*peristalsis*) are typical of its behavior.

Cardiac muscle

Cardiac muscle is the muscle of the heart. Although similar to skeletal muscle in having cross striations, cardiac muscle differs from skeletal muscle in that its fibrous cells branch into one another, producing a syncytium, or multinucleate mass of protoplasm (Fig. 7-2). In contrast to skeletal and smooth muscle, cardiac muscle is characterized by its rhythm and, above all, its automatism.

CONNECTIVE TISSUE

Connective tissue (Fig. 7-3) is the most variable and widespread of all the tissues. Its intercellular substance is highly characteristic and determines whether a particular type of tissue is fluid, jellylike, plastic, or hard. In contrast to epithelial and muscle tissues, the amount of intercellular substance in connective tissue exceeds the cellular elements. Connective tissue houses the internal organs, sheathes the muscles, wraps the joints, and composes the blood and the skeleton. Connective tissue, therefore, may be said to support, protect, or nourish the body.

Connective tissue consists of three basic histologic elements: cells, fibers, and intercellular substance (also called *ground substance* or *matrix*). Because these three elements are the basis for distinguishing one connective tissue from another, we shall say a word or two about each before proceeding to specific types of tissue.

The chief connective tissue cells include the fibroblasts, the macrophages, the mast cells, the plasma cells, and the blood cells. Although fibroblasts do not appear in the blood, they are the most common cellular element in the other varieties of connective tissue. *Fibroblasts* are flat, star-shaped cells (Fig. 7-3) that play the unique role of manufacturing fibers for supportive purposes. *Macrophages* are large, ameboid cells that may be fixed or may wander through the tissues and engulf and destroy microbes, other cells, and foreign particles (phagocytosis). *Mast cells* are mononuclear, irregularly shaped structures with a granular cytoplasm that stores an anticoagulant substance called *heparin. Plasma cells,* usually found in lymphoid and related tissues, are similar to mast cells; they are now generally recognized as the chief producers of antibodies. *Blood cells,* of which there are several types, are discussed in detail in Chapter 13.

There are three types of fibers present in connective tissue: white, elastic, and reticular. White or collagenous fibers* are fine and glistening microscopic strands that occur in interlacing bundles (Fig. 7-3). Elastic fibers, on the other hand, are yellow and coarser and occur as separate strands that branch and join with one another. Reticular fibers are immature strands that compose the netlike supporting framework in lymphoid and myeloid tissues.

Areolar tissue

To the unaided eye, areolar, or loose, connective tissue looks like tissue paper—delicate, thin, and easily torn. Histologically, it represents connective tissue in its truest form: cells and white and elastic fibers scattered through a soft matrix (Fig. 7-3). Areolar tissue forms the basis of subcutaneous tissue and runs between the organs. It principally connects and supports.

Adipose tissue

Human adipose tissue contains fat and looks like the fat on meat. It may be described as a fine meshwork of areolar tissue with fat cells (cells with fat globules) distributed throughout the interspaces (Fig. 7-3). Present under the skin and about the viscera, adipose tissue supports, protects, and insulates, and stores energy as reserve food.

Fibrous tissue

Fibrous connective tissue is of three kinds: white, elastic, and fibroelastic. In white fibrous tissue, white fibers dominate the intercellular space (Fig. 7-3). There are few connective tissue cells and little matrix. This beautiful, pearl-white tissue is flexible but also amazingly strong. It composes organ capsules, deep fasciae, aponeuroses, ligaments, tendons, muscle sheaths, periostea, and the dura mater.

Elastic tissue, often called yellow fibrous tissue because of its color, is characterized by its pronounced elasticity (Fig. 7-3). This tissue composes certain ligaments of the vertebral column (*ligamenta flava*).

---

*These are composed of collagen, the most abundant protein in the body. We know the hydrated form of collagen as gelatin.

Fibroelastic tissue is dense connective tissue containing white and elastic fibers. The large arteries, such as the aorta, are supplied abundantly with this particular variety of tissue.

Cartilage

Cartilage is an opaque, bluish white tissue with the consistency of a hard rubber ball. It is the gristle of meat. Cartilage is composed of cells secluded in widely separated tiny spaces called *lacunae*. The matrix is a dense, translucent, homogeneous substance with or without fibrous elements. Cartilage is covered by a dense membrane, called the *perichondrium*, that serves to nourish and repair the tissue. The principal types of this tough tissue include hyaline, elastic, and fibrous.

Hyaline cartilage is a glasslike cartilage with a translucent, pale blue matrix (Fig. 7-3). It is found at the ends of bones (articular cartilage) and in the larynx; it composes the rings of the trachea and bronchi; and it forms the anterior nasal septum.

Elastic cartilage, or yellow fibrocartilage, is essentially hyaline cartilage with elastic fibers through the matrix. It is present in the external ear, the eustachian cartilage, the epiglottis, and certain arytenoid cartilages of the larynx.

Fibrous cartilage, or white fibrocartilage, is hyaline cartilage with interlacing bundles of white fibers running through the matrix. It is found between the vertebrae and composes the symphysis pubis.

Bone

With the exception of the bones of the face and skull, most bone may be characterized as ossified hyaline cartilage; that is, the semisolid matrix of the cartilage gives way to a hard, dense deposit of calcium salts. In contrast to the apparently inaccessible cells in cartilage, bone lacunae intercommunicate through microscopic channels, called *canaliculi,* arranged concen-

trically about larger channels known as *haversian canals* (Fig. 7-3). Each canal, with its concentric lacunae and *lamellae* (rings of matrix), constitutes a *haversian system.* By this system, the blood supply to bone is distributed to its matrix-locked cells via the canals and the canaliculi. Bone is either compact or cancellous (spongy). In *compact* bone the lamellae of the osseous matrix are put down layer upon layer, whereas in *cancellous* bone the lamellae form a delicate latticework with a spongy appearance.

Reticuloendothelial tissue

Reticuloendothelial tissue, identified by the presence of reticular fibers, composes the reticuloendothelial system: the spleen, the lymph nodes, the bone marrow, and the liver. The system is concerned with blood formation, storage of fatty materials, phagocytosis, and elaboration of antibodies.

Blood

Blood is a connective tissue characterized by a fluid matrix. Its histology and function will be discussed in detail in Chapter 13.

NERVE TISSUE

Nerve tissue composes the brain, the spinal cord, and the nerves. Although nerve cells, or *neurons,* vary considerably in architecture, they are all characterized by an octopus-like arrangement of cytoplasmic projections (*processes*) called *dendrites* and *axons* (Fig. 7-4). In the brain and the spinal cord, there are supporting cells as well as neurons. The supporting cells are characterized by their small oval nuclei and constitute the *neuroglia.* Macroscopically, nerve tissue is white or gray, soft, and friable.

MEMBRANES

Membranes are sheets of tissue that cover or line various parts of the body. They are composed

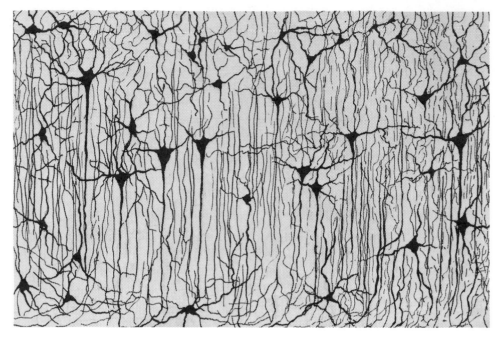

FIG. 7-4

Neurons of cerebral cortex. (From Brooks, S. M.: The V. D. story, New York, 1971, A. S. Barnes & Co., Inc.)

of a layer of epithelium upon a layer of connective tissue. The principal membranes are the mucous, the serous, and the synovial. The skin itself may be considered a membrane; the term is also used to describe various sheets of connective tissue that separate or connect certain structures, for example, the interosseous and the thyrohyoid membranes.

Mucous membranes

Mucous membranes line cavities or passageways leading to the *exterior* of the body. The most extensive include those lining the digestive tract, the genitourinary tract, and the respiratory passageways. The mucous membrane that lines the digestive tract has a stratified squamous epithelium from the mouth to two thirds of the way down the esophagus and a simple columnar epithelium the rest of the way. The mucosa of the respiratory tract is distinguished by its pseudostratified ciliated columnar epithelium. In the genitourinary mucosa, we find both simple ciliated columnar and transitional epithelium.

The term mucous membrane, or mucosa, is derived from the word *mucus,* which means the lubricating and protective slime secreted by the goblet cells in the epithelium. Mucus is essential to life.

Serous membranes

In contrast to mucous membranes, serous membranes line closed cavities. The principal membranes of this type include the pluera, which lines the thoracic cavity, the peritoneum, which lines the abdominal cavity, and the pericardium, which surrounds the heart. The epi-

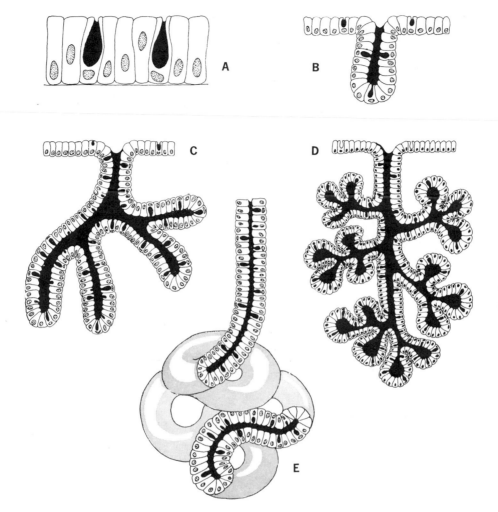

FIG. 7-5
Types of glands. **A,** Goblet. **B,** Simple tubular. **C,** Branched tubular. **D,** Compound tubuloalveolar. **E,** Coiled tubular.

thelium of serous membranes is simple squamous and bears the name mesothelium.

A serous membrane not only lines a particular cavity but also wraps about the organ contained therein. The *visceral* layer refers to the serous membrane about the organ, while the *parietal* layer refers to the layer that lines the cavity. In the close quarters of our internal world the

visceral and the parietal layers are in direct apposition, one rubbing against the other as the organs move about. When we breathe, for example, the visceral pleura about the lung glides along the parietal pleura of the chest wall (p. 64). To reduce the friction arising from this continual movement, nature has placed a small amount of watery fluid (serous fluid) between the two layers. Thus, the body obeys a principle of engineering: Where there are moving parts, there must be lubrication.

Synovial membranes

Synovial membranes line tendon sheaths, bursae, and joint cavities. The epithelium is the same type (mesothelium) that surfaces serous membranes. Synovial membranes secrete a fluid called synovia or synovial fluid, which is similar in composition to serous fluid and serves to lubricate the joints, the tendons, and the bursae.

GLANDS

A gland may be defined as an organ producing a specific secretion. There are two major types: those which secrete via ducts (*exocrine* glands), and those which secrete directly into the blood (ductless or *endocrine* glands). Endocrine glands secrete a variety of potent body regulators (*hormones*) and together constitute the endocrine system. Exocrine glands will be dealt with later as the occasion arises.

Glandular tissue is modified epithelium. The simplest gland is the goblet cell (Fig. 7-5). Multicellular exocrine glands are classified according to the design of the ducts into which the cells pour their secretion (Fig. 7-5). The basic types include the tubular, the saccular (also alveolar or acinous), and the racemose or mixed (combined tubular and saccular structures). These, in turn, are classified as simple or compound, depending upon the complexity of design. The more representative structures are shown in Fig. 7-5.

QUESTIONS

1. Name specific types of tissues you would expect to find in the finger, the stomach, the lung, the liver, and the heart.
2. What is the basic structural characteristic of connective tissue?
3. Name a tissue with a soft matrix; with a stone-hard matrix; with a liquid matrix.
4. Histologically speaking, what does the term "simple" denote?
5. What one word (a noun) characterizes muscle tissue?
6. Distinguish among the terms mucus, mucous, and mucosa.
7. Why is ciliated mucosa especially adapted to the respiratory tract?
8. In a sense the heart is one huge cell. Why?
9. Why is smooth muscle called visceral muscle? involuntary muscle?
10. What general term is applied to cells that engage in phagocytosis?
11. Cite two ways in which the reticuloendothelial system fights infection.
12. Where do we find macrophages?
13. What would you say is the most typical kind of connective tissue?
14. What is the *literal* meaning of canaliculi, lacunae, cancellous, lamellae, matrix, reticular, synovial?
15. What do we call the connective tissue of the nervous system?
16. What is the difference between an axon and a dendrite?
17. Support the view that skin is a membrane.
18. What does macroscopic mean? Give its antonym.
19. The abdominal cavity is often referred to as the peritoneal cavity. Explain.
20. Why, precisely, is pleurisy so painful?
21. See what you can find out about housemaid's knee, especially its association with synovial membranes.
22. Why is the serous membrane about the organ proper referred to as visceral?
23. Are salivary glands exocrine or endocrine?
24. Sweat glands are said to be coiled tubular. What does this mean?
25. How could you tell histologically whether a given tissue, such as a section of heart muscle, was normal or abnormal?

CHAPTER 8

WATER AND ELECTROLYTES

The total body water of the adult male ranges from 50% to 70% of his body weight, that of the adult female, from 45% to 65%. Water content varies with each person and is related mainly to the fat content of the body. There is virtually no water in fat. In obese people (35% fat or more) the water content may drop to as low as 40% of body weight, while in thin people (8% fat or less), water content hovers around 70%. For practical and clinical purposes, let us here consider the human body to be 60% water.

COMPARTMENTS

Physiologists consider body water to be compartmentalized, though this is not exactly correct since water is always on the move. According to this idea, there are two kinds of water within the body: intracellular, or water within the cells, and extracellular, or water outside the cells. Extracellular water is compartmentalized in turn into plasma water and intercellular, or interstitial, water (Fig. 8-1).

As noted, 60% of the body is water: 40% intracellular plus 20% extracellular. Interstitial water and plasma water, in turn, are 15% and 5% of body weight, respectively. According to this breakdown, a man weighing 70 kg (154 pounds) is composed of about 42 kg (0.60×70 kg) of water. Since 1 kg of water is equal to 1 l, this amounts to 42 l of water: 28 l (0.40×70) within the cells and 14 l (0.20×70) outside the cells. Of the latter volume, 3.5 l (0.05×70) is plasma water, and 10.5 l (0.15×70) is interstitial water.

COMPARTMENTAL BALANCE

As indicated, water in the body is not locked up in this or that compartment. On the contrary, it continually is seeping back and forth between compartments. How is it, then, that the percentage of water in a given compartment remains essentially the same, day in and day out? The answer to this question is the backbone of water (fluid) balance. In health, balance is maintained because the water that leaves a compartment is offset by the water that enters the compartment, and vice versa, a case of dynamic equilibrium. The major factors behind this equilibrium are discussed below.

Protein

Plasma proteins (mainly albumin), along with the hydrostatic pressure exerted by the pumping action of the heart, regulate the exchange of water between the plasma and the interstitial fluid compartment; the essential mechanism is depicted in Fig. 8-2. In brief, *blood pressure* (B.P.) at the arteriolar end of the capillary

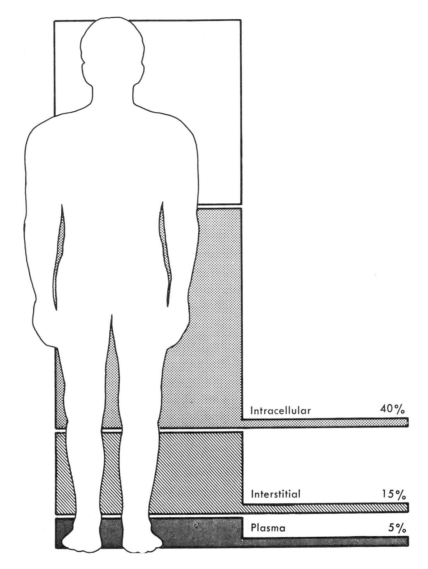

FIG. 8-1
Body water. (From Brooks, S. M.: Basic facts of body water and ions, ed. 3, New York, 1973, Springer Publishing Co., Inc.)

forces water through the permeable walls into the interstitial compartment. Water continues to leave until this pressure is offset by the osmotic (*oncotic*) pull of the protein molecules. At the venular end where the blood pressure drops below the *colloid osmotic pressure* (C.O.P.), water is drawn into the circulation from the interstitial compartment. The amount of water escaping from the capillaries at the arteriolar end is determined by the *filtration*

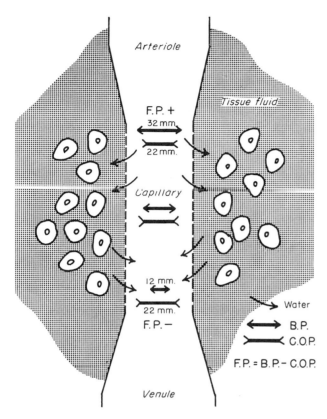

Arteriole

F.P. +

32 mm.

22 mm.

Tissue fluid

Capillary

12 mm.

22 mm.

F.P. −

Water

B.P.

C.O.P.

F.P. = B.P. − C.O.P.

Venule

FIG. 8-2

Fluid balance in tissues. (From Brooks, S. M.: Basic facts of body water and ions, ed. 2, New York, 1968, Springer Publishing Co., Inc.)

pressure (F.P.), or the difference between B.P. and C.O.P.: F.P. = B.P. − C.O.P. But note that at the venular end the F.P. value is negative, accounting for the water that is sucked in rather than squeezed out. Thus, the water lost at one end is balanced by the water gained at the other. This balanced flow of water throughout the tissues is often referred to as *Starling's law of the capillaries*.

Hypoproteinemia, an abnormal decrease in the amount of protein in the blood, leads to *edema,* the accumulation of excessive fluid in the tis-

sues. A decrease in the amount of protein means a decrease in C.O.P., which, in turn, means an increase in F.P. This explains the edema of malnutrition and why "the best remedy is beefsteak." For patients unable to take food by mouth, amino acid solutions are given intravenously.

Sodium

"Water, water everywhere/Nor any drop to drink" bears striking testimony to the role of sodium in compartmental balance. When a person drinks seawater, the kidney is unable to remove the incoming salt fast enough to prevent an abnormal buildup in the interstitial compartment. As a consequence, the osmotic pressure of the compartment intensifies and overpowers the pull of the protein within the cells. Cellular water is lost, and dehydration ensues. For each quart of seawater the castaway drinks, he eliminates 1½ quarts of urine, the difference coming from the cells. In essence, then, excess salt pumps the cells dry.

Lymph vessels

During times of increased metabolic activity, interstitial fluid is produced faster than it can be removed by the capillaries. Were it not for the drainage afforded by the lymphatic system, severe edema would be the rule, not the exception. Under normal conditions, increased interstitial pressure forces tissue fluid into the lymph capillaries (Fig. 8-3), then into the lymph vessels, and finally back into the blood (p. 67). The importance of this escape route is grotesquely demonstrated in filariasis, a disease in which filarial worms get into the lymph vessels and obstruct the flow of lymph. Here the edema is especially pronounced in the lower extremities.

Gastrointestinal secretions

The total volume of juices secreted by the gastrointestinal mucosa and accessory digestive

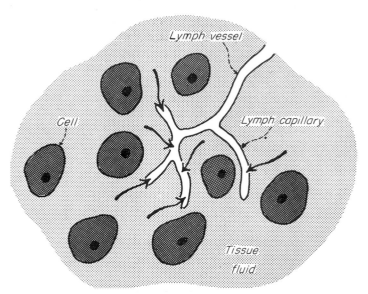

FIG. 8-3
Movement of tissue fluid into lymph capillaries.

organs in 24 hours amounts to about 8 l, over twice the volume of the plasma! All but about 200 ml is reabsorbed by the intestine, particularly by the large intestine. The importance of this resorption immediately becomes apparent in severe vomiting and diarrhea. These situations lead not only to water imbalances and dehydration but also to electrolyte disturbances.

INTAKE VERSUS OUTPUT

Plants and animals vary tremendously in their water requirements. Intake increases with size and, interestingly enough, in a mathematical way. Also of interest is the time required for a given species to imbibe a quantity of water equal to its own weight. A mouse takes 5 days; a cow, 2 weeks; a camel, 3 months; a tortoise, 1 year; a man, 1 month; and a cactus, 29 years! Regardless of amount of water taken in, however, the body normally balances the gain with an equivalent loss. Conversely, the body balances a loss with an equivalent gain. Either way, the central idea is balance. Man possesses phenomenal bal-

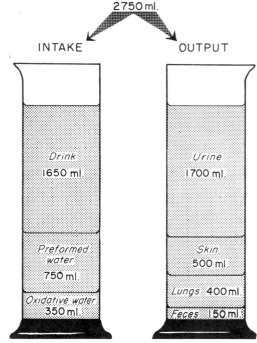

FIG. 8-4
Typical daily intake and output of water. (From Brooks, S. M.: Basic facts of body water and ions, ed. 2, New York, 1968, Springer Publishing Co., Inc.)

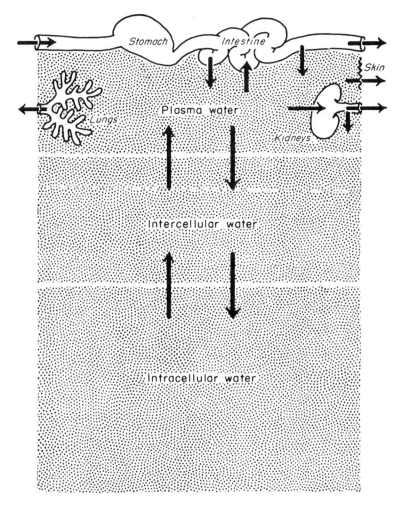

FIG. 8-5

Distribution and movement of body water. (After Gamble; from Brooks, S. M.: Basic facts of body water and ions, ed. 2, New York, 1968, Springer Publishing Co., Inc.)

ance; in a 24-hour period, his weight may vary less than ½ pound.

Intake

On a typical day a man's intake of water is close to 3 l. This water comes not only from imbibed fluids but also from *preformed* water (water trapped in food) and *oxidative* water (water formed as a by-product of oxidation). The respective volumes from these sources are shown in Fig. 8-4. Obviously, food intake has a great bearing on the need for imbibed fluid. The kangaroo rat, a desert creature, never takes a drink as long as it lives, deriving its water solely from solid food! We could do the same if we lived on cucumbers and lettuce.

Output

As noted, the daily output of water balances the daily intake (Fig. 8-4). The kidneys carry the heaviest load, excreting up to about 1.7 l. The other channels, though not as apparent, are no less vital. From the skin and lungs, about 0.5 l and 0.4 l of water are lost, respectively. About 0.2 l leaves the body in the feces. Water escapes from the lungs as vapor. Water lost from the skin is usually vapor (*insensible* perspiration), but it builds up into sweat when the body becomes overheated. Perspiration is about 99% water, with traces of salts (NaCl) and urea. In certain diseases, other constituents may appear, such as bile pigments, albumin, and sugar.

From the foregoing, then, we can see that body water is a matter of compartmental and intake-output balance. This is shown succinctly in Fig. 8-5.

ELECTROLYTES

Physiologically, the term "electrolyte" refers to the ions present in body water. The essential role these charged particles play in the workings of the cell explains why an upset in water balance invariably is associated with an upset in electrolyte balance. In practice, water balance and electrolyte balance generally are looked upon as one and the same. The expression "fluid balance" often is used to ensure the inclusion of both ideas.

The milliequivalent

Because the slightest alteration in electrolytes can often cause havoc—even death—it is imperative that we have the best possible means for expressing ionic concentration. This turns out to be the milliequivalent weight (mEq), which is equal to the atomic weight in grams divided by the product of the valence × 1,000, or

$$mEq = \frac{Atomic\ weight\ (g)}{Valence \times 1,000}$$

TABLE 8-1

Milliequivalent weights of the body's chief ions

Ion	Symbol	Atomic weight	Valence	mEq
Sodium	Na^+	23	1	0.023
Potassium	K^+	39	1	0.039
Calcium	Ca^{++}	40	2	0.020
Magnesium	Mg^{++}	24	2	0.012
Chloride	Cl^-	35	1	0.035
Bicarbonate	HCO_3^-	61	1	0.061
Phosphate	$HPO_3^=$	96	2	0.048
Sulfate	$SO_4^=$	96	2	0.048

The milliequivalent weight of the sodium ion, for example, is

$$\frac{23}{1,000 \times 1} \text{ or } 0.023 \text{ g}$$

For the calcium ion, it is

$$\frac{40}{1,000 \times 2} \text{ or } 0.020 \text{ g}$$

Table 8-1 gives the milliequivalent weights for the body's chief ions.

In actual practice, the electrolyte concentration of a given compartment is expressed as the number of milliequivalents per liter (abbreviate mEq/l). If the concentration of calcium in blood serum is 5 mEq/l, each liter of serum contains 5 × 0.02 g, or 0.10 g, of this electrolyte. On the basis of another method of expressing blood serum concentrations, milligrams per 100 ml, or milligrams percent (mg%), this concentration would be 10 mg%; that is, 10 mg of calcium per 100 ml of blood serum.

Extracellular electrolytes

For all practical purposes, the electrolyte profile of the plasma and the interstitial compartments are the same and are so recognized in Fig. 8-6. The positive ions, or cations, of the extracellular compartment include Na^+, K^+,

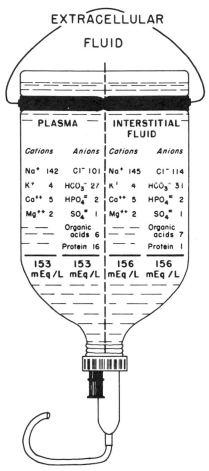

FIG. 8-6
Composition of extracellular fluid. Only major difference between plasma and interstitial (intercellular) fluid relates to protein content. (From Brooks, S. M.: Basic facts of body water and ions, ed. 3, New York, 1973, Springer Publishing Co., Inc.)

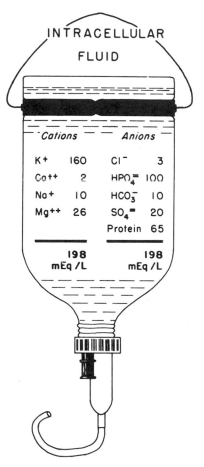

FIG. 8-7
Electrolyte profile of intracellular fluid. Total milliequivalents per liter (mEq/l) of cation equals total milliequivalents per liter of anion. (From Brooks, S. M.: Basic facts of body water and ions, ed. 3, New York, 1973, Springer Publishing Co., Inc.)

Ca^{++}, and Mg^{++}. Sodium, with a concentration of 142 mEq/l, accounts for the bulk of the ions and plays the most influential role in the distribution of body water. The predominating extracellular anions are Cl^- and HCO_3^-. An electrolyte solution is always electrically neutral because the total milliequivalents of cation are always equal to the total milliequivalents of anion. This is interesting physiologically, for regardless of the degree of electrolyte imbalance occasioned by disease, the ions are adjusted to comply with the laws of chemistry (p. 24). A loss of chloride, for instance, is accompanied by a gain in bicarbonate.

Intracellular electrolytes

The electrolyte profile of the intracellular compartment (Fig. 8-7) is just about the reverse of that of the extracellular compartment, particularly in relation to concentrations of Na^+ and K^+. Whereas the extracellular $[K^+]^*$ is only 4 to 5 mEq/l, the intracellular concentration runs as high as 160 mEq/l. Intracellular $[Na^+]$ is about 10 mEq/l. The intracellular ratio of $[Na^+]$ to $[K^+]$ is critical; an increase in one brings about a speedy decrease in the other. This is not peculiar to just this situation, however, for all tissues of the body demand a balanced ionic environment.

ACID-BASE BALANCE

Considering the multitude of diverse substances entering and leaving the various compartments at any given moment, it is remarkable that the body's pH remains so constant. Acid-base balance is a prime example of homeostasis (p. 69). The pH of the blood remains between 7.35 and 7.45; a drop below 6 or a rise above 8 can mean sudden death! The factors responsible for the maintenance of acid-base balance include a variety of chemical buffers as well as compensatory respiratory and renal mechanisms (to be discussed in later chapters).

Buffers

The ability of a solution such as blood serum to resist a change in pH depends upon the presence of *buffer systems*—combinations of weak acids and their salts. The principal buffer system in the body—carbonic acid (H_2CO_3)-bicarbonate (HCO_3^-)—works, like all buffers, in the following way: If strong acid enters the system, the H^+ ions are grabbed up by the HCO_3^- ions and converted to H_2CO_3, a weak acid:

$$HCO_3^- + H^+ \longrightarrow H_2CO_3$$

---

*Brackets, $[\ ]$, around a symbol indicate concentration. Here, $[K^+]$ indicates the concentration of potassium ions.

Conversely, if base is added, it is neutralized immediately by H_2CO_3. The OH^- ion, for example, a strong base, reacts with H_2CO_3 to form water and HCO_3^-, a weak base*:

$$OH^- + H_2CO_3 \longrightarrow HCO_3^- + H_2O$$

By using the proper concentrations of acid and salt, the chemist can produce a buffer system with a desired pH. The body maintains its pH by keeping the ratio of $[HCO_3^-]$ to $[H_2CO_3]$ at 20 to 1. The term ratio cannot be overemphasized, for it is the proportion and not the absolute amounts of HCO_3^- and H_2CO_3 that must remain constant to keep the pH at 7.4. Nevertheless, the body does strive toward the normal concentrations of 1.35 mEq/l for H_2CO_3 and 27 mEq/l for HCO_3^- (Fig. 8-8).

Other important buffers include phosphate and protein, which are particularly critical in control of intracellular acid-base balance.

Acidosis

A decreased pH of the blood, or acidemia, generally is referred to as acidosis. If the altered pH is soon restored by the buffer system, we call the acidosis *compensated*. If the restoration falls short of the normal pH, the acidosis is *uncompensated*. In diabetic acidosis, for example, ketonic acids accumulate in the blood at the expense of HCO_3^-, thereby reducing the $[HCO_3^-]$ to $[H_2CO_3]$ ratio and lowering the pH. The lowered pH stimulates the respiratory center in the medulla, and the lungs are prodded into blowing off more CO_2. Since CO_2 is derived from H_2CO_3, the $[H_2CO_3]$ starts to drop, and the altered ratio is thereby turned toward normal.

Alkalosis

Alkalosis refers to an alkalemia, or increased blood pH. In severe vomiting, for example,

---

*HCO_3^- is a base because it accepts H^+ ions (Brönsted theory, p. 28).

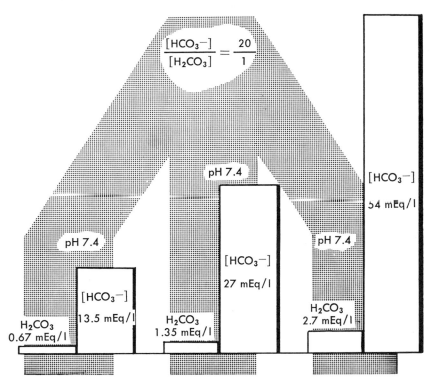

$$\frac{[HCO_3^-]}{[H_2CO_3]} = \frac{20}{1}$$

pH 7.4

$[HCO_3^-]$
54 mEq/l

pH 7.4

$[HCO_3^-]$
27 mEq/l

pH 7.4

$[HCO_3^-]$
13.5 mEq/l

H_2CO_3
0.67 mEq/l

H_2CO_3
1.35 mEq/l

H_2CO_3
2.7 mEq/l

FIG. 8-8

pH and buffer mechanism of blood. As long as ratio of $[HCO_3^-]$ to $[H_2CO_3]$ is 20 to 1, pH remains at 7.4. (From Brooks, S. M.: Basic facts of body water and ions, ed. 2, New York, 1968, Springer Publishing Co., Inc.)

enough hydrochloric acid may be lost to cause a relative increase in base (HCO_3^-) and an altered ratio. To compensate, respiration is diminished, thereby increasing the $[H_2CO_3]$, and the kidneys step up the excretion of HCO_3^-. Both mechanisms working together bring the ratio back toward 20 to 1 and the pH back toward normal.

Another interesting case of alkalosis results from hyperventilation, or excessive breathing. Here the body is plunged into a state of alkalosis because the $[H_2CO_3]$ decreases as CO_2 is blown off, resulting in an increase in the ratio and the pH of the blood.

FLUID IMBALANCES

The loss or retention of abnormal amounts of fluid (water and ions) is a cardinal pathologic feature of many disorders of the body. The more commonly encountered imbalances are touched upon below.

Hemorrhage

The blood, being the smallest of the fluid compartments, can obviously not sustain a great loss. Hemorrhage is an emergency of the first order and demands speedy treatment if a life is to be saved. Since the immediate threat is *shock* due to circulatory collapse, plasma, plasma expanders, or even normal saline solution will suffice to temporarily replace lost blood if whole blood is not available. Though it is quite true that salt and water readily leave the circulation, cardiac output can often be maintained until proper replacement becomes available.

Burns

The burn victim is beset by an upheaval of fluid and electrolytes. Fluid escapes into the interstitial compartment, producing shock and edema; hemoconcentration (due to the fluid loss) withdraws intracellular water and produces dehydration; the accumulation of salt in the edematous area robs the body as a whole of sodium; and the damaged cells release considerable amounts of potassium into the extracellular compartment. Recent data suggest quite strongly that the displacement of sodium and the excess extracellular potassium (hyperkalemia) could very well be the main cause of burn shock.

Burns covering less than 50% of the body offer a favorable prognosis, provided that proper treatment is instituted immediately. In addition to giving drugs to relieve pain and antibiotics to fight infection, the physician must restore fluid balance by accurately computing the amount of fluid lost and then selecting the appropriate intravenous solutions. Whenever possible, however, fluids are given orally.

Vomiting

Severe vomiting results in loss of fluid from the extracellular compartment, producing dehydration and usually alkalosis due to the loss of gastric acid. Treatment centers about the use of water to correct dehydration and the appropriate salt to restore the acid-base balance. If the alkalosis is not pronounced, these objectives can be met by administration of normal saline or Ringer's solution. In severe cases, however, ammonium chloride, an "acid" salt, may be needed.

Diarrhea

Fluid and acid-base imbalance is a common feature in diarrhea, particularly in infants. Since the ion-rich gastrointestinal secretions are swept away before the intestinal wall has a chance to resorb them into the circulation, the body is deprived of water and electrolytes. Because these secretions typically contain more base than acid, diarrhea generally is marked by acidosis. Treatment entails the replacement of water, salt, and potassium and the use of sodium bicarbonate or sodium lactate to correct the acidosis. Sodium lactate is metabolized in the liver into HCO_3^-, the body's major base.

Low-salt syndrome

Perhaps the most frequently encountered fluid imbalance is a self-inflicted variety often called "the low-salt syndrome" or, more popularly, "heat cramps." Marked by nausea, fatigue, and muscle cramps, this condition is caused by excessive sweating. Briefly, the water one drinks in response to the incited thirst lowers the $[Na^+]$ of the extracellular fluid and renders that compartment hypotonic. Consequently, fluid seeps into the more hypertonic intracellular compartment. The signs and symptoms result from a decrease in $[Na^+]$, extracellular dehydration, and too much cellular water. The condition is treated and prevented by taking salt.

Edema

We have mentioned edema, or the accumulation of fluid in the tissues, several times before. There are a variety of etiologic factors. One, hypoproteinemia, was discussed earlier (p. 103). Others include increased capillary permeability (for example, the edema seen in inflammation and allergies), increased hydrostatic pressure (for example, in standing), clogged lymphatics (filariasis), and failing heart. In the latter condition, called cardiac edema, the heart fails to pump out the returning blood fast enough. As a result, the venous pressure becomes elevated and fluid is forced out into the intercellular compartment.

The treatment of edema centers on the use of diuretics and measures to correct the underlying cause—for example, giving digitalis to strengthen the heart.

Acute kidney failure

Acute kidney failure, or renal shutdown, though still an ominous condition, today often yields to the modern principles of electrolyte balance. Its treatment and the use of the artificial kidney are discussed in Chapter 19.

QUESTIONS

1. Discuss the manner in which the water of the body is compartmentalized.
2. Distinguish between plasma water and interstitial water.
3. What is meant by the term "fluid"?
4. What is meant by "water balance"?
5. What effect does hypoproteinemia have upon the filtration pressure in the capillaries?
6. Why is the term "colloid" (for example, colloid osmotic pressure) used in reference to protein solutions?
7. Explain the sequence of events by which the drinking of salt water leads to dehydration.
8. Distinguish between preformed water and oxidative water.
9. What effect does excessive peristalsis have upon water resorption?
10. What is meant by insensible perspiration?
11. Compare the meanings of water balance, electrolyte balance, and fluid balance.
12. What events occur following the ingestion of excess fluid?
13. What is the mg% concentration of sodium in blood serum?
14. Compute the mEq for Mg^{++} (show arithmetic).
15. What is the difference between an atom and an ion?
16. Distinguish between a colloid and a crystalloid.
17. Compare the cations of the extracellular and intracellular compartments.
18. Tissue injury not infrequently leads to hyperkalemia. Explain.
19. Explain how a solution containing citric acid and sodium citrate resists a change in pH.
20. Why is HCO_3^- called a base?
21. Explain how apnea can lead to acidosis.
22. Venous blood has a slightly lower pH than arterial blood. Explain.
23. Discuss the effect of a severe burn upon body water and electrolytes.
24. A solution of $MgCl_2$ contains twice as many Cl^- ions as Mg^{++} ions, and yet it is neutral. Explain.
25. What is the effect upon respiration of the excessive loss of gastric juice?

CHAPTER 9

THE SKIN

The skin, or *integument,* is the body's largest organ. Its functions are numerous, complex, and crucial to survival. Among other things, it shields us from infection and injury, protects us against too little or too much heat loss, informs us of changes in our external environment, and plays a vital role in fluid and electrolyte balance. The skin is composed of two distinct layers, one superficial, the other deep, called the epidermis and the dermis, respectively.

EPIDERMIS AND DERMIS

The epidermis is stratified squamous epithelium (Fig. 9-1). It receives no blood vessels and must rely upon the nutrient fluids derived from the vascularized dermis below. The color of skin depends basically upon its content of *melanin,* the dark pigment formed by special epidermal cells called melanocytes. The skin of the Negro and the darker areas on the white man (for example, about the nipple) contain large amounts of this pigment. In most Caucasians during most of the year, the skin color is most influenced by blood. The white pallor of someone who has fainted is striking.

In the areas where the epidermis is thickest (palms of the hands and soles of the feet), it is made up of five strata. These are, from exterior inward, the stratum corneum, the stratum lu-cidum, the stratum granulosum, the stratum spinosum, and the stratum basale. The stratum *corneum* consists of a couple dozen layers of dead scalelike cells composed of proteins called *keratins;* the most superficial layers are continually flaking off, a process called *desquamation* or *exfoliation.* The stratum *lucidum* consists of three or four layers of translucent cells containing *eleidin,* the precursor of keratin. The stratum *granulosum* consists of two or three layers of cells containing dark-staining granules of *keratohyalin,* the precursor of eleidin. The stratum *spinosum* is composed of several layers of irregularly shaped cells displaying spinelike processes, from which the layer gets its name. The stratum *basale* (or stratum *germinativum*) is composed of a single layer of columnar cells that undergo mitosis. New cells are produced in this deepest stratum at a rate equal to the loss of cells from the stratum corneum. The new cells push upward into each successive layer and eventually flake off.

The dermis, or *corium,* is composed of connective tissue, blood vessels, lymph vessels, nerves, and accessory glandular structures. Characteristically, it thrusts itself upward into peglike structures called *papillae* (Fig. 9-1) that lock together the two layers of skin. The papillae bear a relationship to the fingerprints or *der-*

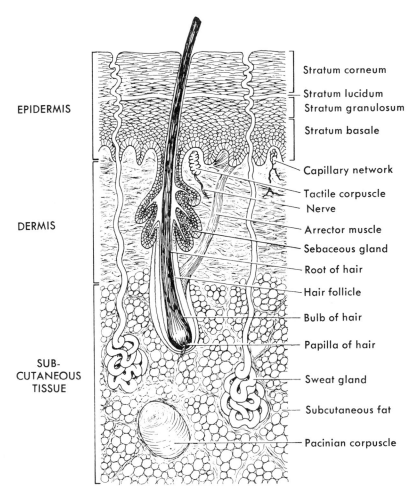

EPIDERMIS

DERMIS

SUB-
CUTANEOUS
TISSUE

Stratum corneum
Stratum lucidum
Stratum granulosum
Stratum basale

Capillary network
Tactile corpuscle
Nerve
Arrector muscle
Sebaceous gland
Root of hair
Hair follicle
Bulb of hair
Papilla of hair
Sweat gland
Subcutaneous fat
Pacinian corpuscle

FIG. 9-1
Skin and subcutaneous tissue. Stratum spinosum not present in hairy skin.

matoglyphics; they directly underly the *sulci* (valleys).

SUBCUTANEOUS TISSUE

Subcutaneous tissue, or *superficial fascia,* anchors the skin to the muscles and the bones. It is composed of areolar connective tissue interlaced with fat, vessels, nerves, receptors, and glands. This tissue nourishes, supports, and cushions the skin, and its fat serves as a food reserve.* Fat is deposited and distributed according to a person's sex, and this, together with muscle and bone development, accounts

---

*Except in the palms and soles where its cushion function supersedes the nutritional one.

for the difference in body shape between the sexes. When we overeat, fat piles up beneath the skin; when we diet, excess fat is burned, the skin resumes its normal texture and contour, and we "get our shape back."

Beneath the superficial fascia lies the *deep fascia*, a thin layer of dense connective tissue without fat that covers the muscles and passes inward to form intermuscular septa. In certain areas the deep fascia thickens to produce ligaments, tendons, and aponeuroses.

HAIR AND NAILS

The accessory organs of the skin include hair, nails, glands, and receptors. Except for the palms of the hands and the soles of the feet, hair adorns the entire body. Over the eyes and in the nose and the ears, it screens out insects, dust, and other airborne debris, and elsewhere it serves as a solar screen and a valuable sensory structure.

A shaft of hair grows upward as a consequence of extensive cellular multiplication at the *papilla*, a structure located at the base of the root (Fig. 9-1). The *root* (the portion of the hair embedded in the skin) and its coat of connective tissue constitute the hair *follicle*. As long as the epithelial cells near the papilla remain alive, a hair will continue to grow and regenerate. As we know, hair stands on end and goose pimples appear in response to fright or to cold. These somewhat primitive responses are caused by contractions of the tiny muscles (*arrectores pilorum*) attached to the hair follicles. Also around the follicle are *sebaceous glands* that oil the hair (Fig. 9-1). The color of pigmented hair is due to the presence of varying amounts of melanin within the shaft; with a decrease in melanin comes gray hair. White hair results from the refraction of light on air spaces in the hair.

The nails are horny distal appendages of the fingers and the toes. They are composed of keratin and develop from the epidermal cells of the stratum lucidum lying under the *lunula,* the white crescent-shaped structure situated at the proximal end of the nail.

GLANDS

For each hair there are at least two sebaceous (oil) glands that secrete about the shaft an oily substance called *sebum.* Modified sebaceous glands lying within the tarsus of the eyelid are referred to as *meibomian glands*. Sebum oils the hair and keeps it supple.

Sudoriferous, or *sweat, glands* (Fig. 9-1) are distributed over the entire body, especially on the palms, the soles, and the forehead, and in the axillary regions. In these areas there are thousands per square inch of skin. Sweat glands soften the stratum corneum and regulate body heat. To a minor degree they help rid the blood of wastes. Sweat is about 99% water but also contains dissolved salts (chiefly NaCl), traces of urea, and miscellaneous other substances. When the body becomes overheated, the sudoriferous glands step up the production of sweat, which evaporates to cool the skin. In the tropics, as much as 10 to 15 l of water may be lost daily through perspiration. The sweat glands are governed by the autonomic nerves, which presumably are under the direction of a "sweat center" in the brain. When the temperature of the blood rises in response to external temperature or muscular activity, the center is triggered and sends impulses to the glands, causing the production of sweat. In addition to sweat, water seeps through the epidermal layers and leaves the surface as water vapor (*insensible* perspiration). On a typical day about 0.5 l is lost this way.

Ceruminous glands are modified sweat glands located in the canal of the external ear. Instead of sweat, they elaborate *cerumen,* a waxlike secretion that aids the hairs in trapping dust, insects, and the like.

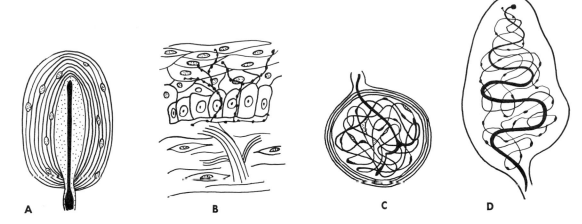

FIG. 9-2
Nerve endings and receptors. **A,** Pacinian corpuscle. **B,** Free nerve endings. **C,** Mucocutaneous end organ (formerly Krause's corpuscle). **D,** Meissner's corpuscle. Mucocutaneous end organs are responsive to mechanical stimuli. See text for function of other receptors. (Redrawn from Francis, C. C: Introduction to human anatomy, ed. 6, St. Louis, 1973, The C. V. Mosby Co.)

RECEPTORS

The skin's extreme sensitivity is effected through sensory nerve endings and specialized structures called *receptors*. These microscopic receivers (Fig. 9-2) are distributed over the entire body, but they are more concentrated in some areas than in others. The back, for instance, has fewer receptors per unit area than the fingertips.

Pain receptors

Pain receptors, or *nociceptors,* are naked nerve filaments found throughout the skin and within the body. They are the most numerous of all the receptors. Temperature and touch are also perceived via free nerve endings.

Meissner's corpuscles

Meissner's corpuscles are encapsulated receptors located in the dermal papillae of the fingers, the toes, the lips, the mammary glands, and the external genitals. They are tuned to the sense of touch.

Pacinian corpuscles

Pacinian corpuscles are ovoid receptors found in the deeper parts of the skin covering the hands and the feet. They also occur throughout the subcutaneous tissue, and in the muscles, the mesentery, and the mesocolon. Each contains a granular central bulb enclosing a single terminal neurofibril sensitive to pressure.

SKIN INFECTIONS

Although the intact skin provides an inimitable barrier against the body's microbial enemies, an injury, however minor, affords a portal of entry. The consequence may be a pimple or, in extreme cases, death. However, an infectious disease involving the skin does not have to attack from without. For instance, smallpox and measles, diseases that have cutaneous manifestations, are caused by inhaled or ingested viruses. The major

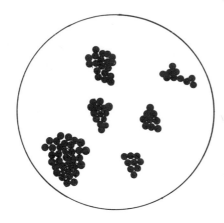

FIG. 9-3
Staphylococcus aureus.

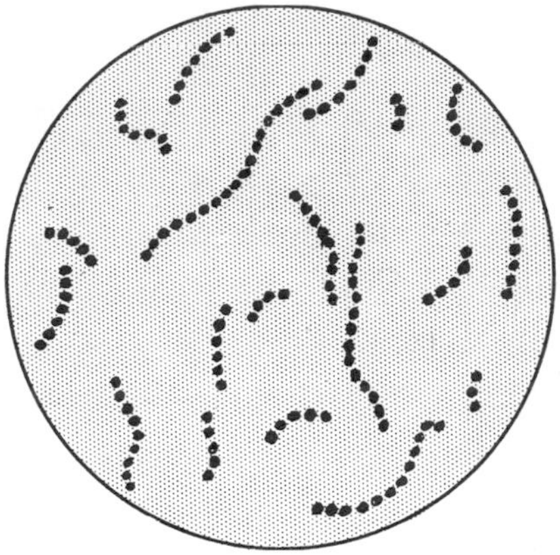

FIG. 9-4
Streptococcus pyogenes.

bacterial, viral, rickettsial, fungal, and ectoparasitic infections (in this order) involving the skin are presented below.

"Staph" infections

Most staph infections are due to *Staphylococcus aureus* (Fig. 9-3). An increasing number of strains of this species are becoming resistant to penicillin and other antibiotic agents. Staphylococci are common causes of sties, furuncles (boils), carbuncles, abscesses, and paronychia. These infections characteristically are marked by considerable pus formation.

"Strep" infections

Another invader of significance is *Streptococcus pyogenes* (Fig. 9-4), commonly referred to as "beta hemolytic strep" or "streptococcus hemolyticus." Among other infections, this pathogen causes cellulitis, erysipelas, septic sore throat, and scarlet fever. The latter disease, characterized by a red rash and a fever, is caused by the exotoxin. Since several conditions mimic scarlet fever, the Schultz-Charlton test is often used to establish a definitive diagnosis. In this

test, a skin dose of scarlet fever antitoxin is injected intradermally into an area of the rash; if blanching occurs (as a result of neutralization of the exotoxin by the antitoxin), the test is positive.

Scarlet fever is treated successfully with penicillin. Patients should be isolated, and exposed susceptible persons should be given the drug for 3 to 5 days. Mass immunization is not recommended because the infection has lost much of its former virulence. An attack of scarlet fever engenders active immunity against the exotoxin, and immune persons react negatively to the *Dick test*. This is performed by injecting a very small amount of scarlet fever exotoxin intradermally into the forearm. In susceptible persons a reaction (a red wheal) occurs within 24 hours at the site of the injection.

Pseudomonas aeruginosa

Pseudomonas aeruginosa is a motile, gram-negative, non-spore-forming bacillus often present in the intestinal tract, in sewage, and in polluted water. Though it does not incite a specific infection, the organism proves a trouble-

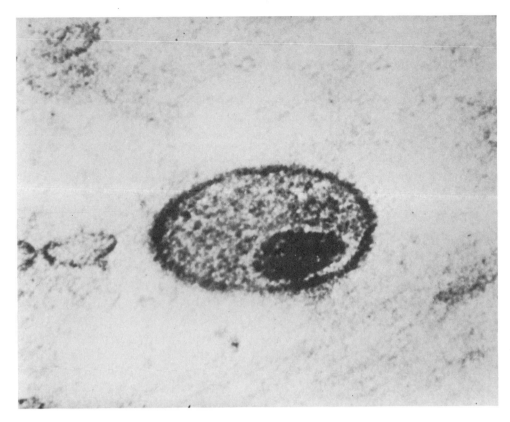

FIG. 9-5
Vaccinia virion. (× 180,000.) (Courtesy Eli Lilly & Co.)

some *secondary invader* and commonly is responsible for abscesses, otitis media (middle ear infection), and infected wounds and burns. (Its involvement in urinary tract infections is discussed in Chapter 19.) Characteristically, it releases a bluish green pigment that tinges the pus of the infection. The antibiotics polymyxin B and gentamicin are used to treat infections caused by this bacterium.

Other bacterial pathogens

Other notable bacterial pathogens associated with the skin include *Bacillus anthracis* (anthrax), *Yersina pestis* (plague), *Francisella tularensis* (tularemia or rabbit fever), *Actinobacillus mallei* (glanders), *Spirillum minus* (rat-bite fever), *Treponema pertenue* (yaws or frambesia), and *Mycobacterium leprae* (leprosy).

Smallpox

With a mortality rate between 10% and 30%, smallpox, or *variola,* is one of the most vicious viral infections. The malady is characterized by a disfiguring vesicular eruption that becomes pustular and finally crusty, leaving pockmarks. The virus is spread from man to man by direct

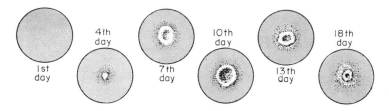

FIG. 9-6

Smallpox vaccination. (From Brooks, S. M.: Basic facts of medical microbiology, ed. 2, Philadelphia, 1962, W. B. Saunders Co.)

contact, and it is spread indirectly via naso-pharyngeal secretions and *fomites*. Smallpox is diagnosed on the basis of clinical appearance and the finding of *inclusion bodies (Guarnieri bodies)* in the epithelial cells.

Since the infection always runs its course, it is highly significant that we possess a valuable vaccine for active immunization. The vaccine, a preparation containing the vaccinia virus* (Fig. 9-5), is usually applied to the outer aspect of the upper arm. In persons who never have been vaccinated or never have had smallpox, an ugly vesicle develops at the site of vaccination (Fig. 9-6). This is an actual attack of vaccinia, and the body responds by elaborating antibodies. These antibodies happen to be effective not only against vaccinia but also against smallpox. Immune persons, those previously vaccinated or those who have had the disease, develop only a red wheal. The absence of any reaction at the site of the vaccination almost always signals an impotent vaccine.

Other prophylactic measures against smallpox include isolation of patients, concurrent and terminal disinfection, and quarantine. Treatment with methiazone, a new drug, may be of some value in the prevention of the disease in exposed persons.

Measles

Measles, or *rubeola*,* is a highly contagious viral disease characterized by skin rash, fever, and acute catarrhal inflammation of the eyes and the respiratory passageways. The virus spreads easily through the air, particularly during the 3 or 4 days preceding eruption of the rash. Contagiousness ceases when the fever drops, and an attack usually provokes permanent immunity. Diagnosis is based upon clinical findings, especially the presence of small white areas inside the mouth called *Koplik's spots*.

Although measles generally is without serious consequence, one must be on guard against such secondary infections as pneumonia, mastoiditis, and otitis media. Should any of these complications occur, antibiotic therapy is indicated. Otherwise, the treatment is symptomatic. Active immunization against measles has been investigated extensively, and workable vaccines are now available. The attenuated or "live" vaccine is now generally considered the preparation of choice. According to recent reports, the disease in time may be completely eradicated. Passive immunization with gamma globulin has been used successfully to treat persons with severe cases and to protect susceptible persons with low resistance.

---

*An attenuated virus related antigenically to the viruses of variola and cowpox. Its origin is not known.

*Not to be confused with rubella, or German measles.

German measles

German measles, or *rubella,* is a highly contagious but mild viral infection marked by a rash and swollen lymph nodes. Usually no treatment is necessary, and an attack almost always produces lasting immunity. Because the virus causes congenital malformations (for example, cataract, microcephaly, deafness, and cardiac defects), pregnant women should avoid all contact with persons who have German measles. This is, indeed, the most important and serious aspect of the disease. An effective attenuated vaccine against the infection is now available.

Chicken pox and herpes zoster

Chicken pox, or *varicella,* is a dermotropic viral infection of childhood characterized by fever and the appearance of vesicles after an incubation period of 1 to 3 weeks. This infection is generally mild, and an attack produces lasting immunity. There are no specific therapeutic or prophylactic measures. Herpes zoster, commonly called *shingles,* is an acute, painful vesicular dermatitis. Since the virus follows the nerve trunks, the location and distribution of the lesions depend upon the particular nerves affected. An attack ordinarily confers lasting

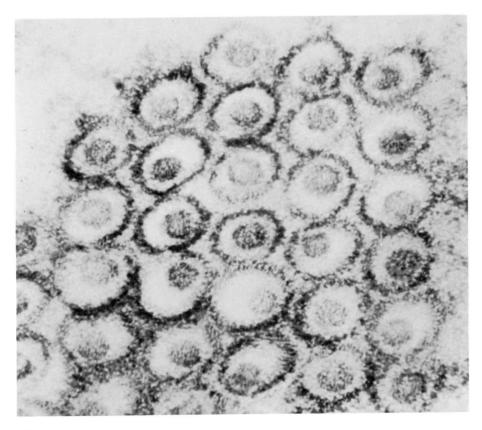

FIG. 9-7
Intranuclear crystal of herpes simplex virus. (× 90,000.) (Courtesy Eli Lilly & Co.)

immunity. The treatment is symptomatic. At present, nothing can be done to prevent herpes zoster.

Shingles and chicken pox are now known to be caused by the same virus. There is considerable difference of opinion among authorities as to the mechanism and explanation behind this rather strange situation, but the general feeling seems to be that the first invasion of the body by the VZ (varicella-zoster) virus results in chicken pox, whereas shingles results from either reinvasion or the activation of a latent virus. Support for this view stems from the inability of patients with shingles to produce a history of prior contact with an external source of the VZ virus and from the fact that certain stimuli and conditions are known to trigger a case of the infection—injury, drugs, leukemia, and hormones, for example.

Herpes simplex

Herpes simplex is an acute vesicular eruption of the skin and mucous membrane caused by *Herpesvirus hominis,* a virus similar to the VZ virus (Fig. 9-7). Characteristically, the vesicles are soft and filled with a watery fluid. Although all areas are subject to attack, the most common site is the lips (the cold sore). Virulent strains of the virus have been known to turn neurotropic and attack the brain. There is no lasting immunity and no specific treatment. There is now strong evidence that herpesvirus type 2 may be a cause of cervical cancer.

Rickettsial infections

Rickettsiae cause infections characterized by fever and a rash. The principal diseases include typhus fever (endemic and epidemic), rickettsialpox, and Rocky Mountain spotted fever. Rocky Mountain spotted fever is the main rickettsial infection in this country, and one of the most severe infectious diseases known. In some regions, the Bitterroot Valley of Montana for

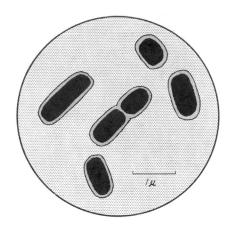

FIG. 9-8

Rickettsia rickettsii, causative agent of Rocky Mountain spotted fever. (From Brooks, S. M.: Basic facts of medical microbiology, ed. 2, Philadelphia, 1962, W. B. Saunders Co.)

example, the mortality rate may reach 90%. The pathogen *Rickettsia rickettsii* (Fig. 9-8) is transmitted by the wood tick (the "western variety" of infection) and the dog tick (the "eastern type"). Diagnosis of these diseases is based on clinical findings, the demonstration of agglutinins in the blood via the Weil-Felix test, and complement-fixation tests. The broad-spectrum antibiotics usually yield excellent therapeutic results.

Fungal infections

Superficial fungal infections of the skin, the hair, and the nails are called *ringworm, tinea,* or more formally, *dermatomycoses;* the pathogens themselves are aptly termed *dermatophytes.* The typical skin lesion is characterized by the formation of ring-shaped, pigmented patches (ringworm) covered with vesicles or scales. Infected nails and hair, unless treated early, are ultimately destroyed. These changes are brought about by the mycelial filaments that gradually creep through the epidermal layers. The more common dermatomycoses, named according to

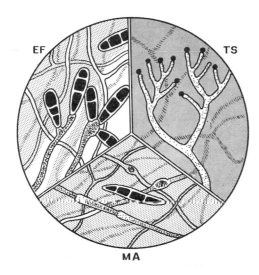

FIG. 9-9
Drawings from micrographs showing three species of dermatophytes (pathogenic fungi that attack the skin): *Epidermophyton floccosum, EF; Trichophyton shoenleini, TS; Microsporum audouini, MA.*

the affected area, include tinea pedis (athlete's foot), tinea unguium (fingernails or toenails), tinea cruris (groin), tinea capitis (scalp), tinea corporis (limbs or trunk), and tinea barbae (face and neck). The diagnosis centers on demonstrating by direct microscopic examination or by cultural methods the presence of the fungal growth (Fig. 9-9) in scrapings taken from the afflicted area. A variety of antifungal agents are available for treating fungal infections, but the most effective are zinc undecylenate (used topically) and the antibiotic griseofulvin (Fulvicin).

Ectoparasites

Although generally not a threat to life, ectoparasites cause uncomfortable and troublesome infestations, often accompanied by secondary bacterial infections. Some ectoparasites are vectors of viral or bacterial infections. The principal invaders include the itch mite *(Sarcoptes scabiei),* the chigger *(Trombicula irritans),* the head louse *(Pediculus humanus* var. *capitis),* the body louse *(Pediculus humanus* var. *corporis),* the crab louse *(Phthirus pubis),* and the human flea *(Pulex irritans).* The itch mite and the chigger burrow into the skin, the flea and the body louse inhabit the surface of the skin, and the head louse and the crab louse infest the hair. Application of gamma benzene hexachloride (Gexane; Kwell) is the treatment of choice for scabies, and DDT preparations are used to destroy lice and fleas.

NONINFECTIOUS DISORDERS

A great many skin conditions are caused by forces other than pathogenic microbes. The etiology of many is unknown, and their treatment necessarily falls short of the mark. Some of the more common and more serious disorders are described briefly here.

Allergy

According to one view, when an antigen-antibody reaction takes place within the cells, the result is *hypersensitivity* or an allergy. Locally, the cells may be injured or destroyed, and systemically, the body may suffer the ill effects of breakdown products released from the cells, such as heparin, serotonin, and histamine. Allergic reactions range in severity from runny noses to fatal anaphylactic shocks. Possibly, an allergy is a kind of weak immunity; whereas microbial antigens stimulate such a high concentration of antibody that the antigens are destroyed or neutralized in the blood before they have a chance to penetrate the tissues, allergy antigens (often called *allergens*) provoke so few antibodies (called *reagins*) that they escape destruction in the bloodstream and enter the cells, where they cause trouble. Almost any substance is a potential allergen. The most common, however, are certain foods (for example, milk, eggs, and strawberries), pollens, dander, and drugs.

At present no one knows for certain why some persons are more susceptible to allergy than are others. One explanation that seems to account for a good many facts is that some individuals, because of their peculiar genetic makeup, possess cells that are damaged easily by cellular allergen-reagin reactions. This does not mean that all the tissues contain such cells, for an allergy is generally confined to localized areas, such as the mucous membranes of the nose and the eyes. Some allergic persons always develop a rash in exactly the same place, perhaps just the place where the cells are most susceptible to the reaction. Until more is known about allergy, or hypersensitivity, the chief control of the condition will continue to center on the use of desensitization and antihistamines.

Allergic dermatitis and allergic eczema are the expressions applied to any inflammation of the skin caused by an allergy. The allergen may come from without or from within. In the former case the skin comes into contact with substances such as sensitizing chemicals or poison ivy; in the latter, certain foods, dust, pollen, molds, and drugs are the culprits. Why some persons are more sensitive than others is still a moot question. The treatment of an allergic dermatitis involves removal of the cause (elimination of certain foods from the diet and avoidance of such allergenic agents as poison ivy), desensitization, and administration of antihistamines and corticosteroids.

Actinic dermatitis

Actinic dermatitis is an inflammatory condition of the skin caused by overexposure to ultraviolet radiation. This condition usually results from staying in the sun too long, but staying under a sunlamp too long can have the same effect. Since sunburn, natural or artificial, can be extremely dangerous, one should always be on guard against it. This is true especially on cloudy summer days when large doses of ultraviolet rays, in contrast to visible light, filter through the clouds.

Psoriasis

Psoriasis is a chronic disorder of unknown cause marked by dry and silvery *papules* and *plaques*. The usual sites are the back and buttocks and the exterior surfaces of the elbows and knees. Some patients develop a joint involvement similar to rheumatoid arthritis. Heredity is a factor in about a third of the patients, and the disease is uncommon in Blacks. Acute attacks usually clear up, but complete and permanent remission is rare. No therapeutic measure assures a cure.

Lupus erythematosus

Lupus erythematosus is a skin condition characterized by disclike patches with raised red edges and depressed centers. These patches are covered with scales that eventually fall off, leaving a white scar. *Disseminated* lupus erythematosus is more serious. It is a chronic and usually fatal disease with systemic repercussions involving *collagen,* the main supportive material of connective tissues. Typically, a morbid redness of the face spreads across the nose in a butterfly pattern (Fig. 9-10). The incidence of the disease is highest in females between puberty and the menopause. Corticosteroids are helpful but not curative.

Scleroderma

In the frequently incurable dermatosis known as scleroderma, the skin becomes thick, hard, and rigid, and is covered with pigmented patches. Often there is involvement of the internal organs. This, too, is a collagen disease of unknown cause. Corticosteroids may be helpful at each stage of the disease. No specific therapy is known.

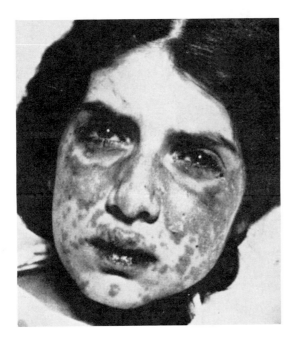

FIG. 9-10
Acute disseminated lupus erythematosus. The "butterfly rash" spreads over nose and cheeks. Lips are swollen and encrusted. (Courtesy Dr. David Omens; from Traut, E. F.: Rheumatic diseases, St. Louis, 1952, The C. V. Mosby Co.)

Pemphigus

Pemphigus, like disseminated lupus erythematosus, involves the general health of the patient and often proves fatal. The dermal features are characterized by successive crops of large blisters that leave deeply pigmented spots following absorption. These lesions often burn and itch. Early recognition is important because corticosteroid therapy may prolong life and produce apparently permanent remission.

Cancer

There are two kinds of tumor: benign and malignant. The latter we usually call cancer. The most common benign tumors of the skin are the various kinds of warts, or *verrucae*, all of which are caused by viruses. Warts and the benign skin tumor *molluscum contagiosum* are about the only human tumors definitely known to have viral etiologies. A variety of skin conditions, although not themselves cancerous, are termed "precancerous" because they so frequently give rise to true cancer. Among these are keratoses, occupational dermatoses, x-ray and radium dermatitis, and leukoplakia.

Skin cancers outnumber all other cancers. The estimated number of new cases per year averages about 120,000, compared to some 70,000 new cases each for cancers of the lung and breast. Sunshine and its ultraviolet rays are considered the major cause of skin cancer. There is much support for this view. In the United States, for instance, skin cancer is three times as common in the southern part of the country as in the northern part. The "big three" among these cancers are basal cell carcinoma, squamous cell carcinoma, and malignant melanoma.

Basal cell carcinoma, which involves the basal cells of the deeper strata of the epidermis, accounts for well over three quarters of all malignancies of the skin but kills few of its victims because it does not *metastasize.* The upper half of the face (Fig. 9-11) and head is the classic site. The early lesion, usually a pale, pearly, raised nodule, slowly enlarges and eventually ulcerates.

Squamous cell carcinoma is a true cancer in every sense of the word—above all, because it metastasizes to parts near and far. The lower lip is a favorite site, and pipe smoking may be a predisposing factor.

Malignant melanoma, or *melanocarcinoma,* is essentially a cancer of the *melanocytes.* It is about twice as common among whites as among blacks, and redheads and blonds are especially susceptible. Typically, the tumor begins as a black to brown nodule surrounded by a reddish

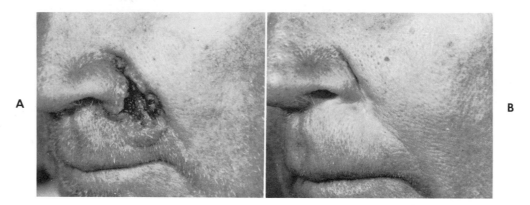

FIG. 9-11

A, Cancer of nasolabial fold. **B,** Following x-ray therapy. (From Ackerman, L. V., and del Regato, J. A.: Cancer: diagnosis, treatment, and prognosis, ed. 4, St. Louis, 1970, The C. V. Mosby Co.)

halo of inflammation; soon small satellite lesions appear about a half inch or so away. The nodule then enlarges, ulcerates, and progressively takes over adjacent tissues via direct extension and lymph channel metastasis. These cancers are unpredictable. Some kill in a matter of months, while others smolder away for years.

Understandably, the diagnosis of skin cancer is a matter of histology. Basal cell carcinoma, for example, shows basal cell involvement, whereas the more devastating squamous cell carcinoma shows an abundance of disorganized squamous cells undergoing rapid division. Treatment centers on surgery and irradiation, with excellent results for both the basal cell and the squamous cell varieties. The outlook for malignant melanoma depends upon whether metastasis has occurred. Without lymph node involvement, well over half of its victims are alive after 10 years have elapsed; with lymph node involvement, only about 10% are still living after this period.

Burns

A severe burn is a medical emergency of the first order, for destruction of the skin in this fashion causes shock, dehydration, electrolyte imbalance, renal damage, and unbelievable agony. Today, burns covering 50% or less of the body generally are not considered fatal, but those involving larger areas often afford a poor prognosis. Obviously, treatment must be multidimensional. Analgesics to relieve pain, antibiotics to fight infection, and plasma and electrolyte solutions to replace lost fluid are all indicated. The correction of water and electrolyte imbalances may spell the difference between life and death. For burns involving more than 15% of body surface area, the aim in the first 24 hours of treatment is to prevent shock with an adequate infusion of fluid without overloading the circulation. According to one scheme, calculation of the proper volume of fluid, in milliliters, is made by multiplying the percent burn (Fig. 9-12) times weight in kilograms times 2, and then adding 2,000. By way of example, a burn involving the head and the front of the trunk of a person weighing 60 kg calls for $60 \times 27 \times 2 + 2,000$, or about 5,000 ml. For burns covering more than 50% of body surface, no more than the amount calculated for a 50% burn is given.

RULE OF NINES

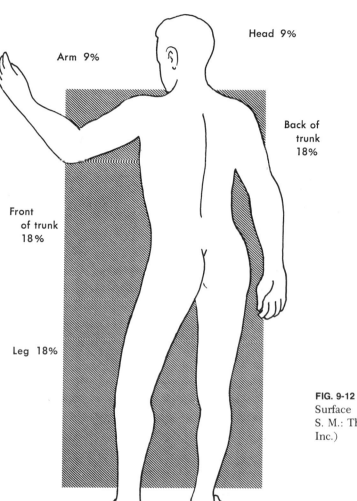

Head 9%

Arm 9%

Back of
trunk
18%

Front
of trunk
18%

Leg 18%

FIG. 9-12
Surface areas of different parts of body. (From Brooks,
S. M.: The sea inside us, New York, 1968, Hawthorn Books
Inc.)

QUESTIONS

1. Explain the role of the skin in the regulation of body temperature.
2. What is meant by the expression "insensible perspiration"?
3. What is the major histological difference between thick skin and thin skin?
4. Medicinal preparations used to remove corns and calluses contain keratolytic agents. What are these?

5. What is albinism?

6. What is the literal meaning of melanocyte?

7. Not forgetting the fact that emotion relates to the nervous system, explain what happens in blushing.

8. A common active ingredient of antiperspirants is aluminum chlorhydroxide. Explain its action.

9. Can a hair really "stand on end"?

10. The epidermis, hair, and nails are said to be "horny." What does this mean?

11. A severe burn may result in kidney damage. Explain.

12. Other things being equal, why do we perspire more when it is humid?

13. Distinguish between sebum and cerumen.

14. Give the derivation of the terms sebum and cerumen.

15. A lesion on or near the lips is especially painful. Why?

16. What is the difference between a receptor and and effector?

17. Skin receptors are sometimes called exteroreceptors. Why?

18. Cite several differencs between the skin of the palm and the skin of the back of the hand.

19. Distinguish among the terms pimples, sties, furuncles, carbuncles, and abscesses.

20. In the Schultz-Charlton test, what constitutes a negative reaction?

21. *Pseudomonas aeruginosa* is a secondary invader. What does this mean?

22. What is the usual etiology of strep infections?

23. What do the infections variola, cowpox, and vaccinia have in common?

24. What are inclusion bodies?

25. In speaking of viral vaccines, "attenuated" and "live" are used synonymously. Why?

26. What is gamma globulin?

27. Does an attack of measles protect us against an attack of German measles?

28. Can you suggest a reason why the same virus can cause shingles on one occasion and chicken pox on another?

29. Does the dog tick cause Rocky Mountain spotted fever? Discuss your answer.

30. "Rocky Mountain spotted fever" is somewhat of a misnomer. Explain.

31. Distinguish among the terms mycosis, dermatomycosis, tinea, and ringworm.

32. How does the laboratory pinpoint the precise cause of a ringworm involvement?

33. The advent of the antibiotic griseofulvin somewhat revolutionized treatment of fungal infections. Explain.

34. What does the term "ectoparasite" tell us?

35. Distinguish between antigen and allergen.

36. What is the relationship between allergen and reagin?

37. Antihistamines are certainly no cure, but they are useful in treating certain allergies. Can you suggest how they work (their "mechanism of action," as the pharmacologists say)?

38. What is the difference between dermatosis and dermatitis, if any?

39. Distinguish between allergic dermatitis and actinic dermatitis.

40. Is it fair to say that the etiologies of most noninfectious diseases of the skin are poorly understood?

41. Lupus erythematosis and scleroderma are among a number of so-called "collagen diseases." What does this mean?

42. Disseminated lupus erythematosis is treated with corticosteroids. Just what are these agents?

43. The immediate threat to life caused by a severe burn is shock. Why shock?

44. What two factors are taken into account in calculating the fluid needed in burn therapy?

45. What is the 24-hour fluid requirement for a burn involving both legs, if the victim weighs 154 pounds?

46. What are "first," "second" and "third" degree burns?

47. What is the basic difference between benign tumors and most malignant tumors?

48. Skin cancers are carcinomas. Why? What are sarcomas?

49. What is the origin of the term "melanocarcinoma"?

50. Skin cancers are most common on exposed and unprotected parts. What is the significance of this fact?

CHAPTER 10

THE SKELETAL SYSTEM

The skeletal system refers to all the bones of the body. This system supports, protects, permits movement, produces blood cells, and plays a key role in mineral metabolism.

TYPES OF BONES

The 206 bones of the body are classified according to their shape into five categories: (1) *long* bones (for example, humerus and femur), (2) *short* bones (for example, carpals and tarsals), (3) *flat* bones (for example, parietal and sternum), (4) *irregular* bones (for example, vertebrae and ethmoid), and (5) *sesamoid* bones (for example, knee cap).

BONE MARKINGS

Learning the name and the position of a given bone is only the beginning. Bones have characteristic markings (determined by location and function), and it is important for the student of anatomy to know the major depressions, openings, and projections (or processes) of each. To this end, the following terms should be committed to memory:

condyle a rounded knob at the end of a bone.
crest generally, a ridge running along the surface of a bone.
foramen (pl., foramina) a hole.
fossa (pl., fossae) a depression or hollow.
head a rounded process at the end of a bone by which it articulates with another bone.

meatus a passageway or tunnel into the interior.
sinus a cavity or spongelike space within a bone.
spine or spinous process a slender, pointed projection of a bone. (The spinal column, or "spine," derives its name from the fact that the majority of the vertebrae possess prominent spinous processes.)
tubercle a small nodule or eminence.
tuberosity a large, broad projection.

BONE STRUCTURE

Bone is composed of about two-thirds inorganic matter and one-third organic matter. The inorganic matter, present chiefly as mineral crystals of hydroxyapatite, occurs in the matrix (p. 97) along with collagen, the body's most abundant protein. The organic matter principally is found in the bone cells, the blood vessels, and collagen.

The manner in which the cells lay down the matrix determines whether bone is compact or cancellous. Compact or hard bone looks and feels like ivory. Cancellous or spongy bone, on the other hand, is light and porous. Compact bone forms the shafts of long bones and the outsides of flat bones. The ends of long bones and the insides of most flat and most irregular bones consist of spongy bone. The spongy bone of the cranial bones is commonly called the diploe.

The histologic features of bone were presented earlier (p. 97). For the gross architecture, let us consider a typical long bone (Fig. 10-1). The

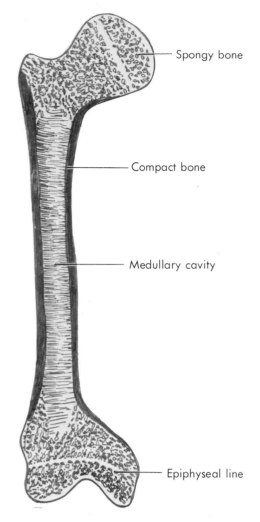

Spongy bone

Compact bone

Medullary cavity

Epiphyseal line

FIG. 10-1
Diagrammatic longitudinal view of typical long bone.

fully stripped away from a fresh bone, tiny beads of blood appear on the bone's surface. These beads represent the entrance of *Volkmann's canals,* the microscopic channels that carry blood from the vessels of the periosteum to the haversian canals (p. 97).

Marrow

Running through the diaphysis is the *medullary cavity,* which is lined with a vascular tissue, the *endosteum,* and filled with a soft, fatty, yellowish substance called *yellow marrow.* At the epiphyses and throughout the cancellous tissue of flat bones and irregular bones, particularly that of the skull, the vertebrae, the ribs, the sternum, and the ilia, we find *red marrow.* In contrast to the yellow variety, red marrow contains little fat and an abundance of marrow cells—erythroblasts, myeloblasts, and megakaryoblasts. The *erythroblasts,* by undergoing a series of transformations, mature into *erythrocytes,* or red blood cells; in a like manner, the *myeloblasts* develop into the *granular leukocytes* (Chapter 13). The *megakaryoblasts,* or giant cells, split up and become the *blood platelets.* Red marrow, particularly that located in the vertebrae and the sternum, is the principal tissue engaged in *hemopoiesis.*

OSSIFICATION

The development from the embryo to the adult skeleton takes about 25 years and is a long, complex process. The "bones" of the embryo actually are not bones at all, but skeletal structures composed of hyaline cartilage and fibrous membranes. As growth continues, the membranes become the flat bones (*intramembranous* ossification), and the cartilages become the long bones (*endochondral* ossification). The soft spots, or *fontanelles,* of the infant's head are fibrous membranes that have not yet undergone ossification.

The ossification of a long bone starts in the diaphysis and the epiphyses and progresses in all

shaft, or *diaphysis,* is composed of hard bone, and the ends, or *epiphyses,* consist of spongy bone. At joints, where the epiphyses meet other bones, they are covered with *hyaline* or articular cartilage. Except at its cartilaginous extremities, a long bone is covered with a tough membrane called the *periosteum.* If this membrane is care-

directions. The term *epiphyseal cartilage* is applied to the cartilage between the diaphysis and the epiphyses. As long as this cartilage remains, a bone continues to grow in length. When the cartilage finally disappears, the remaining external line of juncture is referred to as the *epiphyseal line* (Fig. 10-1).

Osteoblasts and osteoclasts

Histologically and chemically, ossification means the deposition of a bony matrix, a process accomplished through the activities of the *osteoblasts,* the bone cells located in the lacunae and beneath the periosteum. The osteoblasts secrete a protein substance that forms a tough, but not as yet bony, matrix; in this proteinaceous base they bring about the deposition of calcium salts, producing the marblelike apatite referred to earlier. This latter step is carried out largely through the agency of the osteoblastic enzyme *alkaline phosphatase.* By splitting the phosphate radical ($PO_4^{\equiv}$) away from organic phosphates, the enzyme effects the synthesis of Ca^{++}, $CO_3^{=}$, and $PO_4^{\equiv}$ into hydroxyapatite.

Bone continually is being torn down as well as built up, for if bone did not have some way of dissolving itself on the inside, the skeleton would become solid rock, squeezing out the marrow and life itself. But there is another reason for the dissolution of bone: to supply calcium (Ca^{++}) to the extracellular fluids when the concentration of that ion drops below the normal value. Such dissolution, or resorption, of bone is caused by multinucleated giant bone cells called *osteoclasts.* These cells, located in almost all the cavities of bone, secrete an enzyme that digests the protein matrix, thereby releasing Ca^{++} and $PO_4^{\equiv}$ to the blood.

Obviously, osteoblastic and osteoclastic activity must remain in balance if the body is to maintain a healthy skeleton. Stress causes the osteoblasts to become more active, which helps explain why the femur, a bone subjected to great pressure and bending, is so thick and so strong. Paradoxically, but fortunately, crooked leg bones tend to straighten out over the years. Pressure along the inner curvature and stretching along the outer curvature stimulate the osteoblasts and osteoclasts, respectively. As a result, the inner curvature becomes filled with bone, whereas the outer curvature loses bone. Another powerful stimulant of osteoblastic activity is a break in bone. The injured osteoblasts in the vicinity of the break become very active and multiply in all directions, making protein and alkaline phosphatase available for ossification. This accounts for the rapid repair of a broken bone, particularly in the youngster.

Regulating factors

The formation, growth, repair, and metabolism of bone involve several factors, notably calcium, phosphorus, hormones, and vitamins. Briefly, vitamin D (Chapter 18) accelerates the absorption of calcium and phosphate from the gastrointestinal tract; the parathyroid hormone (Chapter 23) stimulates the breakdown or resorption of bone, and the thyroid hormone calcitonin inhibits such resorption. Thus, the two hormones are antagonistic; on the one hand they maintain normal levels of calcium in the blood, and on the other they promote proper bone formation.

ARTICULATIONS

The junction of two or more bones is called a joint, or *articulation*. The articulating surfaces may be separated by a thin membrane, by strong pads of connective tissue, or, in the freely moving joints, by fluid. Joints are perhaps best classified according to the degree and variety of movement they permit.

Synarthroses

In the synarthrotic type of joint there is close contact between the two adjacent bones; there is no joint cavity and there is no movement. *Sutures* are the synarthrotic articulations of the skull characterized by the presence of a thin layer of fibrous tissue uniting the margins of the contiguous bones. A *synchondrosis* is a synarthrosis in which hyaline cartilage separates two bones at their union. Synchondroses are found in growing bones between the diaphysis and the epiphysis. With the cessation of growth, this type of joint disappears.

Amphiarthroses

In the amphiarthrotic type of articulation there is limited movement. The union is effected by fibrocartilage, and the joint is enveloped by ligaments. There may be a joint cavity. Amphiarthroses include the joints between the bodies of the vertebrae, the joints between the sacrum and the ilium, and the joint between the two pubic bones.

Diarthroses

Diarthrotic joints permit variable degrees of movement. At a typical diarthrosis, the articulating surfaces are covered with hyaline cartilage, and the entire joint is invested by a ligament. Characteristically, there is a joint cavity lined by a synovial membrane and filled with synovial fluid (Fig. 10-2); for this reason we often refer to a diarthrosis as a *synovial* joint. Diarthroses are categorized according to degree of movement and include *gliding* joints, where one bone moves or glides across another (such as between carpals and tarsals); *pivot* joints, where movement is in the long axis of a bone (such as between the radius and the ulna); *biaxial* joints, where movement is in two planes at right angles to each other (such as the joint at the base of the thumb); *hinge* joints (such as the joints at the elbows and the knees); and *ball-and-socket* joints, where there is full, free movement (such as the hip and shoulder joint).

Diarthrotic movement. The various possible movements at the diarthroses are given special names:

abduction drawing a part away from the midsagittal plane.
adduction drawing a part toward the midsagittal plane.
circumduction circular movement of a limb at a ball-and-socket joint.
extension increasing the angle of a joint.
flexion decreasing the angle of a joint.
gliding movment at a gliding joint.
rotation movement at a pivot joint.

BURSAE

Where tendons rub against bone ligaments or other tendons, and where the skin moves over a bony prominence, small connective tissue sacs called bursae develop. Lined with a type of synovial membrane and containing a type of synovial fluid, bursae cushion and relieve pressure between moving parts. Those of major clinical importance (because they commonly become inflamed) are situated between the head of the humerus and the acromion process, between the olecranon process and the skin, and between the patella and the skin.

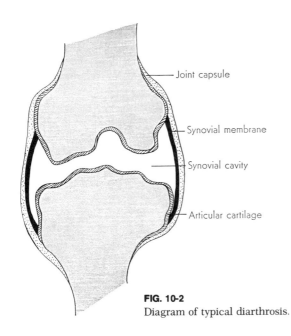

Joint capsule

Synovial membrane

Synovial cavity

Articular cartilage

FIG. 10-2
Diagram of typical diarthrosis.

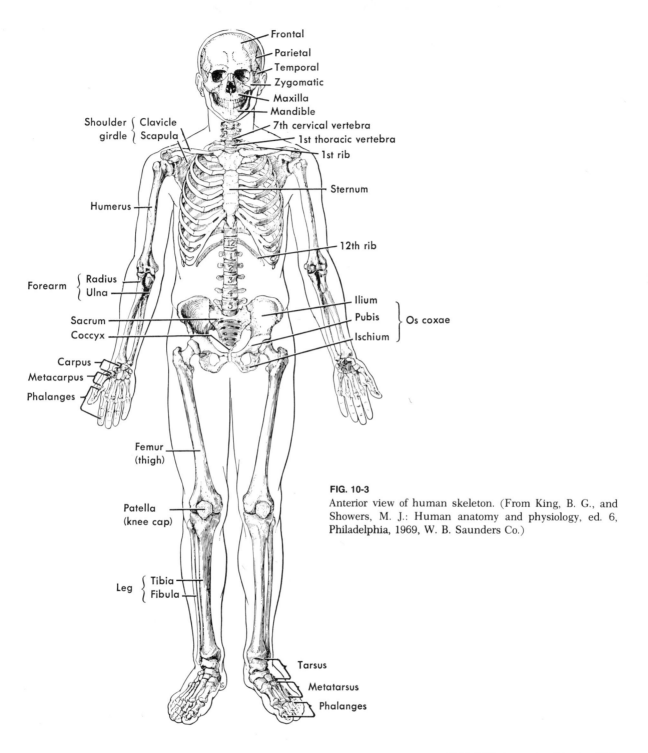

Frontal
Parietal
Temporal
Zygomatic
Maxilla
Mandible
Shoulder girdle { Clavicle Scapula
7th cervical vertebra
1st thoracic vertebra
1st rib
Sternum
Humerus
12th rib
Forearm { Radius Ulna
Ilium
Pubis
Ischium
} Os coxae
Sacrum
Coccyx
Carpus
Metacarpus
Phalanges
Femur (thigh)
Patella (knee cap)

FIG. 10-3

Anterior view of human skeleton. (From King, B. G., and Showers, M. J.: Human anatomy and physiology, ed. 6, Philadelphia, 1969, W. B. Saunders Co.)

Leg { Tibia Fibula
Tarsus
Metatarsus
Phalanges

THE SKELETON

There are two main divisions of the skeleton (Fig. 10-3): the *axial,* including the bones of the head, the neck, and the trunk; and the *appendicular,* including the bones of the extremities. The upper extremity includes the shoulder girdle, the upper arm, the elbow, the forearm, the wrist, and the hand; the lower includes the hip, the thigh, the knee, the lower leg, the ankle, and the foot. The salient features of the axial skeleton will be discussed first.

Skull and cranium

The skull consists of 22 irregularly shaped bones, excluding the ossicles of the ear (Table 10-1), and is divided into two parts, the cranium and the face.

The cranium houses the brain. Its roof (Fig. 10-4) is formed by the frontal, the parietal, and the occipital bones, and its sides, by the temporal bones and the great wings of the sphenoid (Fig. 10-5). These same bones, plus the tiny cribriform plate of the ethmoid bone, also form the floor (Fig. 10-6). The outstanding surface features of the cranium are the foramen magnum, the sutures, the auditory meatus, and the mastoid, styloid, and zygomatic processes. Inside the cranium, looking down at the floor, the chief points of interest include a number of small foramina for nerves and blood vessels, the sella turcica (saddle-shaped depression in the sphenoid bone), and the crista galli (a tiny process of the ethmoid bone). The large opening in the occipital floor—the foramen magnum—permits the spinal cord to join the brain.

Face

The face is said to be made up of 14 bones, but the frontal and ethmoid bones of the cranium are really needed to fill out the complete framework (Fig. 10-5). Like the batlike sphenoid of the cranium, the maxillae serve as the keystones of the face. All bones except the mandible articu-

TABLE 10-1 Bones*

Bone	Single	Paired
Skull CRANIUM		
Frontal	1	
Parietal		2
Occipital	1	
Temporal		2
Sphenoid	1	
Ethmoid	1	
FACE		
Nasal		2
Lacrimal		2
Maxilla		2
Inferior nasal concha		2
Zygoma		2
Palatine		2
Vomer	1	
Mandible	1	
Vertebrae		
Cervical	7	
Thoracic	12	
Lumbar	5	
Sacrum (5 fused)	1	
Coccyx (4 fused)	1	
Thorax		
Ribs		24
Sternum	1	
Upper extremity		
Clavicle		2
Scapula		2
Humerus		2
Radius		2
Ulna		2
Carpus		16
Metacarpus		10
Phalanges of hand		28
Lower extremity		
Hip (3 fused)		2
Femur		2
Patella		2
Tibia		2
Fibula		2
Tarsus		14
Metatarsus		10
Phalanges of foot		28
Miscellaneous		
Ossicles of the ear (3 in each)		6
Hyoid	1	
Total		206

*Adapted from Francis, C. C: Introduction to human anatomy, ed. 6, St. Louis, 1973, The C. V. Mosby Co.

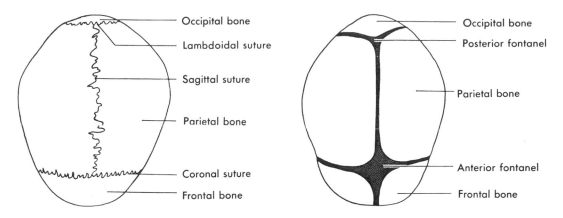

FIG. 10-4
Skull, top view. Left drawing, adult skull. Right drawing, fetal skull.

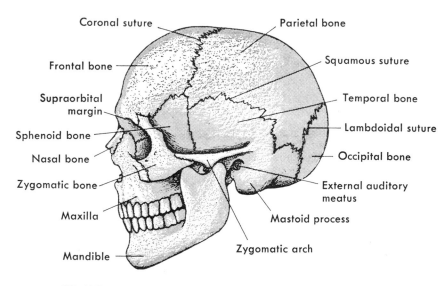

FIG. 10-5
Skull, side view. (Styloid process not shown.)

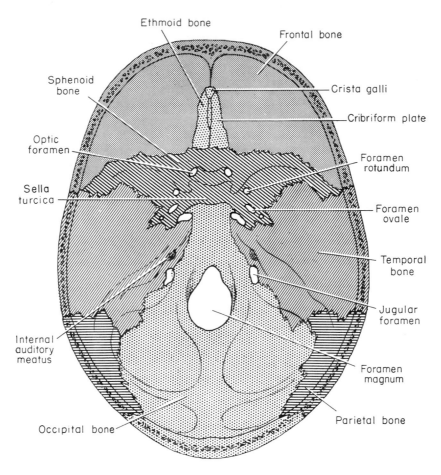

FIG. 10-6

Floor of skull seen from above. (Modified from Francis, C. C: Introduction to human anatomy, ed. 6, St. Louis, 1973, The C. V. Mosby Co.)

late with these two bones. They join together at the midline and form part of the orbits, part of the hard palate, and part of the floor and sidewalls of the nose. In contrast to the two bones (the maxillae) of the upper jaw, the lower jaw consists of a single bone, the mandible, which is the largest, strongest, and most powerful bone of the face. The sockets (alveoli) of the maxillae and the mandible hold the teeth and constitute the alveolar processes of these bones. The cen-

trally located zygomatic (malar) bones, which articulate with the maxillae of the face and with the temporal, frontal, and sphenoid bones of the cranium, form the cheeks. The bony roof of the mouth, or hard palate, is formed anteriorly by the processes of the maxillae and posteriorly by the two palatine bones. These bones are fused together so tightly that the hard palate sometimes looks like a single structure.

The remaining bones of the face—the nasal,

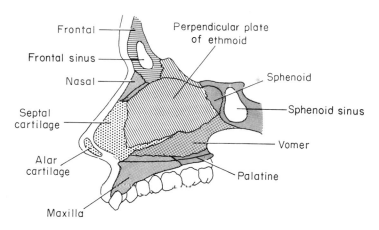

FIG. 10-7

Bones of nasal septum and hard palate. (Modified from Francis, C. C: Introduction to human anatomy, ed. 6, St. Louis, 1973, The C. V. Mosby Co.)

lacrimal, inferior conchae, and vomer—help to form the nose. The key nasal bone, however, is the ethmoid, a cranial bone. This delicate and highly irregular bone forms the upper septum, the sidewalls, and part of the roof of the nose. The vomer, which arises perpendicularly along the midline of the hard palate, meets the perpendicular plate of the ethmoid posteriorly to form the bony nasal septum (Fig. 10-7). The anterior portion of the septum is composed of cartilage. Into each nasal cavity (formed by the septum) project three scroll-shaped structures called conchae, or turbinates. The inferior conchae are separate bones; the middle and superior conchae are processes of the ethmoid. The conchae divide each nasal cavity into passageways, or meati.

Sinuses

The air cavities within the bones of the skull include the paranasal and mastoid sinuses. The paranasal sinuses, which include those of the frontal (Fig. 10-7), sphenoid, ethmoid, and maxillary bones, communicate with the nose; the mastoid sinuses, the tiny cavities within the mastoid process of the temporal bone, communicate with the middle ear. The maxillary sinus, the largest of the skull, is sometimes referred to as the *antrum of Highmore*.

Ear bones

Located within the middle ear cavity in the petrous portion of the temporal bone are three very tiny bones, or ossicles: the hammer (malleus), the anvil (incus), and the stirrup (stapes). These bones, so named because of their respective shapes, participate in the transmission of sound (Chapter 22). We should note that the six ossicles are not counted among the bones composing the skull. They are the only bones in the body that are completely formed at birth and do not grow.

Hyoid

The hyoid is a slender horseshoe-shaped bone situated at the upper border of the larynx. By pressing in with the index finger just above the Adam's apple, one can easily feel it. Suspended from the styloid processes of the temporal bones, the hyoid has the unique distinction of being the only bone in the body that does not articulate with another bone. Attached to the hyoid are the extrinsic muscles of the tongue and a few muscles at the floor of the mouth.

Vertebrae

The vertebrae are the highly irregular bones that form the spinal column. They are separated from each other by discs of fibrocartilage and are bound closely together by ligaments. Each vertebra is equipped with an opening (vertebral foramen), and together these foramina form the vertebral or spinal canal, which houses the spinal cord. The nerves issuing from the cord pass through the intervertebral foramina, the openings formed by the vertebral notches (Fig. 10-8).

The vertebrae differ in size and shape but generally have similar features. A typical vertebra possesses a body (a solid cylinder of spongy bone), articular processes, a spine, and, a foramen. The embryo has 33 or 34 vertebrae, but

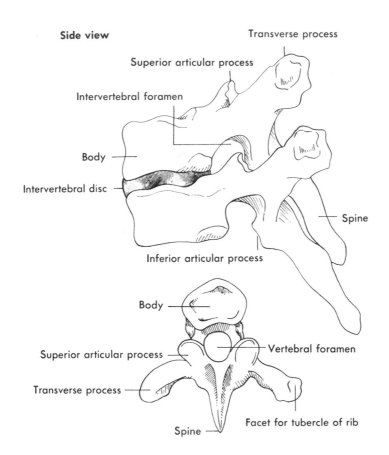

Side view

Transverse process

Superior articular process

Intervertebral foramen

Body

Intervertebral disc

Spine

Inferior articular process

Body

Superior articular process

Vertebral foramen

Transverse process

Spine

Facet for tubercle of rib

Top view

FIG. 10-8
Anatomy and articulation of thoracic vertebrae.

in the adult, the terminal vertebrae are fused, reducing the number to 26. Vertebrae are named according to their location in the spinal column: cervical, thoracic, lumbar, sacral, and coccygeal (Fig. 10-9).

The cervical vertebrae are the seven movable vertebrae of the neck; they are characterized by their small bodies and stubby, spinous processes. The first cervical vertebra, aptly called the atlas, supports the head via two cup-shaped articular depressions that articulate with the two occipital condyles. The second vertebra (called the epi-stropheus or axis) sends up a projection, the odontoid process, or *dens,* around which the atlas rotates, permitting the head to move from side to side.

The twelve thoracic vertebrae are larger and stronger than the cervical vertebrae and possess a well-developed, downward-pointing spinous process. The transverse processes of all except the eleventh and twelfth vertebrae are equipped with facets for articulation with the tubercles of the ribs.

The five lumbar vertebrae, which pass

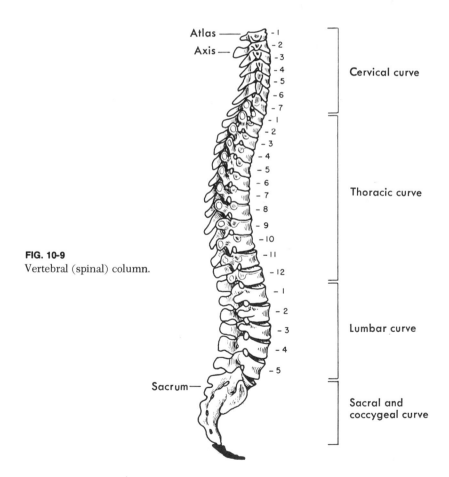

FIG. 10-9
Vertebral (spinal) column.

Atlas — 1
Axis — 2
— 3
— 4
— 5
— 6
— 7

Cervical curve

— 1
— 2
— 3
— 4
— 5
— 6
— 7
— 8
— 9
— 10
— 11
— 12

Thoracic curve

— 1
— 2
— 3
— 4
— 5

Lumbar curve

Sacrum —

Sacral and coccygeal curve

through the loin, are larger and heavier than the thoracic vertebrae. Their spinous processes are characteristically short, thick, and blunt.

The single sacrum of the adult is formed by the fusion of five sacral vertebrae. The first vertebra is the largest, and the succeeding vertebrae become progressively smaller in size. The smooth and concave anterior surface helps form the hollow of the pelvis. In contrast, the posterior surface is rough and uneven to permit attachment to the muscles of the back. The sacrum articulates with the ilia at the sacroiliac joints. The coccyx, or "tail," a small triangular bone at the end of the vertebral column below the sacrum, is formed by the fusion of four (sometimes five) coccygeal vertebrae.

Ribs and sternum

The twelve pairs of ribs (Fig. 10-3), together with the sternum in the front and the thoracic vertebrae in the back, form the skeletal framework of the thorax. A typical rib is a long, slender, curved bone with its head joined to a vertebra and its other end joined to the sternum by a band of cartilage. The first seven ribs are joined to the sternum by separate costal (rib) cartilages, whereas the cartilages of the ninth and tenth ribs are fused with the cartilage of the eighth rib. Accordingly, the first seven pairs often are called "true" ribs, and the remaining five pairs, "false" ribs. The costal cartilages of the eleventh and twelfth ribs do not meet the sternum at all; thus these ribs are called

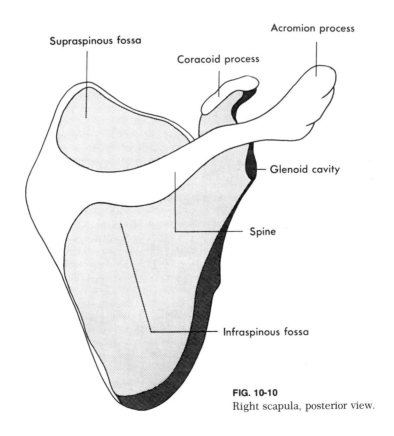

FIG. 10-10
Right scapula, posterior view.

Supraspinous fossa
Acromion process
Coracoid process
Glenoid cavity
Spine
Infraspinous fossa

"floating" ribs. The tips of these ribs may be felt by working the fingers along the lower border of the thoracic cage.

The sternum, or breastbone, is a dagger-shaped bone forming the chest wall at the midline in front. Its upper, handlelike portion is called the manubrium, the middle portion (body) is called the gladiolus, and the tip (unossified in early life) is called the xiphoid process. At the manubrium the sternum articulates with the first rib and the clavicle (collarbone).

Shoulder girdle

The scapula and the clavicle compose the shoulder girdle. Commonly called the shoulder blade, the scapula is a flat, thin, triangular bone possessing a well-developed spine and two prominent processes—the coracoid and the acromion (Fig. 10-10). The latter forms the tip of the shoulder and articulates in front with the lateral end of the clavicle. Below the coracoid process is an oval, shallow fossa, called the glenoid cavity, which accommodates the head of the humerous. The clavicle, or collarbone, forms the bony root of the neck in front. A long, curved, slender bone, the clavicle swings from its articulation with the acromion process to the manubrium at the midline.

Upper arm

The humerus is the bone of the upper arm (Fig. 10-3). One of the best examples of a long bone, the humerus has a well-rounded head at one end and epicondyles at the other. The head articulates with the glenoid cavity of the scapula, and the capitulum and trochlea (rounded surfaces below the epicondyles) articulate with the radius and the ulna, respectively. Between the epicondyles are the coronoid fossa in front and the olecranon fossa behind. The former accommodates the coronoid process of the ulna during flexion, and the latter receives the olecranon process of the ulna during extension.

Lower arm

The ulna and the radius compose the lower arm. The ulna, the longer of the two bones, lies on the side of the little finger. It may be traced by running the fingers from the tip of the elbow to the small projection at the side of the wrist. The ulna is characterized principally by the coronoid and olecranon processes, the latter constituting the elbow. Between these two processes is the semilunar notch which receives the trochlea of the humerus.

The radius runs parallel to the ulna on the outer or thumb side of the lower arm. At the proximal end its head articulates with the ulna and the capitulum of the humerus. At the distal end the radius articulates with the wrist and the head of the ulna.

Wrist

The wrist, or carpus, is composed of eight bones arranged in two rows. (Fig. 10-3). The proximal row includes the navicular, the lunate, the triquetrum, and the pisiform bones; the distal row includes the greater multangular, the lesser multangular, the capitate, and the hamate bones. The prominence at the front of the wrist on the side of the little finger is produced by the pisiform bone.

Hand

The bones of the hand include the five metacarpals of the palm and the fourteen phalanges of the fingers (Fig. 10-3). With the exception of the thumb, which contains two, there are three phalanges in each finger. The metacarpals, beginning with the thumb, are numbered I through V. The phalanges are named the first or proximal phalanx, the second or middle phalanx, and the third or distal phalanx.

Hipbone

The hipbone begins the lower extremity. It is also called os coxae, os innominatum, or, less

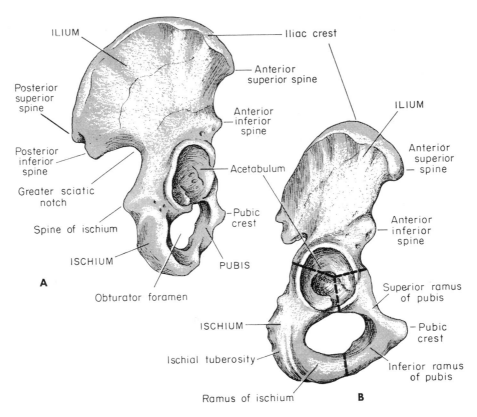

FIG. 10-11

Hipbone. **A,** Lateral view of right hipbone. **B,** Lateral view looking directly into acetabulum. Broken lines indicate where the three bones have joined, usually obliterated by age of 16 years. (Modified from Francis, C. C: Introduction to human anatomy, ed. 6, St. Louis, 1973, The C. V. Mosby Co.)

formally, the innominate bone. Actually, this large, powerful, and irregular bone is not one bone, but three—the ilium, the ischium, and the pubis. (Fig. 10-11). These bones are fused so completely in the adult, however, that it takes a while to detect the lines of juncture. The principal features include the iliac crest, the acetabulum, the obturator foramen, and the greater sciatic notch. The iliac crest borders the ilium at the top and terminates in front and behind as the anterior and posterior superior

spines, respectively. Below these are two other projections of the ilium: the anterior inferior spine and the posterior inferior spine. The acetabulum, a deep socket that receives the head of the femur, is formed by the fusion of the ilium, the ischium, and the pubis. The obturator foramen, the largest foramen or hole in the skeleton, is bounded by the ischium and the pubis.

Strong ligaments unite the two hipbones with the sacrum at the sacroiliac joints to form the pelvis (Fig. 10-3). Anteriorly, the pubic bones articulate with each other at the symphysis pubis to form the pubic arch. The female and male pelves differ in two major respects: in the female the pubic arch is broader, and the pelvic cavity is wider and more capacious.

Thigh and knee

The femur, or thighbone, is the largest and heaviest bone in the body. At its upper end it is equipped with a smooth spherical head that fits neatly into the acetabulum (Fig. 10-11). At the base of its neck are two prominent processes: the greater trochanter, a massive projection lateral to the shaft, and the lesser trochanter, a smaller projection on the inner side. The shaft itself is round and, with the exception of the linea aspera, smooth. The linea aspera is a ridge that runs along the back of the shaft and serves to attach muscles to the femur. At the distal end, the femur forms the lateral and medial condyles, which articulate with the patella and the tibia. Between these processes is the intercondylar fossa.

The patella, or kneecap, is a shell-shaped bone embedded in the tendon at the front of the knee (Fig. 10-3). Because it resembles a grain of sesame, it is often said to be "sesamoid." When the leg is flexed, the patella is forced forward; when the leg is extended, the patella sinks back into the intercondylar fossa.

Lower leg

The tibia and the fibula compose the lower leg. The tibia, or shinbone, is the larger. Its proximal end is thickened and forms two smooth articular surfaces—the lateral and medial condyles— which support the lateral and medial condyles of the femur. At the distal end on the inner side of the ankle is a prominence called the medial malleolus. The fibula is a long slender bone that runs lateral and parallel to the tibia (Fig. 10-3). Proximally, it articulates with the tibia; distally,

it terminates as the lateral malleolus, or the outer prominence of the ankle.

Foot

The bones of the foot (Fig. 10-3) include seven tarsals, five metatarsals, and fourteen phalanges. The metatarsals are numbered I through V, beginning with the big toe. With the exception of the big toe, which has two phalanges, there are three phalanges in each toe; like the phalanges in the hand, they are named the proximal, phalanx, the middle phalanx, and the distal phalanx.

DISEASES OF BONE

Diseases of bone commonly are discussed within the framework of "musculoskeletal and connective tissue disorders," or some similar designation. A great many diseases and disorders are bone associated. Here we shall touch upon certain important conditions that are primarily confined to bone and the skeletal system.

Arthritis

Arthritis is an inflammation of a joint. Since there are so many varieties of the condition and so many ambiguous and confusing clinical terms, we are in need of a better system of classification. Until the causes are known, however, any classification is tentative. For the present, the "categories" presented here will suffice.

Rheumatoid arthritis is a chronic systemic involvement characterized by painful, inflammatory changes in the joints. Deformity and, generally, invalidism occur as the disease progresses. Histologically, the main lesions are lymphocytic infiltration and excess connective tissue in and about the joints. The incidence of the malady in women is two to three times that in men. Rheumatoid arthritis is one of the major causes of disability in this country. Although its etiology remains unknown, the typical case presents some highly interesting and perhaps significant biochemical findings. Treatment centers upon rest, physiotherapy, and drugs (for the most part aspirin and corticosteroids).

Other forms of arthritis relate to injury, degenerative changes, gout, and infectious agents. *Rheumatic fever,* an acute inflammatory complication of group A streptococcal (*Streptococcus pyogenes*) infections, is invariably accompanied by painful and tender joints, especially those of the ankles, knees, elbows, and wrists. After an attack of acute rheumatic fever, continuous antistreptococcal prophylaxis should be maintained to prevent recurrent attacks. The best drug here, on all counts, is penicillin. All strep infections should be treated promptly to prevent rheumatic fever in the first place.

Osteitis deformans

Osteitis deformans, or *Paget's disease,* is a chronic inflammation of bone with marked deformity. Muscle pain, bone pain, and hearing impairment are also commonly experienced. Initially, there is a decalcification and softening of the skeleton that results in the bending and bowing of the weight-bearing bones. In the late stages of this strange disease, there is unbridled recalcification, resulting in a thickened, enlarged, and twisted skeleton. The cause of osteitis deformans is unknown, and its treatment is nonspecific.

Osteitis fibrosa cystica

Osteitis fibrosa cystica, also known as *von Recklinghausen's disease,* is a grotesque bone disorder resulting from *hyperparathyroidism* (Chapter 23). The excess parathyroid hormone washes out the calcium from the skeleton into the blood leaving a flimsy framework that bends, bows, and breaks under the weight of the body. Stones may appear in the kidneys as a result of hypercalcemia. In the early days of medicine, before the cause of this disease was discovered,

the patient eventually grew into an unrecognizable twisted mass of flesh and bone. Today, removal of the hyperplastic or tumorous parathyroid(s) offers an excellent prognosis.

Rickets and osteomalacia

Rickets is caused by a deficiency of vitamin D in growing children, the deficiency arising from inadequate exposure to ultraviolet rays. Under unusual living conditions in temperate climates, it also arises from an inadequate vitamin intake. The pathologic changes essentially

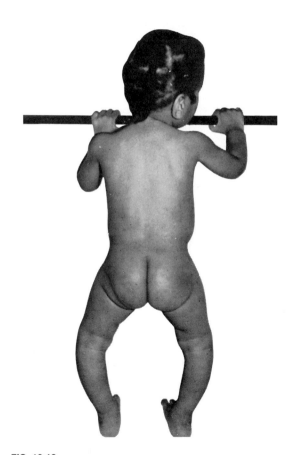

FIG. 10-12

Rickets in young child. (Courtesy Rosa L. Nemire, M.D. and the Upjohn Co.)

consist of defective calcification of bone and hypertrophy of epiphyseal cartilage. This means malformation and skeletal deformities, particularly the bending and bowing of the leg bones (Fig. 10-12). Also, *tetany* may occur, a syndrome manifested by sharp flexion of the wrist and ankle joints, muscle twitchings, cramps, and convulsions. In the adult the counterpart of rickets is osteomalacia, a condition marked by flexible, soft, and deformed bones, and resulting from a lack of vitamin D or calcium or both. The treatment and prevention of rickets and osteomalacia clearly hinges on the administration of vitamin D and calcium.

Osteoporosis

Osteoporosis is a reduction in the amount of calcified bone mass per unit volume of skeletal tissue. It is a matter of decalcification or demineralization. The most common form, *senile* osteoporosis, occurs in the aged, presumably due to a lack of sex hormones. Other etiologic possibilities include an excess of catabolic hormones, as in Cushing's disease (Chapter 23), and long-term administration of corticosteroids in large doses. The chief complications of osteoporosis are bone pain, fracture, kyphosis ("hunch-back") and invalidism. Treatment is supportive and specific. In postmenopausal osteoporosis, for example, estrogens often effectively relieve the pain.

Tumors

Bone tumors are either benign or malignant, and the latter in turn are either metastatic or primary. Metastases are far more common than primary involvements, and just about any cancer in any tissue may travel to the skeleton (Fig. 10-13). The principal primary cancers are the multiple myelomas and osteogenic sarcomas. The diagnosis of these and other bone cancers is made on a basis of x-ray studies, biopsy, and certain chemical tests; treatment centers on various combinations of irradiation and surgery; prognosis ranges from negative to good, depending on the degree of malignancy. There are about 3,000 new cases of bone cancer per year and some 2,000 deaths. Although these are relatively low figures, the agony is especially great, and the prognosis, as indicated, commonly depressing.

Infections

The most important infections of bone are osteomyelitis and tuberculosis. *Osteomyelitis,*

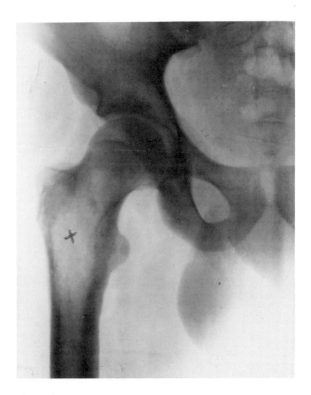

FIG. 10-13
Metastatic cancer (X) of upper femur. Decreased opacity is a result of bone destruction, meaning that this is an osteolytic growth. (Courtesy Charles F. Geschickter, M.D., Georgetown University Hospital, Georgetown University Medical Center, Washington, D.C.)

a bone infection marked by fever, tenderness, and pain over the affected bone, usually arises from an invasion of pyogenic microorganisms, most frequently *Staphylococcus aureus.* These organisms gain access to bone either through the blood as a result of an acute focal infection or directly following compound fracture or other trauma. Before the advent of chemotherapy, osteomyelitis frequently proved fatal. Today, early treatment with the appropriate drug affords a favorable prognosis.

Tuberculosis of the bones and joints occurs as a secondary infection; the pathogen *Mycobacterium tuberculosis,* makes its way to these sites from a focus of infection elsewhere in the body (Chapter 15). The course of the disease is chronic and destructive. Therapy includes a thorough search for the primary infection, the use of appropriate surgical procedures, and the administration of antitubercular drugs.

QUESTIONS

1. Distinguish between the terms foramen and meatus.
2. Distinguish between the terms spine and crest.
3. Compare the substance of the diaphysis and the epiphysis.
4. Compare the terms matrix and intercellular.
5. Follow a red blood cell from a periosteal blood vessel to a bone cell.
6. What is the literal meaning of periosteum? of hemopoiesis?
7. Certain chemicals are toxic to bone marrow. What, therefore, would be the clinical situation in such poisonings?
8. In medicine, what is meant by a "sternal puncture"?
9. With regard to bone, the terms ossification, calcification, and mineralization commonly are used interchangeably. Why?
10. The soft spots of the infants head eventually "turn to bone." By what process?
11. What is the literal meaning of osteoblast?
12. What are the three chief elements present in hydroxyapatite?
13. What is alkaline phosphatase?
14. With regard to bone, the terms dissolution, resorption, and decalcification are often used interchangeably. Explain.
15. For the most part injury to bone stimulates osteoblastic activity. Why doesn't it stimulate osteoclastic activity?
16. Would you say that vitamin D has a direct or an indirect role in ossification? Explain your answer.
17. The parathyroid hormone and calcitonin are physiologically antagonistic. Explain.
18. The parathyroids are triggered to secrete the parathyroid hormone when calcium blood level falls below normal. Does this make sense?
19. For proper bone formation what nutrients are needed in addition to vitamin D?
20. What is the literal meaning of the term articulation?
21. What type of joint(s) are sutures?
22. Distinguish between diarthrosis and diarthroses.
23. Why is joint fluid called synovial fluid?
24. Give the antonyms for flexion and adduction.
25. What is the literal meaning of circumduction?
26. Name the bones associated with the following features: mastoid process, sella turcica, foramen magnum, auditory meatus, and crista galli.
27. What three bones are related to the posterior fontanel?
28. What bones form the hard palate?
29. What is the relationship between the teeth and the alveolar processes?
30. What are the oribts?
31. What bones form the nasal septum?
32. The midline of the hard palate represents the juncture of two pairs of bones. What are they?
33. What is the purpose of all the many and various foramina of the cranial floor?
34. Are the nasal bones considered part of the nasal septum?
35. What purpose do sinuses serve?
36. What is the antrum of Highmore?
37. Distinguish between vertebral foramen and intervertebral foramen.
38. What lies between the bodies of the vertebrae?

39. How many cervical and thoracic vertebrae are there in the infant spinal column?
40. The atlas articulates with one bone above and another bone below. What are they?
41. Distinguish between "false" and "floating" ribs.
42. To what does costal refer?
43. The sacrum and the ilium come together at what articulation?
44. What does manubrium, the upper part of the sternum, mean in Latin?
45. Supply the appropriate anatomical terms for the following: cheek bone, lower jaw, turbinates, hammer (of ear), and collarbone.
46. What is the common name for os coxae?
47. How many individual bones make up the pelvis?
48. Supply the appropriate anatomical terms for the following: shoulder blade, elbow, wrist, thighbone, shinbone, and kneecap.
49. Paradoxically, the foramen magnum is not the largest foramen in the body. What is the largest?
50. Precisely where does the femur articulate with the os innominatum?
51. The anterior superior iliac spine is a process of what bone?
52. Compare the hand and the foot in detail.
53. What is the literal meaning of the expression "intercondylar fossa"?
54. What are the malleoli?
55. What is the literal meaning of the term arthritis?
56. What is the relationship between strep infections and rheumatic fever?
57. Explain the hypercalcemia present in osteitis fibrosa cystica.
58. Explain the kidney stones commonly seen in osteitis fibrosa cystica.
59. Compare rickets and osteomalacia.
60. Explain why inadequate exposure to ultraviolet light results in rickets.
61. Why does a lack of vitamin D often lead to tetany?
62. Compare osteomalacia and osteoporosis.
63. Compare multiple myelomas and osteogenic sarcomas.
64. Why is a compound fracture likely to cause osteomyelitis?
65. What is housemaid's knee?

CHAPTER 11

THE MUSCULAR SYSTEM

The muscular system refers to skeletal muscles, or those muscle masses attached to bone. Movement, maintenance of posture, and heat production are its functions. The musculature contains over one third of all body proteins and constitutes almost half of the body weight.

MUSCLE FIBER

Because muscle cells are so long and narrow, they almost always are referred to as muscle fibers, a given muscle being made up of thousands of these basic units. A single fiber ranges from about 5 to 150 μm in diameter and from 1 to 120 mm in length. As shown in Fig. 11-1, a fiber is composed of *myofibrils* embedded in a semifluid substance known appropriately as *sarcoplasm;* surrounding the fiber is an exceedingly thin elastic sheath, the *sarcolemma,* and just beneath this is situated the numerous nuclei. The myofibrils, in turn, are made up of thick and thin filaments so arranged as to produce dark and light cross striations called *A* (anisotropic) *bands* and *I* (isotropic) *bands,* respectively. We note, too, the less dense midsections, or *H zones,* of the A bands, and the more dense midlines, or *Z lines,* of the I bands. The segment of the myofibril extending from one Z line to the next is called a *sarcomere.* Also embedded in the sarcoplasm is a network of sacs and tubules called the *sarcoplasmic reticulum,* a network similar but not identical to the endoplasmic reticulum of other cells. Finally, a deep folding of the sarcolemma into the interior of the fiber, the *T system,* communicates with the sarcoplasmic reticulum.

Muscle proteins

Proteins unique to muscle are myosin, actin, troponin, tropomyosin, and myoglobin. Myosin molecules compose the thick myofibril filaments; molecules of actin, together with troponin and tropomyosin, compose the thin filaments. As we shall see in just a moment, the interaction of these proteins constitutes muscle contraction. Myoglobin, sometimes referred to as "muscle hemoglobin," is the red pigment of muscle responsible for the transport of oxygen from the capillaries to the mitochrondria. Muscle fibers deficient in myoglobin are pale—hence, the "white meat" of chicken.

CHEMISTRY OF CONTRACTION

Upon arriving at a muscle fiber, a nerve impulse initiates an electrical impulse that travels along the sarcolemma and into the interior, via the T system, to cause the release of calcium ions (Ca^{++}) from the sacs of the sarcoplasmic reticulum. Once released to the sarcoplasm,

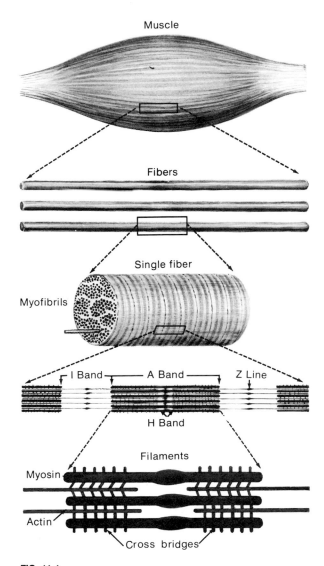

Muscle

Fibers

Single fiber

Myofibrils

I Band — A Band — Z Line

H Band

Filaments

Myosin

Actin

Cross bridges

FIG. 11-1

Striated muscle dissected in schematic fashion. The muscle is made up of many fibers, which appear cross-striated. Single fiber consists of myofibrils, which show alternating dark and light bands. A single sarcomere is the region between two Z lines containing *I* and *A* bands. Bands derive from presence of two sets of interdigitating filaments, thick filaments (myosin) with cross bridges and thin filaments (actin). (From Schottelius, B. A., and Schottelius, D. D.: Textbook of physiology, ed. 17, St. Louis, 1973, The C. V. Mosby Co.)

these ions combine with troponin of the actin filaments and thereby set off an interaction between myosin and actin. The extensions (cross bridges) of the myosin filaments attach to the actin filaments and slide them in toward the midsections of the myosin filaments. The upshot is the shortening, or *contraction,* of the myofibrils and their parent fibers. Based on what is presently known, calcium-free troponin in resting muscle inhibits the interaction of myosin and actin, whereas the calcium-troponin complex is without this effect. Muscle *relaxation* amounts to the breakdown of the complex and the reentry of calcium into the sacs of the sarcoplasmic reticulum. Troponin becomes free once again to exert its inhibitory influence and, by the same token, to react with calcium and cause contraction.

Energy and oxygen debt

The energy needed for muscle contraction is supplied by the breakdown of ATP into ADP and phosphate (p. 79), a reaction triggered by the same impulse that triggers the release of calcium ions from the sacs of the sarcoplasmic reticulum. During periods of relaxation, muscle fibers build up their supply of ATP. They convert some of this ATP energy into reserve creatine phosphate (CP; phosphocreatine)—a high-energy phosphate compound that instantaneously provides additional ATP during periods of strenuous activity. Like most biochemical situations, the relationship between ATP and CP is reversible: during relaxation creatine (C) reacts with ATP to produce CP; during contraction CP reacts with ADP to produce creatine and ATP.

The burning up of ATP and CP in contraction means an increased output of pyruvic acid by glycolysis; if the rate of utilization is high, insufficient oxygen is available for the citric acid cycle to handle all the increased output (p. 82). Consequently, the excess pyruvic acid is reduced to lactic acid, the bulk of which diffuses into the blood and makes its way to the liver, where it is converted to glycogen. The smaller amount of lactic acid remaining in the muscle is eventually oxidized back to pyruvic acid when exercise has decreased sufficiently; the pyruvic acid is then oxidized in the citric acid cycle and the cytochrome system. This demands an amount of oxygen over and above what would be consumed if the excessive muscle contraction had not occurred. This is an "oxygen debt" that must be repaid to reestablish normal conditions in the muscle.

Heat of contraction

From the standpoint of physics, muscle is not efficient in converting energy into work. The average individual has an efficiency of only about 20%, and even the athlete cannot top 35%. The energy not utilized in contraction appears as heat, mainly muscle heat that keeps the body warm and functioning.

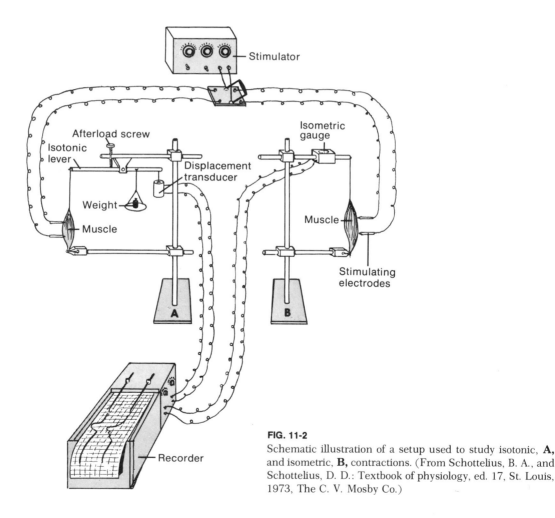

FIG. 11-2

Schematic illustration of a setup used to study isotonic, **A,** and isometric, **B,** contractions. (From Schottelius, B. A., and Schottelius, D. D.: Textbook of physiology, ed. 17, St. Louis, 1973, The C. V. Mosby Co.)

STIMULATION AND CONTRACTION

Electrical stimulation of excised muscles of frogs provides much basic and useful information. The classic preparation is the gastrocnemius muscle; a typical laboratory arrangement of the apparatus used in the student laboratory for the stimulation of this muscle and the recording of its contractile responses is shown in Fig. 11-2.

The stimulus

Electrical stimulation experiments disclose three prime characteristics of a stimulus: its strength, its duration, and its rate of change of intensity.

Other things being equal, the strength of a stimulus delivered to a muscle (in vitro) determines the extent of contraction. The *threshold,* or *liminal,* stimulus is the weakest stimulus that under specified conditions elicits a response. Stimuli of less intensity are said to be *subliminal.*

Stimuli of greater intensity increase the magnitude of contraction up to a point, beyond which no further increase occurs. A stimulus that brings the magnitude of contraction to this point is called the *maximal* stimulus. Stimuli between this and the threshold stimulus are referred to as *submaximal,* and stimuli greater than maximal are said to be *supramaximal.* Under normal conditions, a single muscle fiber responds maximally to all stimuli beyond the threshold stimulus; it gives the fullest response or none at all, a phenomenon known as the *all-or-none law.* Somewhat paradoxically, this explains the greater contraction of the entire muscle to stronger stimulation. Submaximal stimuli of increasing intensity call more and more muscle fibers into play; once the maximal stimulus is reached, once the whole muscle no longer responds with increasing contraction strength to an increase in stimulus intensity, then all the fibers are contracting (Fig. 11-3). The important

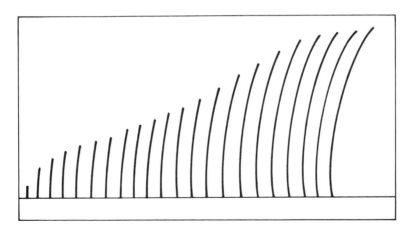

FIG. 11-3

Effect of increasing strength of stimulus on contraction of frog gastrocnemius muscle. Strengths range from threshold (first contraction at far left) to supramaximal at far right. (From The human body: its anatomy and physiology, third edition, by C. H. Best and N. B. Taylor. Copyright © 1932, 1948, 1956, by Holt, Rinehart and Winston, Inc. Adapted and reprinted by permission of Holt, Rinehart and Winston, Inc.)

thing to remember, of course, is that the all-or-none law refers to a muscle fiber, not an entire muscle.

Muscle experiments show that if the length of time during which the stimulus is applied is shortened sufficiently, no contraction results regardless of the strength of the stimulus. In order for a stimulus to produce a response, it must act for a certain length of time. The minimal duration required is called the *excitation time*. Within limits, the stronger the stimulus, the shorter the excitation time. These relationships are portrayed vividly on the strength-duration curve illustrated in Fig. 11-4. Note that on this curve no stimulus below the *rheobase* (minimal stimulus strength required to evoke a

response) is effective regardless of how long it acts. On the other hand, if the duration is too short, no stimulus is effective regardless of its intensity.

The third key characteristic of a stimulus, the rate of change of intensity, can be demonstrated by gradually applying a maximal stimulus to a muscle; that is, gradually building it up from zero to full intensity. Whereas the instantaneous maximal stimulus evokes a vigorous response, the gradual maximal stimulus evokes no response at all because the muscle apparently accommodates itself to the gradual increase in intensity. This relationship between the rate of change of intensity and the response, formally known as the *DuBois-Reymond law,* may be

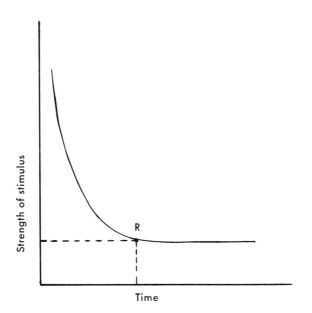

FIG. 11-4
Strength-duration curve showing relationship between strength and duration of stimuli. Stimulus of strength *R* (the rheobase) must be applied for indefinite length of time to elicit response. Stimuli of less intensity are ineffective regardless of how long they act.

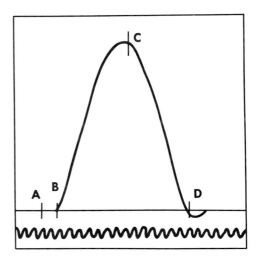

FIG. 11-5
Kymographic record of muscle twitch. *A* to *B,* Latent period. *B* to *C,* Contraction. *C* to *D,* Relaxation. Time indicated by lower tracing. Timer set at 100 vibrations per second; hence, one vibration equals $^1/_{100}$ second. (From The human body: its anatomy and physiology, third edition, by the C. H. Best and N. B. Taylor. Copyright © 1932, 1948, 1956, by Holt, Rinehart and Winston, Inc. Adapted and reprinted by permission of Holt, Rinehart and Winston, Inc.)

stated thus: The efficiency of a stimulus varies with the rapidity of change in strength from zero to maximum intensity.

The muscle twitch

The muscle twitch, the fundamental unit of recordable muscular activity, is a quick, jerky contraction in response to a single stimulus. As shown in Fig. 11-5, a twitch reveals a latent period (or interval between stimulation and response), a period of contraction, and a period of relaxation. The latent period, which is only a fraction of a millisecond, represents the events leading up to the interaction between actin and myosin. Twitches or contractions are isotonic or isometric. An *isometric* twitch is produced by stimulating a muscle suspended between fixed points; the muscle retains its original length but develops tension. An *isotonic* twitch, on the other hand, is a contraction in which the length of the muscle shortens, and the muscle performs mechanical work. The muscles of the body contract both isometrically and isotonically. Lifting the arm, for example, involves isotonic contractions, whereas in standing the muscles of the trunk and legs are contracting isometrically against the force of gravity.

Repetitive stimulation

Thus far we have discussed stimulation in the singular. Repetitive stimulation adds a new dimension. Subliminal stimuli just below threshold intensity can produce contraction if delivered in rapid succession. The first few stimuli in a series, although inadequate, increase the irritability of the muscle to the point where one of the successive stimuli evokes a contraction. This phenomenom is appropriately referred to as *summation* of subliminal stimuli.

A variation of the same theme is to apply a number of maximal stimuli to muscle in such rapid succession that little time is offered for relaxation between successive contractions. The result is the fusion of twitches and the intensification of contractions. As the frequency of stimulation progressively increases, a steady state of contraction is reached showing no individual twitches at all—a state called *tetanus* (Fig. 11-6). Interestingly, tetanus can be predicted from the duration of a single contraction. For example, if it takes a certain muscle $\frac{1}{60}$ second to contract, a stimulating current delivering a volley of 60 or more shocks per second will produce tetanization. Wave summations just below the point of tetanization are referred to as incomplete tetanus. Under normal conditions the muscles of the body perform their duties via a combination of incomplete and complete tetanizations. "Motor units" (p. 153) are fired one at a time but rapidly enough to produce the smooth tetanic type of response rather than jerky twitches.

Another experiment dealing with repetitive stimulation is to subject excised muscle to a *prolonged* period of repeated stimuli. As shown in Fig. 11-7, the first few contractions are greater than the preceding ones, an effect generally attributed to the enhanced irritability occasioned by the accumulation of metabolic waste products and the slight temperature increase. This is called the *treppe,* or *staircase phenomenon.* Beyond the treppe (still referring to Fig. 11-7), the contractions are constant for a while but very soon start to decrease in height. A state of *contracture* appears when the muscle fails to relax to the original baseline between stimuli. In short, this is *muscle fatigue,* the cause of which apparently relates to the accumulation of waste products (CO_2, lactic acid, and the like) and the lack of energy-furnishing substances.

A final matter relating to repetitive stimulation is the *refractory period,* or the time during which muscle fails to respond to an adequate stimulus. In the case of frog muscle, this interval is fantastically short, somewhere in the

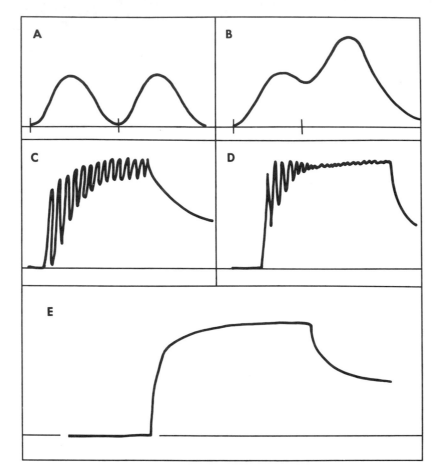

FIG. 11-6
Wave summation and tetanization. **A,** Second stimulus applied after relaxation of muscle. **B,** Second stimulus applied after shorter interval than in **A. C** to **E,** Still shorter intervals. Note complete tetanus in **E.** (From The human body: its anatomy and physiology, third edition, by C. H. Best and N. B. Taylor. Copyright © 1932, 1948, 1956, by Holt, Rinehart and Winston, Inc. Adapted and reprinted by permission of Holt, Rinehart and Winston, Inc.)

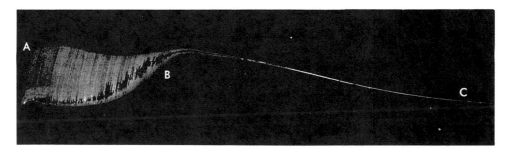

FIG. 11-7
Repeated stimulation of muscle, *A,* leads to incomplete relaxation, *B,* and finally fatigue, *C.* (From Francis, C. C, and Knowlton, G. C.: Textbook of anatomy and physiology, ed. 2, St. Louis, The C. V. Mosby Co.)

vicinity of ½₀₀ second! Thus, a shock falling within this period, say ⅓₀₀ second after the previous shock, fails to evoke a response.

Strength of contraction

The strength of contraction has already been touched upon. In way of review, increasing the strength of the stimulus, up to a point, increases the strength of contraction; and, up to a point, increasing the frequency of stimulation increases the strength of contraction. On the other hand, the accumulation of waste products and the lack of oxygen and of energy-rich substances cause fatigue and lesser contractions. To these factors let us add stretching and the size of the load. Stretching, up to a point, increases the strength of contraction; for each muscle fiber there exists an optimal length that permits the most forceful response. The same applies to the size of the load imposed on muscle.

NERVE SUPPLY

The muscles, like other organs, are supplied with motor nerves whose impulses trigger contraction. Such nerves are composed of fibers (*axons*) that divide into a variable number of branches or *terminal endings,* each of which makes contact with a muscle fiber at the *motor end plate* (neuromuscular junction; myoneural junction). On the average, a single fiber causes the contraction of 150 or so muscle fibers via the release of the neurohormone acetylcholine at the motor end plate (Fig. 11-8). Acetylcholine initiates an impulse that travels along the sarcolemma and thereby sets off the chemical events leading to contraction. In addition to these motor fibers there are a variety of sensory nerve endings attached to special receptors, the purpose being to apprise the central nervous system of "local" conditions. Especially important are the *stretch receptors* (p. 66), which enable the body to maintain muscle tone (tonus) and coordination.

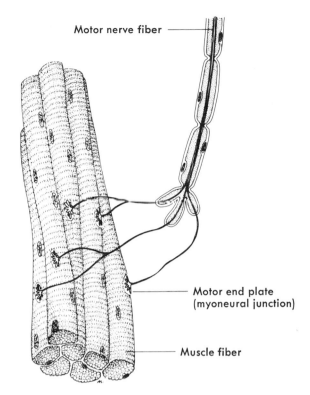

Motor nerve fiber

Motor end plate (myoneural junction)

Muscle fiber

FIG. 11-8
Motor nerve and its end plates.

MUSCLE STRUCTURE

Muscles vary in size and shape and in the arrangement of their fibers (Figs. 11-9 and 11-10). For example, whereas the gluteus maximus forms the bulk of the buttocks, the stapedius muscle of the middle ear is composed of only a few tiny strands. Some muscles are broad, others are narrow; some are long and tapering, others, short or blunt; some are triangular or quadrilateral, others, irregular; some are flat sheets, others are bulky; and so forth. Some fibers run parallel to the long axis; some come together at a narrow attachment; some are oblique; and some are curved or "circular." The direction or arrangement of the fibers tells us

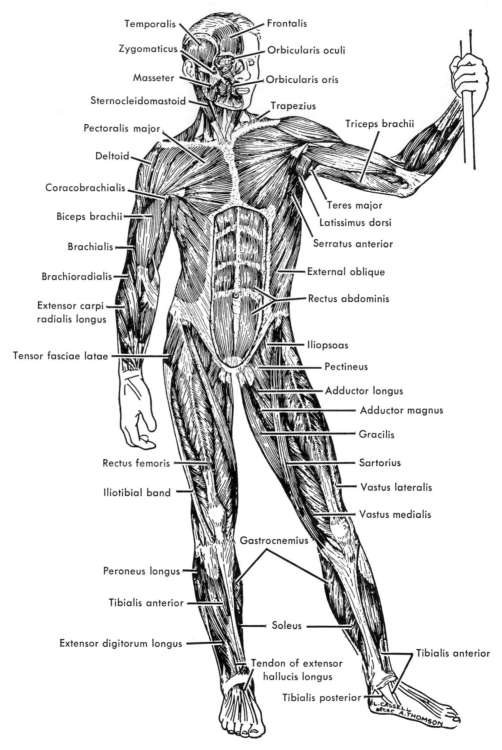

FIG. 11-9

Anterior view of muscles of the body. (From Millard, N. D., King, B. G., and Showers, M. J.: Human anatomy and physiology, ed. 6, Philadelphia, 1969, W. B. Saunders Co.)

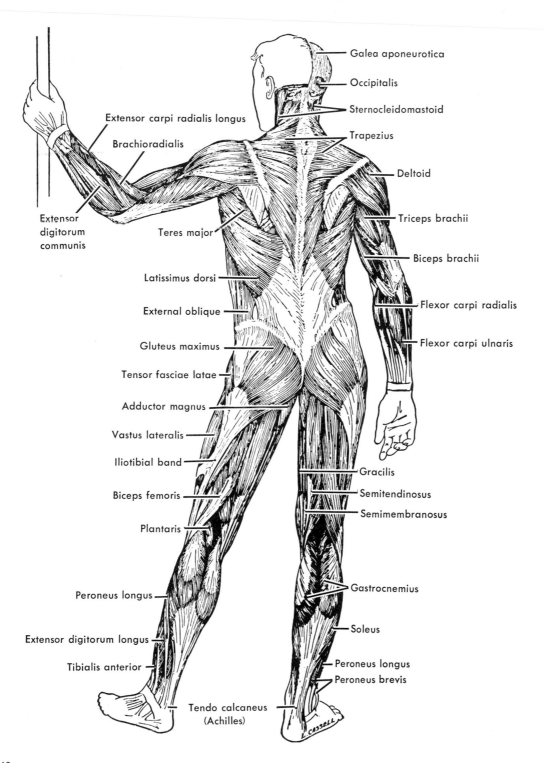

Galea aponeurotica
Occipitalis
Sternocleidomastoid
Extensor carpi radialis longus
Brachioradialis
Trapezius
Deltoid
Triceps brachii
Extensor digitorum communis
Teres major
Biceps brachii
Latissimus dorsi
Flexor carpi radialis
External oblique
Flexor carpi ulnaris
Gluteus maximus
Tensor fasciae latae
Adductor magnus
Vastus lateralis
Iliotibial band
Gracilis
Biceps femoris
Semitendinosus
Semimembranosus
Plantaris
Peroneus longus
Gastrocnemius
Soleus
Extensor digitorum longus
Peroneus longus
Tibialis anterior
Peroneus brevis
Tendo calcaneus (Achilles)

L. CASSELL

FIG. 11-10
Posterior view of muscles of the body. (From Millard, N. D., King, B. G., and Showers, M. J.: Human anatomy and physiology, ed. 6, Philadelphia, 1969, W. B. Saunders Co.)

TABLE 11-1

Major muscles

Muscle	General origin	General insertion	Chief function

Adductors of thigh
  Adductor magnus
  Adductor longus
  Adductor brevis
	Ilium and pubis	Femur and tibia	Adduct thigh

Anterior tibial group	Tibia and fibula	Foot	Dorsiflexes foot
Biceps brachii	Scapula	Radius	Flexes and supinates forearm
Buccinator	—	—	Aids in chewing
Deltoid	Clavicle and scapula	Humerus	Abducts arm
Epicranius	—	—	Wrinkles forehead
External oblique	Lower costal cartilages	Crest of ilium	Flexes trunk
Gastrocnemius	Condyles of femur	Heel bone	Plantar flexes foot
Gluteus maximus	Sacrum and ilium	Femur	Extends thigh

Hamstring group
  Biceps femoris
  Semitendinosus
  Semimembranosus
	Ischial tuberosity	Tibia and fibula	Flexes leg

Iliopsoas	Ilium and lumbar vertebrae	Femur	Flexes thigh
Internal oblique	Iliac crest	Lower costal cartilages	Flexes trunk
Latissimus dorsi	Lower vertebrae and ilium	Humerus	Extends and adducts arm
Orbicularis oculi	—	—	Closes eyes
Orbicularis oris	—	—	Closes mouth
Pectoralis major	Clavicle and chest wall	Humerus	Flexes and adducts arm
Posterior tibial group	Tibia and fibula	Foot	Plantar flexes foot
Quadratus lumborum	Ilium and lumbar vertebrae	Lumbar vertebrae	Flexes vertebrae

Quadriceps femoris
  Rectus femoris
  Vastus intermedius
  Vastus lateralis
  Vastus medialis
	Ilium and femur	Tibia	Extends leg

Rectus abdominis	Pubis	Sternum and costal cartilages	Flexes trunk
Risorius	—	—	Extends corners of mouth
Sacrospinalis	Vertebrae	Vertebrae and ribs	Extends trunk
Sartorius	Anterior superior iliac spine	Tibia	Turns thigh outward
Serratus anterior	Chest wall	Scapula	Draws shoulders forward
Sternocleidomastoid	Clavicle and sternum	Occiput	Flexes head
Tensor fasciae latae	Ilium	Tibia	Abducts thigh
Transversus abdominis	Lower ribs	Linea alba	Compresses viscera
Trapezius	Upper vertebrae	Scapula and clavicle	Braces shoulders
Triceps brachii	Humerus and scapula	Ulna	Extends forearm

much about the function of the parent muscle. Sphincter muscles, for example, which are composed of curved fibers, function to close openings.

Typically, a muscle consists of a main part, the *body,* and two extremities, the *origin* and the *insertion.* The extremities are attached to bone, cartilage, or fascia via extensions of the *epimysium,* the fibrous connective tissue that envelops muscle. When such an attachment takes the form of a strong tough cord, it is called a *tendon;* when it takes the form of a broad, flat sheet, it is called an *aponeurosis.* The epimysium not only covers muscle but also penetrates it to become the partitions between the bundles (the *perimysium*) and individual fibers (the *endomysium*).

CLASSIFICATION AND NOMENCLATURE

According to one source, there are 327 paired and 12 unpaired muscles. Customarily, these are grouped and studied on a basis of their chief function or action. These groups or classes include

flexors decrease the angle of a joint.
extensors increase the angle of a joint.
abductors move a part away from the midsagittal plane.
adductors move a part toward the midsagittal plane.
supinators turn the palm upward.
pronators turn the palm downward.
levators raise a part.
depressors lower a part.
evertors turn a part outward.
invertors turn a part inward.
rotators cause a part to pivot upon its axis.
tensors tense a part.
sphinctors reduce the size of an opening.
dorsiflexors pull the foot backward.
plantar flexors pull the foot downward.

When a muscle contracts, its more fixed attachment is called the origin, and the moveable attachment (where the effects of movement are produced) is called the insertion (Table 11-1). The muscle responsible for a given movement is called the *agonist,* or *prime mover,* and its opposing muscle is called the *antagonist.* When the agonist contracts, the antagonist relaxes, and vice versa. The agonist rarely acts alone; for the sake of smoothness, neighboring muscles, called *synergists,* contract at the same time.

A muscle is named according to one or more features, including size, shape, number of divisions, direction of fibers, location, points of attachment, and basic action. The deltoid is triangular; the triceps brachii is a three-part muscle acting on the arm; the transversus abdominis is an abdominal muscle with transverse fibers; the adductor magnus is a large adductor; and so on.

MUSCLES OF FACIAL EXPRESSION

As a group, the muscles of facial expression, are complex and clearly peculiar to man. Those of special interest include the frontalis (raises eyebrows and draws scalp forward), the corrugator (wrinkles forehead vertically), the orbicularis oculi (closes eye), the orbicularis oris (close lips), the zygomaticus major (draws angles of mouth upward), the risorius (draws angles of mouth outward), the platysma (draws corners of mouth down), and the buccinator (compresses cheek). (Refer to Fig. 11-11 and Table 11-1.)

MUSCLES OF MASTICATION

The muscles of mastication pass from the skull to the lower jaw and act upon the temporomandibular joint. The temporalis and the masseter close the jaws, and the pterygoids grate the teeth.

MUSCLES ACTING ON HEAD

Head movements involve flexion, extension, rotation, and lateral bending; flexion and extension relate to the atlanto-occipital articulation (Fig. 11-12). The principal muscles are

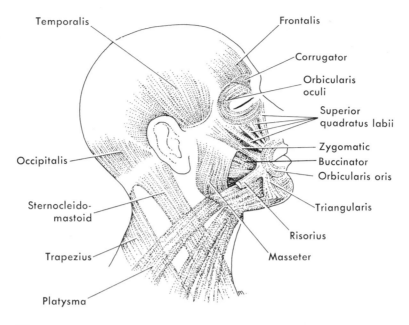

FIG. 11-11
Muscles of facial expression and mastication.

the sternocleidomastoid, the semispinalis capitis, the splenius capitis, and the longissimus capitis.

MUSCLES ACTING ON SHOULDER (SCAPULA)

The trapezius, the serratus anterior, and the pectoralis minor are the principal muscles acting on the shoulder. The trapezius raises or lowers shoulders (and shrugs them); the pectoralis minor pulls the shoulder down and forward; and the serratus anterior pulls the shoulder forward and adducts and rotates it upward (Figs. 11-9 and 11-10).

MUSCLES ACTING ON TRUNK (SPINE)

Although there is little flexion or extension of the spine in the thoracic region, there is free movement in the lumbar region, especially between the fourth and fifth lumbar vertebrae (Fig. 11-13). The principal muscles include the sacrospinalis, which extends the spine, maintains erect posture of the trunk, and, acting singly, abducts and rotates the trunk; the quadratus lumborum, which extends the spine and acting singly, abducts the trunk; and the iliopsoas and the rectus abdominis, both of which flex the spine.

ABDOMINAL MUSCLES

The sides of the abdominal wall are formed by three sheets of muscle: the external oblique, the internal oblique, and the transversus abdominis. The wall has great strength because the fibers of these muscles run in three different directions. The front, central wall of the abdomen is formed by the rectus abdominis and

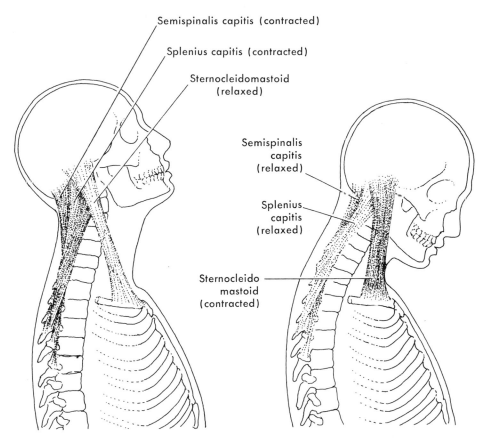

Semispinalis capitis (contracted)

Splenius capitis (contracted)

Sternocleidomastoid (relaxed)

Semispinalis capitis (relaxed)

Splenius capitis (relaxed)

Sternocleido mastoid (contracted)

FIG. 11-12
Flexion and extension of head.

the aponeuroses of the muscles at the sides (Figs. 11-9 and 11-14). The rectus abdominis arises from the pubis and passes upward as a strong column of muscle to be inserted into the xiphoid process and the fifth, sixth, and seventh costal cartilages. Cutting across the rectus abdominis transversely are three or four fibrous bands called inscriptiones tendineae which, together with the linea alba, produce the grid-like appearance of the skinned abdomen (Fig. 11-9). The linea alba (white line) results from the meeting of the aponeuroses of the external oblique muscles at the midline. The posterior abdominal wall is formed by the bony vertebral column and the quadratus lumborum, a strong column of muscle that arises from the iliac crest and the lower lumbar vertebrae and inserts into the twelfth rib and the upper lumbar vertebrae. As already noted, the quadratus lumborum extends the spine; the other muscles compress the abdomen and thereby aid in straining, defecation, forced expiration, and childbirth.

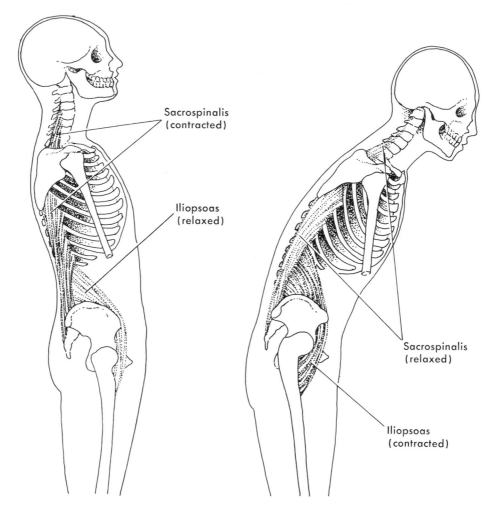

FIG. 11-13
Extension and flexion of spine.

Inguinal canal

The inguinal canal is a channel, 1 inch or so in length, that tunnels obliquely through the lower abdominal wall just above the inguinal ligament. In the male the spermatic cord passes through the inguinal canal on its way to the testis. In the female the canal anchors the round ligament of the uterus. The canal is the weakest site in the abdominal wall, which accounts for the common occurrence of inguinal hernias. In this type of hernia in the male, the parietal peritoneum makes its way through the canal and

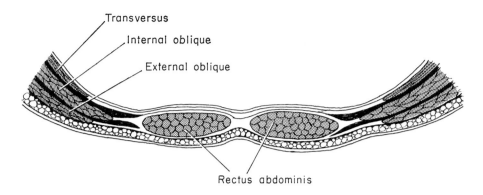

Transversus

Internal oblique

External oblique

Rectus abdominis

FIG. 11-14

Abdominal muscles in cross section. (Modified from Anthony, C. P., and Kolthoff, N. J.: Textbook of anatomy and physiology, ed. 8, St. Louis, 1971, The C. V. Mosby Co.)

finally into the scrotum. As this outpouching continues, a portion of the abdominal contents, such as the omentum or a loop of intestine, may follow and aggravate the condition. Other weak areas in the wall occur about the umbilicus and along the linea alba above the umbilicus.

MUSCLES OF RESPIRATION

Respiration involves inspiration (due to an increase in size of the chest cavity) and expiration (a decrease in size of the cavity). The muscles of quiet inspiration include the external intercostals, situated between the ribs, and the diaphragm, the muscular sheet between the thoracic and abdominal cavities. Contraction of the external intercostals elevates the ribs, and contraction of the diaphragm increases the vertical dimension of the chest cavity. The relaxed diaphragm is domeshaped with the convex surface forming the floor of the thorax. Quiet expiration involves the relaxation of these muscles and a moderate contraction of the abdominal muscles (see above). Forced expiration calls for a strong contraction of the abdominal muscles and the contraction of the internal intercostals. Forced inspiration calls for the contraction of

not only the diaphragm and the external intercostals, but also the sternocleidomastoid, the sacrospinalis, the trapezius, and the pectoralis major, among others.

MUSCLES ACTING ON PELVIC FLOOR

The levator ani and coccygeus muscles together form the floor of the pelvic cavity. The levator ani draws the rectum forward and upward, and the coccygeus draws the coccyx forward after it has been pressed back during defecation. If these muscles are torn at childbirth or become too relaxed, the bladder or uterus may "drop out."

MUSCLES ACTING ON UPPER ARM (HUMERUS)

The shoulder joint, the most freely moveable in the body, involves the head of the humerus and the glenoid cavity of the scapula. Abduction, adduction, flexion, extension (Fig. 11-15), rotation, and circumduction of the upper arm are all permitted. The principal muscles (Figs. 11-9 and 11-10) are the pectoralis major (flexion, adduction), the latissimus dorsi (extension, adduction), the deltoid (adduction, flexion), the

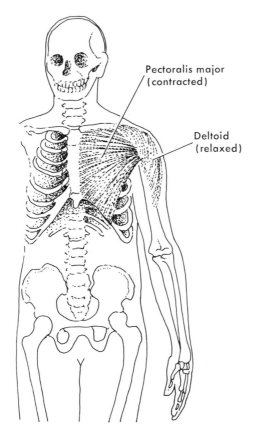
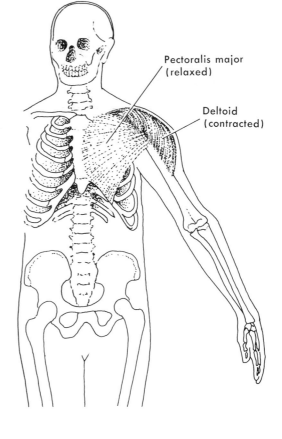

FIG. 11-15
Abduction and adduction of upper arm.

coracobrachialis (adduction, flexion, extension), the supraspinatus (abduction), the teres major (extension, adduction, rotation), the teres minor (rotation), and the infraspinatus (rotation).

MUSCLES ACTING ON LOWER ARM

Movements of flexion and extension of the lower arm (forearm) take place at the elbow between the ulna and the humerus. The brachialis, the brachioradialis, and the biceps brachii are the flexors, and the triceps brachii is the extensor (Figs. 11-16 and 11-17). Supination is

effected by the supinator, and pronation, by the pronator teres and pronator quadratus.

MUSCLES ACTING ON HAND

A complex grouping of muscles act upon the hand. Those of special note include the biceps brachii (supination of forearm and hand), the supinator (supination of forearm and hand), the pronator teres (pronation of forearm and hand), the pronator quadratus (pronation of forearm and hand), the flexor carpi radialis (flexion of hand), the flexor digitorum super-

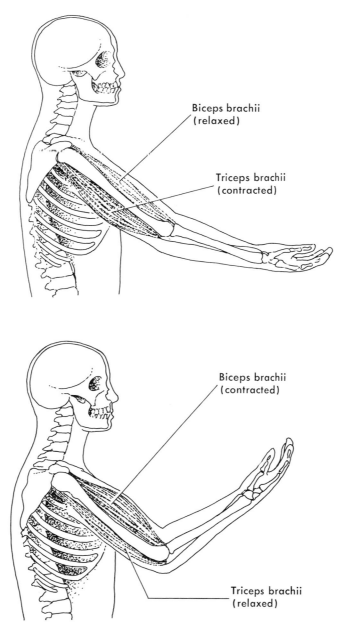

Biceps brachii
(relaxed)

Triceps brachii
(contracted)

Biceps brachii
(contracted)

Triceps brachii
(relaxed)

FIG. 11-16
Extension and flexion of supinated forearm.

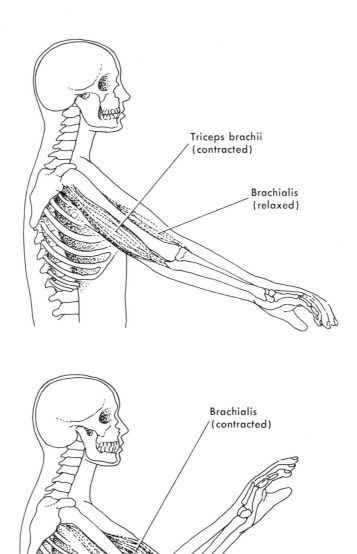

Triceps brachii
(contracted)

Brachialis
(relaxed)

Brachialis
(contracted)

Triceps brachii
(relaxed)

FIG. 11-17
Extension and flexion of pronated forearm.

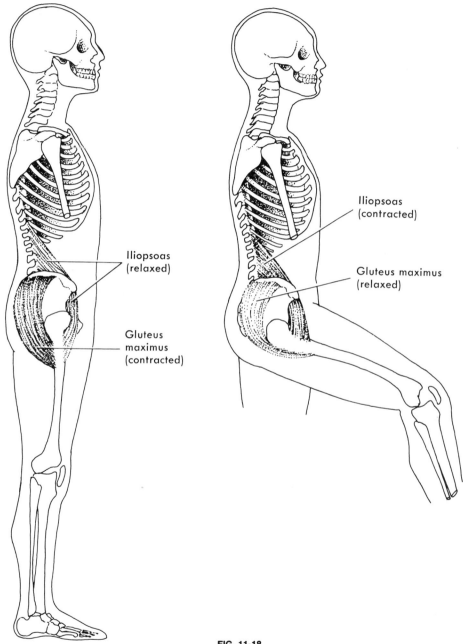

Iliopsoas
(relaxed)

Gluteus
maximus
(contracted)

Iliopsoas
(contracted)

Gluteus maximus
(relaxed)

FIG. 11-18
Extension and flexion of thigh.

ficialis (flexion of second, third, fourth, and fifth middle phalanges upon proximal phalanges), the flexor digitorum profundus (flexion of distal phalanges upon proximal), the flexor pollicis longus (flexion of thumb), the extensor carpi radialis longus (extension of hand), the extensor carpi radialis brevis (extension of hand), the extensor carpi ulnaris (extension of hand), the extensor digitorum (extension of fingers), the extensor pollicis longus and brevis (extension of thumb), the abductor pollicis (abduction of thumb).

MUSCLES ACTING ON THIGH (FEMUR)

Movement of the thigh (femur) occurs at the hip joint and is free in all directions. The flexors include the iliopsoas and the rectus femoris; the extensors include the gluteus maximus and the piriformis; the abductors include the gluteus medius, the gluteus minimus, and the tensor fasciae latae; and the adductors include the adductor brevis, the adductor longus, the adductor magnus, and the gracilis (Figs. 11-9, 11-10, and 11-18). Additionally, the gluteus maximus and gluteus medius rotate the thigh outward, and

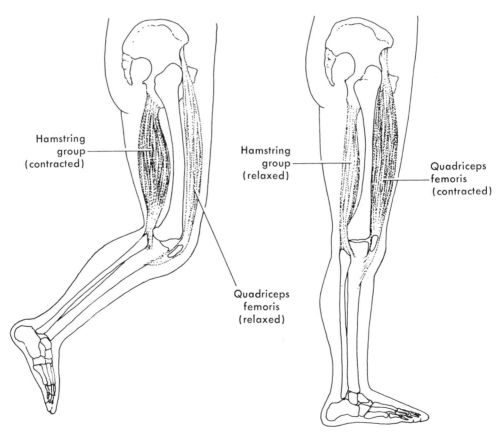

Hamstring group (contracted)

Hamstring group (relaxed)

Quadriceps femoris (contracted)

Quadriceps femoris (relaxed)

FIG. 11-19
Flexion and extension of lower leg.

the gluteus minimus rotates the thigh inward. When the femur acts as the origin, the iliopsoas, which includes the iliacus and psoas major, flexes the trunk.

MUSCLES ACTING ON LOWER LEG

Muscles acting on the lower leg (Figs. 11-9 and 11-10) operate via the knee joint, which is a modified hinge joint involving the patella, the condyles and the patellar surface of the femur, and the upper surface of the tibia. Flexion and extension are the principal movements (Fig. 11-19). The flexors include the sartorius, which is the longest muscle in the body, the gracilis, and the "hamstring group" (semitendinosus, semimembranosus, and biceps femoris); the extensors include the rectus femoris, the vastus medialis, the vastus lateralis, and the vastus intermedius, the four together composing the

"quadriceps femoris." Additionally, the rectus femoris and vastus lateralis flex the thigh, the sartorius adducts the leg, and the biceps femoris aids in extending the thigh.

MUSCLES ACTING ON FOOT

Movements of the foot relate mainly to the ankle joint (Fig. 11-20), and include dorsiflexion (bending of the foot toward the anterior part of the leg), plantar flexion (extension or straightening out of the foot or leg), inversion (turning the foot in), and eversion (turning the foot out). The chief muscles are the tibialis anterior (dorsiflexion and inversion), the gastrocnemius (plantar flexion and flexion of the leg), the soleus (plantar flexion), the peroneus longus (plantar flexion and eversion), the peroneus brevis (dorsiflexion and eversion), the tibialis posterior (plantar flexion and inversion), and the peroneus

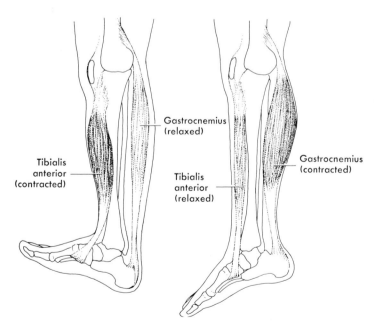

Tibialis anterior (contracted)

Gastrocnemius (relaxed)

Gastrocnemius (contracted)

Tibialis anterior (relaxed)

FIG. 11-20
Dorsiflexion and plantar flexion of foot.

tertius (dorsiflexion and inversion). Among the many muscles acting on the toes are the extensor digitorum longus (extension of toes), the extensor hallucis longus (extension of big toe), the flexor hallucis longus (flexion of big toe), and the flexor digitorum longus (flexion of toes).

MUSCULAR ATROPHIES AND DYSTROPHIES

In the language of clinical medicine, muscular *atrophy* refers to muscular weakness and wasting secondary to some pathologic involvement of the central nervous system or peripheral nerves. A number of muscular atrophies are recognized. One important example is *infantile spinal progressive muscular atrophy*, or *Werdnig-Hoffman disease*. This disease appears in infants and is frequently familial. It is characterized by weakness due to the degeneration of motor nerve cells in the spinal cord and the medulla oblongata.

By contrast, *myopathy* is muscular weakness and wasting without evidence of neural involvement or degeneration; it may be primary or secondary to some distant biochemical derangement. The best known myopathies are the *muscular dystrophies*, a class of hereditary diseases marked by progressive weakness due to degeneration of the muscle fibers. The underlying biochemical defect is unknown. Two major varieties are recognized: the Duchenne form, which affects only boys, and the Landouzy-Dejerine form, which affects both sexes.

There is no specific treatment or cure for any of the atrophies or dystrophies. Physiotherapy and understanding care are essential.

MYASTHENIA GRAVIS

Myasthenia gravis is marked by fatigue and exhaustion of the muscles and is almost always improved by cholinergic drugs (drugs that mimic the action of acetylcholine). Characteristically, there is progressive paralysis without atrophy or sensory disturbance. Although any muscle may be affected, those of the head and the neck are most commonly involved. This accounts for the expressionless face and drooping eyelids of the person who has this disorder. For some reason, the nerve impulse fails to cross the myoneural junction. Present evidence indicates the cause of this defect to be some sort of "autoimmune disease" involving the thymus gland (p. 219). About one third of the patients with myasthenia gravis have either an abnormal persistence of the gland, which normally disappears during childhood, or a thymoma (thymic tumor). Also, about one half of the patients with thymoma display the signs and symptoms of myasthenia gravis. In such cases thymectomy (removal of the gland) may afford some relief. For the most part, though, the treatment centers on the use of neostigmine and related cholinergic drugs.

TETANY

Tetany, a disorder of neuromuscular tissue, is marked by muscle twitchings, cramps, and convulsions. Any situation that produces a drop in blood calcium (hypocalcemia) can cause tetany. The usual etiologic factors include alkalosis (p. 109), vitamin D deficiency (Chapter 18), and hypoparathyroidism (Chapter 23).

GAS GANGRENE

Gas gangrene is a highly fatal disease caused by *Clostridium perfringens* and several other species of the genus. These bacteria are grampositive, sporulating, anaerobic bacilli that thrive when introduced into the tissues, especially muscle. This explains the common occurrence of the infection in dirty, lacerated wounds. As indicated by the name, there is death of cells en masse (gangrene) and considerable amounts of gas in the affected area. Both effects result from the elaboration of toxins and enzymes by the bacilli. Treatment and prevention include thorough cleansing of wounds and administration of gas gangrene antitoxin and penicillin.

TRICHINOSIS

Trichinosis is an infestation caused by the nematode *Trichinella spiralis,* one of the smallest of parasitic worms (Fig. 11-21). In brief, when contaminated pork is eaten, the encysted larvae are released in the intestine and at maturity copulate. The female later gives birth to about a thousand new larvae that migrate by the lymph and the bloodstream to the skeletal muscles, where they encyst and eventually become encapsulated and calcified. The early manifestations of the infestation (the "intestinal phase") include fever, nausea, abdominal pain, and di-

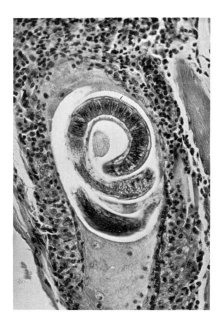

FIG. 11-21
Microscopic section of muscle showing embedded *Trichinella spiralis.* (From Ivey, M.: Helminths. In Frankel, S., Reitman, S., and Sonnenwirth, A.: Gradwohl's clinical laboratory methods and diagnosis, ed. 7, St. Louis, 1970, The C. V. Mosby Co.)

arrhea. Later, when the larvae take up residence in the muscles, the patient experiences stiffness, pain, and swelling; insomnia is also a prominent feature. Treatment is largely symptomatic, and the prognosis is good in most cases. The drug thiabendazole is highly effective against the parasite in host animals, but in humans the response has been variable. Still, most doctors feel the drug is worth a try in the acute stages of the disease. The answer to trichinosis, of course, is prevention, which here means thoroughly cooking all pork products and sterilizing all garbage fed to hogs.

QUESTIONS

1. A man of lean build weighs 80 kg. About how many pounds of muscle does this represent?
2. A muscle fiber is 100 mm in length. What is this in inches?
3. Put the following in order of increasing size: muscle, filament, fiber, myofibrils.
4. In the instance of a muscle cell, or fiber, the cell membrane and the cytoplasm have special designations. What are they?
5. Why are the anisotropic bands of a muscle fiber darker than the isotropic bands?
6. Myoglobin transports oxygen from the capillaries to the mitochondria. Precisely, what does this tell us about the role of the mitochondrion?
7. Can you account for the white meat of fowl?
8. Based on present knowledge, what is the role of troponin?
9. How does the answer to question 8 fit in with the role of Ca^{++} in muscle contraction?
10. Distinguish between myosin and myoglobin.
11. What is the function of the T system?
12. What are the "horizontal units" of a myofibril called?
13. In muscle relaxation do the H zones narrow or widen? Why?
14. What is the role of phosphocreatine in muscle contraction?
15. What is meant by the expression "high-energy phosphate"?
16. In excess amounts, pyruvic acid is reduced to

lactic acid. What do we call the reverse of this reaction?

17. Why do we continue to pant after strenuous exercise? Explain fully.
18. What eventually happens to the pyruvic acid entering the citric acid cycle?
19. Of what value is the heat of contraction?
20. Compare the following terms: threshold, liminal, maximal, subliminal, supramaximal.
21. The all-or-none law applies only to muscle fibers, and not to muscles. Explain.
22. Distinguish between rheobase and chronaxie.
23. When a maximal stimulus is very gradually applied to a muscle, no response is elicited; but when such a stimulus is suddenly removed, a response is elicited. Explain.
24. What are the three phases of a muscle twitch?
25. Is weight-lifting an isotonic or an isometric exercise?
26. What is "isometrics"?
27. How can subliminal stimuli be made to elicit a response?
28. How do you account for your answer to question 27?
29. What is the relationship among the terms summation, complete tetanus, and incomplete tetanus?
30. If a certain muscle takes 0.01 second to contract, what frequency of shocks must be applied to produce tetanization?
31. Why is the staircase phenomenon also called the treppe?
32. What, if any, is the relationship between the "warm-up" in athletics and the "treppe"?
33. Distinguish between the latent period and the refractory period.
34. Name five basic factors affecting the strength of muscle contraction.
35. Acetylcholine is often referred to as a "transmitter substance." Why?
36. Curare, the "arrow poison," produces paralysis of skeletal muscle. Direct electrical stimulation of the poisoned muscle, however, provokes normal contractions. With the hint that curare *does not* interfere with the passage of the nerve impulse over the nerve fibers, suggest the poison's site and mechanism of action.

37. In curare poisoning, what is the actual cause of death?
38. Distinguish among the terms epimysium, perimysium, and endomysium.
39. Distinguish among the terms tendon, ligament, and aponeurosis.
40. A muscle is an agonist at one time and an antagonist at another. Explain.
41. For some movements a certain site is an origin at one time and an insertion at another. Explain.
42. What is the literal meaning of synergists?
43. For any given movement, the body typically employs a number of muscles when possibly one or two would suffice. Why?
44. What is meant by "muscle tone" (or tonus)?
45. Cite two examples each for muscles named for their size, shape, number of divisions (heads), direction of fibers, location, points of attachment, major function.
46. Name the agonists and the antagonists in flexion of the head, extension of the spine, forced expiration, abduction of the upper arm, extension of the lower arm, pronation of the hand, adduction of the thigh, flexion of the lower leg, dorsiflexion of the foot.
47. In the case of the upper arm, what is the meaning of flexion and extension?
48. Distinguish among atrophy, dystrophy, and myopathy.
49. Muscular weakness that responds to neostigmine and certain other cholinergic drugs points to a diagnosis of myasthenia gravis. Explain.
50. Does the fact that hypocalcemia causes tetany relate in any way to the chemistry of contraction?
51. *Clostridium perfringens* is characterized as a "gram-positive, anaerobic bacillus." Elaborate on this quote.
52. The fact that *Clostridium perfringens* is an anaerobe ties in very well with gas gangrene. Why?
53. Account for the stiffness and pain of trichinosis.
54. The text states that all pork should be "thoroughly cooked." Investigate to see what two or three reliable cookbooks have to say about this.

CHAPTER 12

THE CIRCULATORY SYSTEM
THE HEART AND VESSELS

The cells of the body demand homeostasis (p. 69) for survival; the tissue fluid that surrounds them must supply nutrients, hormones, and the like on the one hand, and remove metabolic wastes on the other. In turn, this fluid must be adjusted continuously to maintain homeostasis. This is where the circulatory system comes into the picture. For example, blood takes oxygen and nutrients to the tissues. In this fashion the circulation sustains all the cells and keeps us alive. Circulation is made possible by the pumping action of the heart. The heart, the blood, and the blood vessels together constitute the *circulatory system*. In this chapter we shall direct our attention to the heart, the vessels, and the actual circulation; the blood itself is best treated as a separate and special subject (Chapter 13).

THE HEART

The heart (Fig. 12-1) is a sac-enclosed, muscular pump located in the lower mediastinum directly behind the sternum. It is about the size of the individual's fist and somewhat pear shaped, with the apex directed to the left, and between the fifth and sixth intercostal spaces (about 3 inches from the sternum).

Pericardium

The sac that surrounds the heart, the pericardium, consists of an external layer of dense, fibrous tissue and an inner, serous layer that surrounds the heart directly (visceral pericardium) and then reflects over the inner surface of the fibrous coat to form the parietal pericardium. The base of the pericardium is attached to the diaphragm. The space between the visceral and parietal layers is filled with a thin serous liquid.

Chambers

The heart is a mass of specialized muscle, the myocardium (p. 95), organized into four chambers: two atria and two ventricles. These chambers are lined inside and out by the endocardium and epicardium, respectively. (The epicardium is another term for the visceral pericardium.)

The atria (sing., atrium), the two upper chambers, are much smaller than the ventricles and have relatively thin walls. Characteristically, each has an ear-shaped or auricular appendage, hence the use of the word auricle for atrium. In the wall between the two atria is a small oval depression, the fossa ovalis, which marks the site of an opening, the foramen ovale, in the fetal

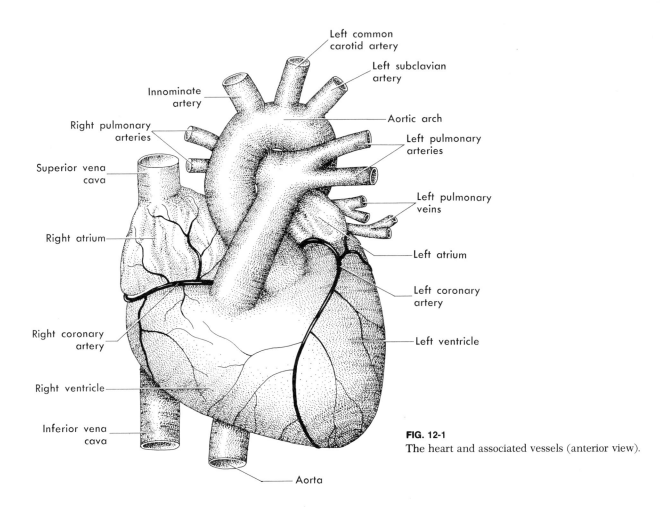

Left common
carotid artery

Left subclavian
artery

Innominate
artery

Aortic arch

Right pulmonary
arteries

Left pulmonary
arteries

Superior vena
cava

Left pulmonary
veins

Right atrium

Left atrium

Left coronary
artery

Right coronary
artery

Left ventricle

Right ventricle

Inferior vena
cava

Aorta

FIG. 12-1
The heart and associated vessels (anterior view).

heart (Chapter 24). The right atrium receives the venae cavae (the great veins that return blood from all parts of the body) and the coronary sinus (the heart vein returning blood from the heart muscle itself). Below, the right atrium leads into the right ventricle through an opening guarded by the right atrioventricular valve. Because this valve has three tapering projections (cusps), it is referred to commonly as the tricuspid valve. The left atrium receives the four pulmonary veins from the lungs. Below, it leads into the left

ventricle through the left atrioventricular valve, also called the bicuspid (two cusps) or mitral valve (Fig. 12-2).

The inside walls of the right ventricle are formed into interlacing bundles of muscle called trabeculae carneae and finger-shaped projections called papillary muscles. Strong white cords, the chordae tendineae, run from the apices of the papillary muscles to the edges of the tricuspid valve, serving to prevent the valve from being swept up into the atrium during con-

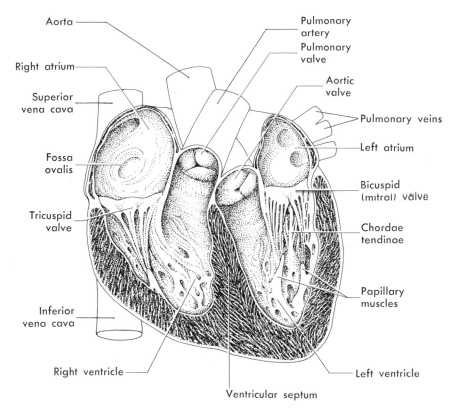

FIG. 12-2

Anatomic highlights within the heart. Pulmonary and aortic valves are semilunar valves.

traction. Blood leaves the right ventricle via the pulmonary artery. To prevent the blood from flowing back into this chamber once it has been pumped out, the opening is guarded by a semilunar valve (Fig. 12-2). When the blood bumps back against this valve during ventricular relaxation, its three flaps fill out and close the opening.

The left ventricle is the largest chamber and forms the apex of the heart. Its thick, powerful walls must develop great pressure to pump out the blood, via the aorta, to all parts of the body. Like the right chamber, its internal architecture is characterized by trabeculae carneae, papillary muscles, and chordae tendineae. The aortic

opening is equipped with a semilunar valve similar to the one in the pulmonary artery.

Purkinje system

In watching the heart beat, one gets the impression that it is contracting en masse. Unless the muscle fibers contract almost simultaneously, the heart loses its compression and pumping power. However, cardiac fibers would not contract as they do if it were not for the Purkinje system. This system, composed of modified cardiac fibers (*Purkinje's fibers*), transmits electrical impulses to all areas of the ventricles four times as fast as the regular muscle. As shown in

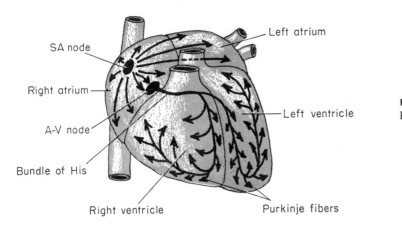

SA node

Left atrium

Right atrium

A-V node

Bundle of His

Right ventricle

Left ventricle

Purkinje fibers

FIG. 12-3
Purkinje system.

Fig. 12-3, these fibers begin at the tiny atrioventricular node (AV node) in the right atrial wall and continue on as the bundle of His, the latter eventually giving way to countless individual Purkinje fibers.

CARDIAC CYCLE

The heart is characterized by automatic and rhythmic beating. The sinoatrial node (SA node; pacemaker), a bit of nervous tissue located in the right posterior atrial wall just below the opening of the superior vena cava, emits electrical impulses at the rate of about 72 per minute. These impulses flash over the atria and then over the ventricles via the Purkinje system. As a consequence, the chambers are caused to contract and keep time with the SA node. The *cardiac cycle,* which lasts about 0.8 second, starts with the initiation of an impulse at the SA node and ends following ventricular relaxation. The spread of the actual nerve impulse takes about 0.2 second!

Electrocardiogram

As the electrical impulses pass through the cardiac muscle, very weak currents are transmitted to the surface of the body. Since these currents are a reflection of the electrical happenings in the heart, they have proved of great value to the researcher and the physician. To pick up, amplify, and record these currents, recording leads from an instrument called an electrocardiograph are connected to the body via smooth metal plates moistened with a conductive paste. The various sites of the body to which the leads may be attached yield characteristic electrical wave patterns, and so it is essential to specify the lead number: lead I, right arm–left arm; lead II, right arm–left leg; lead III, left arm–left leg; precordial (or chest) leads, V_1, V_2; and so on. Since heart disease is generally accompanied by abnormal electrical events, the electrocardiogram (ECG) is of singular diagnostic importance. A normal electrocardiogram for lead I is shown in Fig. 12-4. The *P wave* corresponds to the impulse passing through the atria; the *QRS complex* corresponds to the impulse passing through the ventricles; and the *T wave* represents the reestablishment of the original electrical properties of the ventricular muscle cells.

Heart sounds

During the cardiac cycle, the stethoscope discloses two sounds that resemble the syllables "lub" and "dub." The first sound (lub) is caused

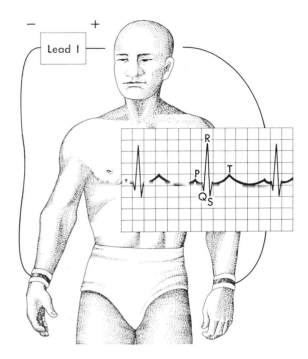

FIG. 12-4
Normal electrocardiogram (ECG) for Lead I.

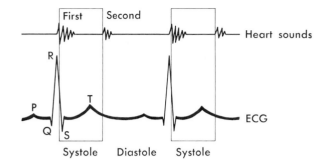

FIG. 12-5
The two heart sounds and their relation to the ECG.

by the sudden closing of the atrioventricular valves when the ventricles contact; the sudden closing creates vibrations in the blood and heart walls that are transmitted to the surface of the chest. The second sound (dub) is caused by the blood bumping back against the semilunar valves at the end of contraction. Once again, this sets up vibrations in the blood and walls of the heart.

Systole and diastole

Unless qualified otherwise (that is, atrial systole), *systole* means contraction of the ventricles. Likewise, *diastole* means relaxation of the ventricles. The systolic and diastolic periods can be noted by either the stethoscope or the electrocardiograph. Mechanical systole starts with the first sound and ends with the second, whereas electrical systole starts with the beginning of the QRS complex and ends with the T wave (Fig. 12-5). By the same token, mechanical diastole starts with the second heart sound and continues until the first, and electrical diastole starts with the end of the T wave and lasts until the outset of the QRS complex.

Starling's law

Perhaps the most basic fact relating to the action of the heart is *Starling's law,* which states that the volume of blood pumped by the heart is normally determined by the volume of blood returned to the heart. Thus, up to a point, cardiac muscle contracts with greater force the more it is stretched. The diseased heart, however, frequently does not follow Starling's law. Indeed, in congestive heart failure, the heart fails to pump effectively even normal volumes of blood.

Nervous control

Though the heart continues to beat when cut off from its nerve supply, optimum performance depends upon regulatory impulses from the autonomic nervous system. Parasympathetic stimulation inhibits the SA node and thereby slows the heart. Conversely, sympathetic stimulation causes the heart to beat faster and harder. Parasympathetic control prevails during periods of quietude, while sympathetic impulses are called forth in stressful situations when the body demands a more rapid flow of blood.

The pump in action

Blood returns to the heart via the venae cavae entering the right atrium. The atrium then contracts, sending the blood through the tricuspid valve into the relaxed right ventricle. Ventricular systole follows, and blood is forced out through the semilunar valve into the pulmonary artery leading to the lungs. Here, the blood is aerated, then returned to the left atrium via the pulmonary veins. Atrial systole now forces the blood through the bicuspid valve into the left ventricle, where it is ejected under great pressure into the systemic circulation via the aorta.

BLOOD VESSELS

The vessels through which the blood flows include the arteries, the arterioles, the capillaries, the venules, and the veins, in this order.

Arteries

An artery is a vessel that carries blood away from the heart. Its thick, tough wall is composed of three coats: the tunica intima, the tunica media, and the tunica externa (Fig. 12-6). The tunica intima, the smooth inner lining of endothelium, is unique in that it runs uninterrupted throughout the vascular system; it forms the lining of all the vessels. The tunica media, or middle coat, contains variable amounts of elastic tissue and smooth muscle. The aorta, for example, contains a great deal of elastic tissue and almost no muscle. The tunica externa, or outer coat, is composed of connective tissue. Interestingly, the larger arteries have within their walls smaller vessels, known as vasa vasorum, to nourish the thick tunics. Generally, the arteries and accompanying veins run along the flexor side of the extremities so as to be well protected against injury and stretching during movement. Also, in most cases the artery is placed more deeply than the vein.

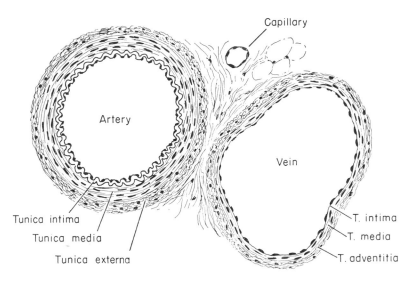

FIG. 12-6
Artery, vein, and capillary in cross section. (The terms tunica externa and tunica adventitia are used interchangeably.)

The role of the arteries in circulation is by no means a passive one. Their elastic tissue permits them to give a little during systole, and their smooth muscle, under the direction of nerve impulses, permits them to constrict or dilate. Indeed, nerve fibers form extensive plexuses about the arteries. As we shall see, these features have a life-and-death bearing on blood pressure.

Arterioles

As indicated by the suffix, arterioles are the smallest arteries. Because of the relatively large amount of smooth muscle in their walls, they are capable of being constricted or dilated well beyond the limits of other vessels. This feature plus their great number account for the fact that, next to the heart, they play the most vital role in the control of blood pressure. By the laws of physics, the arterioles increase the pressure when constricted and lower it when dilated.

Veins

A vein is a vessel that carries blood toward the heart. Like an artery, it has the three tunics (Fig. 12-6), but the vein wall is not so thick, so strong, or so elastic as that of an artery. Unlike arteries, veins collapse when empty; many of the larger veins are equipped with valves. These valves are semilunar folds of the intima, with their free ends pointed toward the heart so as to prevent backflow. Valve-equipped veins are particularly common in the lower extremities.

FIG. 12-7

Capillary bed of striated muscle. (× 200.) (Courtesy Dr. Benjamin W. Zweifach, University of California, San Diego.)

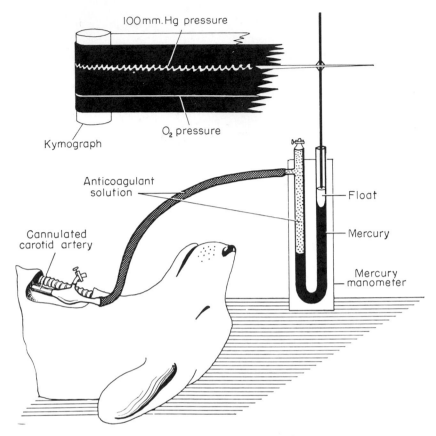

FIG. 12-8

Laboratory setup for taking blood pressure of anesthetized dog. (Modified from Guyton, A. C.: Function of the human body, Philadelphia, 1959, W. B. Saunders Co.)

Venules

As the name indicates, venules are the smallest veins. They deliver the blood from the capillaries to the veins proper.

Capillaries

Capillaries are the endothelial, one-cell thick, microscopic blood channels (Fig. 12-7) omnipresent throughout the body; they are essentially an interlacing network of tunica intimas. In a sense, the capillaries are the key to the circulatory system, for it is through their walls that nutrients and wastes diffuse between the blood and the tissue fluid. Also, the capillaries are a keystone in the maintenance of fluid balance. According to the *law of the capillaries* (p. 103), the volume of fluid leaving the capillaries through the walls is balanced by the fluid returning through the walls. In this fashion, the volume of blood and the volume of intercellular fluid are kept at constant values.

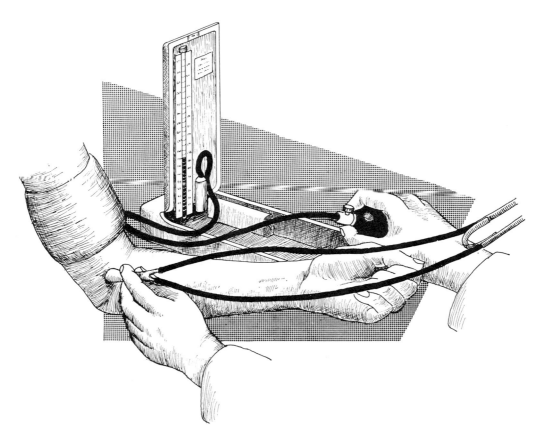

FIG. 12-9

Sphygmomanometer. (Redrawn from Schottelius, B. A., and Schottelius, D. D.: Textbook of physiology, ed. 17, St. Louis, 1973, The C. V. Mosby Co.)

BLOOD PRESSURE

The blood pressure reaches its highest point during systole and its lowest point during diastole. In normal young adults the systolic pressure is about 120 mm Hg (millimeters of mercury) and the diastolic pressure about 80 mm Hg. By convention, this is expressed as 120/80.

Measuring blood pressure

The classic and still most commonly used device for measuring blood pressure in the laboratory is the mercury manometer. As shown in Fig. 12-8, a rubber tube leads an anticoagulant solution from the left arm of the instrument to a glass cannula inserted into the animal's carotid artery. Thus, the pressure of the blood is transmitted directly to the mercury. By convention, the pressure is taken to be the difference, in millimeters, between the levels in the two arms. For example, if the difference is 100 mm, the *mean* pressure (p. 180) is said to be "100 mm Hg."

For the purpose of "putting the pressure on paper," a recording arm is floated on the mercury and adjusted to write on the kymograph. As the level of the mercury bobs up and down in concert with the beating of the heart, the kymograph records not only the pressure but also affords a vivid picture of the heart in action.

In the doctor's office a special type of manometer called the *sphygmomanometer* records blood pressure indirectly (Fig. 12-9). An inflatable cuff, connected to the manometer by a rubber tube, is placed about the upper arm and pumped up by means of a rubber bulb to cut off the blood supply to the lower arm. At all times the pressure within the cuff is registered on the manometer. A stethoscope is placed on the flexor surface just below the cuff, which is slowly deflated. At the instant the operator hears the first thumping sounds, he notes the reading on the manometer. This is taken as the highest or *systolic* pressure. The sound is caused by the spurts of blood overcoming the pressure in the cuff. The pressure in the cuff is further reduced until the sounds suddenly become dull and muffled. The pressure is noted and taken as the lowest or *diastolic* pressure. The reason the sounds disappear, of course, is that the blood, no longer obstructed by the cuff, flows along smoothly.

Pulse pressure

The pulse pressure is the difference between the systolic and diastolic pressures. Thus, a blood pressure of 120 over 80 (120/80) produces a pulse pressure of 40 mm Hg. The chief factors affecting the pulse pressure are the cardiac stroke volume and arterial elasticity.

The cardiac *stroke volume* is the volume of blood ejected by the heart per beat. It averages about 70 ml and ranges anywhere from 10 to 160 ml. The greater the stroke volume, the greater the systolic pressure and, in turn, the greater the pulse pressure. The greater the

stroke volume, the slower the heart has to beat to maintain adequate circulation. The heart rate of the well-trained athlete, for example, is often below the normal rate. In contrast, the weak heart must increase its rate to compensate for an inadequate output.

The more elastic the arteries, the more easily the arterial system accommodates the stroke volume; by giving a little, the systolic and pulse pressures are lessened. Advancing age brings about a rise in the pulse pressure because of the gradual decrease of elastic tissue in the arterial wall. In *arteriosclerosis*, or hardening of the arteries, the pulse pressure may go as high as 100 mm Hg.

Pulse wave

When a stone is thrown into the middle of a quiet pond, a wave travels out in all directions, losing its vigor the farther it goes. When this wave hits the shore, it will be reflected backward, and its contour will be changed. This situation is very much like the transmission of a pulse wave. With each beat, a wave of blood spreads out through the arteries and, upon reaching the smaller arteries, is reflected backward. What happens to the pulse pressure and the pulse wave from the aorta to the venae cavae is shown in a most vivid fashion in Fig. 12-10. As the blood travels away from the heart, there is not only a dampening of the pulse pressure and the pulse wave but also a fall in arterial pressure, which drops almost to zero in the great veins. This is easy to understand if we recall the total cross section of the vessels increases tremendously in passing from the aorta to the capillaries. Since the overall resistance to the flow of blood progressively decreases, the pressure decreases.

Mean pressure

The arterial blood pressure throughout the cardiac cycle is not even, jumping from some

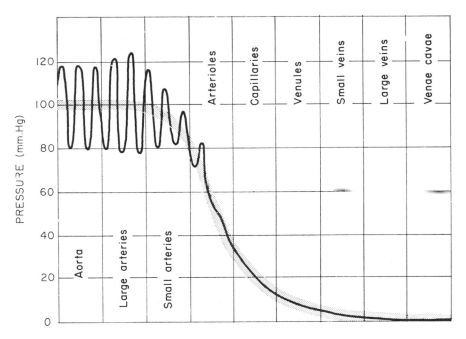

FIG. 12-10

Blood pressure curve showing pressure gradients in different divisions of circulatory system.

80 mm Hg to 120 mm Hg. Physiologically, the mean pressure is more significant than the systolic or the diastolic pressure because it is the pressure that determines the average rate at which blood flows through the body. The mean pressure, found by averaging the pressures at all stages of the cardiac cycle, is about 100 mm Hg in the resting condition. Generally, this figure can be approximated quite closely by averaging the systolic and diastolic pressures.

Factors affecting mean arterial blood pressure. The mean arterial blood pressure depends upon cardiac output, total peripheral resistance, blood volume, and arterial elasticity (p. 176). *Cardiac output,* or the rate at which blood is pumped, relates to heart rate and stroke volume (p. 180), the latter in turn depending upon the force of the heart and the inrush of blood to be pumped. *Peripheral resistance,* or resistance to

blood flow in the vessels, depends principally upon the caliber of the arterioles, the number of open capillaries, and the blood viscosity. Vasoconstriction increases blood pressure; vasodilatation decreases it. The number of open capillaries (Fig. 12-11)—the more open, the lower the pressure—depends upon such factors as temperature, metabolites, and blood flow through the arterioles supplying the capillary bed.

Nervous regulation of blood pressure

A variety of nervous reflexes regulate heart rate and peripheral resistance, and thereby exert a profound influence on blood pressure. These reflexes operate through the *vasomotor center* situated in the medulla oblongata of the brain. This center has four areas that cause vasoconstriction, vasodilatation, cardioacceleration (increased heart rate), and cardioinhibition (de-

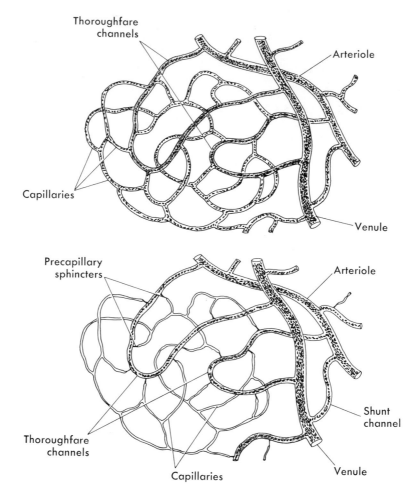

FIG. 12-11
Blood flows into capillary bed through arterioles and out of it through venules; between these vessels it flows through thoroughfare channels and capillaries. Flow of blood into capillaries (upper illustration) is regulated by precapillary sphincters, contraction of which cuts off blood supply (lower illustration).

creased heart rate) when stimulated. The majority of impulses from these areas are transmitted down the spinal cord and over the sympathetic nervous system to the heart and the blood vessels. To a lesser degree, impulses are carried by the parasympathetic system.

The vasomotor center is stimulated directly and most powerfully by medullary *ischemia* (deficiency of blood) and by an increase in the blood concentration of carbon dioxide. The vasoconstrictor and cardioaccelerator areas are caused to step up the transmission of impulses to the blood vessels and to the heart, respectively. The result, of course, is an increase in arterial pres-

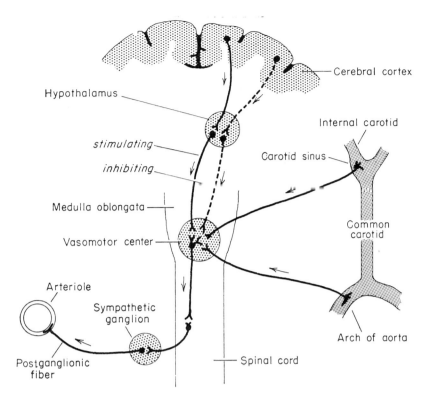

FIG. 12-12
Nervous regulation of blood pressure via vasomotor center.

sure. In this fashion, the body rids its tissues of accumulated waste products, of which carbon dioxide is only one, and otherwise remedies an embarrassed circulation.

The vasomotor center is affected indirectly by specialized sensory nerve endings, known as pressoreceptors or baroreceptors, that are located in the walls of the aortic arch and the carotid sinuses (Fig. 12-12). When for some reason the arterial pressure starts to rise, these receptors are stimulated to step up their transmission of impulses to the center and thereby cause the vasodilator and cardioinhibitor areas to lower arterial pressure. Conversely, when the arterial pressure drops below normal, the cardioaccelerator and vasoconstrictor areas become stimulated. This reciprocal relationship between blood pressure and heart rate is called *Marey's law,* after the man who first described it.

The higher brain centers, as indicated in Fig. 12-12, may influence blood pressure. Emotion, rage, and nervousness elevate blood pressure by stimulating the cardioaccelerator and vasoconstrictor areas via the hypothalamus.

Capillary regulation of blood pressure

When for some reason the blood volume becomes too great, the concomitant rise in capillary pressure forces more fluid through the capillary walls into the tissues, thereby lowering the blood volume and the blood pressure. Conversely, loss of blood from the circulation lowers the blood pressure and decreases capillary filtration. As a result, the relatively greater osmotic pressure of the blood will draw water from the tissues into the circulation (p. 103). The shift of fluid across the capillary wall plays an important role in regulating the blood volume and thereby the blood pressure.

Kidney regulation of blood pressure

Like the capillaries, the kidneys regulate arterial pressure by acting upon the blood volume. After an excessive intake of fluid, the kidneys step up urine production; following hemorrhage, they decrease it, thereby allowing ingested fluid to remain in circulation. The mechanisms involved in this renal control are the blood pressure itself and two hormones, the antidiuretic hormone and aldosterone (Chapter 23). In brief, an increase in blood volume enhances glomerular filtration, the first step in the production of urine. A drop in pressure has the opposite effect. The antidiuretic hormone and aldosterone decrease the output of urine by acting upon the renal tubules (Chapter 19). Both hormones are released when there is a drop in blood volume.

Venous blood pressure

In a standing position, blood pressure in the veins varies from about −10 mm Hg in the head to +90 mm Hg in the feet. When a person is moving about, however, muscle contraction milks the blood along, greatly reducing the hydrostatic pressures in the legs and the feet. This action depends upon the proper functioning of the valves. If they are faulty, the blood slips back, thereby elevating the hydrostatic pressure and producing varicose veins.

Venous pressure throughout the body is determined mainly by the pressure in the right atrium.

Since this pressure is just about zero, venous blood always flows toward the heart. A significant rise in atrial pressure interferes with venous return and the circulation. Increased atrial and venous pressures are the cardinal features of heart failure. Normally, as pressure within the right atrium increases, certain pressoreceptors in the great veins and the atrial wall are stimulated, and this causes an increase in heart rate via the vasomotor center. The opposite occurs when atrial pressure decreases. This response of the heart to changes in atrial-venous pressure is called the *Bainbridge reflex*.

BLOOD FLOW

Circulation effects a flow of blood through the tissues. At rest the normal heart pumps about 5,000 ml of blood per minute. However, during exercise the flow may reach the phenomenal value of 35,000 ml per minute! The output is equal to the stroke volume times the heart rate. For example, with a stroke volume of 70 ml and a rate of 72, the output is 5,040 ml per minute.

Of the some 5,000 ml of blood leaving the heart per minute, about 25% (1,250 ml) flows through the muscles, 25% through the kidneys, 15% (750 ml) through the abdominal region, 10% (500 ml) through the liver, 8% (400 ml) through the brain, 4% (200 ml) through the coronary vessels, and 13% (650 ml) through the remaining areas.

The flow of blood through any given vessel of the body can be expressed according to *Poiseuille's law:*

$$\text{Blood flow} = \frac{\text{Pressure} \times \text{Radius}}{\text{Length} \times \text{Viscosity}}$$

which states that blood flow is directly proportional to blood pressure and the size of the vessel and inversely proportional to the length of the vessel and the viscosity of the blood. Of these factors, vessel size is usually the most important because of the rapidity with which the autonomic system can produce vasoconstriction or vasodilatation.

Exercise

During exercise the muscles demand extra oxygen and nourishment. In an attempt to meet this demand, the arterioles dilate to enhance the blood flow. However, since vasodilatation reduces blood pressure, the vessels in other areas of the body (the skin, kidneys, liver, and gastrointestinal tract) must be constricted. The body shunts blood from its "reservoirs" to where the blood is needed most.

Temperature control

The shifting of blood to and from the skin plays a vital role in temperature control. When the body temperature rises, the vasomotor center dilates the arterioles of the skin and thus shunts more warm blood to the surface. This increases heat loss and restores the temperature to normal. When the temperature starts to drop, the arterioles are constricted, and more blood is shunted from the skin to the interior. In this manner, heat is conserved, and temperature maintained.

Hemorrhage

When large amounts of blood are lost, the blood pressure falls to dangerously low levels. The body defends itself by constricting the vessels in the less vital areas so as to assure an adequate flow of blood through the brain and the heart muscle.

Reactive hyperemia

Reactive hyperemia is a classic example of the ability of the body to solve a vital problem in a simple way. As the metabolism in a given tissue increases, the need for more blood to supply oxygen, nutrients, and remove wastes is met simply and effectively by local vasodilatation, or reactive hyperemia. Conversely, in a state of low metabo-

lism the vessels constrict, thereby reducing the blood flow. In this way the tissues are supplied with no more and no less than they need. The precise mechanism responsible for reactive hyperemia is not known. Possibly vessels dilate because the muscles in their walls relax due to lack of nutrients; or possibly they dilate in response to waste products such as lactic acid or carbon dioxide.

CIRCULATION

The general path the blood follows in its circulation is depicted in Fig. 12-13; major arteries are shown in Figs. 12-14 and 12-15, and major veins in Figs. 12-16 and 12-17. Blood flows from the heart via the arteries and the arterioles and returns via the veins and the venules; between the arterioles and the venules lie the vast capillary networks. The highlights of the circulation relating to special areas are discussed below.

Coronary circulation

Immediately above the aortic valve the left and right coronary arteries (Fig. 12-18) take leave of the great vessel and run along the surface of the heart, branching as they go and finally penetrating the myocardium. The blood is directed from the venules within the muscle into the coronary veins, then into the right atrium via the coronary sinus. The flow through the coronary arteries depends upon cardiac activity and sympathetic control. As in other tissues, increased activity speeds up the flow via reactive hyperemia. Also, movement of the beating heart massages the vessels, thereby moving along the blood inside. Sympathetic stimulation improves coronary flow directly by vasodilatation and indirectly by increasing the heart rate. Interestingly, the coronary blood flow is greater during diastole than during systole. This is in stark contrast to the circulation in other areas of the body. (The reason, of course, is that during

Text continued on p. 192.

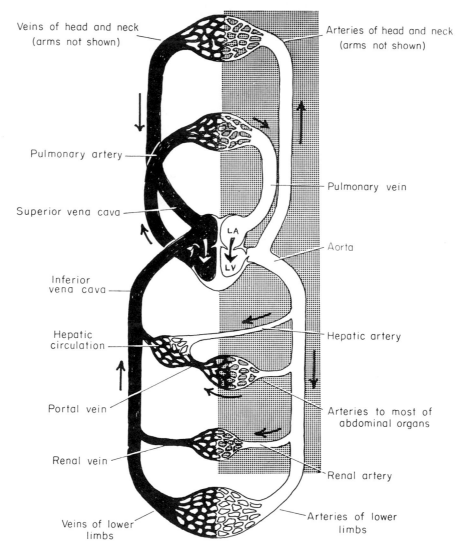

Veins of head and neck
(arms not shown)

Arteries of head and neck
(arms not shown)

Pulmonary artery

Pulmonary vein

Superior vena cava

LA

LV

Aorta

Inferior
vena cava

Hepatic
circulation

Hepatic artery

Portal vein

Arteries to most of
abdominal organs

Renal vein

Renal artery

Veins of lower
limbs

Arteries of lower
limbs

FIG. 12-13
General circulation.

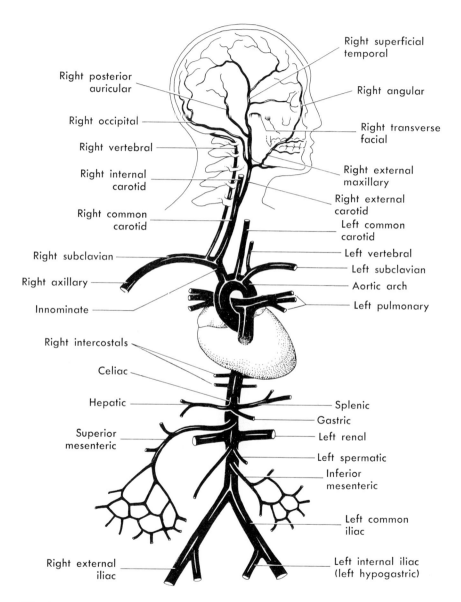

FIG. 12-14
Major arteries of head and trunk.

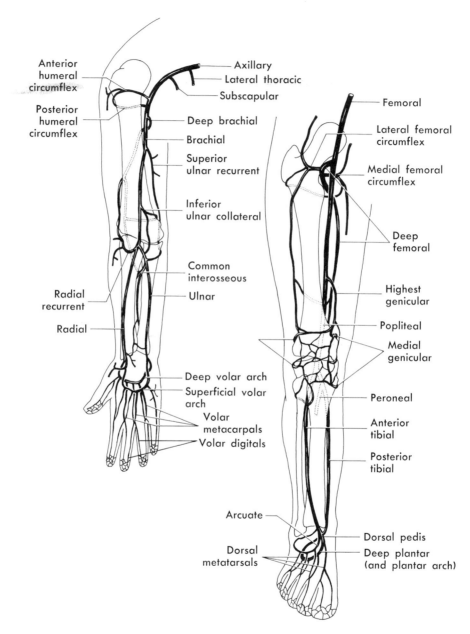

Anterior humeral circumflex
Posterior humeral circumflex
Radial recurrent
Radial

Axillary
Lateral thoracic
Subscapular
Deep brachial
Brachial
Superior ulnar recurrent
Inferior ulnar collateral
Common interosseous
Ulnar
Deep volar arch
Superficial volar arch
Volar metacarpals
Volar digitals

Femoral
Lateral femoral circumflex
Medial femoral circumflex
Deep femoral
Highest genicular
Popliteal
Medial genicular
Peroneal
Anterior tibial
Posterior tibial

Arcuate
Dorsal metatarsals

Dorsal pedis
Deep plantar (and plantar arch)

FIG. 12-15
Major arteries of upper and lower extremities (anterior view).

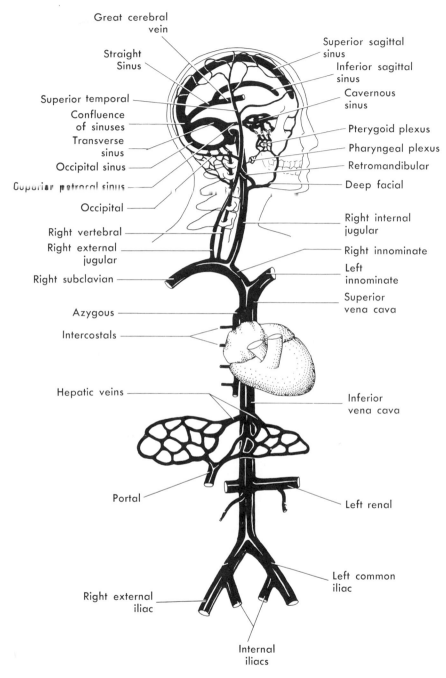

Great cerebral
vein

Straight
Sinus

Superior temporal

Confluence
of sinuses

Transverse
sinus

Occipital sinus

Superior petrosal sinus

Occipital

Right vertebral

Right external
jugular

Right subclavian

Azygous

Intercostals

Hepatic veins

Portal

Right external
iliac

Superior sagittal
sinus

Inferior sagittal
sinus

Cavernous
sinus

Pterygoid plexus

Pharyngeal plexus

Retromandibular

Deep facial

Right internal
jugular

Right innominate

Left
innominate

Superior
vena cava

Inferior
vena cava

Left renal

Left common
iliac

Internal
iliacs

FIG. 12-16
Major veins of head and trunk.

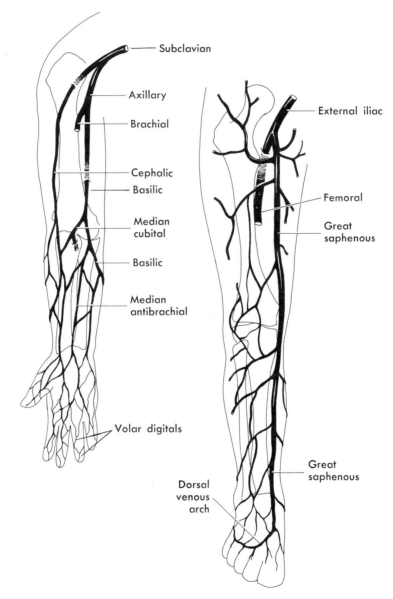

FIG. 12-17

Major veins of upper and lower extremities (anterior view). Femoral vein becomes anterior tibial and popliteal, and the latter in turn becomes posterior tibial and peroneal.

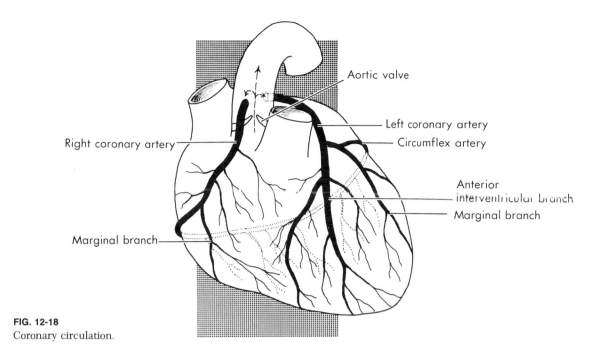

FIG. 12-18
Coronary circulation.

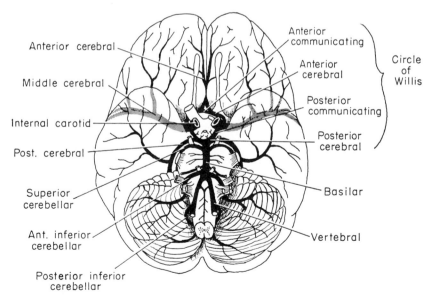

FIG. 12-19
Cerebral circulation. (Modified from Francis, C. C, and Farrell, G. L.: Integrated anatomy and physiology, ed. 3, St. Louis, 1957, The C. V. Mosby Co.)

systole the vessels are squeezed and occluded.) Consequently, an adequate coronary flow is more dependent upon diastolic pressure than upon systolic pressure.

Pulmonary circulation

The pulmonary circulation refers to the flow of blood between the heart and the lungs (Fig. 12-13) for the purpose of taking in oxygen and releasing carbon dioxide. The blood leaves the right ventricle via the pulmonary artery (guarded by a semilunar valve) and returns to the left atrium via the four pulmonary veins. In the lungs the capillaries surround the alveoli, making possible the exchange of gases (Chapter 15). The resistance to the flow of blood in the pulmonary circuit is so low that the arterial pressure in this area averages from about 15 to 85 mm Hg less than the mean systemic arterial pressure! The vessels of the lung dilate so easily that the pulmonary pressure does not rise greatly even during exercise, when the heart is pumping out several quarts of blood a minute. In general, the pulmonary flow follows the systemic flow; the greater the flow through the systemic vessels, the greater the flow through the pulmonary vessels, and vice versa.

Cerebral circulation

Relatively speaking, the brain receives a very rich blood supply. Blood is pumped to the undersurface through the internal carotid and the vertebral arteries; the former join the circle of Willis and the latter join the basilar artery (Fig. 12-19). The veins of the brain are highly characteristic. In contrast to the usual architecture, they are actually sinuses channeled in the fibrous lining, or dura mater, of the skull. These sinuses drain into the jugular veins.

The flow of blood through the brain depends upon reactive hyperemia and sympathetic control. Although activity or inactivity can bring about a change in blood flow in a localized area, the overall cerebral flow changes very little. This is fortunate, for nervous tissue is easily excited by sudden changes in blood flow.

Portal circulation

The major anatomic features of the portal system are shown in Fig. 12-20. The system essentially amounts to the path by which blood flows from the spleen and the intestines, through the liver, and eventually into the inferior vena cava. Intestinal blood is heavily laden with microbes from the gastrointestinal tract; if it were to pass directly into the general circulation, infection would be the rule, not the exception. To remove these microbes and other foreign debris, the minute blood sinuses of the liver are equipped with miraculous phagocytizing scavengers called *Kupffer's cells*. In addition to providing this defense mechanism, the liver detoxifies an endless variety of ingested poisons and toxins, again preventing insult to the general circulation. The liver also removes and stores nutrients from the intestinal blood.

Muscular circulation

Blood flow through the muscles is proportional to the amount of work being performed. As contraction increases, metabolic wastes cause vasodilatation and an increased flow via reactive hyperemia. Also, contraction puts more carbon dioxide into the circulation, thereby increasing the arterial pressure via the vasomotor center (p. 183). During exercise the latter mechanism is especially important in bringing blood to the muscles.

Cutaneous circulation

The flow of blood through the skin serves not only to supply nourishment but also to regulate body temperature and blood pressure. An extensive plexus of venous sinuses lies everywhere underneath the skin; whenever the body temperature becomes too high, blood is allowed to flow

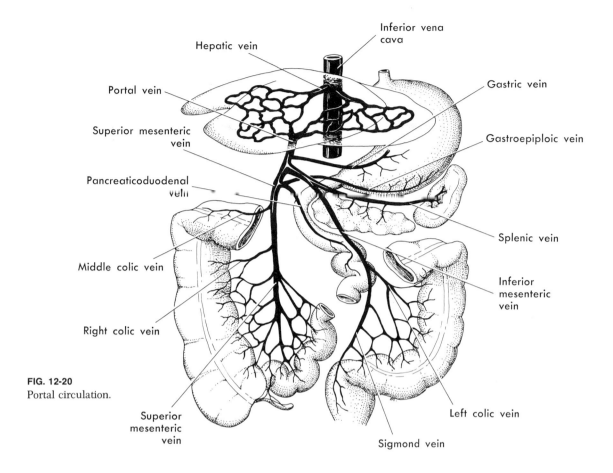

Inferior vena cava

Hepatic vein

Portal vein

Gastric vein

Superior mesenteric vein

Gastroepiploic vein

Pancreaticoduodenal vein

Splenic vein

Middle colic vein

Right colic vein

Inferior mesenteric vein

FIG. 12-20
Portal circulation.

Superior mesenteric vein

Left colic vein

Sigmond vein

rapidly into them via the arteriovenous anastomoses. The result is a warming of the skin and loss of heat. On the other hand, when the internal temperature starts to drop, the arteriovenous anastomoses constrict and shunt blood away from the skin. These same vessels also are caused to constrict if blood pressure drops, the shunted blood serving to bolster volume and pressure.

CARDIOVASCULAR DISEASE

Cardiovascular disease, the nation's number one cause of death, kills about a million Ameri-cans a year. Of this million, some 675,000 are victims of heart disease. But the great killer—the one behind what we call the "heart attack"—is atherosclerosis.

Atherosclerosis

Various diseases affecting the blood vessels fall under the generic designation *arteriosclerosis*. By far the most important involvement is atherosclerosis, an arterial lesion characterized by thickening of the tunica intima due to localized accumulations of lipids. Such accumulations, known as *atheromas,* are composed mainly of cholesterol, cholesterol esters, phospholipids, and neutral fat in association with protein. These compounds, collectively referred to as β-lipoproteins, are the hallmark of atherosclerosis; indeed, the disease cannot be experimentally produced in animals unless the β-lipoprotein concentration is elevated greatly, either by diet or by hormones. The great mystery and debate, however, is where the β-lipoproteins fit into the overall picture. All clinical evidence points to a multidimensional interplay of environmental and hereditary factors. Cigarette smoking, obesity, lack of exercise, anxiety, and a diet leading to elevated blood lipids all appear to promote and accelerate atherosclerosis; at the very least they appear to promote the disease in most individuals. Some of us are more susceptible than others; this is where heredity has to be taken into account. For example, β-lipoproteins are synthesized in the liver, and the enzymes needed to do the job are under genetic control. Atherosclerosis is an insidious disease. For years it is silent; then all at once something happens: an outpouching of an arterial wall (*aneurysm*), a hemorrhage (due to a ruptured aneurysm), a blood clot (*thrombosis*), or a heart attack. Not

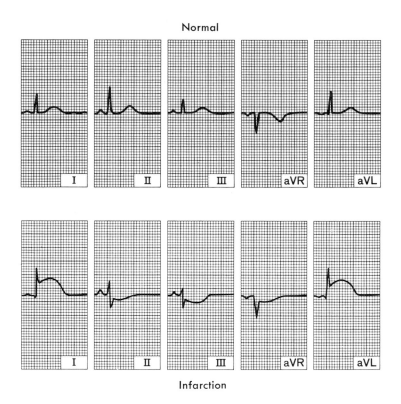

FIG. 12-21

ECG taken a few hours after myocardial infarction ("heart attack"). Note elevated ST segment in *I* and *aVL,* and depressed ST segment in other leads. (Courtesy Merck, Sharp, and Dohme.)

uncommonly, such an event is sudden, sharp, and severe; not uncommonly, there is sudden death.

Coronary artery disease

Narrowing—or *occlusion*—of the coronary arteries results in an imbalance between blood supply and myocardial demands; the underlying cause is usually atherosclerosis. The result is myocardial ischemia (deficiency of blood to heart muscle) and commonly, damage, the severity of which depends upon the size and location of the heart region involved, the rate of development, and the duration of the deficiency. A short period of relatively mild ischemia results in angina pectoris, while a severe and prolonged ischemia results in myocardial infarction, the classic "heart attack" (Fig. 12-21).

The events leading to the heart attack are briefly as follows: As atherosclerosis advances, a fatty deposit may rip through the tunica intima and trigger a clot or thrombus ("coronary thrombosis"); a protruding fat deposit may break off; or a deposit may erode a vessel of the vasa vasorum, producing within the wall a tiny hemorrhage that pushes out the tunica intima, thereby blocking the vessel. The area of the heart supplied by the blocked or occluded vessel dies from a lack of oxygen. This we call *infarction*. Whether the victim survives the attack depends upon the size of the vessel involved and the speed of the occlusion. In the event of recovery, the heart is weakened for several months and often for life. Treatment centers on the use of vasodilators, such as nitroglycerine, to enhance collateral circulation, anticoagulants to prevent the development or enlargement of clots, digitalis to strengthen heart muscle, diuretics to rid the body of excess salt and water, and proper attention to rest, exercise, diet, and tobacco, and so on. In selected cases, surgery may be lifesaving. The three main surgical approaches are providing the heart muscle with an alternate supply of blood, repairing the diseased artery, and repairing the diseased heart.

Cardiac arrhythmias

Cardiac arrhythmias are changes in heart rhythm (Fig. 12-22) caused by disturbances of the pacemaker, by abnormal automatic mechanisms replacing normal rhythm, and by disturbances in the Purkinje system. Some of the more common arrhythmias include *paroxysmal tachycardia* (heart rate suddenly increasing to 100 or more per minute), *atrial flutter* (atria contracting regularly at rates between 200 and 400 per minute), *atrial fibrillation* (atria con-

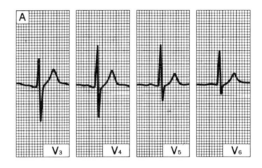

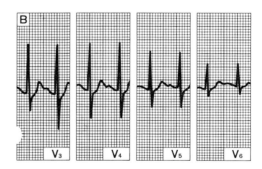

FIG. 12-22
Cigarette smoking and ECG. **A,** Normal resting pattern. **B,** After smoking. Rate has increased from 80 to 130, and there is depression of ST segment. (Courtesy Merck, Sharp, and Dohme.)

tracting regularly at rates between 400 and 1,000 per minute), and *ventricular fibrillation* (convulsive movements of the ventricles).

In ventricular fibrillation, the most ominous of the arrhythmias, the blood pressure falls to zero, and the outcome is commonly sudden death. The causes include coronary occlusion, overdosage of digitalis and certain other drugs, cyclopropane anesthesia, and certain surgical procedures, notably those involving the heart. Ventricular fibrillation may be the cause of death in electrocution. The treatment is largely preventive and consists of using every precaution in giving drugs known to incite arrhythmias. If fibrillation begins, electrical defibrillation by external electric countershock may be lifesaving. If this is not available, intravenous chemical defibrillators—lidocaine, procainamide, or diphenylhydantoin—are given. Artificial respiration and external cardiac massage are also essential to maintain cardiac output.

Rheumatic heart disease

The relationship between rheumatic heart disease (rheumatic fever) and streptococcal infections has now been established. The essential lesion in this condition concerns the valves, generally the mitral and aortic. Damage to these structures leads to either stenosis or regurgitation. In mitral or aortic stenosis, the narrowed opening causes damming of blood in the lungs, with the result that the elevated pressure in the capillaries forces fluid into the alveoli. In the end, the patient drowns in his own water. On the other hand, if the valves fail to close properly because of erosion, there is regurgitation, or the leaking backward of blood. The heart, in attempting to correct itself by beating harder and faster, becomes strained or weakened. The prompt use of penicillin in all streptococcal infections is now recognized as the best defense against rheumatic heart disease. Treatment is continued until the last threatening or-

ganism has departed, and then a short time more.

Congenital heart disease

In contrast to the high incidence of acquired heart disease, congenital diseases are relatively, and fortunately, uncommon. These conditions are caused by failure of normal development in utero. Since cyanosis in many instances is the first symptom to appear (because of poor oxygenation), we often hear the expression "blue baby." Clubbing of the fingers frequently accompanies congenital heart disease. Some of the more common and important congenital disorders of the heart are pulmonary stenosis, atrial-septal defect, patent interventricular septum, and tetralogy of Fallot. The first condition, a narrowing of the opening between the pulmonary artery and the right ventricle, decreases the flow of blood through the lungs, on the one hand, and the return of blood to the heart, on the other. The "patent" defects embarrass the circulation by permitting the blood to flow directly between the atria and between the ventricles. The tetralogy of Fallot is characterized by four lesions—pulmonary stenosis, patent interventricular septum, enlargement of the right ventricle, and displacement of the aorta to the right. Other congenital defects relate to faulty valves and absence of valves.

Patent ductus arteriosus is a congenital defect that indirectly involves the heart. When the ductus arteriosus fails to close after birth, the heart has the added burden of pumping more blood than is needed through the lungs. This abnormal situation strains the heart and in later life results in lung damage because of the high pressure in the pulmonary vessels.

In no other area of cardiovascular research have there been more fruitful or more dramatic results than in the surgical management of congenital heart disease. The surgeon can now ligate a patent ductus arteriosus, repair and re-

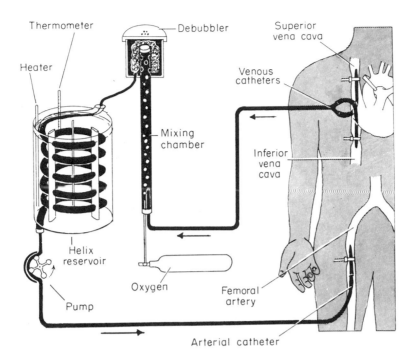

FIG. 12-23

Heart-lung machine. Blood flows from two catheters inserted in venae cavae to bottom of mixing chamber. Bubbles of oxygen rising through chamber are removed in debubbler. Blood then flows into helix reservoir and is pumped into femoral artery of patient.

place bad valves, and patch openings between the chambers. Indeed, even the tetralogy of Fallot has yielded to the needle and scalpel. Such procedures demand an artificial heart and an artificial lung to keep the body alive during the operation. The essentials of the heart-lung machine are shown in Fig. 12-23.

Congestive heart failure

The term congestive heart failure is applied to the inability of the heart to pump at top efficiency. Although the condition can result from more blood returning to the heart than the heart can pump out (for example, overloading the circulation with intravenous fluids), the usual cause stems from heart damage; either the valves are faulty, as in rheumatic heart disease,

or the muscle has been weakened by coronary disease or other injury.

The signs and symptoms of congestive heart failure are referable to poor circulation. A rapid, weak pulse, bluing of the skin, shortness of breath, fatigue, and edema are the classic features. Edema results from the increased venous pressures forcing fluid through the capillaries into the tissues and from poor filtration through the kidneys. Also, the presence of the adrenal hormone aldosterone (Chapter 23) stimulates the kidney to retain additional water and salt. Since salt "holds" water, this is an especially aggravating feature. The treatment of congestive heart failure centers about the use of digitalis preparations to strengthen the myocardium and diuretics to flush out the water and salt held in the tissues, and the implementation of a salt-free diet.

Endocarditis

Endocarditis, a dangerous and often fatal heart disease marked by inflammation of the endocardium and valves, is usually caused by either streptococci or staphylococci. One of the more serious and common forms of the disease is subacute bacterial endocarditis. The etiologic microbe here is generally *Streptococcus viridans,* alone or in association with other microbes. Administration of pencillin and related antibiotics is the preferred treatment.

Pericarditis

Pericarditis, inflammation of the pericardium (the sac around the heart), is characterized by fever, pain over the heart, rapid pulse, cough, and labored breathing. The condition usually follows in the wake of rheumatic fever, pneumococcal infection, and tuberculosis. Treatment centers on chemotherapy.

Hypertension

Hypertension, or high blood pressure, is another ominous medical enigma. It occurs in about one out of every five persons, causing misery and not infrequently death. Hypertension kills by rupturing a vessel in a vital organ or by causing the heart or kidneys to fail. About 90% of all cases of the disease are characterized either as primary or as essential—"cause unknown." However, possible causes have been postulated. Some researchers, for example, believe that essential hypertension may be caused by a genetic defect in the areas where the sympathetic fibers innervate the arterioles, causing excessive vasoconstriction. This attractive hypothesis is supported by the strong tendency of essential hypertension to be inherited and by certain recent laboratory developments. Treatment centers on the use of vasodilators and diuretics (to rid the body of excess sodium). Known

causes of hypertension include arteriosclerosis, certain brain abnormalities, kidney failure (Chapter 19), and an excessive output of adrenal hormones (Chapter 23).

Phlebitis

Phlebitis, or inflammation of the veins, is a potentially dangerous condition marked by infiltration of the vein walls and formation of blood clots. The disease is accompanied by redness, edema, stiffness, pain, and sometimes infection in the affected area. Vasodilators, anticoagulants, and antibiotics commonly are prescribed for this condition to enhance the flow of blood in the affected part (usually the leg), to prevent enlargement of the clot, and to fight infection, respectively.

Varicose veins

Varicose veins are veins that are weak and dilated. Although factors that elevate the venous pressure, such as continuous standing and congestive heart failure, are usually associated with the disease, there may be an inherited weakness of the walls. Since varicose veins bleed easily, a slight injury to the affected areas may incite formation of a clot. Treatment is directed against the underlying cause; surgical intervention and anticoagulant drugs, if needed, are employed.

Thromboembolism

Thromboembolism, or *embolism,* is the blocking of a blood vessel with a thrombus that has broken loose from its site of formation. If the lung is affected, death is often instantaneous. Embolism is always a threat in such peripheral vascular diseases as phlebitis and varicose veins. The treatment centers on the use of anticoagulants.

Shock

Shock is perhaps the most baffling medical emergency. The ominous signs (hypotension, pallor, clammy skin, feeble and rapid pulse, decreased respiration, anxiety, and often unconsciousness) are frightening even to the seasoned physician. Morbidly interesting, shock is triggered not by one situation but by many. A severe blow, a burn, a hemorrhage, a heart attack, an unpleasant experience, or even a bee sting may result in shock and death. Shock relates to the circulation, and most forms can be explained, at least in part, on the basis of circulatory dynamics. In a severe hemorrhage or burn, for example, the reduced blood volume leads to reduced venous return and, in turn, to reduced cardiac output. A severe heart attack causes a drastic cut in cardiac output because the pump is damaged.

In mild shock, the body does a remarkable job of protecting itself by constricting the vessels and the venous reservoirs. Thus, even though blood may be lost, the constricted vessels effect a normal venous return and cardiac output. If more than a quart of blood is lost, however, this mechanism cannot cope with the task, and cardiac output starts to fall. The treatment demands speed and proper choice of restorative measures. To correct a volume deficiency, whole blood, plasma, or a plasma expander (for example, dextran) is used. If one of these is not available, electrolyte solutions or 5% glucose may be employed. Digitalis is indicated if the heart has been weakened. Vasopressors, such as levarterenol and metaraminol, may be necessary in the emergency treatment of shock from any cause when systolic pressure is below 80 mm Hg, but these drugs should not be used until blood volume has been restored.

QUESTIONS

1. Discuss the derivation of the term homeostasis.
2. Distinguish among the terms epicardium, visceral pericardium, and parietal pericardium.
3. Compare the terms auricle and atrium.
4. Distinguish between fossa ovalis and foramen ovale.
5. What are the venae cavae?
6. What is the purpose of atrioventricular valves?
7. Distinguish among the terms trabeculae carneae, papillary muscles, and chordae tendinae.
8. Locate and compare the two semilunar valves.
9. What is the purpose of the Purkinje system?
10. Why is the SA node commonly called the pacemaker?
11. Account for the two heart sounds disclosed by the stethoscope.
12. Distinguish between systole and diastole.
13. Other things being equal, mechanical systole and electrical systole are just about the same. Why?
14. Does Starling's law apply in a general way to skeletal muscle?
15. Under normal conditions, a good venous return ensures a good cardiac output. Explain.
16. Discuss the nervous control of the heart.
17. What is the best example to refute the often-made statement that "arteries carry oxygenated blood"?
18. Refute the statement that "veins carry oxygen-poor blood."
19. What is the purpose of the semilunar valves?
20. Compare the walls of an artery, a vein, and a capillary.
21. Do arteries and veins have valves?
22. The "whole purpose" of the circulatory system is manifested in the capillaries. Explain in full.
23. What is the law of the capillaries?
24. What does a pressure of 150 over 90 mean?
25. Why is the systolic pressure greater than the diastolic pressure?
26. Mention several factors that could possibly produce false values in taking the blood pressure.
27. What volume of blood, in liters, is pumped per minute when the stroke volume is 60 ml, and heart rate, 72?
28. What is the mean arterial pressure with systolic reading of 115 and diastolic of 75?
29. Which vessels have greatest effect upon peripheral resistance?
30. What is the relationship of the following to the blood pressure: cardiac output, peripheral resistance, elasticity, blood volume?
31. What bearing does blood pressure have upon capillary filtration?
32. What happens to the output of urine in shock? Why?

33. Give two reasons why the blood pressure drops as the blood moves away from the heart.

34. Cite several factors that have a bearing on the number of open capillaries.

35. It makes good physiologic sense for medullary ischemia to stimulate the vasoconstrictor and cardioaccelerator areas. Why?

36. Carbon dioxide stimulates the vasoconstrictor and cardioaccelerator areas. This might be due in great part to the drop in pH. Explain.

37. Why are pressoreceptors also called baroreceptors?

38. Applying a little pressure at just the right spot in the neck slows the heart. Why?

39. What would be the effect of cutting the sympathetic fibers going to the heart?

40. Compare Marey's law and the Bainbridge reflex.

41. In the standing position, what is the difference in pressure between the feet and the head?

42. Other things being equal, what are the effects of the following on the blood flow in a given vessel: a decrease in blood pressure, a decrease in caliber, an increase in length, a decrease in blood viscosity?

43. Why does rubbing redden the skin?

44. Discuss in detail the various effects of exercise on the circulatory system.

45. Discuss the role of reactive hyperemia during periods of high metabolic activity.

46. Discuss the role of the circulation and the skin in temperature control.

47. A certain amount of blood can be lost without a significant drop in pressure. Explain.

48. What is the influence of sympathetic stimulation upon the coronary circulation?

49. The relatively low blood pressure in the pulmonary circuit underscores the influence of peripheral resistance. Discuss.

50. What is the relationship of the internal carotid arteries to the circle of Willis?

51. What is the relationship of the blood sinuses to the dura mater?

52. What is the central purpose of the portal system?

53. Naming all the major vessels, trace a red cell from the right foot to the left ear.

54. Naming all the major vessels, trace a red cell from the right hand to the left hand.

55. Trace a molecule of glucose from the intestine to the brain.

56. Compare the terms arteriosclerosis and atherosclerosis.

57. What does sclerosis mean?

58. It is chemically incorrect to call cholesterol a fat. Why?

59. What are lipoproteins?

60. What were the findings of the now famous "Framingham (Mass.) study"?

61. Discuss the role of "polyunsaturates" in the prevention of atherosclerosis.

62. Compare the following terms: hyperlipoproteinemia, hyperlipemia, hypertriglyceridemia, and hypercholesterolemia.

63. A popular drug in the management of hyperlipoproteinemias is clofibrate. What is its mechanism of action and how effective is it?

64. The severe, classic heart attack carries the medical name "acute myocardial infarction." Precisely what does this mean?

65. Discuss the treatment and prognosis of acute myocardial infarction.

66. Distinguish among the terms coronary artery disease, coronary occlusion, and coronary thrombosis.

67. Why does atherosclerosis predispose to thrombosis?

68. Cite possible causes of congestive heart failure.

69. Congestive heart failure is often characterized more specifically as "left side failure" or "right side failure." Explain.

70. The term digitalis is imprecise when it comes to writing the prescription. Check with your local pharmacist and find out why.

71. Distinguish between congenital and acquired heart disease.

72. Two features of tetralogy of Fallot are pulmonary stenosis and enlargement of the right ventricle. Are these related and, if so, why?

73. In the diagnosis of hypertension, what pressure is considered the more diagnostic—systolic or diastolic?

74. What is the purpose of giving lidocaine to the heart attack victim?

75. Distinguish between hypovolemic shock and cardiogenic shock.

CHAPTER 13

THE CIRCULATORY SYSTEM
THE BLOOD

Blood, the viscous red fluid of the body, is composed of cell-like bodies* suspended in plasma. This scarlet humor nourishes the tissues, takes away the wastes, fights infection, and plays a key role in the adjustment of pH, fluid balance, and temperature; clearly, it is the central element in homeostasis. The volume of blood in individuals of normal build is estimated to be about 70 ml/kg of body weight. Blood has an average specific gravity of 1.055, an average pH of 7.4, and a viscosity some five times that of water. By means of a centrifuge (*hematocrit*), blood cells and plasma can be separated easily and their volume percents determined (Fig. 13-1). Normal values for plasma average about 55% (of the blood volume), and those for the cellular elements about 45%. Blood cells fall into three categories: red cells, white cells, and platelets.

PLASMA

Plasma is a complex mixture of water (90%) and a galaxy of solutes, including proteins, inorganic salts, lipids, glucose, wastes, vitamins,

---

*Variously referred to as "cells," "corpuscles," and "formed elements."

gases, enzymes, hormones, and antibodies (Table 13-1). With the exception of proteins, whose molecules are largely confined to the circulation, the other constituents diffuse easily into the intercellular fluid. By the same token, intercellular constituents diffuse into the circulation. Thus, except for the presence of protein, plasma and intercellular fluid are practically identical. Indeed, they are lumped together as the extracellular fluid.

Plasma proteins fall into three types: albumins, globulins, and fibrinogen. *Albumins,* which make up the bulk of the 6% to 8% protein concentration in plasma, are responsible for almost all the osmotic force of the blood. The *globulins* are concerned with immunity; almost all antibodies are gamma globulins (immunoglobulins). *Fibrinogen* participates in the vital process of clotting. Whereas albumin and fibrinogen are manufactured principally in the liver, globulins are formed by the plasma cells of the reticuloendothelial system.

Serum

When freshly drawn blood is placed in a tube and allowed to coagulate, a clot forms that gradually draws away from a clear liquid called the

TABLE 13-1
Blood chemistry*

Constituents	Concentration
Alcohol (ethyl)	< 1.5 mg/100 ml
Amylase	4–25 units/ml
Bilirubin, total	0.1–1.5 mg/100 ml
Calcium	9–11.5 mg/100 ml
Carbon dioxide pressure (Pco_2) (arterial blood)	35–45 mm Hg
Chloride (Cl^-)	98–108 mEq/l
Cholesterol, total	150–280 mg/100 ml
Corticosteroids	6–25 μg/100 ml
Creatine	0.2–0.6 mg/100 ml (♂)
	0.4–1.0 mg/100 ml (♀)
Creatinine	0.7–1.5 mg/100 ml
Estrogens	< 0.2 μg/100 ml
Fatty acids	250–390 mg/100 ml
Fibrinogen	0.2–0.4 g/100 ml
Gamma globulin	0.7–1.3 g/100 ml
Glucose	65–100 mg/100 ml
Lipase	0–1.5 units/ml
Lipids	450–1,000 mg/100 ml
Magnesium (Mg^{++})	1.3 mg/100 ml
Nitrogen, total	900–1,350 mg/100 ml
Nonprotein nitrogen (NPN)	20–35 mg/100 ml
Oxygen pressure (Po_2) (arterial blood)	80–100 mm Hg
Potassium	3.5–5.6 mEq/l
Protein, total	6–8 g/100 ml
Sodium	138–148 mEq/l
Urea	20–40 mg/100 ml
Uric acid	3–7 mg/100 ml (♂)
	2–6 mg/100 ml (♀)
Vitamin A	10–60 μg/100 ml
Vitamin B_1	1–9 μg/100 ml
Vitamin B_2	2.3–3.7 μg/100 ml
Vitamin B_6	1–18 μg/100 ml
Vitamin B_{12}	14–98 μg/100 ml
Vitamin C	0.1–2.5 mg/100 ml
Vitamin D	1.7–4 μg/100 ml
Vitamin E	0.4–0.8 mg/100 ml

*Suggested values most likely to be found in health. Material analyzed is serum except for glucose, CO_2, and O_2 (whole blood).

serum (Fig. 13-1). Serum differs from plasma in only one significant respect; it contains no fibrinogen. Although the two expressions are often used as synonyms, there is this difference. If plasma is desired, therefore, blood must be treated with an anticoagulant (to prevent coagulation) and then centrifuged.

RED BLOOD CELLS

In each cubic millimeter of blood, there are about 5 million red cells, or *erythrocytes*, in the male and 4.5 million in the female. Since the red cells outnumber the white cells by about 500 to 1, the *cellular volume percent*, or *hematocrit*, is essentially a reflection of the erythrocyte content. The red blood cell is a biconcave disc, about 8 μm in diameter, marked by the absence of a nucleus and the presence of an iron-bearing red protein called *hemoglobin*. Hemoglobin makes up approximately 33% of the cell and averages between 14 and 16 g per 100 ml of whole blood. Other components of the red cell include proteins and lipids, which make up the internal framework, and an assortment of minerals and enzymes. The principal function of the red cell is to transport oxygen and carbon dioxide. Since this role is in the province of respiration, we shall reserve the details for Chapter 15.

Formation

In the fetus, red cells are manufactured by the liver, the spleen, and the bone marrow. By the time of birth, the job has been taken over exclusively by the red marrow. The flat and irregular bones produce more cells than the long bones do; although there is a gradual decrease in blood cell production in all the bones with advancing age, the marrow of the long bones is engaged in this chore only during preadolescence.

There are several stages in the development of a red cell, beginning with the large *hemocytoblast* and ending with the erythrocyte, or mature

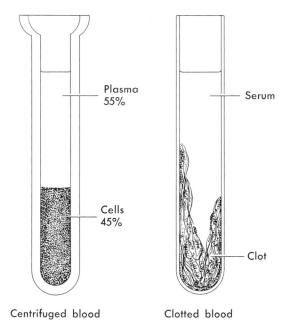

FIG. 13-1
Hematocrit (left) and clotted blood.

Plasma 55%

Cells 45%

Centrifuged blood

Serum

Clot

Clotted blood

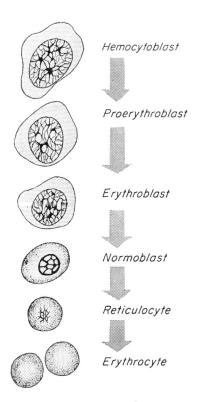

Hemocytoblast

Proerythroblast

Erythroblast

Normoblast

Reticulocyte

Erythrocyte

FIG. 13-2
Development of red cell. (Modified from Guyton, A. C.: Function of the human body, Philadelphia, 1959, W. B. Saunders Co.)

red cell (Fig. 13-2). Once formed, the red cells squeeze their way into the capillaries and enter the circulation at large. Sometimes, however, when they are produced at an accelerated rate, *reticulocytes* and even *normoblasts* enter the circulation before they have fully matured. The number of reticulocytes in the circulation is indicative of hemopoietic activity. Special agents, particularly vitamin B_{12} and folic acid, are needed for the proper development of red cells. If either of these two vitamins is lacking, the *erythroblasts* fail to mature; instead, they grow into oversized cells called *megalocytes*. The presence of this particular type of cell in the blood is a sign of megaloblastic anemia.

Destruction

It has now been established that the red cell has a life-span of about 120 days. At the end of this time it becomes fragile and breaks down. The fragments are phagocytized in the bone marrow, liver, and spleen. The hemoglobin is set free and eventually degraded to globin, the bile pigment bilirubin, and certain forms of iron. The liberated iron may be used immediately to form new red cells or united with a protein to form *ferritin* and stored as this compound in the spleen and the liver until needed.

Oxygen regulation

The production of red cells is controlled chiefly by the oxygen in the tissues. If the concentration drops, erythropoietic activity in the marrow increases; if the concentration rises, erythropoietic activity decreases. Persons living at high altitudes or engaged in hardy exercise may have red blood cell counts as high as 8 million or more. In contrast, the sedentary, sickly person may have a count as low as 3 million. The mechanism of this regulation is as yet unknown. However, recent investigations indicate that diminished oxygen concentration triggers the release of some hormone that stimulates the bone marrow.

Hemolysis and crenation

The destruction of red cells and the release of hemoglobin is called hemolysis. *Chemical* hemolysis is brought about by any number of agents, including certain antibodies (*hemolysins*), certain poisons, animal venoms, fat solvents, and so on, that act directly on the cell membrane. *Osmotic* hemolysis is caused by the excessive uptake of water in *hypotonic* solutions; enough water enters the cells to cause rupture. In a 0.9% (*isotonic*) solution of NaCl red cells retain their shape, but as the concentration becomes progressively lower, they begin to swell and rupture; hemolysis is complete once the 0.3% level is reached. In *hypertonic* solutions, on the other hand (for example, in a NaCl solution above the 0.9% level) red cells shrivel and shrink. This is called *crenation*. Understandably, solutions given intravenously should be as nearly isotonic as possible.

Anemias

Anemia may be defined as a deficiency in red cells or hemoglobin or both; quite understandably, the hematocrit is low. Anemia causes damage to the body in two ways. The most obvious insult is a lack of oxygen in the tissues due to the impaired ability of the blood to carry that gas. As a result, the cells degenerate, especially those of the nervous system. This explains such early symptoms as disinterest, fatigue, and loss of energy. The second way in which anemia damages the body has to do with viscosity. Decreased hematocrit values mean decreased viscosity, which in turn enhances the blood flow. This results in excessive return of blood to the heart, causing overwork and, not infrequently, heart failure. There are various anemias with various causes. Some are treated easily, and some lead to early death. The major forms are discussed below.

Blood loss anemia is a consequence of hemorrhage or the chronic loss of blood. Although the marrow may be able to maintain a nearly normal red blood count, the iron stores of the body become progressively decreased. Accordingly, this type of anemia is marked by not only a mild drop in the number of red cells but also a severe drop in hemoglobin.

Hemolytic anemias involve excessive destruction of red cells. The more common forms include sickle cell anemia, erythroblastosis fetalis (Rh incompatibility), and poisoning due to drugs and chemicals. Sickle cell anemia is a hereditary disease marked by distorted red cells (Fig. 13-3). Since the membranes of these cells are fragile, they are easily damaged and destroyed. In erythroblastosis fetalis (p. 211), the pregnant woman builds up destructive antibodies against the red cells of the fetus. Consequently, the baby is born with a severe anemia.

Iron-deficiency anemia relates to the insufficient intake of iron. Although the cell count is usually near normal, the concentration of hemoglobin in each cell is reduced greatly—hence, the expression *hypochromic* anemia (Fig. 13-3). This condition is quite common and treated successfully with iron compounds such as ferrous sulfate.

Pernicious anemia is a megaloblastic anemia that proves fatal if not treated. The highly characteristic blood picture is shown in Fig. 13-3. Pernicious anemia results from a deficiency of an

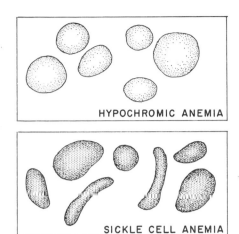

HYPOCHROMIC ANEMIA

SICKLE CELL ANEMIA

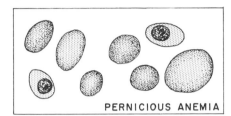

PERNICIOUS ANEMIA

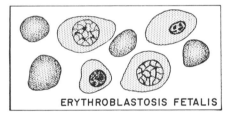

ERYTHROBLASTOSIS FETALIS

FIG. 13-3
Red cells in four types of anemia.

intrinsic factor present in the mucosa of the stomach and duodenum. Since this factor serves to enhance the absorption of vitamin B_{12} (the *extrinsic factor*) into the circulation, a deficiency will severely curtail the manufacture of red cells in the marrow. The treatment of choice is parenteral administration of vitamin B_{12}.

Aplastic anemia is a fatal condition stemming from destruction of the bone marrow. The usual causes are drugs, poisons, and overexposure to X rays, gamma rays, or other ionizing radiations.

Polycythemia

An increased red cell count, or polycythemia, is a normal response during strenuous exercise and at high altitudes. In polycythemia vera (or *erythemia*), however, the elevated count results from tumorous bone marrow. A tremendous number of red cells are poured into the circulation, and the hematocrit and the viscosity are thereby dangerously increased (p. 201). The blood "flows like molasses"; there is a tendency to thrombosis. The treatment includes judicious use of blood letting and administration of anticoagulants and radioactive phosphorus.

WHITE BLOOD CELLS

There are approximately 7,000 white cells, or *leukocytes,* per cubic millimeter of blood. These cells range in diameter from 9 to 25 μm, and are called "white" simply because they are colorless. They are divided into granulocytes and agranulocytes (Fig. 13-4). The *granulocytes,* named for the granules in their cytoplasm, include the neutrophils, the eosinophils, and the basophils. The *agranulocytes,* which contain no specific granules, include the lymphocytes and monocytes.* All leukocytes originate in the bone marrow. Unlike red cells, leukocytes can leave the circulation in great numbers by squeezing through the pores in the capillary wall; in this manner they infiltrate the tissues and aid in fighting infection. It has been demonstrated in the laboratory that the average white cell remains in the blood for no longer than 3 to 4 days. Its life-span, however, may be almost a year.

Neutrophils

Neutrophils are among the most spectacular cells in the human body. They move about like amebas and carry on extensive phagocytosis, especially during inflammation. The inflam-

---

*All leukocytes may contain azure granules, but these are not of the type that characterize the granulocytes.

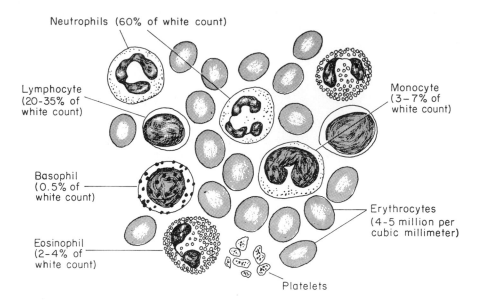

Neutrophils (60% of white count)

Lymphocyte (20-35% of white count)

Basophil (0.5% of white count)

Eosinophil (2-4% of white count)

Monocyte (3-7% of white count)

Erythrocytes (4-5 million per cubic millimeter)

Platelets

FIG. 13-4
Cellular elements of blood. (Platelets average about 300,000 per cubic millimeter.)

matory response runs essentially as follows: Damaged tissue releases a substance, called *leukotaxine,* that increases the permeability of the capillaries in the affected area. The neutrophils line up along the inside wall *(margination)* and slowly squeeze through the enlarged pores *(diapedesis).* Once through, they head for the center of trouble, apparently in chemotaxic response to leukotaxine. Here they phagocytize microbes (Fig. 13-5) and tissue debris. A single neutrophil may devour as many as 50 or more bacterial cells before falling victim to its meal. Pus, the product of inflammation, is composed of dead neutrophils that have fallen in defense of the body.

Damaged tissue also releases into the blood a leukocyte-promoting factor that stimulates the the marrow. This brings about an increase in

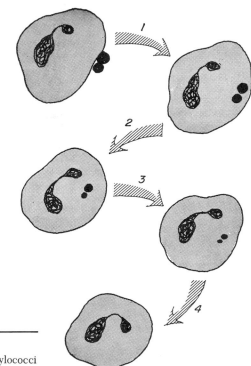

FIG. 13-5
Phagocytosis. Progressive engulfment and destruction of two staphylococci by phagocytic leukocyte.

the number of all white cells in the blood (leuko-cytosis), particularly the number of neutrophils (neutrophilia). Pronounced neutrophilia, there-fore, is an indication that all is not well. Because of the increased production, many neutrophils leave the marrow in an immature form. The more commonly encountered early forms include the juvenile and stab cells.

Eosinophils

Eosinophils, though generally quite similar to neutrophils, are distinguished easily by their larger granules, which take acidic dyes such as *eosin*—hence their name. Since these cells move about and scavenge to a much lesser extent than do neutrophils, one wonders what role they play in the defense of the body. They could, of course, perform some vital activities unknown to the researcher. It is certainly quite significant, for in-stance, that eosinophils increase greatly (*eosino-philia*) in cases of hypersensitivity and parasitic infestation.

Basophils

The basophil has a rather obscure, S-shaped nucleus, and the large granules of the cytoplasm stain with basic dyes. Basophils show little move-ment and are limited in their phagocytic powers. Basophil granules contain *heparin* (an antico-agulant) and *histamine* (a vasodilator), both of which may well serve a purpose during the heal-ing process of inflammation.

Lymphocytes

Lymphocytes have a nongranular cytoplasm, and the nucleus is large, single, generally spherical, and sharply defined. Small cells often are more numerous than large ones. Lympho-cytes are manufactured in the bone marrow and multiply by cell division in the lymph nodes, the spleen, and the thymus gland (Chapter 14). These cells pervade most of the tissues, to which they are delivered by the bloodstream. About half of the lymphocytes, the *T cells*, pass through

the thymus gland on their way to the tissues; the other half, the *B cells,* do not. Foreign sub-stances (*antigens*) stimulate B cells to grow, change in structure, and divide, eventually giving rise to a large number of *plasma cells,* lymphocytes specialized in the rapid synthesis and secretion of antibody molecules (p. 59). Each cell is committed in advance to the produc-tion of one specific antibody. T cells can destroy other cells (cancer cells, transplanted tissues) and can also affect the functioning of B cells. Depending on circumstances, they either sup-press B cells or assist them in the output of antibody molecules.

Monocytes

The monocyte is the largest leukocyte, with an average diameter of about 25 μm. The nucleus is single and kidney or horseshoe shaped, and stains blue; the cytoplasm is abundant, stains a grayish blue, and may contain azure granules. Monocytes migrate readily through capillary walls and carry on extensive phagocytosis. They are able to mop up tissue debris that proves in-digestible to the neutrophils, and in chronic infections, they are the body's major cellular defenders.

Leukopenia

Leukopenia is a decrease in the number of white cells in the blood. The most serious form is *agranulocytosis,* or a lack of granulocytes. This condition results when the bone marrow is damaged by poisons, certain drugs, or radiation. Since the body is deprived of one of its major defenses against bacterial infection, the result is ulceration of the gastrointestinal and respira-tory tracts. Unless anti-infective drugs are ad-ministered, the patient usually dies in a day or two.

Leukemia

Leukemia is the lawless proliferation of white cells. These cells are immature, abnormal, and oftentimes literally flood the bone marrow and lymphatic tissues. Paradoxically, the white cell count of the circulating blood is not always elevated and upon occasion may actually be down! The number of new victims for 1972 was estimated to be about 20,000, the number of deaths, 15,000. The disease accounts for about half of all cases of cancer in persons between the ages of 3 and 14. The precise etiology is unknown, but two facts stand out: certain viruses cause animal leukemia; radiation is definitely one cause of human leukemia.

Leukemias are classified on a basis of the speed of the disease (*acute* or *chronic*) and the type of cell chiefly involved. Leukemias in which the granulocytes predominate are labeled *myelocytic;* those involving lymphocytes and monocytes are labeled *lymphocytic* and *monocytic,* respectively. In sum, then, we have acute and chronic myelocytic leukemias; acute and chronic lymphocytic leukemias; and acute and chronic monocytic leukemias. For the acute form, chemotherapy is often spectacularly effective, and cures have been reported. Among the drugs (*antineoplastics*) of major value are 6-mercaptopurine (6-mp), methotrexate (amethopterin), and corticosteroids.

PLATELETS

Platelets, or *thrombocytes,* are minute, granular, disc-shaped bodies in the blood that number about 300,000 per cubic millimeter. Not really cells, they are believed to be fragments of the giant megakaryocytes present in the red bone marrow; based on present knowledge platelets have a life-span of about a week. The role they play in blood coagulation is presented below.

HEMOSTASIS AND COAGULATION

At the site of injury, the first response to stem the loss of blood is local vasoconstriction (the *vascular phase* of hemostasis), followed within seconds by the adherence of the platelets to the surface of the vessels and to one another (*platelet*

phase). Then commences the critical *coagulation phase,* the details of which are still shrouded in much biochemical mystery. Dozens and dozens of reactions are known to be involved, and there is every reason to believe that many more await discovery. One thing is certain: Coagulation is a cascade affair with each step activating the next. In each step, a coagulating *factor* is converted from an *inactive* to an *activated* enzymatic form, which in turn activates the next factor, and so on. The naming of the various *plasma* factors has now been standardized. Each is designated by a Roman numeral, and the activated form is specified by the suffix *a. Platelet* factors are assigned Arabic numerals.

According to one scheme, the overview of coagulation runs thus: The contact between shed blood and an extravascular surface converts factor XII to factor XIIa, the latter cascading to factor IXa as shown in Fig. 13-6. Factor IXa now reacts with factor VIII and factor 3 in such a way as to convert factor X to factor Xa. Factor Xa interacts with factor V and factor 3 to form thromboplastin (factor III). From here on, the process is more familiar and better understood. *Thromboplastin* converts *prothrombin* (factor II) to the active enzyme *thrombin,* which in turn converts soluble *fibrinogen* (factor I) to insoluble, tangled threads of *fibrin.* Blood cells catch in this sticky meshwork to produce the *clot.* Contributing to these events are the *tissue factors* released from the injured cells at the site of the injury. As shown in Fig. 13-6, these factors interact with factor VII to form a product also capable of converting factor X to factor Xa, and so on. In most of these steps, calcium (factor IV) is essential. Also critical is the indirect role of vitamin K, which the liver needs to manufacture prothrombin.

Coagulation tests

A number of tests are employed in the diagnosis of coagulation abnormalities. The most useful for screening purposes include bleeding

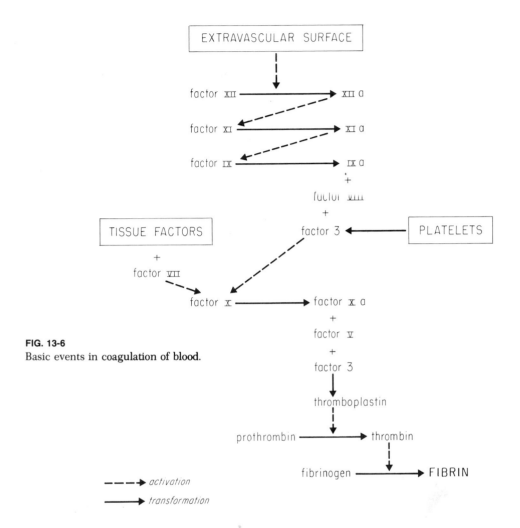

FIG. 13-6
Basic events in coagulation of blood.

EXTRAVASCULAR SURFACE

factor XII ⟶ XII a

factor XI ⟶ XI a

factor IX ⟶ IX a
+
factor VIII
+

TISSUE FACTORS   factor 3 ⟵ PLATELETS
+
factor VII

factor X ⟶ factor X a
+
factor V
+
factor 3

thromboplastin

prothrombin ⟶ thrombin

fibrinogen ⟶ FIBRIN

----→ *activation*
⟶ *transformation*

time, coagulation time, partial thromboplastin time (PTT), prothrombin time, and platelet count. *Bleeding time* is determined by puncturing the skin, applying to it at regular intervals a piece of absorbent paper, and noting the length of time elapsing before the paper is no longer stained with blood. Normally, this is less than 6 minutes. *Coagulation,* or *clotting, time,* on the other hand, is the length of time required for the coagulation of shed blood. Normally, this is less than 15 minutes. The *platelet count* requires a

phase contrast microscope or automated equipment, and the *PTT* and *prothrombin* tests are basically biochemical procedures. The bleeding time and platelet count are tests of the vascular and platelet phases of coagulation, whereas the other three procedures are tests of the coagulation phase.

Antithrombins and fibrinolysin

The coagulation mechanism does not get started in normal vessels because the smooth and nonwettable endothelial lining is a surface to which platelets do not adhere. Other deterrents to intravascular clotting are antithrombins and fibrinolysin. *Antithrombins,* as the name indicates, oppose or inactivate thrombin. The best known, heparin, is present in most tissues. Somewhat paradoxically, the concentration of heparin in the blood is too low to have much influence on coagulation; presently its physiological role is not understood. Therapeutically, however, it enjoys wide use as an anticoagulant in the management of thrombosis and embolism. Widely used, too, are the synthetic anticoagulants bishydroxycoumarin (Dicumarol) and warfarin (Coumadin).

Fibrinolysin (or *plasmin*) is important in dissolving clots. Once formed, a clot by some unknown mechanism triggers the release of this enzyme from a plasma precursor called *plasminogen*. Occasionally, fibrinolysin is used as an adjunct to anticoagulant therapy.

Thrombosis

A common cause of death is thrombosis, or the formation of an intravascular clot. Such a clot, or *thrombus,* may remain where it is, or it may be swept away by the blood. In the latter event, the clot is called an *embolus,* and the condition is termed *embolism.* Death results from the obstruction of blood to a vital area or organ. The classic example is coronary thrombosis (p. 195). Thrombosis of the peripheral vessels is most often traceable to sluggish blood flow, particularly in the legs. Since some of the platelets have a tendency to adhere to the walls of the veins anyway, they are even more likely to do so in areas of poor circulation. In so doing, the platelets rupture, and clotting ensues. Other predisposing factors include atherosclerosis, phlebitis, and varicose veins. As noted above, the management of thrombosis and embolism centers on the use of heparin and other anticoagulants.

Hemorrhagic disorders

Hemorrhagic disorders are diseases characterized by an abnormal tendency to bleed. The various causes may be categorized under vascular, platelet, and coagulation disorders. Examples of *vascular* disorders include allergic purpura, scurvy, and hereditary hemorrhagic telangiectasia. *Platelet* disorders relate to the various forms of *thrombocytopenia. Coagulation* disorders typically relate to a deficiency of some plasma factor or factors involved in the actual coagulation mechanism. Major *hereditary* disorders of this kind are hemophilia and von Willebrand's disease; *acquired* disorders of this kind include, among many others, deficiency of vitamin K–dependent coagulation factors, defibrination syndrome, multiple myeloma, and kidney disease. Treatment of a particular hemorrhagic disorder can sometimes be directed at the underlying cause with signal success. For example, bleeding arising from a deficiency of vitamin K–dependent coagulation factors is rapidly managed by injections of phytonadione (vitamin K).

BLOOD GROUPS

The reason one person's blood may prove fatal to another person relates to blood grouping. A given group, or type, is characterized by the presence of a certain protein factor in the red cells called *agglutinogen*. Three agglutinogens—A, B, and Rh—are the most likely to cause difficulties with transfusions.

ABO groups

Persons with agglutinogen A red cells are type A (41% of the population); persons with agglutinogen B red cells are type B (10%); the presence of both agglutinogens constitutes type AB (4%); the absence of both, type O (45%). Further, there are two kinds of antibodies present in the plasma, anti-A agglutinin and anti-B agglutinin. Anti-A agglutinin agglutinates (clumps) only red cells containing agglutinogen A, and anti-B agglutinin agglutinates only red cells containing agglutinogen B. Both agglutinins are present in type O blood; both are absent in type AB blood. Type A and type B blood contain anti-B agglutinin and anti-A agglutinin, respectively. This is the way it has to be, for otherwise plasma would agglutinate its own red cells. Agglutinated red cells are dangerous because they become trapped in the smaller vessels and interfere with the circulation. Moreover, they soon disintegrate and release massive amounts of hemoglobin, enough to plug the kidney tubules and cause renal failure.

The critical point in regard to transfusions is the effect of the recipient's agglutinin on the donor's agglutinogen. The reverse is not ordinarily of concern because the donor's blood is diluted to such an extent that its agglutinating power becomes negligible. Theoretically, type O individuals are universal donors, and type AB individuals are universal recipients. Due regard, of course, must be paid also to the Rh factor (below).

Rh factors

There are at least a dozen Rh factors (agglutinogens) in countless combinations, but only about a half dozen combinations are common. About 85% of the population possess Rh factors, and these individuals are said to be *Rh-positive;* the absence of Rh factors makes an individual *Rh-negative*. Rh-positive blood transfused into an Rh-negative individual sensitizes the latter's immunity system to such an extent that a subsequent transfusion will provoke dangerous levels of anti-Rh agglutinin, enough to cause mass agglutination and a severe reaction, perhaps death. The classic Rh incompatibility is erythroblastosis fetalis, which occurs in about 1 out of every 50 babies born to Rh-negative mothers and Rh-positive fathers. If the fetus inherits the Rh factor from the father and becomes Rh positive, its red cells will provoke the output of anti-Rh antibodies when they enter the

mother's circulation. This is most likely to happen at the time of delivery. The first child is usually unaffected, but because the plasma cells are now sensitized, subsequent pregnancies shoot the antibody concentration higher and higher, the result being a severe case of erythroblastosis fetalis. Rather recently a way was found to circumvent the dilemma by giving the mother an injection of anti-Rh antibody at the time of delivery at *each* pregnancy. Such an injection neutralizes the Rh factor in the event red cells have escaped into the maternal circulation and prevents the factor from sensitizing the plasma cells.

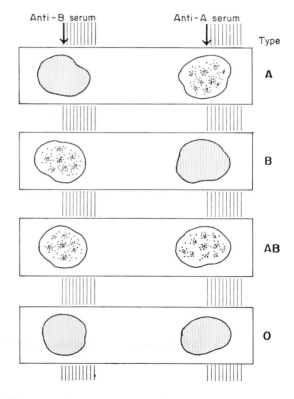

FIG. 13-7
Typing of blood. (From Brooks, S. M.: Basic facts of medical microbiology, ed. 2, Philadelphia, 1962, W. B. Saunders Co.)

M and N factors

M and N are red cell agglutinogens giving rise to three blood groups: M, N, and MN. Fortunately no agglutinins are involved, and consequently no transfusion problems are encountered. These groups are often of great value in the establishment of parentage. For example, if a child is type N, the mother and father must be N or MN; the father cannot be type M; and so on.

Blood typing

To determine the particular blood group to which an individual belongs, the blood is mixed with the appropriate agglutinin(s). In ABO typing, for example, anti-A agglutinin ("anti-A serum") and anti-B agglutinin ("anti-B serum") are used (Fig. 13-7). AB blood is agglutinated by both agglutinins; O blood is *not* agglutinated by either; A blood is agglutinated only by anti-A serum; B blood is agglutinated only by anti-B serum.

BACTEREMIA

The presence of bacteria in the blood is called bacteremia. *Transient* bacteremia occurs frequently in various infections but invariably clears up once the primary infection (pneumonia, for example) is brought under control. *Unresponsive* bacteremia, however, notably staphylococcal bacteremia, is generally fatal. Treatment centers on parenteral antibotics and intensive supportive measures; the choice of antibiotics depends on the sensitivity of the organism involved. Lifesaving drugs include, among others, gentamicin, oxacillin, and cephalothin.

MALARIA

Malaria, the number one infection of man, is a protozoiasis associated with the blood. The precise nature of malaria depends upon which of the four species of the genus *Plasmodium* is involved. Benign tertian malaria, the most common form of the disease, is caused by *Plasmodium vivax*.

The other three forms are malignant tertian malaria (*Plasmodium falciparum*), quartan malaria (*Plasmodium malariae*), and an uncommon form confined to East Africa and South America caused by *Plasmodium ovale*. In benign tertian and quartan malaria, the paroxysms of fever, which occur every third and fourth day, respectively, are followed by chills; in malignant tertian malaria, chills may or may not occur. In each instance fever and anemia result from the destruction of red cells.

Plasmodia have an alternation of generations, with the sexual stage involving insects and the asexual stage involving vertebrates. In the case of man (Fig. 13-8), the situation begins when an infected *Anopheles* mosquito bites him. Asexual spores (*sporozoites*) from the saliva of the mosquito enter the liver cells and develop into *merozoites;* the latter in turn invade the red blood cells and multiply. Eventually, the red cells rupture and pour forth countless merozoites, which thereupon invade other red cells, and so on. In time, some merozoites develop into *microgametocytes,* and others, into *macrogametocytes.* If these enter the body of a biting mosquito, they develop into *microgametes* (sperm) and *macrogametes* (eggs). At this point in the cycle, man has infected the mosquito. The sperm now fertilize the eggs, and from the resulting zygotes, or *ookinetes,* develop *oocysts* containing numerous sporozoites. Upon rupture of the oocysts, the released sporozoites wend their way into the salivery glands of the mosquito to await injection.

THE BLOOD IN DIAGNOSIS

The value of the blood in the diagnosis of disease is obvious. Conceivably, all abnormalities are in some fashion reflected in the blood, and the trick is to devise a test or tests to detect this reflection. "High sugar" plus "high acetone" plus "low pH," for example, spells diabetic acidosis. The number of tests is always on the rise,

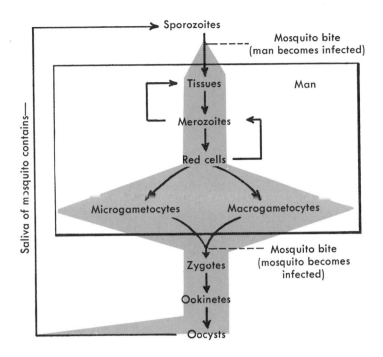

FIG. 13-8

Life cycle of malarial parasite at a glance. (From Brooks, S. M.: Basic facts of pharmacology, ed. 2, Philadelphia, 1963, W. B. Saunders Co.)

and for the most part, the tests grow more sophisticated. In practice, most hospital laboratories categorize blood tests as hematology, serology, and chemistry. In this sense, *hematology* refers to the gross features of blood, namely, cell counts, hematocrit, bleeding time, and so on; *serology* refers to the detection of telltale antibodies; and *chemistry* refers to the endless solutes dissolved in the plasma, ranging from acetone to uric acid. Some tests are routine, others are special, but clearly what is special today may very well be routine tomorrow.

QUESTIONS

1. What is the blood volume of a 154-pound man of normal build?
2. A given sample of blood has a specific gravity of 1.055. What is the weight of 30 ml of this blood?
3. What are the two meanings of hematocrit?
4. Distinguish between plasma and serum.
5. Compare the compositions of plasma and interstitial fluid.
6. Discuss the roles of plasma proteins.
7. In regard to blood cell counts, a common error among students is to say "cc" instead of the correct cubic millimeter. Compare these two units.
8. How many atoms of iron are there per molecule of hemoglobin?
9. Distinguish between hemopoiesis and erythropoiesis.
10. Why does a "black eye" turn yellowish?
11. Anemia is not always characterized by a low red cell count. Explain.
12. In the treatment of pernicious anemia, why is vitamin B_{12} given by injection for best results?
13. Distinguish between polycythemia and erythrocytosis.

14. Explain the elevated red cell count among those living at high altitudes.
15. In the red blood cell "fragility test," hemolysis normally begins at 0.45% (salt solution concentration) and is complete at 0.39%. What does it mean if these values are increased (to say, 0.60% and 0.50%, respectively)?
16. Normal saline solution (0.9% NaCl) is "isotonic." Isotonic with what?
17. Which possess the greater danger in intravenous therapy—hypotonic or hypertonic solutions?
18. Leukocytes are not white. Why then are they called leukocytes?
19. What are polymorphonuclear leukocytes?
20. How do we distinguish among the three kinds of granulocytes?
21. Compare the lymphocyte and the monocyte.
22. Compare agranulocytosis and leukopenia.
23. What is meant by a "differential count"?
24. Discuss the role of the blood in the body's immunity system.
25. What were the major findings of the clinical studies done on the survivors of the atomic bomb?
26. Specifically, what does "chronic myelocytic leukemia" tell us?
27. What role do the blood vessels play in hemostasis?
28. What is meant by the "platelet phase" of hemostasis?
29. Dozens and dozens of plasma factors are involved in coagulation, but the lack of only a single factor can result in bleeding. Why?
30. What role do injured tissues play in coagulation?
31. Platelets play both a physical and chemical role in hemostasis. Explain.
32. What is the relationship between thrombin and prothrombin?
33. Distinguish between fibrin and fibrinogen.
34. The newborn bleeds very easily. Why?
35. How do you explain the fact that liver disease can result in bleeding?
36. The prolonged use of certain antibiotics causes bleeding. Explain.
37. Citrates and oxalates prevent blood from clotting. Explain.
38. The administration of anticoagulants is obviously not without danger. How is the dosage regulated?
39. What is the specific antidote for dicumarol poisoning?
40. Why is the bleeding time shorter than the coagulation time?
41. Specifically, what does the prothrombin time measure?
42. Describe the procedure for doing the prothrombin time.
43. Why are there so many coagulation tests?
44. Compare the action of anticoagulants with that of fibrinolysin.
45. Isotonic sodium chloride is a 0.9% solution, whereas isotonic glucose is a 5% solution. Explain.
46. Give the signs and symptoms, the diagnosis, and the treatment of idiopathic thrombocytopenic purpura.
47. What is hereditary hemorrhagic telangiectasia?
48. What is von Willebrand's disease?
49. What coagulation factors are involved in the most common forms of hemophilia?

50. Discuss the treatment and prognosis of the hemophilias.
51. Distinguish between antigen and agglutinogen.
52. Distinguish between agglutinins and agglutinogens.
53. Distinguish between coagulation and agglutination.
54. Individuals with type O blood are universal donors even though type O plasma agglutinates red cells of types A, B, and AB blood. Explain.
55. Is it safe to give Rh-negative blood to an Rh-positive individual?
56. The first time Rh-positive blood is given to an Rh-negative individual no transfusion reaction occurs. Explain.
57. Erythroblastosis fetalis can be prevented. How?
58. Is it safe to give M blood to an individual with type N blood?
59. Given two samples of blood, one type A, the other type B, tell in detail how you would use these two bloods to determine the type of an unknown blood.
60. Why is it important to know the type of malaria a patient has?
61. The treatment of malaria has four aspects: termination of acute attacks, curative therapy, suppression, and gametocidal therapy. Discuss.
62. "The value of the blood in the diagnosis of disease is obvious." Discuss and elaborate on this statement.
63. The laboratory report reads coagulation, 7 min.; triglycerides, 100 mg/100 ml; hematocrit 40%; and VDRL, negative. Classify these values on a basis of hematology, chemistry, and serology.

CHAPTER 14

THE LYMPHATIC SYSTEM

Increased metabolic activity fosters an excess of tissue fluid that must be drained away and returned to the blood. A major function of the lymphatic system is to effect this drainage via a vast network of vessels embedded in the tissues throughout the body. What we call *lymph* is the tissue fluid within these vessels. Additionally, the system encompasses a variety of structures, glands, and organs that play a vital role in protecting the body against infection.

LYMPHATICS

Lymph vessels (lymphatics) originate among the cells as microscopic channels, or lymph capillaries. The capillaries come together to form the progressively larger lymphatic vessels, much as twigs on a tree coalesce into branches (Fig. 14-1). The vessels join together to form the main channels—the thoracic duct and the right lymphatic ducts. The *thoracic duct* is about 16 inches in length and originates in the lumbar region as a dilated structure called the *cisterna chyli*. It runs up the trunk and finally empties its lymph into the bloodstream at the juncture of the left internal jugular and subclavian veins. The right lymphatic ducts (sometimes there is just one) join the venous system at the juncture of the right internal jugular and subclavian veins. Except for the right arm, the right upper chest, and the right side of the head, which are drained by the right lymphatic ducts, all lymph flows into the thoracic duct. The lymphatics originating in the *villi* of the small intestine are called *lacteals,* and their milky lymph is called *chyle*.

Tissue fluid, which is mainly water and escaped blood protein, enters the lymph capillaries and becomes lymph when it is in excess and its pressure rises. Lymph flows slowly and steadily along in this drainage system as a result of breathing movements and the massaging action of contracting skeletal muscles. Lymphatics are equipped with numerous valves, much like those of veins, that prevent a backflow. The rate of flow varies greatly, depending upon the degree of physical activity, but the volume "pumped" is always fantastically low. Compared to the 5 to 6 liters of blood passing through the heart per minute, only a fraction of an ounce of lymph makes its way through the lymphatics.

LYMPH NODES

Lymph nodes are whitish, bean-shaped structures composed of dense masses of lymphocytes (p. 207) and anastomosing sinuses lined with phagocytic reticuloendothelial cells called macrophages. Varying in size, from a pinhead to an inch or so in diameter, they are scattered along

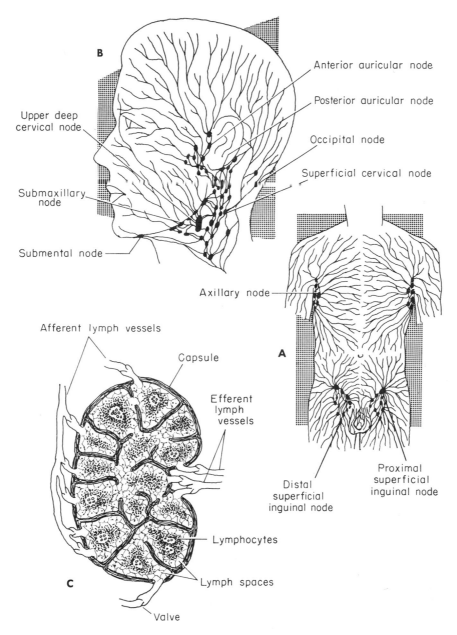

FIG. 14-1

Lymphatic system. **A,** General distribution of lymphatic vessels of trunk. **B,** General distribution of lymphatic vessels of head. **C,** Structure of lymph node. (**A** and **B** after Sappey; modified from Francis, C. C: Introduction to human anatomy, ed. 6, St. Louis, 1973, The C. V. Mosby Co.)

the course of the lymphatics, the most conspicuous groupings occurring in the cervical, axillary, and inguinal regions. As shown in Fig. 14-1, lymph moves into the nodes via two or more afferent lymphatics, then makes its way through the sinuses, and finally emerges via a single (usually) efferent lymphatic. The *cervical* nodes drain lymph from the head and neck; the *axillary* drain the arm and upper part of the chest wall; and the *inguinal* drain the legs and external genitals.

Lymph nodes aid the body in its defense against infection. Macrophages engulf and destroy microorganisms and other tissue debris, and lymphocytes differentiate into plasma cells, the all-important producers of *circulating* antibodies (p. 207). Unfortunately, some lymphocytes synthesize *cellular* or *tissue-rejecting*

antibodies, a major concern in organ transplantations. Unfortunately, too, lymph nodes screen out dislodged cancer cells and thereby set up new growths, or metastases. Indeed, the lymphatic system has much to do with making cancer the great killer that it is.

SPLEEN

The spleen (Fig. 14-2) is situated directly below the diaphragm in the left hypochondrium. It is ovoid and in the adult is from 5 to 6 inches long and 4 inches wide. Histologically, the organ resembles a lymph node and is especially characterized by numerous venous sinuses. During times of quietude and plenty these sinuses can collectively accommodate a liter of blood and then release it to the circulation in times of stress and deprivation. And aside from this reservoir role, the spleen serves the body in other ways, notably in processing lymphocytes and monocytes. It also manufactures red cells in the fetus, and via the macrophages lining the venous sinuses, it destroys microorganisms, platelets, and worn out red cells. Interestingly, the iron and globin released from the hemoglobin are salvaged and used again in hemopoiesis.

TONSILS

The tonsils are masses of lymphoid tissue located in the mouth and throat. The *faucial* or *palatine* tonsils are embedded in the lining of the throat between the arches of the fauces; the *adenoids* or *pharyngeal* tonsils lie in the posterior wall of the nasopharynx; and the *lingual* tonsils are situated at the base of the tongue. The basic histologic features of these organs are the deep pits (crypts) reaching into the interior and the abundance of lymphocytes, mast cells, and plasma cells occurring in the connective tissue. Unlike lymph nodes, which they resemble, tonsils do not possess sinuses and do not filter lymph; their only supposed function relates to lymphocytes and body defense. Typically, ton-

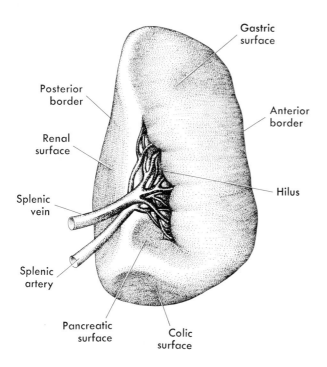

FIG. 14-2
The spleen and associated vessels (medial aspect).

sils reach their highest state of development in childhood and involute with age.

THYMUS

The thymus is a flat, pinkish gray, two-lobed gland lying high in the chest behind the sternum. Large in relation to the rest of the body in fetal life and in early childhood, by the age of puberty it has stopped growing and begins to atrophy. The thymus is the source of T-cell lymphocytes, the cells involved in delayed hypersensitivity and the rejection of skin grafts. This is underscored by the fact that infants born without the thymus develop no delayed hypersensitivity in response to various antigens and sometimes completely fail to reject a graft of skin from an unrelated donor. However, these infants are not deficient in antibodies, which indicates that the lymph node plasma cells are not processed in the thymus. Thus, there appears to be two separate immunologic systems, one dependent upon the thymus and the other independent of it. The plasma cells, as discussed earlier (p. 207), develop from antigen-stimulated B-cell lymphocytes.

LYMPHATIC DISEASES

As the body's drainage system, lymphatic tissue is clearly a prime target for infection and cancer; also, like all tissues, it gives rise to trouble in its own right. The highlights of some of the more common and important lymphatic diseases follow.

Lymphadenitis

The inflammation of one or more lymph nodes is called lymphadenitis. Just about any organism can be responsible, but the usual pathogens are staphylococci and streptococci. The involvement may be local and minimal or widespread and severely toxic. The nodes may enlarge greatly and become extremely painful and tender. The surrounding tissues often become inflamed and give way to abscesses. Treatment centers on the removal of the underlying cause, and the use of the appropriate antimicrobials.

Lymphangitis

Inflammation of the lymphatic vessels, or lymphangitis, may be caused by any pathogen, but the usual agents are staphylococci and streptococci. As the infection creeps along the vessel from the portal of entry, its path becomes red, swollen, and painful, and there is almost invariably lymphadenitis. Cellulitis and ulcers may develop along the path of the infection, and the pathogen may even enter the blood to cause bacteremia and septicemia ("blood poisoning"). Fever, chills, headache, malaise, and generalized aching are prominent systemic features, even without blood poisoning.

Tonsillitis

Tonsillitis is inflammation of the palatine tonsils. The acute form, usually caused by certain strains of nonhemolytic streptococci, can be a severe infection in which the tonsils become red, swollen, and painful, and the crypts filled with noxious debris and pus. The throat is severely sore and swallowing is agonizing. Systemically, the repercussions include muscular pains, malaise, fever, and chills. Bed rest, aspirin, and antibiotics are the generally prescribed therapeutic measures. Although tonsillectomy was formerly the fashion of the day, this is no longer the case. Since the tonsils are apparently a natural defense mechanism (perhaps much more so than has been realized), many authorities believe that they should be removed only in cases in which they are a chronic focus of infection.

Infectious mononucleosis

Infectious mononucleosis is an acute disease characterized by the constitutional symptoms of an infection and the generalized involvement of

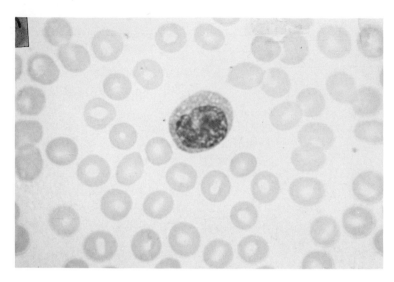

FIG. 14-3

Blood picture of patient with infectious mononucleosis. Abnormal lymphocyte stands out in the center of photo. (From Miale, J. B.: Laboratory medicine: hematology, ed. 4, St. Louis, 1972, The C. V. Mosby Co.)

lymphatic tissue. Sporadic attacks occur chiefly between the ages of 15 and 30, and epidemics occur for the most part in children. College students appear likely candidates and "deep kissing" is often cited as the reason. More particularly, the etiology is now generally thought to be the EB virus—the same virus thought to cause Burkitt's lymphoma (p. 221). Infectious mononucleosis is extremely varied in its severity. The incubation period may run from a few days to several weeks. Typically, the initial features include fever, malaise, sore throat, and headache. In almost all cases the lymph nodes are enlarged, especially those of the neck and arm-pit regions. In about half of the cases the spleen is tense and swollen, and the liver is enlarged. Signs and symptoms persist for 2 to 8 weeks and sometimes for several months. Rarely death may occur as a result of an overwhelming infection, a ruptured spleen, a damaged heart, or suffocation from swollen cords obstructing the windpipe. A defini-tive diagnosis is made on a basis of the presence in the blood serum of atypical lymphocytes (Fig. 14-3) and peculiar immune bodies called *heterophil agglutinins*. Therapy is symptomatic.

Filariasis

Filariasis is a tropical infestation caused by the filarial worm (Fig. 14-4), two species of which are pathogenic to man, *Wuchereria bancrofti* and *Wuchereria malayi*. The adult worms, which are threadlike in shape and measure 1 to 3 inches in length, seek out the lymph vessels and nodes, particularly those of the pelvis and groin. Obstruction of the flow of lymph and elephantiasis (tremendous swelling in the legs) result. The life cycle of the filarial worm involves certain species of mosquitoes that inject the larval stage into the blood upon biting, the mosquito itself having been infected by another stage of the worm (*microfilaria*) present in the bloodstream of infected persons. The chief prophylactic measure

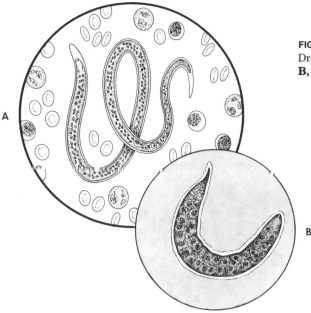

FIG. 14-4
Drawings of filarial worm larva. **A,** In human blood.
B, In mosquito. (About × 1,000.)

in this helminthiasis is mosquito control. The
diagnosis of filariasis is made by the finding of
microfilariae in blood smears, and treatment
consists chiefly in administering anti-infective
agents, especially Hetrazan.

Lymphomas

Lymphomas are tumors of lymphatic tissue.
Most are cancers, and the yearly death toll is
about 20,000. The cause or causes are unknown,
but there is much evidence that at least one kind
of lymphoma (*Burkitt's* lymphoma) relates to
the EB virus (p. 220). The major lymphoma is
Hodgkin's disease. The victim characteristically
is a young adult, and in the classic case the cancer begins in a lymph node in the neck (Fig.
14-5). Soon adjacent nodes are involved, then

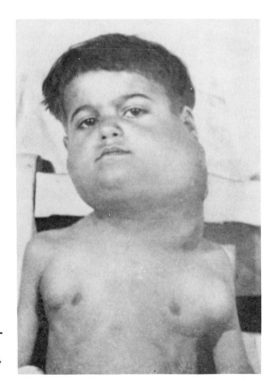

FIG. 14-5
Hodgkin's disease. (From Goodale, R. H.: Nursing pathology, ed. 2,
Philadelphia, 1956, W. B. Saunders Co.)

adjacent nodal regions, then the nodes behind the kidney, then the nodes in the groin, and finally nonlymphoid tissues. Ultimately, all organs are involved. Treatment centers on irradiation and antineoplastics. In the very early stages of the disease, cures are possible.

QUESTIONS

1. Distinguish among intercellular fluid, interstitial fluid, tissue fluid, and lymph.
2. What is meant by the expression "lymphatic system proper"?
3. What is meant by lymphatics?
4. What feature do the components of the lymphatic system have in common?
5. What is the function of lymph vessels?
6. Compare lymph capillaries and blood capillaries.
7. Are the lymphatics more like arteries or veins?
8. Compare the function of the thoracic duct and the right lymphatic ducts.
9. What causes tissue fluid to enter the lymph capillaries?
10. Explain how breathing aids the movement of lymph.
11. Explain the swelling distal to a tight garter.
12. What are the functions of a lymph node?
13. What are macrophages?
14. What is meant by the expression "anastomosing sinuses"?
15. Distinguish between efferent and afferent lymphatics.
16. What is meant by the expression "circulating antibodies"?
17. Why can a ruptured spleen cause sudden death?
18. Is the spleen essential to life?
19. Why are the palatine tonsils also called faucial?
20. Why are the pharyngeal tonsils called adenoids?
21. What role does the thymus play in the body's immunity system?
22. What is meant by the expression "delayed hypersensitivity"?
23. What does lymphadenopathy mean?
24. Lymphadenitis and lymphangitis are relatively common. Why?
25. An infected finger left unattended can easily lead to axillary lymphadenitis. Why?
26. What are buboes?
27. Explain the role of the lymphatic system in the metastasis of cancer.
28. A standard treatment for cancer of the breast is radical mastectomy. Why radical?
29. Distinguish between bacteremia and septicemia.
30. At one time tonsillectomies were fashionable. This is clearly not true today. Why has there been this change in attitude?
31. To what facts does the disease infectious mononucleosis owe its name?
32. Describe the heterophil agglutination test for infectious mononucleosis.
33. Elephantiasis is "pathological proof" for at least one function of the lymphatic system. Which one?
34. Strictly speaking, lymphoma means tumor (of lymphatic tissue) not cancer. Explain.
35. Biopsy is the mainstay in the diagnosis of Hodgkin's disease, as it is in most cancers. What specific histologic feature distinguishes this lymphoma from the others?
36. What is blood poisoning?

CHAPTER 15

THE RESPIRATORY SYSTEM

Respiration encompases the exchange of gases in the lungs, the exchange of gases in the tissues, and the transport of gases between the lungs and the tissues. The exchange between blood and air that occurs in the lungs is called external respiration, and the exchange between blood and the fluid in the tissues is called internal respiration. The system proper includes the nose, the pharynx, the larynx, the trachea, the bronchi, the lungs, and the respiratory muscles (p. 161). We shall discuss first the structure and function of the various parts of the respiratory system and then pass on to the actual mechanics of breathing and the exchange and transport of gases.

NOSE

The nose is divided into left and right cavities by the bony partition (septum) that arises perpendicularly from the roof of the mouth (hard palate). Each nasal cavity is further divided into three passageways by the conchae (turbinates) that arise from the lateral walls of the internal nose (Fig. 15-1). The superior and middle conchae are processes of the ethmoid bone (p. 135), whereas the inferior conchae are separate bones. The passageways lead from the nostrils, or anterior nares, through the posterior nares, to the nasopharynx. These passageways, or meati,

take their names from the conchae: superior meatus, middle meatus, and inferior meatus. The nasal cavities, and most other structures along the respiratory tract, are lined with pseudostratified ciliated columnar epithelium (p. 94). This tissue not only secretes mucus but also propels it along in one direction ("outward") by the lashlike motion of the microscopic cilia (not to be confused with the macroscopic hairs of the nose). In this manner foreign particles trapped on the sticky surface are removed to the outside. Taken en masse, the cilia are unbelievably forceful.

By means of the extensive mucosal surface, the nose filters, warms, and moistens the air while leading it to the nasopharynx. Olfactory receptors embedded in the mucosa lining the roof of the nasal cavities make the nose the organ of smell. The nose also aids in speaking, or phonation. When the vocal cords vibrate, the air within the nasal cavities, like the air in the other cavities and sinuses of the head, is caused to vibrate, or resonate. This vibration or resonance imparts a certain quality to the voice that a "stuffy nose" can radically alter.

PHARYNX

The pharynx, or throat, is a tubelike structure, about 5 inches in length, that passes immedi-

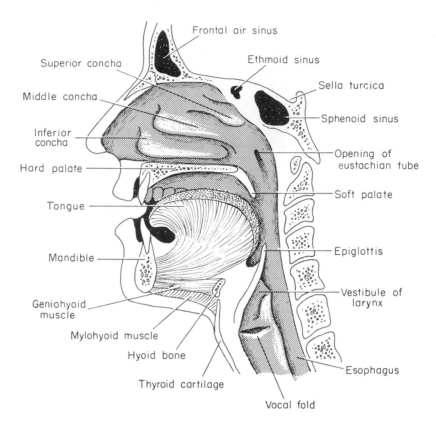

FIG. 15-1
Sagittal section of head and neck. (Modified from Francis, C. C: Introduction to human anatomy, ed. 6, St. Louis, 1973, The C. V. Mosby Co.)

ately before the cervical vertebrae from the base of the skull to the esophagus (Fig. 15-1). Its walls are well supplied with skeletal muscle, and with the exception of the oropharynx (the part behind the fauces), the lining is ciliated epithelium. The area behind the nose is referred to as the nasopharynx, and that behind the larynx the laryngopharynx. In all, the pharynx has seven openings: two from the eustachian tubes, two from the nose (posterior nares), one from the mouth (inner fauces), one into the larynx, and one into the esophagus. The pharynx is also characterized by the presence of lymphoid struc-

tures called tonsils (p. 218). The pharyngeal tonsils (adenoids) are located on the posterior wall of the nasopharynx, opposite the posterior nares, and the faucial or palatine tonsils are located in the oropharynx (see Fig. 16-2). The lingual tonsils are located at the base of the tongue.

The pharynx serves as a passageway for air and food and plays an important part in phonation. The pharynx can change its shape in a split second and gives pitch and quality to the sound waves produced in the larynx. Learning to speak largely lies in training the pharyngeal muscles.

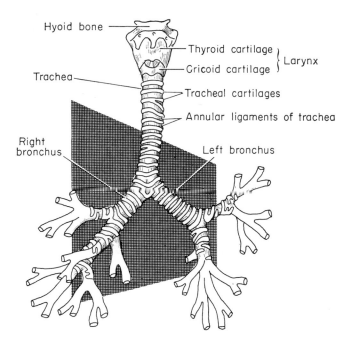

FIG. 15-2
Larynx and trachea. (Modified from Francis, C. C: Introduction to human anatomy, ed. 6, St. Louis, 1973, The C. V. Mosby Co.)

The eustachian tubes are nature's unique way of protecting the eardrums against rupture from the impact of powerful sound waves; they equalize the air pressure on either side of the eardrum. The snapping and crackling sensation in the ears as one ascends in an elevator is due to relatively greater air pressure in the middle ear. Swallowing relieves the condition because air from the middle ear is sucked through the eustachian tubes. The tubes remain closed except during yawning, swallowing, and yelling.

LARYNX

The larynx (Fig. 15-2), aptly called the voice box, is situated at the top of the trachea and conjoins the trachea with the pharynx. It is composed of nine pieces of cartilage and abundant skeletal muscle woven together by connective tissue into a tough, wedge-shaped box. The thyroid cartilage forms the anterior portion of the larynx and gives it its characteristic shape (Adam's apple). Generally, the larynx is larger in men than in women. The epiglottis, a tonguelike structure attached along one edge of the thyroid cartilage, prevents food and drink from going down the "wrong way." The cricoid cartilage, which forms the bottom border of the larynx, is attached to the trachea by connective tissue. The larynx is lined with ciliated mucosa thrown up in two prominent horizontal folds, the so-called false vocal cords. The true vocal cords, situated below these, are composed of two fibrous bands strung horizontally. The opening between them is called the glottis.

Sound is produced by a substance in vibration (p. 15). In the case of phonation, sound is made by ejecting bursts of air through the glottis, thereby causing the vocal cords to vibrate. In turn, this causes the air throughout the respiratory passageways, cavities, and sinuses of the head to resonate. Like other sounds, the human voice is characterized by loudness, pitch, and quality. The pitch of the human voice is controlled by the tension placed on the vocal cords by the laryngeal muscles. Tense cords yield high-pitched notes, relaxed cords, low-pitched notes, as may be simply demonstrated by plucking a rubber band under varying degrees of tension. Formally stated, the pitch of a musical string varies directly with tension and inversely with its length and diameter. The quality of the human voice depends upon the number and nature of the overtones (the tones of a higher pitch than the fundamental) that combine with the fundamental to produce a complex musical sound. Overtones, in turn, relate to the structure of the nose, mouth, pharynx, and sinuses, and the position of the tongue.

TRACHEA

The trachea, or windpipe, is a 5-inch cartilaginous and membranous cylindrical tube descending in front of the esophagus from the larynx to the bronchi (Fig. 15-2). It averages about 1 inch in diameter. Like the other respiratory passageways, the trachea is lined with ciliated epithelium. The cartilage is arranged in C-shaped rings embedded at equal intervals in fibrous tissue. The open ends of the rings, which face the posterior surface, are filled in with bands of smooth muscle. These rings prevent the trachea from collapsing and shutting off vital air. The trachea leads air into the lungs and in the process filters, warms, and moistens it via the ciliated mucosa.

BRONCHIAL TREE

At the level of the fifth thoracic vertebra, the trachea divides, or bifurcates, into the bronchi

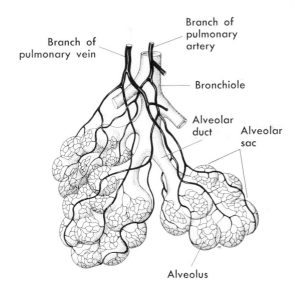

FIG. 15-3

Respiratory bronchiole and associated alveoli. Note especially the blood supply.

(Fig. 15-2). The right bronchus is shorter, wider, and more vertical than the left. Both bronchi structurally resemble the trachea. As the bronchial tubes divide and subdivide into smaller and smaller structures, they gradually lose their cartilage and fibrous tissue until finally all that remains is a microscopic tube (the *bronchiole*) composed of only a thin layer of smooth muscle and elastic fibers. The entire bronchial tree from the primary bronchi to the bronchioles is lined with ciliated epithelium. The bronchioles terminate as "pulmonary units." As shown in Fig. 15-3, a respiratory bronchiole enters an alveolar duct, which leads into an alveolar sac, a structure that somewhat resembles a bunch of hollow grapes. Each "grape" of the sac is called an alveolus. Because the alveolar wall is only one cell thick, gases easily diffuse across it between the air and the blood. This, of course, is the whole purpose of external respiration.

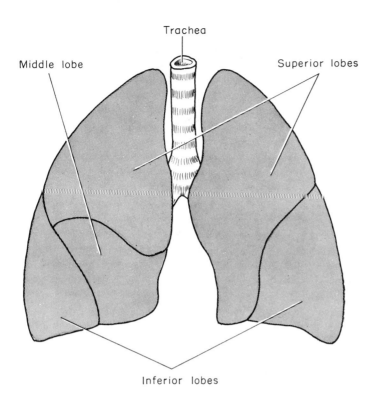

Trachea

Middle lobe

Superior lobes

Inferior lobes

FIG. 15-4
Lobes of the lung (anterior view).

LUNGS

The lungs are light, spongy, cone-shaped organs (Fig. 15-4) that fill the lateral cavities of the chest. They are separated from each other by the heart and the other mediastinal structures. Their bases rest on the diaphragm, and their apices extend slightly above the clavicles. To accommodate the heart, the medial aspect of each lung has a concave surface; the left lung has a greater concavity, or "notch," than the right. Other significant gross features include the hilus, the lobes, and the pleura. The *hilus* (or *hilum*) is a slit in the medial surface through which the primary bronchus and the pulmonary blood vessels, wrapped in a sheath of connective tissue, penetrate the lung. The *lobes* are the sections of the lung formed by fissuring;

the left lung partially divides into two lobes and the right lung into three. The *visceral pleura* is a thin serous membrane that envelops the lungs, and the *parietal pleura* is a similar membrane that lines the chest wall (p. 64). The two are always in immediate contact and move against each orther in breathing. To facilitate this movement and to reduce friction, they are moistened with a serous fluid.

According to one source, the lung is made up of some 200 million or so alveoli (Fig. 15-3). There is a strong tendency for these microscopic structures to collapse because of the elastic fibers surrounding them and the potential surface tension of the thin film of fluid covering their inside linings. These forces account for the lung's elastic recoil in the expiratory phase of breathing (below). Total collapse, however, must be and is prevented by one or more *surfactants* present in the fluid lining, which act to reduce surface tension. As the alveoli become smaller during expiration, the concentration of these agents increases, making them most effective when they are most needed.

THORAX

The thorax, or chest, is defined by the sternum, the ribs, the thoracic vertebrae, and the diaphragm; within, the pleurae describe the mediastinum and pleural cavities (p. 64). The parietal pleura covers the diaphragm and lines the inside surface of the chest. When the diaphragm and the external intercostal muscles (p. 161) contract, the thorax is increased in all dimensions, thereby increasing the volume of the cavity. In this manner the *intrapleural pressure** is lowered and the lungs "sucked out," causing air to be inspired. To perform a normal expiration, the body merely relaxes these muscles; the diaphragm returns from the stretched-out horizontal condition to its rounded dome shape and the intercostal muscles drop the ribs. This causes the cavity to decrease in all dimensions, the intrapleural pressure to increase, and the lungs to deflate by elastic recoil. In forced breathing, other muscles are called into play. To take in the greatest possible volume of air, the sternocleidomastoids, the rhomboids, and the levator scapulae muscles, among others, aid the external intercostals in enlarging the thorax. Forced expiration involves the internal intercostals and the abdominal muscles.

EXCHANGE OF GASES

As the volume of the thoracic cavity increases as a consequence of diaphragmatic and intercostal contraction (p. 161), the intrapleural

---

*The intrapleural (or *intrathoracic*) pressure refers to the pressure in the potential space between the visceral and parietal pleura. It does not refer to pressure within the lungs (*intrapulmonic* pressure).

pressure decreases in accordance with Boyle's law. The drop in pressure averages about 4 mm Hg, going from about 756 mm at the end of expiration to about 752 mm at the end of inspiration (Fig. 15-5). This reduced pressure caused the lungs to expand, with the concomitant decrease in the *intrapulmonic* pressure, which goes from atmospheric (760 mm) to about 757 mm Hg. As a result, air from the respiratory passageways moves into the alveoli, and, in turn, fresh air enters through the nose.

External respiration

At the end of inspiration the air pressure within the alveoli is atmospheric. Of this pressure, 100 mm is due to oxygen; in accordance with Dalton's law, we say that oxygen has a partial pressure (P_{O_2}) of 100 mm Hg. Since the P_{O_2} of venous blood entering the pulmonary capillaries is only about 37 mm Hg, oxygen diffuses into the blood. Carbon dioxide diffuses into the alevoli as a consequence of the difference in partial pressures of that gas; precisely, the P_{CO_2} of venous blood is 46 mm Hg versus about 40 mm Hg in the alveoli. Expired air is about 4% carbon dioxide and about 16% oxygen.

Internal respiration

Tissue fluid and arterial blood entering the systemic capillaries have P_{O_2}'s of about 37 mm Hg and 100 mm Hg, respectively. Consequently, oxygen diffuses into the tissue fluid. Carbon dioxide, on the other hand, diffuses into the blood because the arterial P_{CO_2} is less (40 mm Hg) than that of tissue fluid (46 mm Hg). Venous blood, the blood returning to the lung, has a P_{O_2} of 37 mm Hg and a P_{CO_2} of 46 mm Hg.

OXYGEN AND CARBON DIOXIDE TRANSPORT
Lung phase

As the blood spills into the vast expanse of the pulmonary capillaries, reduced hemoglobin (here abbreviated HHb) combines with oxygen to form

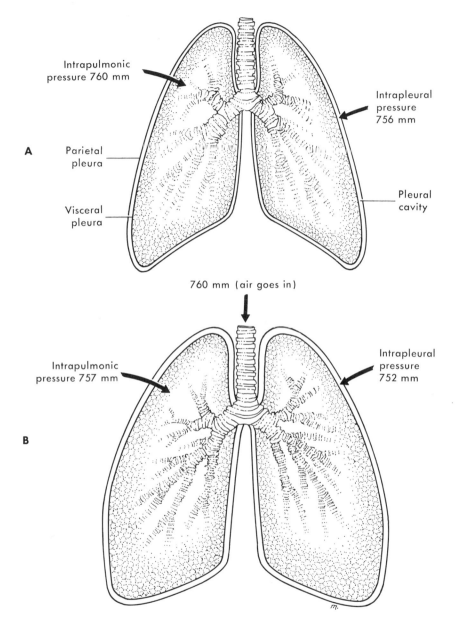

Intrapulmonic
pressure 760 mm

Intrapleural
pressure
756 mm

Parietal
pleura

Visceral
pleura

Pleural
cavity

A

760 mm (air goes in)

Intrapulmonic
pressure 757 mm

Intrapleural
pressure
752 mm

B

FIG. 15-5
Intrapleural and intrapulmonic pressures at the end of expiration, **A,** and inspiration, **B.** Air moves into lungs, **B,** as a result of lowered intrapulmonic pressure.

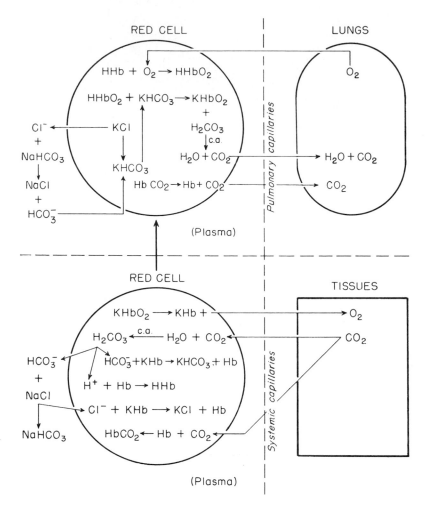

FIG. 15-6

Diagrammatic summary of chemical events in oxygen and carbon dioxide transport by the blood. Red cell in tissues (lower) releases O_2 and returns to lungs with its hemoglobin reduced (HHb); plasma returns to lungs with charge of $NaHCO_3$. In lungs, HHb reacts with O_2 to yield $HHbO_2$, which in turn reacts with $KHCO_3$ to yield $KHbO_2$ and H_2CO_3; and so on. (From Brooks, S. M.: Basic facts of general chemistry, Philadelphia, 1956, W. B. Saunders Co.)

oxyhemoglobin ($HHbO_2$) (Fig. 15-6). Oxyhemoglobin reacts with potassium bicarbonate (also present in the red cell) to form potassium oxyhemoglobin ($KHbO_2$) and carbonic acid (H_2CO_3), according to the reaction:

$$HHbO_2 + KHCO_3 \longrightarrow KHbO_2 + H_2CO_3$$

In the presence of the enzyme carbonic anhydrase, H_2CO_3 is split into CO_2 and H_2O and given off to alveolar air. The loss of $KHCO_3$ causes an imbalance between the bicarbonate ion (HCO_3^-) in the red cell and the plasma. As a result, Cl^- diffuses out of the red cell into the plasma (*chloride shift*), where it reacts with $NaHCO_3$ to form HCO_3^-:

$$NaHCO_3 + Cl^- \longrightarrow NaCl + HCO_3^-$$

HCO_3^- diffuses into the red cell to replace Cl^- and reacts with KHb to form $KHCO_3$, which, as explained above, reacts with $HHbO_2$, to form H_2CO_3, and so on. This exchange of Cl^- for HCO_3^- continues until the plasma's "charge of CO_2" (carried as $NaHCO_3$) is blown off.

What about oxygen? As shown, oxygen has gone through two steps to become $KHbO_2$, the form in which it is carried to the tissues. A very small amount of oxygen is carried in simple solution. On the average, 100 ml of normal blood transports 20 ml of oxygen, 19.5 ml as $KHbO_2$ and 0.5 ml in simple solution.

Tissue phase

When arterial blood enters the capillaries, O_2 is released from the unstable $KHbO_2$:

$$KHbO_2 \longrightarrow KHb + O_2$$

At the same time, CO_2 diffuses into the red cells, where, in the presence of the enzyme carbonic anhydrase, most of it reacts with H_2O, to form carbonic acid (H_2CO_3). H_2CO_3 dissociates into H^+ ions and HCO_3^- ions. Most of the HCO_3^- ions diffuse into the plasma, where they react with NaCl to form $NaHCO_3$ and Cl^-; the H^+ ions left behind react with hemoglobin (Hb) to produce HHb (reduced Hb):

$$H_2CO_3 \longrightarrow H^+ + HCO_3^-$$
$$HCO_3^- + NaCl \longrightarrow NaHCO_3 + Cl^-$$
$$H^+ + Hb \longrightarrow HHb$$

The Cl^- formed in the plasma diffuses into the red cell, balancing the loss of HCO_3^- (the "chloride shift" again). All of the HCO_3^- ions, however, do not leave the cell. Some react with KHb to form Hb and $KHCO_3$. Also, some of the original CO_2 entering the red cells combines directly with hemoglobin, forming carbaminohemoglobin ($HbCO_2$).

CO_2, then, is carried by the blood in three forms: $NaHCO_3$, $HbCO_2$, and $KHCO_3$. Two thirds of the total blood CO_2 is carried in the plasma as $NaHCO_3$, and one third is carried in the red cells as $HbCO_2$ and $KHCO_3$.

RESPIRATORY CONTROL

The respiratory center located in the medulla oblongata of the brain is believed to wield the most influence over the depth and frequency of respiration. It does so via nerve impulses sent to the muscles of respiration. Although the pneumotaxic center in the pons of the brain also shares in this regulation, we shall speak only of medullary control. In order for the respiratory center to operate effectively, it must receive peripheral information regarding the body's oxygen requirements (Fig. 15-7). For example, if we breathe harder during exercise than at rest, it is because the respiratory center has been apprised of the body's need and has reacted accordingly.

Hering-Breuer reflex

The presence of *stretch receptors* among the alveoli prevents overinflation and aids in maintaining the basic rhythm of respiration. These receptors, as their name implies, are stimulated by expansion of the alveoli during inspiration, sending impulses to the medulla via sensory nerves. This causes inhibition of the center and reflex relaxation of the inspiratory muscles. As the lungs deflate, the pressure upon the receptors is reduced; the inhibitory impulses flowing to the medulla cease, causing the respiratory center to release impulses to the muscles of inspiration. Once again, the lungs inflate, the stretch receptors are stimulated, and so on. This feedback series of events is known as the Hering-Breuer reflex.

Carbon dioxide and pH

A moderate increase in carbon dioxide results in faster breathing. As its partial pressure in the

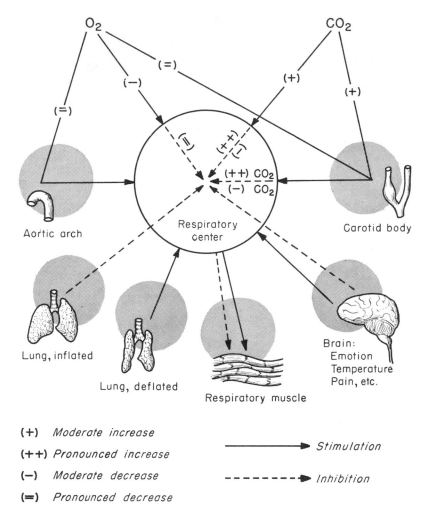

O₂   CO₂

(=)   (+)
(−)   (+)

(=)

Aortic arch   (++) CO₂
                     (−) CO₂   Carotid body

Respiratory
center

Lung, inflated

Lung, deflated   Brain:
                                  Emotion
          Respiratory muscle   Temperature
                                  Pain, etc.

(+) *Moderate increase*
(++) *Pronounced increase*   ——————————▶ *Stimulation*
(−) *Moderate decrease*
(=) *Pronounced decrease*   - - - - - - -▶ *Inhibition*

FIG. 15-7
Summary of mechanisms involved in control of respiratory rate. Note that CO_2 and O_2 act on respiratory center directly and indirectly.

blood rises above 40 mm Hg, CO_2 stimulates the respiratory center directly.* Indirectly, and to a much lesser degree, it steps up respiration by stimulating the chemoreceptors in the carotid

bodies and aortic arch. At high concentrations, however, carbon dioxide depresses these areas. Fortunately, this does not occur under ordinary conditions.

A moderate decrease in the pH of arterial blood stimulates the respiratory center both indirectly, via the chemoreceptors, and directly.

———————
*Mostly as H⁺ ions from H_2CO_3 ($CO_2 + H_2O \rightarrow H_2CO_3$).

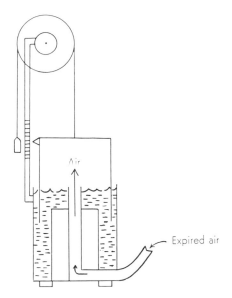

FIG. 15-8

Spirometer. Volume of air entering or leaving bell is recorded on scale.

Oxygen

A decrease in blood P_{O_2} stimulates the respiratory center directly. This is not true in severe hypoxic situations, however, because nervous tissue cannot function properly when starved of oxygen. Indeed, the nervous system is the most sensitive of all systems to oxygen deprivation. A pronounced decrease in arterial P_{O_2} reflexly stimulates respiration via chemoreceptors in the carotid body and aortic arch. However, since the oxygen concentration of the blood never falls to such a level under normal conditions, this particular regulatory mechanism operates only in hypoxic emergencies.

Other factors affecting respiration

Respiration is altered by countless stimuli other than the Hering-Breuer reflex and the influence of oxygen, carbon dioxide, and pH. The physician spanks the newborn infant to induce breathing. Irritant chemicals, strong light, sudden pain, and cold are other examples of such stimuli. Few of us can forget the distressing experience of having our breath "shut off" following a plunge into cold water. Many of us may recall the dramatic response that attends administration of smelling salts to a victim of fainting. There are also psychological influences. It is common experience, for instance, that when one is spellbound or frightened one "forgets" to breathe. The sign following such an experience is quite real; the period of apnea produces enough carbon dioxide to trigger an audible respiratory sound. Other emotional varieties of respiration include laughing, crying, and yawning.

AIR VOLUMES

Using an appratus called a *spirometer* (Fig. 15-8), we can easily measure the intake and output of the lungs.

Tidal air

Tidal air is the volume of air that is carried to and fro in respiration. It is the volume of air normally exhaled following inspiration. In most persons it averages about 500 ml at rest.

Inspiratory reserve and expiratory reserve

Inspiratory reserve is that volume of air in excess of tidal air that can be drawn into the lungs by maximal inspiration; expiratory reserve is that volume which can be expelled from the lungs in excess of the tidal air. Inspiratory reserve is normally about 3 l, expiratory reserve, about 1 l.

Vital capacity

Vital capacity may be defined as the greatest volume of air that one is able to expel after the greatest possible inhalation. This is equal to tidal air plus inspiratory reserve plus expiratory reserve, about 4.5 l.

Residual and minimal air

Regardless of how hard one tries, one cannot expel all air from the lungs. The amount that remains in the alveoli after the greatest possible expiration is called residual air (about 1.2 l). When the thorax is opened, the lungs collapse and most, but not all of the residual air is forced out. The tiny bronchioles collapse and in so doing cut off the escape of a small amount of air from the alveoli. This permanently trapped air is referred to as minimal air. This explains why the lungs, unlike other soft tissue, do not sink when placed in water. The lungs of stillborn infants, which contain no air, are an exception. Advantage is taken of this fact in medicolegal work.

Dead air

Well over a quarter of the air taken in at each breath fails to enter the alveoli and participate in the exchange of gases. This so-called "dead air" fills the nose, throat, and other respiratory "dead spaces." External respiration specifically concerns alveolar air, the volume of which is equal to the volume of tidal air minus the volume of dead air.

NORMAL BREATHING

Normal quiet breathing, called *eupnea,* may be costal, diaphragmatic, or both. In the costal type, the upper ribs move first and the abdomen second, with the elevation of the ribs being the more noticeable of the two movements. In the diaphragmatic type, the abdomen bulges outward first, followed by a less energetic movement of the thorax. Diaphragmatic respirations are the deeper of the two. Costal breathing may be caused by tight clothing that obstructs diaphragmatic action.

Respiratory rate

Eupnea in the adult runs from 8 to 20 respirations per minute, but during exercise or emotiotional tension the rate in healthy persons increases substantially (*polypnea*). Age has an important bearing on respiration. During the first year of life, the respiratory rate averages about 50 per minute, and by the age of 5 years the rate has dropped to 25. The rate continues to decrease, reaching the normal adult level sometime during the late teens or early twenties.

ABNORMAL BREATHING

There are a number of breathing abnormalities. Some of the principal forms include the following:

apnea temporary cessation of respirations.[*]
dyspnea difficult or labored breathing due to conditions that interfere with the free passage of air into and out of the lungs.
orthopnea inability to breathe in the horizontal position.
tachypnea excessive rapidity of respiration.
Cheyne-Stokes respiration a type of respiration characterized by rhythmical variations in tidal volume, occurring in cycles; typically, each cycle consists of a dyspneic phase followed by an apneic period lasting from 5 to 30 seconds (Fig. 15-9); Cheyne-Stokes respiration is usually seen in coma and often precedes death.

ACID-BASE AND WATER BALANCE

The loss of CO_2 from the lungs relates to acid-base balance because the loss in effect amounts to a loss of H_2CO_3 (acid) in the blood (p. 108). Accordingly, a decrease in pH (acidosis) stimulates respiration and an increase in pH (alkalosis) inhibits respiration; the respiratory system compensates for metabolic acidosis and metabolic alkalosis. On the other hand, the respiratory system can cause an acid-base imbalance, in which case we employ the labels *respiratory* acidosis and *respiratory* alkalosis. Hyperventilation leads to respiratory alkalosis; hypoventilation leads to respiratory acidosis. About 400 ml of water is lost from the lungs as vapor in a 24-hour period (insensible water loss). This

---

[*] May be normal, as after a sigh.

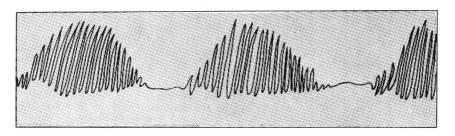

FIG. 15-9
Stethographic tracing of Cheyne-Stokes breathing. (From Halliburton, W. D.: Handbook of physiology, ed. 16, Philadelphia, 1923, P. Blakiston's Son & Co.)

relates to the body's water balance and must be repaid.

HYPOXIA

Hypoxia means a reduction of oxygen in the body tissues to below physiologic levels. If this relates to interference with the source of oxygen, the expression *hypoxic* hypoxia is used. Etiologic possibilities include high altitudes, low oxygen mixtures, respiratory depression, and mechanical obstruction (namely, tracheal tumor, asthma, foreign bodies, and pulmonary edema). *Anemic* hypoxia results from inadequate oxygen being carried from the lungs to the tissues because of a decreased capacity of blood to transport the gas. The etiologic factor may be anemia, methemoglobinemia, or carbon monoxide poisoning. *Histotoxic* hypoxia is caused by certain poisons and drugs that interfere with the utilization of oxygen. Still other forms are *anoxia neonatorum* and *stagnant* hypoxia. The former is anoxia of the newborn, and the latter results from failure of the circulation to move blood fast enough through the vessels. The clinical management of hypoxia depends upon the type and the cause. It may involve surgery (for example, tracheotomy or removel of tumors), drugs (for example, epinephrine for asthma, vitamin B_{12} for anemia, and vasodilators for gangrene), or antidotes against histotoxic poisons. In most hypoxic emergencies, of course, oxygen is administered.

PNEUMONIA

Pneumonia is an inflammation of the small bronchioles and the alveoli. These spaces become filled with mucous exudate, and hypoxic hypoxia results. Pneumonia is often characterized as lobar or lobular. *Lobar* pneumonia, which involves one or more lobes of the lung, is caused in most instances by *Diplococcus pneumoniae*, a gram-positive, encapsulated microorganism (Fig. 15-10) that gains entrance to the body via the nose or mouth. Although it may be trans-

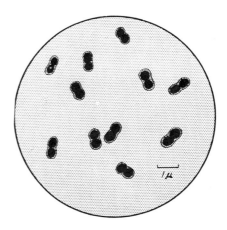

FIG. 15-10
Diplococcus pneumoniae. (From Brooks, S. M.: Basic facts of medical microbiology, ed. 2, Philadelphia, 1962, W. B. Saunders Co.)

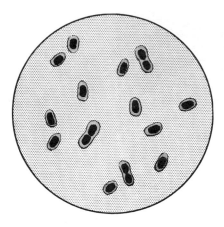

FIG. 15-11

Klebsiella pneumoniae. (From Brooks, S. M.: Basic facts of medical microbiology, ed. 2, Philadelphia, 1962, W. B. Saunders Co.)

mitted indirectly by food or fomites, most cases result from droplet infection. The diagnosis is usually established by demonstration of the organism in the sputum. Another pathogen that causes lobar pneumonia is *Klebsiella pneumoniae,* or *Friedländer's bacillus* (Fig. 15-11), an encapsulated, non-spore-forming, gram-negative organism that also is capable of infecting the nasal sinuses, bronchi, and middle ear. *Lobular* pneumonia and *broncho*pneumonia are generally secondary infections scattered throughout the lung tissue and caused by staphylococci, streptococci, and a variety of other organisms. Treatment of all pneumonias centers on the administration of antibiotics and oxygen, and complete bed rest.

THE COMMON COLD

The signs and symptoms of the common cold (*acute coryza*) are well known, as is the cause, a special class of microbes called *rhinoviruses.* Hundreds of these viruses have been identified. The infection may be limited to the upper respiratory tract or may extend to the larynx, trachea,

and middle ear. Often, however, the latter involvements are bacterial complications. The treatment of the common cold remains purely symptomatic, and its prevention is still an unsolved problem. Because of the multiplicity of pathogens, the prospects for an effective vaccine are not especially bright at this time.

INFLUENZA

On a world-wide basis, influenza, or the "flu," is man's most deadly enemy. In the great pandemic of 1918, the death toll was close to 20 million! Lesser pandemics occurred in 1957 ("Asian flu"), 1968 ("Hong Kong flu"), and 1973 ("London flu"). Epidemics occur every 1 to 4 years and develop rapidly since the incubation period is only 1 to 3 days. Viruses (Fig. 15-12) belonging to types A, B, and C cause the infection. Type A viruses continue to be the major problem because of their propensity to mutate. The signs and symptoms of influenza include inflammation of the respiratory mucous membranes, aches and pains, fever, and prostration. In severe cases, fever (as high as 104° F) may last 4 or 5 days, and fatigue, weakness, and sweating may last 4 or 5 weeks. Treatment is purely symptomatic except for the use of antibiotics in the event of bacterial complications (namely, pneumonia). Vaccination is effective against prevalent strains but clearly affords little or no protection against major antigenic mutations.

TUBERCULOSIS

Although chemotherapy has drastically reduced the death rate, tuberculosis still remains a major threat to man's welfare, particularly in underprivileged areas. The infection is caused by the actinomycete *Mycobacterium tuberculosis,* a slender, nonmotile, non-spore-forming acid-fast bacillus (Fig. 15-13). Like other members of the genus, this species is often observed in the branching and filamentous state. Tuberculosis

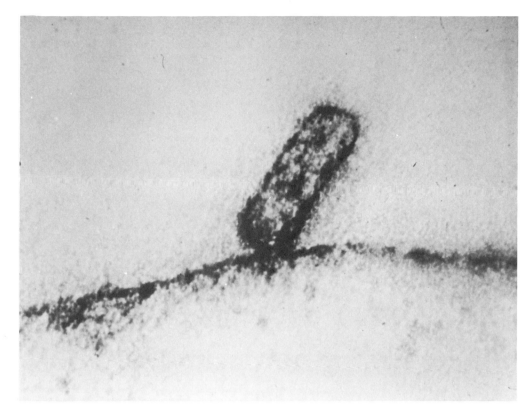

FIG. 15-12
Influenza virus showing a single virion attached to cell membrane. (× 90,000.) (Courtesy Eli Lilly & Co.)

is usually spread via contact with "open cases" through coughing, sneezing, and kissing. The organism may also enter the body via milk or other contaminated foods. Pasteurization and the removal of infected cattle, however, have practically eliminated, in the United States at least, milk-borne tuberculosis. The clinical picture depends upon the tissues involved and the host's resistance. The principal lesions occur in in the kidneys, bones, meninges, and lungs. Pulmonary (lung) tuberculosis is the most common form of the disease. From the primary infection, the tubercle bacilli migrate throughout

the body and, if unchecked, may overrun the body's defenses. Treatment, centers on the use of drugs, including isoniazid (INH), streptomycin (SM), para-aminosalicylic acid (PAS), and ethambutol. Prevention of the spread of tuberculosis centers on early diagnosis through chest X ray and the tuberculin test. The tuberculin test is based upon the fact that a person previously or presently infected with the tubercle bacillus becomes sensitized to tuberculin, an extract derived from the tubercle bacillus. Since a positive reaction can mean an inactive as well as an active focus of infection, it can never be

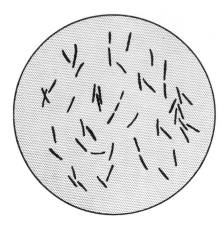

FIG. 15-13
Mycobacterium tuberculosis. (From Brooks, S. M.: Basic facts of medical microbiology, ed. 2, Philadelphia, 1962, W. B. Saunders Co.)

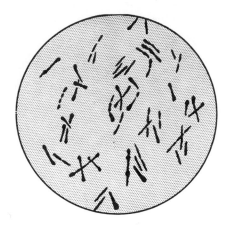

FIG. 15-14
Corynebacterium diphtheriae. (From Brooks, S. M.: Basic facts of medical microbiology, ed. 2, Philadelphia, 1962, W. B. Saunders Co.)

interpreted as unequivocal evidence of clinical tuberculosis except in infants and young children. Vaccination with BCG (vaccine) has been recommended for tuberculin nonreactors who have been heavily exposed to tuberculosis. For chemoprophylaxis, INH has proved effective.

DIPHTHERIA

Diphtheria is a severe acute infection generally confined to young children. It is marked locally by inflammation of the upper respiratory tract and systemically by damage to the heart, kidneys, and nerves. The pathogen, *Corynebacterium diphtheriae* (often called the Klebs-Loeffler bacillus after its discoverers), is a pleomorphic, non-spore-forming, nonmotile, nonencapsulated, gram-positive bacillus. Characteristically, the bacilli stain irregularly and are frequently banded or beaded with metachromatic granules (Fig. 15-14). The principal threat to life stems from the powerful exotoxin released by the organism. The concomitant use of diphtheria antitoxin and penicillin affords a good treatment if given early. The antitoxin neutralizes the toxin and the penicillin

attacks the bacilli. Active immunization with the toxoid is the most important prophylactic measure; when vaccinated in infancy, almost all persons are endowed with potent immunity that lasts from 2 to 5 years. For immediate prophylaxis, however, the antitoxin must be given (Fig. 15-15). It is recommended that the first shot of toxoid be given at the age of 4 to 6 months and that a booster be given before the child enters school. To determine whether a person is susceptible to diphtheria, a very small quantity of the exotoxin is injected into the skin. Susceptible persons develop a pink or red wheal at the site of the injection, but immune persons show no reaction. This is called the Schick test.

PSITTACOSIS

Also called "ornithosis" and "parrot fever," psittacosis is an infection of the respiratory tract marked by fever, chills, headache, sore throat, nausea, and vomiting. The possibility of developing a severe type of pneumonia or involving the central nervous system is the main danger. Until rather recently, the etiologic agent was classed

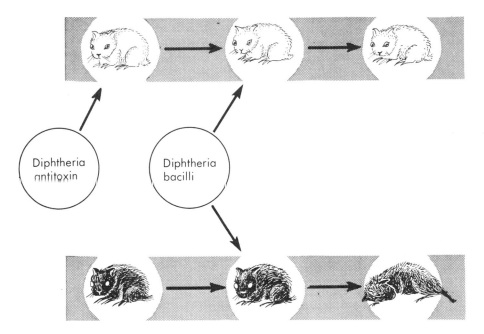

FIG. 15-15

Antitoxin affords excellent protection against diphtheria when administered without delay. In guinea pig, injection of antitoxin affords absolute protection against an otherwise lethal dose of diphtheria bacilli.

as a virus, but the facts now point to a rickettsial category. The reservoir of infection includes pigeons, ducks, chickens, turkeys, and most especially psittacine birds (parrots, parakeets, and the like). Most cases of psittacosis are contracted by handling sick birds and inhaling dust charged with bird feces. Treatment centers on the use of antibiotics, namely the tetracyclines; prevention entails avoiding undue handling of all birds and destroying those that are diseased.

HISTOPLASMOSIS

Histoplasmosis is a systemic fungal infection caused by the pathogenic yeast *Histoplasma capsulatum* (Fig. 15-16). It is characterized by emaciation, leukopenia, anemia, fever, and ulceration of the nasopharynx, intestine, liver, and spleen. A mild form of the disease, common

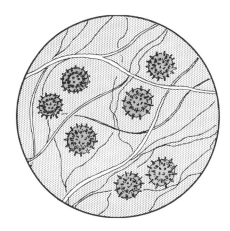

FIG. 15-16

Histoplasma capsulatum (in spore stage). (From Brooks, S. M.: Basic facts of medical microbiology, ed. 2, Philadelphia, 1962, W. B. Saunders Co.)

along the Mississippi basin, attacks the lungs and closely parallels early tuberculosis. Cases of histoplasmosis have often been misdiagnosed as tuberculosis. The antibiotic amphotericin B yields excellent results in the progressive disseminated form of the disease. Other notable fungal infections involving the respiratory tract and the lungs include blastomycosis, coccidioidomycosis, and aspergillosis.

EMPHYSEMA

Pulmonary emphysema is a disease of the lungs characterized by abnormal enlargement of the air spaces and by destructive changes in the alveolar walls. Histologically, the alveoli are distended and disrupted, with a decrease in the amount of elastic tissue and in the number and size of vascular channels. Wheezing, chronic coughing, and labored breathing are prominent features. In the United States, the incidence of emphysema has increased alarmingly. According to one survey, 27% of men over 40 years old demonstrate evidence of chronic obstructive disease of the airway. The incidence of the disease is four to five times greater among men than among women. Heavy cigarette smoking is the usual history. Symptomatic benefit and measurable improvement in pulmonary function usually follow intensive therapy for airway obstruction. This is accomplished best by the use of bronchodilators and intermittent positive pressure breathing. Above all, the patient should not smoke and should avoid all respiratory irritants.

BRONCHIAL ASTHMA

Bronchial asthma is a distressing disease marked by recurrent attacks of paroxysmal dyspnea, wheezing, and coughing due to spasmodic contraction of the bronchi and bronchioles. An acute attack leads to hypoxic anoxia and often sudden death. The cause usually can be attributed to such external factors as pollens, animal danders, lint, and the like ("allergic asthma") or to respiratory infections (infective allergy). Secondary factors that greatly influence the frequency and severity of attack include temperature, humidity, noxious fumes, fatigue, endocrine changes, and emotional stress. Specific treatment is directed to the basic cause, and symptomatic treatment includes the use of bronchodilators (epinephrine, isoproterenol, aminophylline, and others), corticosteroids, cough depressants, and sedatives.

ATELECTASIS

Atelectasis is the collapse, or failure of expansion, of a part or all of the lung. The condition may be acquired or congenital. Acquired atelectasis may result from intrapulmonary obstruction by secretions accumulated postoperatively or by tumors of the bronchi. Also, outside pressure (as from pleural effusions or pneumothorax) may cause atelectasis. The signs and symptoms depend upon the rapidity of onset; in acute episodes, dyspnea and cyanosis are cardinal signs. Treatment includes giving oxygen and correcting the underlying cause.

PLEURISY

Pleurisy is inflammation of the pleura commonly arising from some unrecognized subpleural infection; the condition often accompanies pneumonia. In the acute form, the membranes become reddened and often covered with a serous exudate; the inflamed surfaces tend to unite by adhesions. The symptoms are a "stitch" in the side, chills, fever, and dry cough. Chronic pleurisy is long continued and characterized by dry surfaces, purulent exudation, adhesions, and even calcification. Treatment is both specific (when the cause is known) and symptomatic (mainly relief of pain).

PNEUMOCONIOSES

Pneumoconioses are chronic fibrous reactions in the lung provoked by inhalation of dust particles. The lungs characteristically become hard and pigmented, but the precise clinical picture depends upon the nature of the dust. The more important dusts are coal, silica, asbestos, and beryllium. For the most part, treatment is symptomatic.

LUNG CANCER

Lung cancer is the leading cause of male cancer deaths, the rate now being 15 times what it was at the turn of the century. In round figures, there are about 75,000 new cases per year and some 70,000 deaths (55,000 among men). The statistical evidence against cigarettes is overwhelming, and at the laboratory level researchers have induced lung cancer in dogs using methods simulating human smoking. Common early features of this morbid involvement are coughing (often with blood), wheezing, hoarseness, and persistent fever; advanced features basically relate to the outright destruction of lung tissue and above all to metastases (regional lymph nodes, liver, bones, kidneys, brain, or heart). The diagnosis encompasses x-ray studies (Fig. 15-17), cytology, bronchoscopy, thoracotomy, and always biopsy. The only curative treatment—in curable cases—is the removal of a lobe or an entire lung. Radiation is helpful preoperatively and in the management of the inoperable patient. Only about 5% of all cases are curable because of the advanced standing of the tumor upon diagnosis; metastases underscore a negative prognosis. Cancers confined to the lung are more hopeful; almost half the patients who have such cancers and are operated on live up to 5 years longer, or more.

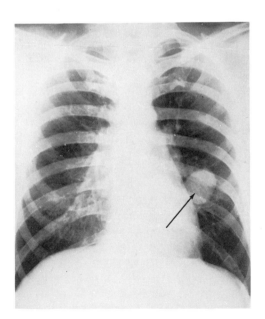

FIG. 15-17
Well-delineated lesion (arrow) of lung that proved to be cancerous. (From Maier, H. C., and Fischer, W. W.: Adenomas arising from small bronchi not visible bronchoscopically, J. Thorac. Surg. **16**:392-398, 1947).

QUESTIONS

1. Discuss the various meanings of respiration.
2. What bones compose the nasal septum?
3. Distinguish among conchae, turbinates, meati, and nares.
4. Precisely, why does a stuffy nose change the voice?
5. What function does respiratory mucus serve?
6. A severe pharyngitis may lead to otitis media. Explain.
7. What "mechanical problem" can enlarged adenoids cause?
8. In ordinary usage, what particular glands do "the tonsils" refer to?
9. As one ascends in an elevator, do the eardrums tend to push in or push out? Why?
10. Is the windpipe anterior or posterior to the esophagus?
11. Distinguish between glottis and epiglottis.
12. Compare the false vocal cords and the true vocal cords.
13. Discuss the role of the vocal cords, nose, nasal sinuses, and pharynx in phonation.
14. What is the length of the trachea, in centimeters?

15. The rings of the trachea are composed of cartilage. Why cartilage?
16. An object swallowed into the trachea is more likely to end up in the right bronchus than the left. Why?
17. Compare the trachea and the bronchi.
18. Describe the bronchial tree.
19. Distinguish between tracheostomy and tracheotomy.
20. In going from air to the blood, oxygen must pass through what two barriers?
21. Distinguish between alveolus and alveolar sac.
22. Would a decrease in the surface tension of mucus tend to increase or decrease the size of an alveolus?
23. The lungs may be regarded as "great ramifications of the bronchi." Explain and elaborate.
24. What are the pleural cavities?
25. "The intrapleural pressure refers to the pressure in the potential space between the visceral and parietal pleura." Discuss the meaning of "potential" here.
26. Distinguish between intrapleural pressure and intrapulmonic pressure.
27. Upon inspiration, the intrapleural pressure decreases. Is this the same as saying that the intrapleural pressure becomes more negative?
28. What two gas laws are involved in external respiration?
29. What is pneumothorax?
30. What are reduced and oxidized hemoglobin?
31. Write the equation for the reaction between carbon dioxide and water.
32. Write the equation for the reaction between reduced hemoglobin and oxygen.
33. What anions are involved in the chloride shift?
34. The partial pressure of oxygen averages about 100 mm Hg in alveolar air, compared to about 160 mm Hg in atmospheric air. Explain this difference.
35. Based on the figure of 100 mm Hg (above), what is the percent oxygen in alveolar air?

36. The actual transport of CO_2 and O_2 is more a matter of chemistry than of physics. Discuss.
37. The pressure of CO_2 is higher in tissue fluid than in blood plasma. Why?
38. About what percent of the CO_2 in the blood is carried by the red cells?
39. When HCl is added to blood plasma, CO_2 is liberated. Write the equation for this reaction.
40. What muscles are involved in the Hering-Breuer reflex?
41. What are chemoreceptors?
42. Locate and describe the carotid body.
43. In regard to cause, there are two basic kinds of respiratory paralysis: peripheral and central. Illustrate through the examples of barbiturate poisoning on one hand and curare poisoning on the other.
44. Explain the use of succinylcholine and other curare-like drugs in surgery.
45. What is anoxemia?
46. Hypoxia and anoxia are often used synonymously. Discuss.
47. What is hypercapnia?
48. Tell precisely how you would measure your tidal air, inspiratory reserve, and expiratory reserve using the spirometer.
49. At the slaughterhouse, the lungs are referred to as "lights." Why?
50. What is Valsalva's maneuver?
51. What effect does Valsalva's maneuver have upon the circulation?
52. "The loss of CO_2 in effect amounts to a loss of H_2CO_3." Illustrate this point by means of a chemical equation.
53. In respiratory acidosis, hypoventilation is primary; in metabolic alkalosis, hypoventilation is secondary. Elaborate.
54. In respiratory alkalosis, hyperventilation is primary; in metabolic acidosis, hyperventilation is secondary. Elaborate.
55. Indicate the effects of respiratory acidosis and respiratory alkalosis on the following blood (laboratory) values: pH, CO_2 content, and Pco_2.

56. The daily water requirement is about 1 l over and above the volume of fluid needed to balance the urinary output. Account for this figure.

57. The laboratory based its diagnosis of pneumococcic pneumonia on a sputum specimen. Explain what the laboratory did to arrive at the diagnosis.

58. What is bronchopneumonia?

59. What is the immediate threat to life in all pneumonias?

60. The common cold is caused by a virus. Elaborate.

61. There is no specific treatment of the common cold and no means of protection. Elaborate fully.

62. "Type A viruses continue to be the major [influenza] problem because of their propensity to mutate." Discuss.

63. In flu epidemics, what is the usual immediate cause of death?

64. Fully discuss influenza vaccination. Be sure to include the composition of the vaccine and the duration of protection.

65. *Mycobacterium tuberculosis* is an acid-fast bacillus. What does this mean?

66. What is the major disadvantage of all drugs used in the treatment of tuberculosis? How is this, to some degree, circumvented?

67. How is the tuberculin test done, and what is meant by a positive and by a negative reaction?

68. *Corynebacterium diphtheriae* is a pleomorphic, non-spore-forming, nonmotile, encapsulated, gram-positive bacillus. Discuss this statement.

69. What is diphtheria toxoid?

70. Compare the Schick and the Dick tests.

71. For the immediate protection of exposed, diphtheria-susceptible individuals, the antitoxin is used. Why not the toxoid?

72. Compare the etiologies and treatments of psittacosis and histoplasmosis.

73. Discuss in some detail the treatment of emphysema and bronchial asthma.

74. The most common lung cancer is squamous cell carcinoma. Briefly discuss this type malignancy.

75. What is anthracosis?

CHAPTER 16

THE DIGESTIVE SYSTEM

The digestive system (Fig. 16-1) includes the gastrointestinal tract and associated organs and glands; its purpose is to convert carbohydrates, fats, and proteins into molecules capable of being absorbed into the blood. We shall discuss the system's structure, the digestive process proper, and certain major disorders and diseases affecting the system.

MOUTH

The mouth (Fig. 16-2) begins the gastrointestinal tract. It is bounded on the sides by the cheeks, above by the hard and soft palates, and below by the tongue. In front, the mouth is opened and closed by the lips, and in the back it opens into the pharynx. Characteristically, the soft palate ends behind as a free medial projection called the *uvula*. On the outside, the lips and the cheeks are covered with skin; on the inside, they are lined with mucous membranes. The palatine or faucial tonsils are situated laterally in crypts formed by the glossopalatine and pharyngopalatine arches, muscular folds continuous with the soft palate. The arches aid the soft palate in closing off the mouth from the pharynx during mastication.

Tongue

The tongue, a highly agile muscular organ, is attached in back to the hyoid bone and the epiglottis, and underneath to the floor of the mouth by an anteroposterior mucosal fold called the *frenulum linguae*. The tongue is covered by a mucous membrane that is continuous with the lining of the mouth and thrown up into a number of papillae on the upper surface. Most of these papillae are slender, threadlike projections called filiform papillae. Some are knoblike, or fungiform, and the eight or so arranged on the back part of the tongue in a V-shaped line are called vallate papillae. Distributed over the surface of the tongue, but most conspicuously on the sides of the vallate papillae, are microscopic, barrel-shaped structures, called *taste buds,* that serve as the receptors of taste. The nerve fibers supplying the buds carry sensory impulses to the taste areas in the brain via the seventh and ninth cranial nerves. In order for a substance to trigger a taste bud, it must diffuse through the taste pore in solution, which explains in part why such readily soluble foods as sugar and salt produce a most profound effect. Each taste bud is sensitive to one of the primary taste sensations: bitter, sweet, sour, and salty, and, according to some authorities, alkaline and metallic.

Teeth

The teeth are the bony structures of mastication. The deciduous or milk dentition consists of 20 teeth, 10 in each jaw; the permanent denti-

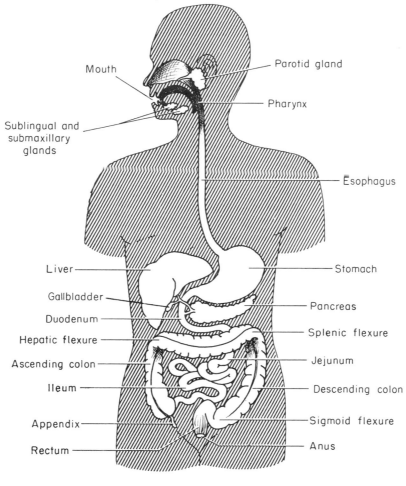

FIG. 16-1
Digestive system.

tion contains 32 teeth, 16 in each jaw. The teeth are divided into four groups: incisors, canines, premolars, and molars. The incisors, with their sharp, chisel-shaped crowns, are the cutting teeth. The canines, with their sharp, pointed crowns, are used in tearing; the premolars and molars are grinders. Whereas the incisors and canines each have a single root, the molars have two or three. The premolars usually have but one root. Table 16-1 shows the number and average eruption time of the four types of teeth in the two dentitions.

The basic anatomy of a tooth is shown in Fig. 16-3. The crown, the portion above the gum, is coated with enamel, the hardest substance in the body. Beneath this, and making up the bulk of the tooth, is the dentin, a dense, yellow-white, hard, striated material. The cavity within the

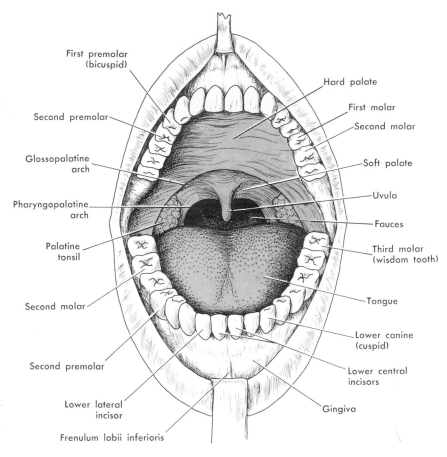

First premolar (bicuspid)

Second premolar

Glossopalatine arch

Pharyngopalatine arch

Palatine tonsil

Second molar

Second premolar

Lower lateral incisor

Frenulum labii inferioris

Hard palate

First molar

Second molar

Soft palate

Uvula

Fauces

Third molar (wisdom tooth)

Tongue

Lower canine (cuspid)

Lower central incisors

Gingiva

FIG. 16-2
Oral cavity.

TABLE 16-1
Time of eruption of teeth*

Deciduous teeth	Months	Permanent teeth	Years
Lower central incisors	6- 8	First molars	6
Upper central incisors	9-12	Central incisors	7
Upper lateral incisors	12-14	Lateral incisors	8
Lower lateral incisors	14-15	First premolars	9-10
First molars	15-16	Second premolars	10
Canines	20-24	Canines	11
Second molars	30-32	Second molars	12
		Third molars	17-18

*From Francis, C. C: Introduction to human anatomy, ed. 6, St. Louis, 1973, The C. V. Mosby Co.

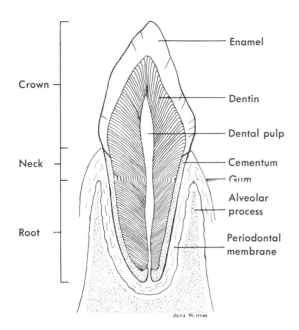

FIG. 16-3

Basic anatomy of a tooth. (From Francis, C. C: Introduction to human anatomy, ed. 6, St. Louis, 1973, The C. V. Mosby Co.)

dentin, the pulp chamber, contains connective tissue, nerve endings, and blood vessels. The root, the portion of the tooth below the gum, is anchored to the alveolar process of the jawbone by the alveolar periosteum or periodontal membrane. The cementum is a layer of modified bone forming a sheath for the root.

Like bone, the teeth are composed of calcium carbonate and calcium phosphate bound together into a hard crystalline substance. The deposition and the resorption of this chemical complex occur almost exclusively in the dentin and cementum. The mineral turnover in the enamel is almost nil. The speed of growth and the rate of eruption of the teeth are stimulated by the thyroid and growth hormones (Chapter 23). Deposition of the calcium salts in the dentin depends upon the supply of calcium, phosphorus, fluorine, vitamin D, and the parathyroid hormone. A deficiency of one or more of these leads to defective ossification and poor teeth.

SALIVARY GLANDS

The salivary glands are exocrine glands arranged in pairs about the face (Fig. 16-1). They include the parotid, the sublingual, and the submaxillary glands. The parotid gland lies immediately below and in front of the ear. Its duct, known as Stensen's duct, pierces the buccinator muscle and opens in the mouth at the level of the upper second molar. The submaxillary, a smaller gland than the parotid, is situated below the mandible about midway between the angle and the end of the lower jaw. Its secretions enter the mouth via Wharton's duct, which opens in the floor of the mouth near the midline. The sublingual glands lie in the floor of the mouth on either side of the tongue. Each gland communicates with the mouth via several small ducts that open into the floor behind Wharton's ducts.

Saliva

Saliva, the clear, colorless, sticky juice secreted by the salivary glands, facilitates mastication and swallowing by its moistening and lubricating effects; it also begins the digestion of carbohydrate. Saliva is composed of water, mucin, ptyalin, and certain inorganic salts. Its pH averages between 6 and 7.

Mucin, the characteristic constituent of mucus and digestive juices, is an amazing substance; it not only serves as an excellent lubricant but also protects the gastrointestinal mucosa against the eroding action of the potent digestive enzymes. Ptyalin, or salivary amylase, is the only enzyme present in saliva. Its action will be discussed in connection with the digestive process.

The secretion of saliva is provoked by the presence of food in the mouth (gustatory stimuli) and by seeing, smelling, and thinking about food (psychic stimuli). Such stimuli trigger the superior and inferior salivatory nuclei in the brain which, in turn, via the autonomic nervous system, stimulate the glands.

PHARYNX

The pharynx, or throat, a musculomembranous structure connecting the nose, the mouth, and the esophagus, serves as a passageway for air and food. It leads air into the larynx and food into the esophagus. The structure of the pharynx was discussed in conjunction with the respiratory system (p. 223).

ESOPHAGUS

The esophagus is a muscular tube, 10 to 12 inches long, running behind the trachea from the lower pharynx (at the level of the sixth cervical vertebra) to the cardiac orifice of the stomach. On the way, it passes through the posterior mediastinum and pierces the diaphragm just before the cardiac orifice. The esophagus, as well as the rest of the alimentary tract, is composed of four coats. From inside out, these include the mucosal, submucosal, muscular, and serous coats. The muscular coat has an inner and an outer layer of fibers arranged in a circular and a longitudinal fashion, respectively. In the

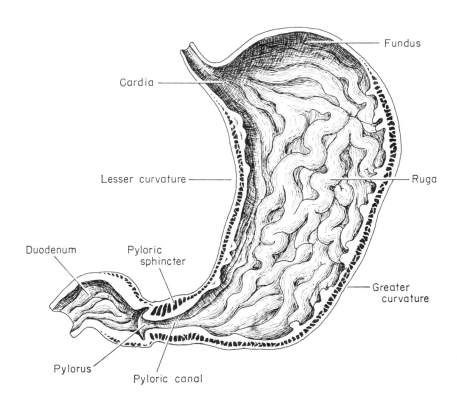

FIG. 16-4

The stomach. (Modified from Francis, C.C: Introduction to human anatomy, ed. 6, St. Louis, 1973, The C. V. Mosby Co.)

upper esophagus, the muscle is of the skeletal type; in the lower esophagus, it is smooth. In cross section, the lumen of the esophagus has a stellate outline as a result of the partial contraction of its musculature. During swallowing, however, the puckered wall is distended. Mucus is the only secretion of the esophagus.

STOMACH

The stomach is an ovoid, musculomembranous pouch connected above to the esophagus and below to the duodenum. As shown in Fig. 16-4, the major areas include the cardia, the fundus, the body, and the pylorus. The cardia and the pylorus lead into the esophagus and the duodenum, respectively. The fundus is the dome-shaped portion above the cardia, and the body is the portion between the cardia and the pylorus. Along the lesser curvature or concave margin of the stomach (Fig. 16-4) is attached the lesser omentum, a peritoneal fold running to the transverse fissure of the liver. Along the greater curvature is attached the greater omentum, an apron-shaped double fold of peritoneum that hangs down loosely over the intestines. Within, the healthy empty stomach has a small cavity and a thick, tough wall thrown up into deep wrinkles, or *rugae*. This telescoped surface permits great distention; the adult stomach can accommodate a meal with a volume of 3 to 4 l. The stomach wall has four coats: an outer peritoneal or serous coat, a muscular coat, a submucosal coat (submucosa), and a mucous coat (mucosa) that lines the inside. The muscular coat is composed of an outer layer of longitudinal fibers, a middle layer of circular fibers, and an inner layer of oblique fibers; circular fibers compose the pyloric sphincter.

Gastric juice

Gastric juice contains water, mucin, salts, enzymes, and hydrochloric acid, and is secreted by three types of mucosal cells working in con-cert. The parietal cells, located in the fundus and the body of the stomach, secrete hydrochloric acid; the mucous cells, located throughout the stomach, secrete mucin; and the zymogenic, or chief, cells, also found throughout the stomach, secrete the enzymes. During fasting, the juice has a pH of about 3.5, but following stimulation of secretion by food, it has a pH of about 1. The principal enzyme is pepsin. It is not secreted as such but as a zymogen, called pepsinogen, that becomes activated by hydrochloric acid. Gastric lipase and rennin, two other gastric enzymes, are relatively unimportant. The former has a slight action upon fat, and the latter causes milk protein to coagulate and therefore to stay in the stomach longer. Rennin is almost absent from adult gastric juice.

The flow of gastric juice is regulated by neurogenic and hormonal mechanisms. Secretory signals are transmitted reflexly to the gastric glands from the brain as a consequence of both psychic and physical factors. Meat and other protein foods when in the stomach provoke the release of a hormone called *gastrin* that stimulates the secretion of a highly acidic gastric juice. Interestingly, histamine, an intracellular chemical agent associated with allergy (p. 121), also stimulates the output of gastric juice.

Mixing and emptying

In stark contrast to the idealized architecture shown in Fig. 16-4, the stomach can contort itself into an endless variety of shapes in the course of mixing a swallowed meal; at any given moment, its shape depends upon the contents, the muscle tone, and the position of the body. Mixing is performed by closing the pyloric sphincter in the face of the forward movement of powerful peristaltic waves. In the presence of copious amounts of gastric juice, the meal is transformed into a semifluid mush called chyme. Chyme enters the duodenum in intermittent squirts, the frequency of which depends upon

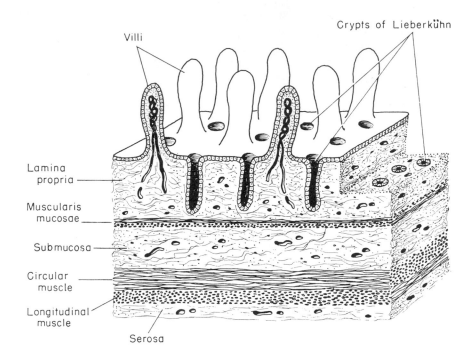

Villi

Crypts of Lieberkühn

Lamina propria

Muscularis mucosae

Submucosa

Circular muscle

Longitudinal muscle

Serosa

FIG. 16-5
Three-dimensional view of wall of small intestine.

the acidity and the volume ejected. As the volume of chyme builds up in the intestine, the walls become distended, and the myenteric receptors are thereby stimulated; reflexly, this causes contraction of the pyloric sphincter and closing of the pylorus. In time, when distention has subsided and the acid has become neutralized, the reflex becomes nonoperative, thus allowing the pylorus to open and more chyme to be ejected. Also of importance in the emptying process is the chyme-stimulated release by the duodenum of enterogastrone, a hormone that inhibits gastric motility. Fat in the chyme is considered the most potent stimulant of enterogastrone.

SMALL INTESTINE

The small intestine is the musculomembranous tube (Fig. 16-1) that extends from the pylorus to the ileocecal valve. It has three some-

what arbitrary divisions: the duodenum, the jejunum, and the ileum. The intestine begins with the duodenum, a C-shaped tube about 10 inches in length that curves toward the right and in its descending portion receives the ducts from the liver and the pancreas. The jejunum and the ileum form the main body of the small intestine and measure about 8 and 11 feet, respectively. The jejunum occupies the upper abdominal cavity; the ileum fills the lower right abdominal cavity and the upper pelvic cavity.

The small intestine, like the other portions of the digestive tract, has four basic coats. The inside coat, or mucosa (Fig. 16-5), consists of a surface layer of simple columnar epithelium supported by a thin layer of connective tissue, called the lamina propria, and below this a very thin layer of smooth muscle called the muscularis mucosae. The mucosa characteristically sends up an astronomical number of fingerlike

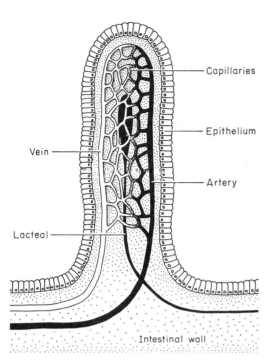

FIG. 16-6

The villus. (From Brooks, S. M.: Basic facts of general chemistry, Philadelphia, 1956, W. B. Saunders Co.)

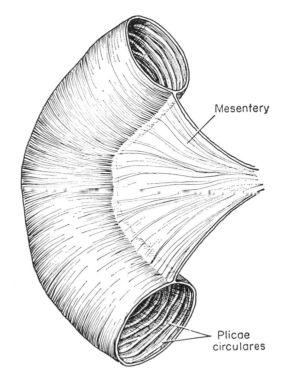

FIG. 16-7

Segment of small intestine.

projections called villi, structures that give the inside surface its soft, velvety texture. Each villus is supplied with a capillary network and a large lymphatic vessel called a lacteal (Fig. 16-6). The entire mucosa is thrown up into a large number of circular folds called plicae circulares (Fig. 16-7) that, like the villi, gradually decrease toward the end of the intestine. Large, oval, elevated areas (Peyer's patches) in the mucosa of the lower ileum are composed of lymph nodules closely packed together. In contrast, the nodules in the other sections of the intestine are solitary.

Beneath the mucosa lies the submucosa, which is composed of a layer of strong con-nective tissue. Next comes the muscular coat composed of an inner circular layer and an outer longitudinal layer of smooth muscle. The outer-most coat, the serosa, is made up of transparent tissue continuous with the mesentery (Fig. 16-7), a fan-shaped fold of peritoneum that attaches the entire length of the jejunum and ileum to the posterior abdominal wall. Between the two layers of the mesentery run blood vessels, lymphatic vessels, and nerve fibers to the intestine.

Intestinal juice

Among the villi are openings that lead into microscopic tubular structures called the glands (or crypts) of Lieberkühn (Fig. 16-5). The modified epithelium lining these glands secretes the intestinal juice, or succus entericus. This secretion contains water, mucin, carbohydrases, peptidases, and enterokinase. Brunner's glands, lying deep in the duodenal mucosa, secrete large amounts of mucus to counteract the highly acidic chyme. In a 24-hour period, total intestinal secretion amounts to about 3 liters of fluid! This volume, plus about 5 liters of juice contributed by the mouth, stomach, liver, and pancreas, is almost entirely reabsorbed before reaching the large intestine.

Intestinal juice is secreted in response to a number of factors, but the myenteric reflexes, triggered by distention and acidity, are considered the most significant. Hormonal agents, collectively called enterocrinin, are thought to be important in determining the character of the juice; protein increases the concentration of peptidases, and carbohydrate increases the concentration of carbohydrases. Secretion of intestinal juice is also weakly enhanced by parasympathetic stimulation.

Peristalsis

Peristalsis is the wormlike movement of the gastrointestinal tract by which ingested material is moved along from the pharynx to the anal opening. This movement results chiefly from a slowly advancing circular constriction. Intestinal peristalsis does more than propel, however, for by swishing the chyme back and forth, it brings about thorough mixing.

Throughout the gastrointestinal tract, the musculature is supplied with ganglia (groups of nerve cells) and nerve fibers. This myenteric plexus controls practically all peristaltic movements. The effects may be local (for example, irritation along a small segment of the intestine causes contraction and secretion only in that area) or generalized due to impulses arriving at the ganglia over the autonomic nervous system. Parasympathetic stimulation increases peristalsis and relaxes the sphincters guarding the pyloric and ileocecal valves, whereas sympathetic stimulation decreases peristalsis and constricts the sphincters.

LARGE INTESTINE

The large intestine is a musculomembranous tube, 5 feet or so in length, which runs up, across, and down the lower abdominal cavity. In order, its various sections include the cecum, the ascending colon, the hepatic flexure, the transverse colon, the splenic flexure, the descending colon, the sigmoid flexure, the rectum, and the anal canal (Fig. 16-1). In contrast to the small intestine, the large intestine has no villi and an incomplete layer of longitudinal smooth muscle arranged in three ribbonlike strips called taeniae coli. Characteristically, its walls are puckered into incomplete sacculations called haustra (sing., haustrum). These characteristic features are shown clearly in Fig. 16-8.

Cecum and appendix

The cecum is the dilated intestinal pouch into which open the ileum and the vermiform appendix (Fig. 16-8). The ileocecal valve guards the opening of the ileum. When a peristaltic wave reaches the valve, the sphincter is relaxed reflexly, and the chyme is pushed on through. The valve is constructed in such a way that fecal matter is kept in the colon. The vermiform appendix is the blind wormlike tube branching from the lower cecum. It averages about ¼ inch in diameter and varies anywhere from 3 to 6 inches in length. Microscopically, the structure resembles the rest of the intestine. Presumably, the appendix serves little useful purpose. It is regarded generally as a vestigial organ.

Rectum

The terminal part of the large intestine, extending from the sigmoid flexure to the anus, is called the rectum. In the adult it measures from

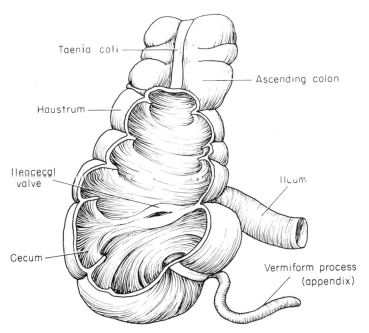

Taenia coli

Ascending colon

Haustrum

Ileocecal valve

Ileum

Cecum

Vermiform process (appendix)

FIG. 16-8
The cecum and associated structures.

6 to 8 inches in length. In the terminal inch, referred to as the anal canal, the mucosa is arranged in vertical folds known as rectal columns, each of which is supplied with an artery and a vein. The anus, or external opening, is guarded by two sphincters, an internal one composed of smooth muscle and an external one composed of skeletal muscle.

Feces and defecation

The major role of the large intestine is to absorb water and electrolytes from the soupy chyme. As the chyme is propelled along, it becomes progressively less watery and eventually becomes a semisolid material called feces, an excrement composed of food residues (for example, cellulose), bacteria, and intestinal secretions. The characteristic chocolate color of feces is due to the pigments formed by the action of the digestive enzymes on bile.

Defecation, the discharge of feces, is initiated by the passage of fecal material into the rectum. Sensory impulses are relayed to the spinal cord and peristalsis is stimulated; simultaneously, the internal anal sphincter is relaxed to aid passage through the anal opening. The external sphincter is then relaxed voluntarily to complete the act. The abdominal muscles, too, play a role in defecation, for when contracted, they compress the intestinal contents downward against the rectum and stimulate the sensory endings in the rectum. In this manner, defecation is both initiated and promoted.

LIVER

The liver (Fig. 16-9), the largest organ in the body, is situated beneath the diaphragm and occupies almost all the right hypochondrium and the upper part of the epigastrium. It is dark red, has a plasticlike texture, and is divided by the falciform ligament into left and right lobes. The right lobe in turn is divided into the right

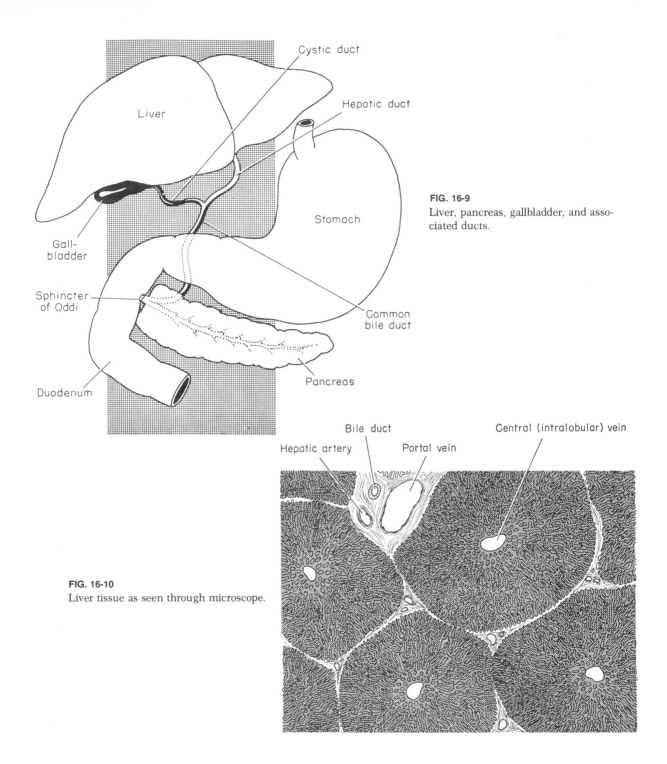

FIG. 16-9
Liver, pancreas, gallbladder, and associated ducts.

Cystic duct

Hepatic duct

Liver

Stomach

Gall-bladder

Sphincter of Oddi

Common bile duct

Duodenum

Pancreas

Bile duct

Hepatic artery

Portal vein

Central (intralobular) vein

FIG. 16-10
Liver tissue as seen through microscope.

lobe proper, the caudate lobe, and the quadrate lobe. Histologically, the liver is made up of anatomic units called hepatic lobules, each characterized by a central or intralobular vein (a branch of the hepatic vein) surrounded by radiating columns of hepatic cells (Fig. 16-10). Surrounding each lobule are tiny arteries (branches of the hepatic artery), interlobular veins (branches of the portal vein), and interlobular bile ducts (branches of the hepatic bile duct). Irregular blood channels, or sinusoids, extending between the radiating columns of hepatic cells connect the interlobular veins with the central vein.

Duct system

The interlobular bile ducts within the liver come together to form two large ducts that emerge from the undersurface and immediately join to form the hepatic duct. This duct and the cystic duct from the gallbladder then merge to form the common bile duct, which opens into the duodenum at the papilla, or ampulla, of Vater, a small raised area about 3 inches below the pylorus. The cardinal features of this duct system are shown clearly in Fig. 16-9.

Liver function and bile

Chemically, the liver is perhaps the most amazing organ in the body. It plays a key role in the metabolism of carbohydrates, fats, and proteins; it stores vitamins and minerals; it detoxifies harmful substances; it produces large amounts of body heat (second only to skeletal muscle); it manufactures, among other things, antibodies, blood proteins, and bile. The phagocytes (Kupffer cells) that line the sinusoids are actively engaged in the destruction of microbes and other foreign matter.

Bile, the product of the liver concerned with digestion, is a yellowish green secretion containing minerals, mucin, cholesterol, bile salts, and the pigments bilirubin and biliverdin. Bilirubin, a red pigment, is the catabolic product resulting from the breakdown of hemoglobin; biliverdin, a green pigment, is a derivative of bilirubin. Secretin, a hormone released into the blood by the duodenal mucosa in response to the presence of chyme, weakly stimulates the production of bile but has a more pronounced effect on the pancreas. The role of bile in the digestive process is to emulsify fat into microscopic globules. This action, due entirely to the presence of the bile salts, allows fat to be acted upon more readily by pancreatic lipase.

GALLBLADDER

The gallbladder, a pear-shaped pouch 3 to 4 inches long and about 1 inch wide, is partially embedded in the undersurface of the liver (Fig. 16-9). Its walls are composed of smooth muscle, and it is lined on the inside with a mucosa thrown up into rugae. Bile enters the gallbladder via the cystic duct (Fig. 16-9) and is stored and concentrated. When chyme containing fat enters the duodenum, it causes the release of a hormone called cholecystokinin, which makes its way to the gallbladder via the blood and there brings about the ejection of bile by stimulating the contraction of the muscular wall. At the same time, peristalsis, enhanced by the presence of chyme, stretches the sphincter of Oddi (at the ampulla of Vater) and allows the passage of bile into the duodenum.

PANCREAS

The pancreas is a pinkish, fish-shaped gland running behind the stomach and horizontally across the posterior abdominal wall (Fig. 16-1). Its head and neck are nestled in the C-shaped curve of the duodenum, and its tail just touches the spleen. The organ measures about 9 inches in length and weighs about 3 ounces. Upon careful dissection and microscopic examination, the pancreas is found to be divided into lobes and lobules that are composed of enzyme-secreting cells arranged about microscopic ducts. These ducts unite into larger ducts that in turn join the duct of Wirsung, the chief pancreatic duct, which extends the entire length of the gland and empties into the duodenum at the sphincter of Oddi (Fig. 16-9). Another duct, the duct of Santorini, is usually found arising from the head of the pancreas and entering the duodenum about 1 inch above the papilla.

Scattered throughout the pancreas and among the enzyme-secreting cells are a million or so cell clusters situated about blood capillaries. These areas, called the islands (or islets) of Langerhans, are seen easily under the microscope. The beta cells of these islets secrete the hormone insulin; the alpha cells secrete the hormone glucagon (Chapter 23). Whereas the lobule cells secrete into ducts, the alpha and beta cells have no ducts and secrete directly into the blood. Thus, the pancreas is both an exocrine and an endocrine gland.

The pancreas secretes about a liter of juice per day into the duodenum. The juice contains principally pancreatic amylase for digesting starches, pancreatic lipase for digesting fats, trypsin and chymotrypsin for digesting proteins, and variable amounts of sodium bicarbonate for neutralizing the hydrochloric acid from the stomach. The acidic chyme causes the release of the hormone secretin from the duodenal mucosa. Secretin is taken via the blood to the pancreas, where it stimulates a copious flow of juice containing large amounts of sodium bicarbonate. Also, chyme causes the mucosa to release pancreozymin. Unlike secretin, this hormone stimulates the secretion of a pancreatic juice containing little bicarbonate but high concentrations of digestive enzymes.

CARBOHYDRATE DIGESTION

In order for carbohydrates to pass into the villi and enter the circulation, they must be broken down into simple sugars or monosaccharides. Since polysaccharides (namely, starches,

dextrins, and pectins) are polymers of the monosaccharide glucose, they yield glucose as the end product of digestion. Sucrose, the principal sugar of the diet, and lactose, the sugar in milk, also yield glucose. Clearly, then, glucose is the chief end product of carbohydrate digestion. On the average, about 80% of the monosaccharide output in the digestive mill is glucose; the remainder is made up almost entirely of fructose (from sucrose) and galactose (from lactose).

Salivary digestion

In the mouth, food is mixed with saliva, and the digestive process begins, with salivary amylase, or ptyalin, the only enzyme present in saliva, working on the starches. Although most starches are not substantially altered during their short stay in the mouth, the continued action of the enzyme in the stomach converts about 40% of their molecules into the disaccharide maltose. The starch molecule is successively split apart by water (hydrolysis); the enzyme catalyzes the reaction. Before maltose is reached, however, the molecule yields intermediates called dextrins. The salivary digestion of starch, then, may be summarized thus:

$$(C_6H_{10}O_5)_x + H_2O \xrightarrow[\text{amylase}]{\text{Salivary}} \text{Dextrins} \xrightarrow[\text{amylase}]{\text{Salivary}} C_{12}H_{22}O_{11}$$
Starch   Maltose

Gastric digestion

There are no gastric enzymes that act upon carbohydrates, but hydrochloric acid no doubt converts some starch into maltose. Like salivary amylase, it does this by catalytic action. In the presence of heat, hydrochloric acid is extremely adept at completely breaking down the starch molecule into glucose. This can be demonstrated easily in the laboratory.

Intestinal digestion

The bulk of carbohydrate digestion is carried out in the intestine. Starches and dextrins are broken down into maltose by pancreatic amylase, and all disaccharides are converted into monosaccharides by the appropriate carbohydrases. Maltase splits a molecule of maltose into two molecules of glucose; sucrase splits a molecule of sucrose into one molecule of glucose and one molecule of fructose; lactase splits a molecule of lactose into one molecule of glucose and one molecule of galactose. These reactions may be summarized as follows:

$$(C_6H_{10}O_5)_x + H_2O \xrightarrow[\text{amylase}]{\text{Pancreatic}} \text{Dextrins}$$
$$\text{Starch}$$

$$(C_6H_{10}O_5)_x + H_2O \xrightarrow[\text{amylase}]{\text{Pancreatic}} C_{12}H_{22}O_{11}$$
$$\text{Dextrins} \qquad\qquad \text{Maltose}$$

$$C_{12}H_{22}O_{11} + H_2O \xrightarrow[\text{Lactase}]{\substack{\text{Maltase}\\\text{Sucrase}}} 2C_6H_{12}O_6{}^*$$

Maltose	Glucose
Sucrose	Fructose
or	or
Lactose	Galactose

FAT DIGESTION

Fats and oils are glyceryl esters of fatty acids. The most important of these acids are stearic ($C_{17}H_{35}COOH$), palmitic ($C_{15}H_{31}COOH$), oleic ($C_{17}H_{33}COOH$), linoleic ($C_{17}H_{31}COOH$), linolenic ($C_{17}H_{29}COOH$), and arachidonic ($C_{19}H_{31}COOH$). The diet contains a small amount of other lipids, the most important of which are perhaps the fatty acid esters of certain sterols, especially cholesterol. Before pancreatic lipase (steapsin) can effectively act upon fats and oils, they first must be emulsified by bile into tiny globules to increase their surface area. In the absence of emulsification, most of the fats and oils leave the body in an undigested state.

---

*These sugars are isomers and thus have the same empirical formula.

Once emulsified, fats and oils are split by lipase into glycerol, fatty acids, and glycerides. Only about half of the fat molecules are converted completely into glycerol and fatty acids (the "true" end products); the rest are converted into glycerides. A glyceride is a molecule with one or two fatty acid residues and may be considered a "partial fat." The glycerides are apparently absorbed with almost the same speed as glycerol and the free fatty acids.

Using glyceryl tristearate (tristearin) as a typical example, the complete digestion of fat may be expressed as shown below:

$$
\begin{array}{l}
\phantom{C_{17}H_{35}COO-|-}3H \;\vdots\; 3OH \\[4pt]
C_{17}H_{35}COO-|-CH_2 \qquad\qquad C_{17}H_{35}COOH \quad HO-CH_2 \\
\phantom{C_{17}H_{35}COO-}| \\
C_{17}H_{35}COO-|-CH + 3H_2O \xrightarrow{\text{Steapsin}} C_{17}H_{35}COOH + HO-CH \\
\phantom{C_{17}H_{35}COO-}| \\
C_{17}H_{35}COO-|-CH_2 \qquad\qquad C_{17}H_{35}COOH \quad HO-CH_2
\end{array}
$$

Tristearin • Stearic acid • Glycerol

Note especially that one molecule of completely digested fat yields one molecule of glycerol and three molecules of fatty acid.

PROTEIN DIGESTION

During the process of digestion, proteins are broken down into amino acids, the building blocks of the protein molecule. Before these end products are reached, however, the various enzymes successively chop off a great many intermediate products of decreasing complexity: proteoses, peptones, and polypeptides, in that order. The simplest possible molecule before the amino acid stage is the dipeptide, one molecule of which, upon digestion, yields two molecules of amino acid. Generically, this may be expressed as shown at the top of p. 258.

Protein digestion begins in the stomach. Secreted pepsinogen is activated by hydrochloric acid into pepsin, which then converts protein into proteoses, peptones, and some polypeptides. In addition to activating pepsinogen, hydro-

$$R - \underset{\underset{H}{|}}{\overset{\overset{NH_2}{|}}{C}} - \overset{\overset{O}{\parallel}}{C} \mathrel{\vdots} \underset{\underset{H}{|}}{\overset{\overset{H}{|}}{N}} - \underset{\underset{H}{|}}{\overset{\overset{\overset{OH \mathrel{\vdots} H}{}}{R}}{C}} - \overset{\overset{O}{\parallel}}{C} - OH + H_2O \xrightarrow{\text{Dipeptidase}} R - \underset{\underset{H}{|}}{\overset{\overset{NH_2}{|}}{C}} - \overset{\overset{O}{\parallel}}{C} - OH + R - \underset{\underset{H}{|}}{\overset{\overset{NH_2}{|}}{C}} - \overset{\overset{O}{\parallel}}{C} - OH$$

Dipeptide   Amino acid   Amino acid

chloric acid provides the proper pH for optimum digestion. Upon entering the small intestine, the protein intermediates are acted upon by trypsin and chymotrypsin* from the pancreatic juice and converted into polypeptides of low molecular weights. In the final step, the intestinal peptidases† (namely, the dipeptidases and the aminopolypeptidases) split the peptides into amino acids.

ABSORPTION

Water, minerals, vitamins, and the products of digestion are absorbed into the intestinal villi by both passive (diffusion, osmosis, and the like) and active means. According to one view, in active transport the substance to be absorbed combines with some carrier agent in the epithelial cells and is transported in this form from the lumen into the villi. Once inside, it breaks away from the carrier and enters the capillaries.

Fats are absorbed as glycerol, fatty acids, glycerides, and, to some extent, undigested fat. Once through the mucosa, some of these molecules reunite into neutral fat, which then enters the lacteals. The release of a hormone called villikinin from the intestinal mucosa stimulates rhythmic contractions of the villi. The contractions massage the fat-laden lymph into the lymphatic vessels, and the lymph eventually enters the blood.

Chyle

The milky, fat-laden lymph emerging from the lacteals after digestion is called chyle. When examined under the microscope, chyle is seen to be a fine emulsion of fat globules, or *chylomicrons,* dispersed in lymph. Chylomicrons are also found in the blood for about 3 hours following a fatty meal, but normally will have disappeared in 8 to 10 hours unless additional food has been taken.

DENTAL CARIES

Dental caries, or tooth decay, is a common and, in many respects, major problem. The pathologic process involves the gradual disintegration and dissolution of the enamel, the dentin, and the pulp, in that order. Acid-producing microorganisms are chiefly responsible (including *Lactobacillus acidophilus* and certain anaerobic streptococci), and fermentable carbohydrates serve as the substrates for their enzymatic activities. Brushing the teeth immediately after eating effectively minimizes microbial action, as does drinking fluoridated water. For maximum benefit, the fluoride ion should be at a concentration of 1 ppm (one part per million parts water) and consumed during the time the teeth are being calcified. Fluoride-containing dentifrices are also beneficial. The protective action of fluoridation relates to the incorporation of the ion into the apatite crystal to form the less soluble hydroxyapatite. Excessive fluoride,

---

*Like pepsin, trypsin and chymotrypsin are secreted in inactive forms, called trypsinogen and chymotrypsinogen, respectively. The intestinal enzyme enterokinase activates trypsinogen to trypsin, and trypsin itself activates chymotrypsinogen to chymotrypsin.

†Formerly referred to as "erepsin."

however, is harmful to the teeth and causes mottling of the enamel.

MALABSORPTION SYNDROME

Impaired absorption of nutrients from the small intestine results in a varied and complex array of signs and symptoms, ranging from minor and subtle complaints like lassitude and fatigability, to major ones like bleeding and convulsions; hence, the expression malabsorption syndrome. Most cases, though, are characterized by a specific finding, the abnormal loss of fecal fat. Possible underlying causes of the syndrome are manifold. Established etiologies include, among many others, radiation injury, hormonal disorders, infection, liver disease, enzyme deficiencies, and a lack of the intrinsic factor. Treatment is symptomatic and, whenever possible, specific.

PEPTIC ULCER

Peptic ulcers are erosions of the mucosa exposed to pepsin and hydrochloric acid; most occur on the lesser curvature of the stomach and the first portion of the duodenum. The cause is obscure, but hypersecretion of gastric juice and emotional tension are clearly involved. The signs and symptoms include pain, heartburn, nausea, vomiting, anorexia, weight loss, diarrhea, and anemia. Perforation, massive hemorrhage, or obstruction (of the outlet of the stomach) arc grave complications. Medical treatment centers on mental and physical rest and the suppression of hyperacidity and gastric activity. If these measures fail, surgery must be considered.

DIVERTICULOSIS

Diverticulosis refers to the outpouchings (diverticula) that are prone to develop in weakened areas of the intestinal wall. The sigmoid colon is especially susceptible to the condition. When fully developed, diverticula are spherical pouches connected with the intestinal lumen by narrow necks; commonly these necks become obstructed and infected, giving rise to diverticulitis. Diverticulosis is often of little consequence and no treatment is required. Diverticulitis, on the other hand, calls for antibiotics, and in the event of complications, surgery is indicated. Common complications are abscess formation, intestinal obstruction, and hemorrhage. A congenital sacculation, located 1 to 5 feet proximal to the ileocecal valve, is called Meckel's diverticulum. This may become infected (Meckel's diverticulitis) and rupture (simulating acute appendicitis) or continue in a chronic form and cause obstruction. The treatment is surgical.

COLITIS

The term colitis is applied to inflammatory diseases of the colon; the major forms of the condition included amebic, bacillary, and ulcerative. Chronic ulcerative colitis involves the large intestine, the rectum and sigmoid being affected earliest and most severely; in one third or less of cases, the ileum also may be involved. First there is inflammation and edema of the mucosa, followed by necrosis and ulceration. Hemorrhage and perforation (with peritonitis) account for about 40% of the deaths. The incidence of colorectal cancer among patients with colitis is about ten times that of the rest of the population. The cause of ulcerative colitis is unknown, and there is no specific treatment. Resting the bowel as much as possible and providing supportive therapy is the best that can be done. If these measures fail, or if the condition grows progressively worse, the removal of the rectum and the colon (total proctocolectomy) is recommended.

CIRRHOSIS

Cirrhosis refers to a fibrosis or scarring of the liver that is progressive and not simply the stationary, healed, end stage of an injury. It is a chronic disease, and all parts of the liver are involved. There are several varieties of cirrhosis, which differ in etiology, nature, form, and effects. Accordingly, classifications and terminologies abound, and there is a fair amount of confusion. Generally agreed upon, however, are three histological or morphological forms: postnecrotic, biliary, and portal. Postnecrotic cirrhosis results from hepatotoxic drugs and chemicals and from viral hepatitis; biliary cirrhosis is associated with gallbladder disease; and portal cirrhosis, the best known form of the disease, is generally associated with alcoholism and malnutrition. The effects and complications of the disease include faulty liver function, obstruction of the portal circulation, gastrointestinal hemorrhage, and possibly cancer. Treatment is essentially symptomatic.

HEMORRHOIDS

Of the several annoying and painful anorectal disorders, hemorrhoids (piles) are the most common and perhaps the most excruciating. Hemorrhoids are varicosities of the veins situated beneath the mucosa of the anus and lower rectum; these vessels act as collateral connections between the systemic venous system and the portal system. Because the portal system is without valves, increased intra-abdominal pressure is transmitted through the system to these dependent veins—hence, the varicosities. Precipitating causes include pregnancy, straining at the stool, obesity, coughing, sneezing, portal hypertension (as in cirrhosis of the liver), and abdominal tumors. Also, as in all varicosities, there is apparently a hereditary element of susceptibility.

CANCER

Cancers of the gastrointestinal tract and associated organs and glands are a common threat to life. Colorectal cancers number about 75,000 new cases per year, and colorectal cancer deaths,

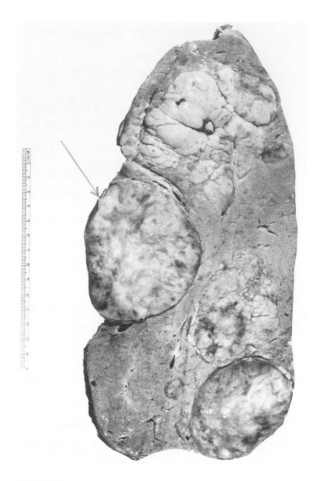

FIG. 16-11

Metastatic involvement of liver. Arrow points to one of four secondary growths. (Courtesy Dr. W. A. D. Anderson, Miami, Fla.)

some 50,000. About half of these cancers involve the rectum, and about a quarter, the sigmoid colon. Stomach cancer averages about 16,000 new cases per year, about half of what it was at the turn of the century. Other malignancies of special note involving the digestive system are cancers of the mouth and throat, the pancreas, the liver, and the gallbladder. In the United States, liver cancers are predominantly

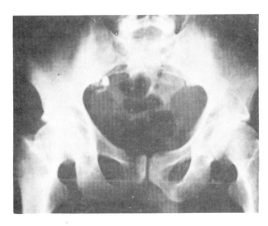

FIG. 16-12

X-ray film showing .22-caliber bullet lodged in appendix. Patient, who accidentally swallowed bullet, demonstrated signs and symptoms of appendicitis. (Courtesy Dr. M. W. H. Friedman and Dr. Walter C. MacKenzie.)

secondary or metastatic (Fig. 16-11). The causes are essentially unknown, but there is much evidence that tobacco is responsible for oral cancer, and the chemical benzpyrene, for stomach cancer. Benzpyrene, a proved carcinogen in animals, has been isolated from smoked meat and fish, which are dietary staples in Iceland and Japan, the countries leading the world in stomach cancer. Treatment of all the cancers cited centers for the most part on irradiation and surgery.

APPENDICITIS

Appendicitis, or inflammation of the vermiform appendix, varies from mild involvements to those with serious and fatal results. The full etiology is not known, but there is a general agreement that obstruction of the opening between the appendix and the cecum (Fig. 16-12) is the underlying cause. In most cases, the obstruction is thought to be a fecalith, or dried fecal concretion. According to the "obstruction theory," appendiceal mucus is prevented from emptying into the cecum, resulting in the build-

up of pressure of sufficient intensity to squeeze the walls and the blood vessels within and thereby cut off the blood supply. This, of course, lowers the resistance of the tissues and sets the stage for attack by fecal bacteria. Gangrene, inflammation, perforation (rupture), and peritonitis follow, typically in that order. Peritonitis is the classic cause of death. An acute attack of appendicitis calls for an immediate appendectomy.

FOOD POISONING

The most common cause of food poisoning is the enterotoxin produced by certain strains of *Staphylococcus aureus*. The usual events, which begin 2 to 4 hours following ingestion of the contaminated food, include nausea, severe vomiting, abdominal cramps, diarrhea, and prostration. Despite the violence of an acute attack, the symptoms usually subside after about 6 hours. Symptomatic treatment, with replacement of fluids, is generally all that is required. Inasmuch as "staph" food poisoning is due to a toxin, most authors nowadays refer to it as staph intoxication. The most common type of food poisoning due to the activities of living organisms (food infection) is *salmonellosis*, a gastroenteritis involving dozens of species of the genus *Salmonella*, the same genus to which the pathogens of typhoid and paratyphoid fever belong. Because many of these species are natural pathogens of domestic animals, meats, milk, eggs, and meat, milk, and egg products are the usual culprits in salmonellosis. Also, foods may become contaminated during processing by human carriers and by infected rats and mice that inhabit food plants, warehouses, and kitchens. The clinical picture is severe gastroenteritis; treatment involves replacement of water and electrolytes and administration of antibiotics.

The most deadly food poisoning, and fortunately the least common, is *botulism*. Botulism is caused by *Clostridium botulinum*, an anaerobic spore-forming bacillus that produces a highly lethal, paralytic exotoxin. Gastrointestinal effects are slight, and some victims experience none at all; the usual immediate cause of death is respiratory depression. The fatality rate may run over 50%, even with the best treatment. The antitoxin is of limited value once the signs and symptoms appear, but highly effective prophylactically. All exposed persons should be given trivalent antiserum immediately. Problem foods include sausages, smoked fish, mushrooms, and home-canned vegetables. Unless all spores are destroyed at the time of processing, they may give rise to the vegetative cells that in turn give rise to the exotoxin in the anaerobic environment of the can.

TYPHOID FEVER

Typhoid fever is a severe bacterial infection that starts in the intestinal mucosa and, if untreated, progressively infiltrates the tissues throughout the body. It is characterized by severe diarrhea (often with bloody stools), fever, abdominal pain, and prostration. The most serious aspect of the infection is the possibility of intestinal perforation, for once the organism enters the abdominal cavity, a fulminating peritonitis ensues. When this occurs, the prognosis is poor indeed. The etiologic microbe, *Salmonella typhosa,* is a motile, non-spore-forming, gram-negative bacillus that enters the body through the mouth via contaminated food or water. Treatment of typhoid fever entails good nursing care and the use of antibiotics. At present, chloramphenicol (Chloromycetin) is the drug of choice. For prevention, vaccination (using a "killed vaccine") is a must in endemic and disaster areas. Meticulous attention must be paid to the handling of food, water, garbage, and sewage if the malady is to be eradicated. Such measures in the United States have reduced the typhoid fever death rate from 40 per 100,000 in 1900 to less than 0.1 per 100,000 today.

Paratyphoid fever, an enteric infection similar to typhoid fever but much less severe, is caused by three species of *Salmonella* that are often spoken of collectively as *Salmonella paratyphi.* The modes of infection and the diagnosis, prevention, and treatment are about the same as those cited for typhoid fever.

BACILLARY DYSENTERY

The term dysentery is applied to a number of enteric inflammations that are marked by abdominal pain and severe diarrhea, often with stools streaked with blood and mucus. Bacillary dysentery, or shigellosis, is the type caused by several species of the genus *Shigella.* These microbes are nonmotile, non-spore-forming, gram-negative bacilli that are cultured easily on most ordinary laboratory media. Like the typhoid and paratyphoid organisms, the shigellae find their way into the body via contaminated food and water. Since they appear in the feces during the first few days of the infection, stool specimens afford a means of diagnosis. Treatment centers on fluid replacement, and for the most part, antibiotics are unnecessary. Prophylactically, good sanitation is the most effective way to control the infection.

AMEBIC DYSENTERY

Amebic dysentery, or amebiasis, is caused by the protozoan *Entamoeba histolytica.* The ingested cysts develop into trophozoites that subsequently burrow their way into the intestinal wall (Fig. 16-13). The invasion soon results in ulceration, pain, and diarrhea. In the event the trophozoites or amebas enter the blood, abscesses may be produced in the liver, lungs, and brain, and for this reason amebic dysentery is extremely dangerous. Although one may be prone to think that amebiasis is confined to tropical and subtropical regions, the incidence of the malady reportedly approaches 20% of the population in certain areas of the United States

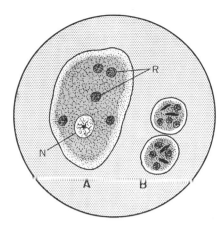

FIG. 16-13
Entamoeba histolytica. **A,** Trophozoite. **B,** Cyst stage.
N, Nucleus; *R,* red cells.

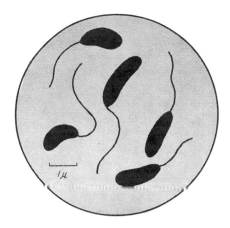

FIG. 16-14
Vibrio cholerae.

and Canada. Diagnosis is made by the demonstration of cysts or trophozoites in stool specimens. A wide variety of drugs have been used to manage the intestinal and systemic stages. The more successful agents include metronidazole, broad-spectrum antibiotics, fumagillin, emetine, and chloroquine. Persons with diagnosed cases must be kept under close surveillance to prevent them from contaminating food and drink. Good sanitation, of course, is a prime concern.

CHOLERA

Cholera is a vicious enteric infection that has plagued the peoples of Asia for centuries—hence, the term Asiatic cholera. The etiologic agent, *Vibrio cholerae,* a short spirillum with one or two polar flagella (Fig. 16-14), is a motile, non-spore-forming, gram-negative organism that grows well on most ordinary artificial media. The classic signs of cholera are profuse rice-water stools, severe vomiting, general emaciation, prostration, and collapse. A definitive diagnosis is made by isolating the pathogen from the feces of patients and carriers. Fluid replace-

ment is essential to recovery, and tetracycline is helpful in inhibiting the organism. Prophylactically, stringent sanitary codes are a must, and active immunization employing killed cholera spirilla has been shown to offer protection in most persons. All those going into cholera areas should be vaccinated.

YELLOW FEVER

The discovery by Reed that the mosquito (*Aedes aegypti*) transmits yellow fever was one of the great moments in medicine, a discovery that led to the conquest of the disease in Panama and made possible the construction of the Panama Canal. Yellow fever is an acute viral disease marked by high fever, pain, hemorrhage, and jaundice. The jaundice results from the damaging effect of the virus upon the liver. Since there is no specific therapy available, the emphasis is upon prevention: eradication of mosquitoes and active immunization with a vaccine prepared from the attenuated virus. The immunity engendered by this vaccine is good for about 6 years. An actual attack of yellow fever usually confers life-long immunity.

VIRAL HEPATITIS

Of the various types of hepatitis (inflammation of the liver), the viral forms are of most concern for the general population. There are two viral forms, infectious hepatitis and serum hepatitis. Although the signs and symptoms are similar and often indistinguishable, the causative viral agents are quite distinct, as evidenced by the shorter incubation period of serum hepatitis, the different modes of transmission, and the different responses to immune serum globulin. Infectious hepatitis is usually spread in a fecal-oral fashion, whereas blood transfusions and inadequately sterilized needles and syringes are responsible for most cases of serum hepatitis (hence its name). Typical signs and symptoms for both forms include lassitude, weakness, drowsiness, nausea, abdominal discomfort, headache, and fever; jaundice is common in infectious hepatitis, but uncommon in serum hepatitis. Treatment is symptomatic. Prevention centers on paying strict attention to sanitation, to sterilization procedures, and to "donor quality" at blood banks. Immune serum globulin is effective in aborting infectious hepatitis in susceptible contacts.

MUMPS

Mumps, or parotitis, is an acute viral infection, usually afflicting children, characterized by swelling of the salivary glands. It is not considered serious except after puberty when sterility may develop in both sexes as a result of orchitis and oophoritis. Other complications include facial paralysis and encephalitis. There is no specific form of therapy, but gamma globulin has proved of considerable prophylactic value; when gamma globulin is administered 7 to 10 days after exposure, a substantial portion of susceptible persons are protected. An attack of mumps typically engenders a permanent immunity; an effective vaccine is available for artificial active immunity.

ENTEROVIRUSES

Enteroviruses, so designated because they commonly occur in the intestinal tract, fall into three major groups: Coxsackie viruses, ECHO viruses, and polioviruses (Chapter 20).

Coxsackie viruses comprise two clinical groups, A and B. Viruses of group A cause herpangina (an illness marked by fever and eruption of blisters) and aseptic meningitis (a nonbacterial meningitis often indistinguishable from nonparalytic poliomyelitis). All five viruses of group B have been found to be associated with human illnesses, the most important of which are aseptic meningitis, pleurodynia (painful disease of the muscles of the chest and diaphragm), fatal inflammation of the heart in newborn infants, and certain polio-like diseases. It is now apparent that many illnesses diagnosed as nonparalytic poliomyelitis may actually be Coxsackie-virus infections.

ECHO virus stands for *enteric cytopathogenic human orphan*. Orphan refers to the fact that in many cases there has been no association with disease. Other enteric viruses of a similar nature have been isolated from monkeys (ECMO), cattle (ECBO), and swine (ECSO). It is now recognized that certain ECHO viruses cause disease. Their association with outbreaks of diarrhea, particularly in hospital nurseries, has recently been proved. In one study, ECHO viruses were isolated in 31% of cases. They have also been associated with febrile infections of the respiratory and enteric tracts, paralytic aseptic meningitis, and skin eruptions.

TRENCH MOUTH

Also known as *Vincent's angina*, trench mouth is an ulcerative infection of the mouth and throat marked by the presence of a pseudomembrane. Stained smears from the ulcers disclose two microbes, *Borrelia vincentii*, a spirochete, and *Fusobacterium fusiforme*, a bacillus, that may or may not be the etiologic factors.

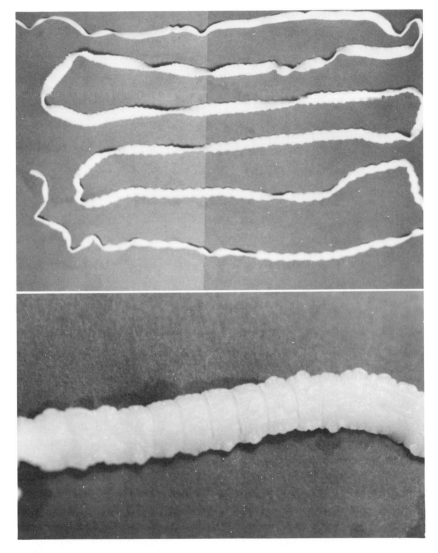

FIG. 16-15

Beef tapeworm *(Taenia saginata)* recovered from stool of patient following administration of quinacrine (the anthelmintic of choice in this condition). Top photo shows entire worm (6 feet in length) except for neck and head. Bottom photo (×2) underscores segmentation.

Trench mouth is no longer considered contagious. Precipitating causes include poor oral hygiene, malnutrition, debilitating diseases, worry, and heavy smoking. Treatment consists of débridement, implementation of a bland diet, and administration of peroxide rinses, analgesics, vitamins, and, in the event of high fever, antibiotics.

WORM INFESTATIONS

The presence of parasitic worms in the intestine is the most common form of helminthiasis. Such infestations harass the host by causing intestinal obstruction and inflammation, faulty digestion, and anemia. The most commonly encountered intestinal nematodes in the United States are the roundworm (*Ascaris lumbricoides*), hookworm (*Ancylostoma duodenale* and *Necator americanus*), threadworm (*Strongyloides stercoralis*), whipworm (*Trichuris trichiura*), and pinworm (*Enterobius vermicularis*); the most common cestodes or tapeworms (Fig. 16-15) are the dwarf (*Hymenolepsis nana*), beef (*Taenia saginata*), pork (*Taenia solium*), and fish (*Diphyllobothrium latum*). Helminthiases involving these worms are diagnosed by the finding of eggs (ova) in stool specimens. Definitive treatment entails the use of the appropriate anthelmintic. Piperazine is effective against pinworm and roundworm; tetrachloroethylene, against hookworm; thiabendazole, against threadworm and whipworm; and quinacrine, against all tapeworms.

QUESTIONS

1. Describe the anatomy of the mouth.
2. What is the function of the frenulum linguae?
3. Name the three types of papillae found on the surface of the tongue.
4. Discuss the mechanism of taste.
5. Compare the deciduous and the permanent teeth.
6. Describe the anatomy of the tooth.
7. Discuss the structure and formation of the teeth and the role played therein by calcium, vitamin D, parathyroid hormone, and fluorine.
8. Discuss the pros and cons of fluoridation.
9. Discuss the anatomy and physiology of the salivary glands.
10. Describe the acts of mastication and deglutition.
11. Describe the structure of the esophagus.
12. Describe the stomach wall.
13. Discuss the neurogenic and the hormonal stimulation of gastric juice.
14. Describe the mixing and emptying of the stomach contents.
15. Describe the gross anatomy of the large intestine.
16. What are the three somewhat arbitrary divisions of the small intestine?
17. Compare the walls of the small and large intestines.
18. What are Peyer's patches?
19. Give the precise location of the vermiform appendix.
20. Distinguish between the villi and the plicae circulares.
21. What is the lamina propria?
22. Describe the structure and function of the mesentery.
23. Discuss the hormonal activation of the glands of Lieberkühn.
24. Describe peristalsis and its regulation.
25. Describe in detail the structure of the rectum and the anal canal.
26. Describe the mechanism of defecation and the composition of the feces.
27. What prevents the passage of bile into the duodenum during fasting?
28. What are the composition and the function of the bile?
29. Briefly discuss the various functions of the liver.
30. What role does secretin play in the digestive process?

31. What is the relationship of the gallbladder to the liver?
32. What is the function of cholecystokinin?
33. What does it mean to say that the pancreas is a "dual gland"?
34. Locate the duct of Wirsung and the duct of Santorini.
35. What is the function of pancreozymin?
36. When saliva is added to a dilute starch solution that has been colored blue by the addition of a drop of iodine, the color slowly fades away. Explain.
37. What is the chief function of the mouth in the digestive process?
38. The presence of excessive gastric juice in the stomach when food is not present is a serious matter. Explain.
39. The stomach plays a relatively minor chemical role in the digestive process. Explain.
40. Write the equation for the reaction that occurs when baking soda enters the stomach.
41. What are antacids?
42. What is pepsinogen?
43. Compare the digestion of maltose, sucrose, and lactose.
44. Name the two most important amylases in the digestive process.
45. Compare the action of trypsin and erepsin.
46. Glucose, fructose, and galactose are isomers. Explain.
47. What is a polypeptide?
48. Write the balanced equation for the enzymatic hydrolysis of glyceryl tripalmitate into palmitic acid and glycerol.
49. Write the equation for the enzymatic hydrolysis of glycylglycine.
50. State the function of enterokinase.
51. State the function of villikinin.
52. Distinguish between chyle and chyme.
53. Explain the common occurrence of serum hepatitis in adults.
54. Discuss the etiology and treatment of Vincent's angina.
55. Discuss the etiology, clinical picture, treatment, and prevention of typhoid fever.
56. In light of the fact that abstainers develop portal cirrhosis, is it fair to say that alcoholism is a cause of the disease? Discuss.
57. Distinguish between bacillary and amebic dysentery.
58. Discuss the etiology, clinical picture, treatment, and prevention of Asiatic cholera.
59. What role does "plaque" play in dental caries?
60. What do the various forms of the malabsorption syndrome have in common relative to etiology?
61. What is steatorrhea?
62. What types of drugs are used in the treatment of a peptic ulcer to control hyperacidity and gastric activity?
63. Distinguish between diverticulosis and diverticulitis.
64. What possible causes have been suggested for colitis?
65. Hemorrhoids are common in pregnancy. Account for this fact.
66. The drastic decline in the incidence of stomach cancer cannot be explained by better diagnosis and better treatment. What, then, are some other possibilities?
67. What are "metastatic tumors"?
68. The incidence of oral cancer is higher in men than in women. Account for this fact.
69. Why, specifically, does a ruptured appendix generally lead to peritonitis?
70. In the event of a ruptured appendix, the greater omentum may well be life saving. Explain.
71. It is possible that we ingest "food-poisoning" strains of *Staphylococcus aureus* daily with impunity. Discuss.
72. What, if any, is the relationship of ptomaines to food poisoning?
73. Shellfish is often associated with infectious hepatitis. Why?
74. Enteroviruses belong to a family called picornaviruses. What is the derivation of this name?
75. Describe in detail the treatment of cestodiasis.

CHAPTER 17

METABOLISM

Metabolism, as noted earlier, is the complex of processes involved in the maintenance of life; in the more restricted and usual sense, it deals with the chemical pathways taken by carbohydrates, lipids, and proteins following absorption into the blood. These materials supply the cells with energy and are used by them to synthesize new structural components and to build storage compounds to be used at a later time (for energy or synthesis). For the most part, the major pathways are present in every cell, but not to the same degree. Many tissues become highly specialized in their functions; adipose cells, for example, are chiefly concerned with the synthesis of fat. There is considerable metabolic integration and a great many compounds actually share the same pathway(s). This, of course, affords efficiency and adaptability.

CARBOHYDRATE METABOLISM

Carbohydrates are the body's major source of energy and can be derived from proteins and fats when the occasion arises (Fig. 17-1). "Carbohydrate" enters the cell as the monosaccharides ($C_6H_{12}O_6$) glucose, fructose, and galactose, the end products of carbohydrate digestion, all three of which must be converted to glucose-6-phosphate or fructose-6-phosphate before they can be catabolized for energy or converted to glycogen and fat for future use. The catabolism of glucose and its release of energy (as ATP) was discussed earlier (p. 79). In the first phase (glycolysis), the sugar begins as glucose-6-phosphate and ends as pyruvic acid; in the second phase (citric acid cycle), pyruvic acid* is degraded to CO_2 and H_2O. Both phases are interlocked with the cytochrome system, in which pathway the bulk of energy is released. The overall reaction is this: one molecule of glucose plus six molecules of oxygen yields six molecules of carbon dioxide, six molecules of water, and 38 molecules of ATP or:

$$C_6H_{12}O_6 + 6O_2 \longrightarrow 6H_2O + 6CO_2 + 38ATP$$

The reaction is the same for the isomers fructose and galactose. Fructose enters the glycolysis pathway as fructose-6-phosphate and galactose as glucose-6-phosphate (to which form it is converted as discussed below).

Glycogen

Glucose, fructose, and galactose are converted into glycogen in the process of *glycogenesis*. The major steps are shown in Fig. 17-2.

---

*During strenuous exercise, of course, an "oxygen debt" develops, and much pyruvic acid is converted into lactic acid.

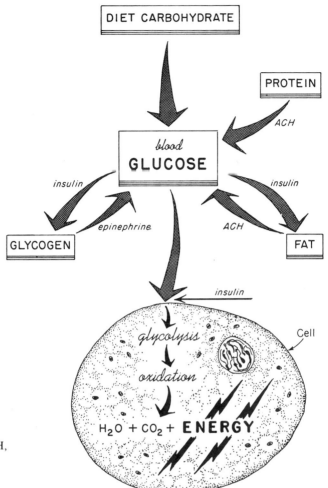

FIG. 17-1
Metabolism at a glance. (ACH, adrenocortical hormones.)

When the need for energy arises, glycogen is broken down (*glycogenolysis*) by the enzyme phosphorylase to glucose-1-phosphate, which in turn is transformed to glucose-6-phosphate by phosphoglucomutase. Glucose-6-phosphate then enters the glycolysis pathway, and so forth. Complete glycogenolysis, or the conversion of glycogen to glucose, occurs only in the liver through the action of glucose-6-phosphatase on glucose-6-phosphate.

Fat

If there is more sugar than the body needs, the sugar is converted into fat and stored in adipose tissue. In brief, the metabolic pathway runs as follows. Pyruvic acid, the end product of glycolysis, reacts with coenzyme A to form acetyl-coenzyme A, which thereupon undergoes reverse beta oxidation to produce fatty acids. Fats (triglycerides) are then formed when these fatty acids combine with glycerol. The source of

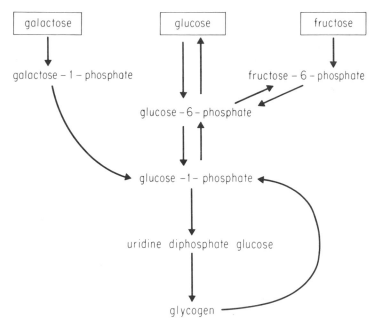

FIG. 17-2

Basic steps in synthesis of glycogen (glycogenesis).

glycerol is phosphoglyceraldehyde, the fourth step in glycolysis.

Hormonal control

A number of hormones (Chapter 23) have a bearing on carbohydrate metabolism; the most critical is insulin. Among other things, insulin accelerates glucose transport through cell membranes and enhances the activity of the enzyme glucokinase, which is critical in initiating the breakdown of glucose in the glycolysis sequence. Other hormones of special note are glucagon, epinephrine, and thyroxine. Glucagon and epinephrine stimulate the breakdown of glycogen, and thyroxine stimulates glucose catabolism. Acting to increase the amount of glucose in the blood in times of stress are the glucocorticoids of the adrenal cortex, hormones that promote the metabolism of protein into glucose (*gluconeo-*

genesis). Acting indirectly are the thyroid-stimulating hormone (thyrotropin) and the adrenocorticotropic hormone (ACTH). The thyroid-stimulating hormone serves to decrease blood glucose, and the adrenocorticotropic hormone serves to increase blood glucose.

LIPID METABOLISM

The lipids of chief metabolic concern are fats (triglycerides), phospholipids, and sterols. Fats are a valuable energy source and are stored in adipose tissue as energy reserves; phospholipids, along with certain proteins, form the structural basis for the cell membrane; and cholesterol, the chief sterol, is the chemical precursor of vitamin D, bile salts, and a number of hormones. Body lipids are derived directly from the diet and indirectly from carbohydrates and proteins (Fig. 17-3).

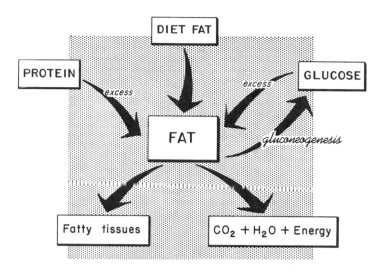

FIG. 17-3

Diagrammatic summary of fat metabolism.

Lipid catabolism

Within the tissues, fats are broken down into glycerol and fatty acids, just as they are in the intestinal tract. Glycerol is converted to phosphoglyceraldehyde and in this form enters the glycolysis pathway to be catabolized. Fatty acids demand a much more complex handling via the pathway known as *beta oxidation,* which in essence amounts to the successive splitting off of two-carbon fragments through the agency of many enzymes, including the crucial coenzyme A. The final product is the two-carbon acetyl-coenzyme A, which begins the citric acid (Krebs) cycle. Thus, the catabolism or oxidation of fats is a matter of converting them into metabolites that enter the same pathways as glucose—namely, glycolysis and the citric acid cycle.

Ketone bodies

The last metabolite before acetyl-coenzyme A in the beta oxidation of a fatty acid is acetoacetyl-coenzyme A, most of which, as already indicated, breaks down into two molecules of acetyl-coenzyme A and enters the Krebs cycle. Normally, a small amount of acetoacetyl-coenzyme A is converted to acetoacetic acid, which is eventually oxidized in the Krebs cycle. Under certain abnormal conditions, however, such as starvation and diabetes mellitus, the body employs fat as the chief fuel, and more acetoacetic acid is formed than can be catabolized. Consequently, there is a buildup of this acid and of its breakdown products, acetone and β-hydroxybutyric acid, all three compounds customarily being grouped together under the heading "ketone bodies." Such buildup is dangerous because the presence of these compounds upsets the acid-base balance and causes acidosis.

Lipid anabolism

Fatty acids can be synthesized by an anabolic pathway that is essentially the reverse of beta oxidation. Acetyl-coenzyme A molecules (the majority derived from the glycolysis of excess glucose) are added to smaller, preformed fatty acids to yield the fatty acids of higher molecular weight needed for making the essential fats and phospholipids. The phospholipids are formed when fatty acids combine with glycerol in the form of glycerophosphate: one molecule of glycerophosphate reacts with two molecules of fatty acid to yield a phosphatidic acid, which then goes on to yield a triglyceride (fat) or a phospholipid. The synthesis of other lipids involves different pathways, the majority of which are exceedingly complex and have not been worked out in complete detail. One of the most brilliant biochemical achievements was the establishment of the pathway of formation of cholesterol, a pathway commencing with two carbons (acetyl-coenzyme A) and ending with the complex, 27-carbon molecule.

Hormonal control

Lipid metabolism is regulated by several hormones, including insulin, growth hormone, glucocorticoids, and ACTH (Chapter 23). Whereas insulin promotes anabolism, the other hormones bring about an increase in fat catabolism (Fig. 17-1). When carbohydrate is in ample supply, for example, these hormones are secreted in such a way as to step up the storage of fat. On the other hand, a decrease in carbohydrate in the diet promotes the catabolism of fats in order to meet the body's energy needs.

PROTEIN METABOLISM

Proteins constitute the basic material of protoplasm, and the manner in which they are synthesized from amino acids was discussed in some detail in Chapter 6. Actually, the subject of the "genetic code" (p. 85) is fundamentally concerned with protein anabolism. Let us now complete the picture by taking a look at protein catabolism and the "nonprotein" fate of amino acids.

Protein catabolism

Through the catalytic agency of intracellular enzymes called cathepsins, proteins constantly are broken down into amino acids for metabolic service elsewhere. The quantity of protein in a given cell is determined by the balance between synthesis and degradation. Under normal conditions, the body needs about 30 g of new protein daily to replace the protoplasm destroyed through wear and tear and to fortify the reserves. Amounts over and above this figure are converted to other compounds or oxidized in the Krebs cycle. These events relate to amino acid metabolism (below).

Amino acid metabolism

The amino acids absorbed from a digested meal (plus those derived from protein catabolism) are either polymerized into new proteins or else metabolized by a variety of other mechanisms, including deamination, transamination, and decarboxylation (Fig. 17-4). *Deamination* involves the removal of the amino group (NH_2) through the agency of specific enzymes, the most important of which are amino acid oxidases present in liver and kidney tissues. This reaction leads to the formation of the corresponding α-keto acid and ammonia. The ammonia is converted to urea and excreted, and the keto acid is either oxidized in the Krebs cycle to supply energy or converted to glucose or fat, the particular pathway taken depending on the metabolic situation at the time. *Transamination* also involves the removal of the amino group, but in this case the group is transferred to a keto acid, with the resulting formation of a new amino acid and new keto acid. These reactions are catalyzed by specific enzymes called transaminases. *Decarboxylation* refers to the removal of the carboxyl group by a specific decarboxylase to yield CO_2 and an amine.

Through deamination, transamination, decarboxylation, and other mechanisms, the tissues

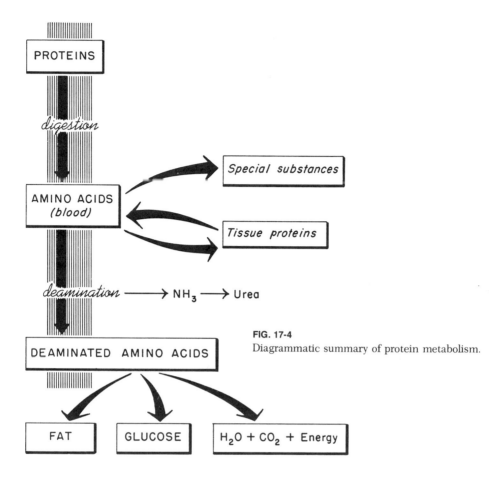

FIG. 17-4
Diagrammatic summary of protein metabolism.

are able to transform amino acids into essentially any substance the body needs, nonprotein as well as protein. Nonprotein products include a number of hormones, purines, pyrimidines, porphyrins, and creatine. Purines and pyrimidines are used in the synthesis of nucleic acids; protoporphyrin IX serves as the nucleus for the heme group in hemoglobin and cytochrome enzymes; and creatine phosphate serves as a reservoir of energy in muscle contraction. The waste products of protein metabolism include carbon dioxide, water, urea, excess creatine,

and creatinine. Carbon dioxide and water are released in the Krebs cycle; urea is formed in deamination; creatine is synthesized from the amino acids arginine, glycine, and methionine; creatinine is formed from creatine; and uric acid is an indirect product arising from the catabolism of purine nucleotides.

Hormonal control

Like carbohydrate and lipid metabolism, protein metabolism is mainly under hormonal control. The growth hormone and testosterone (the male hormone) promote protein anabolism, and glucocorticoids and ACTH speed up the mobilization and catabolism of tissue protein. Thyroid hormones also have a considerable effect on protein metabolism, but the direction of the effect depends on the availability of carbohydrates and fats. When these nutrients are in amply supply, thyroid hormones promote protein anabolism; when in short supply thyroid hormones promote protein mobilization and catabolism.

METABOLIC RATE

Since the energy released in the body appears largely as heat, the rate at which the heat is produced provides the most accurate measure of metabolism. Depending on the degree of physiologic activity, the metabolic rate may run from about 50 Calories (Cal) per hour up to 2,000 Cal per hour.

Factors affecting rate

Obviously, any factor that increases the liberation of energy will increase the metabolic rate. Muscular activity yields tremendous amounts of heat. Strenuous exercise can push the metabolic rate almost 50% above normal! Next to muscular activity in effect on metabolic rate stands the environmental temperature. As the temperature drops, the metabolic rate must increase to balance the increased loss of heat. Having a marked influence, too, is the extra heat associated with the consumption of food. This so-called "specific dynamic action of food" (SDA) is now thought to arise from the deamination of amino acids. Proteins cause an increase of about 30% over and above their basic caloric content, and carbohydrates and fats about 6% and 4%, respectively.

Basal metabolic rate

The metabolic rate is the most meaningful single statistic relating to body chemistry and has considerable diagnostic value to the physician. However, a person's metabolic rate has little or no meaning unless it is taken under basal conditions, that is, at complete physical and mental rest. The basal metabolic rate, or BMR, is a measure of inherent cellular chemistry.

The obvious way to take the basal metabolic rate would be to place the subject in a giant calorimeter and measure the number of calories given off in a given period of time. More commonly, however, indirect calorimetry is employed; the subject's oxygen consumption (Fig. 17-5) is measured over a short period of time (10 minutes or so), and from this data calories can be calculated. The consumption of 1 l of oxygen corresponds to the release of 4.825 Cal. Since the basal metabolic rate has been shown to vary in proportion to the body surface area, the latter must be found and incorporated into the BMR. For this purpose, use is made of the chart shown in Fig. 17-6. By way of illustration, suppose a 30-year-old man weighing 75 kg and measuring 160 cm in height consumed 2.5 l of oxygen in 10 minutes. Thus, in 1 hour he consumes 15 l of oxygen and releases 15 times 4.825, or 72.38 Cal. Now, since the surface area of his body, according to the chart, is 1.8 m², his BMR is 72.38 divided by 1.8, or 40.2 Cal per square meter per hour (40.2 Cal/m²/hr). This value, however, is relatively useless unless we compare it with the normal BMR, which for this particular individual is 37.2 Cal/m²/hr (Fig. 17-7). This we do on a percentage basis:

$$\frac{40.2 - 37.2}{37.2} \times 100 = 8\%$$

Clinically, this is expressed as +8, meaning 8% above normal. On the other hand, had the subject's oxygen consumption been say 34.2 Cal/m²/hr, his rate would be −8, or 8% below normal. In cases of extreme hyperthyroidism, the BMR may climb as high as +100, and in

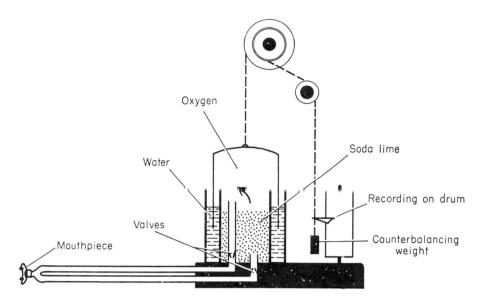

FIG. 17-5

Diagrammatic representation of oxygen-filled spirometer used to measure basal metabolism. Subject (with nose closed) breathes in and out through mouthpiece. As oxygen is consumed, metal can descends, thereby recording on drum volume of oxygen used. (After Guyton, A. C.: Function of the human body, Philadelphia, 1959, W. B. Saunders Co.)

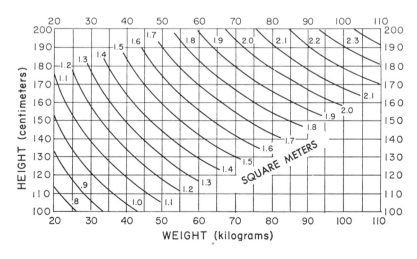

FIG. 17-6

BMR chart for determining body surface area (in square meters) from weight and height. (After Dubois; from Guyton, A. C.: Function of the human body, Philadelphia, 1959, W. B. Saunders Co.)

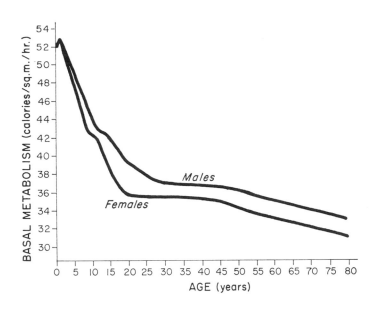

FIG. 17-7
Normal basal metabolic rates of males and females at different ages. (After Guyton, A. C.: Function of the human body, Philadelphia, 1959, W. B. Saunders Co.)

severe hypothyroid states, the BMR may drop to a low of −50. Generally, the range of the normal BMR extends from +15 to −15.

RESPIRATORY QUOTIENT

Another interesting metabolic statistic is the respiratory quotient (RQ), which is the ratio of the volume of carbon dioxide expired to the volume of oxygen inspired in a given time. A person on a pure carbohydrate diet has an RQ of 1, a fact that is explained by the following equation:

$$C_6H_{12}O_6 + 6O_2 \longrightarrow 6CO_2 + 6H_2O$$

$$RQ = 6/6, \text{ or } 1$$

In the catabolism of fats and proteins, however, proportionately less carbon dioxide is given off, making the RQ less than 1. Fats produce an RQ of about 0.7, and proteins, an RQ of about 0.8. A person on a mixed diet has an RQ in the vicinity of 0.85.

METABOLIC DISEASES

One is hard put to name a body abnormality that is devoid of biochemical ramifications; the expression "metabolic disease" becomes more common with the passing of time. By and large, most textbooks of medicine employ the expression to denote the pathology arising from a faulty metabolic pathway, especially an inherited fault, or "inborn error of metabolism." Such abnormalities follow genetic patterns and accordingly may be dominant or recessive (Chapter 25). They may be mild or fatal. Here we shall confine our remarks to the more important and serious inborn errors.

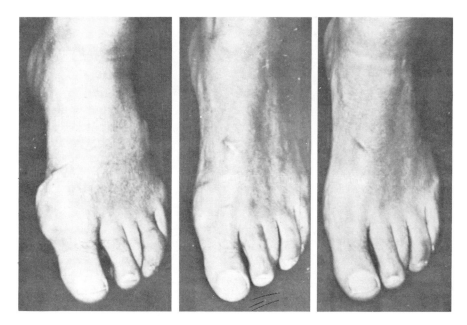

FIG. 17-8

First case demonstrating dramatic action of treating gout with drugs that increase excretion of uric acid. Left, December, 1950; center, September, 1951; right, January, 1953. (From Gutman and Yü, J.A.M.A. **157:**1096, 1955.)

Gout

Gout is a recurrent acute arthritis of the fingers, toes, and knees that may become chronic and deforming. Usually an affliction of middle-aged males, gout results from the deposition of monosodium urate crystals in and about the joints and tendons. In an acute attack, the joint (classically the big toe) suddenly becomes swollen and extremely painful and tender, and the overlying skin becomes tense and shiny. Such attacks come and go but in the main become progressively more frequent. Chronic gout is characterized by joint deformities and subcutaneous nodules (tophi) of the urate deposits; enlarging tophi may erupt and discharge their chalky contents. The kidneys, too, suffer from urate deposits, accounting for the fact that renal failure is the most common cause of death in the disease. Although it is generally agreed that the underlying cause in the majority of patients arises from excessive purine synthesis and the consequent uric acid production (p. 273), the precise metabolic defect remains a mystery. Colchicine is the drug of choice in the treatment and prevention of acute attacks, and drugs to prevent and resolve tophi are now available. Allopurinol, for example, blocks uric acid production, and probenecid increases uric acid excretion (Fig. 17-8).

Porphyria

Porphyria refers to a group of diseases characterized by abnormal porphyrin metabolism; porphyrin refers to any one of a group of iron-free or magnesium-free pyrrole pigment derivatives present throughout the tissues.

In congenital, or erythropoietic, porphyria, a disease inherited as a dominant trait, the onset of signs and symptoms occurs in early life. The urine is colored Burgundy red, skin lesions develop on exposed surfaces, there is abnormal sensitivity to light, and there may be hemolytic anemia and enlargement of the spleen. Treatment centers on avoidance of sunlight, application of corticosteroids, and removal of the spleen in the event of anemia. Even with the best treatment few patients live to middle age.

Hepatic porphyria, the other well known form of the disease, is inherited as a dominant trait and arises from the overproduction of porphyrin precursors. These precursors and the resulting porphyrins accumulate in the liver, hence the term "hepatic." The disease typically occurs in later life, affects women more than men, and is characterized by intermittent acute attacks. Colic pain, vomiting, diarrhea, anxiety, delirium, hallucinations, and convulsive seizures constitute the usual manifestations. A high-carbohydrate, high-protein diet and tranquilizers provide symptomatic relief. Barbiturates and alcohol are to be avoided because they may provoke acute attacks.

Phenylketonuria

Phenylketonuria (PKU) is an inborn error of metabolism passed along by a recessive gene, characterized by high levels of phenylalanine in the blood, and commonly associated with mental retardation. The cause is a deficiency of phenylalanine hydroxylase, the liver enzyme needed to convert phenylalanine (an amino acid) into tyrosine (another amino acid). The disease has an incidence of about one case per 25,000 births and is rare among Jews and Blacks. During the first few weeks of life, the usual features include irritability, epileptic seizures, dry skin, and certain neurological abnormalities; two thirds of those afflicted will go on to show evidence of mental retardation. Treatment cen-

ters on limiting phenylalanine in the diet from the first few weeks of life up until the age of 5 years. Such a diet is of little value, however, in cases of retarded children past the age of 3 years.

Galactosemia

Galactosemia is a hereditary disease of carbohydrate metabolism marked by vomiting, diarrhea, jaundice, poor weight gain, cataracts, cirrhosis, and mental retardation. The defect arises from a deficiency of galactose-1-phosphate uridyltransferase, the enzyme needed to convert galactose to glucose. The disease is diagnosed easily (urine is positive to Benedict's test but negative to glucose oxidase test) and is treated by excluding lactose and galactose from the diet for at least 3 years while the central nervous system is developing.

Hyperlipemia

The major classes of blood-serum lipids are the triglycerides, phospholipids, cholesterol, cholesterol esters, and free fatty acids. These lipids are bound to proteins to form macromolecular complexes called lipoproteins. Inborn diseases of lipid metabolism marked by abnormally high levels of lipoproteins in the blood or tissues are known as hyperlipoproteinemia, or the less elongated "hyperlipemia." Five distinct types are recognized.

Type I hyperlipemia (exogenous hypertriglyceridemia), transmitted as a recessive trait, is caused by a deficiency of an enzyme needed for assimilation of triglycerides. The major features include a marked increase in triglycerides, yellowish fat deposits in the skin (xanthomas), and pancreatitis. A diet restricting all the common sources of fat is highly effective in the management of the involvement.

Type II hyperlipemia (familial hypercholesterolemia) is characterized by high blood cholesterol, tender xanthomas, atherosclerosis, and early death from myocardial infarction. A diet that completely eliminates cholesterol and saturated fats and replaces them with unsaturated fats is effective in reducing serum cholesterol, but apparently fails to prevent the development of heart disease.

Type III hyperlipemia ("broad beta disease") is thought to be a recessive trait. Here both cholesterol and triglyceride levels are elevated. The clinical features include eruptive xanthomas and a premature atherosclerosis. Simultaneous restriction of cholesterol and saturated fats, weight loss, and administration of the drug clofibrate dramatically resolve the xanthomas but have an unknown effect on the progression of the atherosclerosis.

Type IV hyperlipemia (endogenous hypertriglyceridemia) is a common condition that appears to be transmitted as a dominant trait. It is marked by high levels of serum triglycerides and by a pronounced predisposition to obesity and cardiovascular disease. Treatment centers on weight reduction and low-carbohydrate diet.

Type V hyperlipemia (mixed lipemia) is marked by an increase in triglyceride levels and quite possibly represents a genetic variant of the type IV disease; frequently it is associated with diabetes mellitus. The risk of atherosclerosis is not known, but obesity and enlargement of the liver are well-recognized complications. Treatment centers on weight loss and the use of clofibrate to reduce the lipemia.

Hypolipemia

Hypolipemia, or *acanthocytosis*, is a rare congenital disease in which the levels of serum lipids are markedly reduced. The earliest finding is malabsorption of fat, which appears in infancy and persists to a varying extent throughout life. Defective vision (due to retinitis pigmentosa) appears at about age 8 to 10, and often progresses to blindness. Also, there is progressive brain damage, which usually manifests itself in ataxia, slurred speech, and muscular atrophy. There is no treatment.

Lipid-storage diseases

A normal person has a large variety of enzymes that cause the breakdown of excess or unwanted lipids. When such an enzyme is inactive or absent, the result is a lipid-storage disorder, commonly marked by enlargement of the spleen and liver and by mental retardation. These disorders are hereditary, and most are fatal. Tay-Sachs disease, the classic lipid-storage disorder, is characterized by a lack of the enzyme hexosaminidase A and by the accumulation of ganglioside G_{M2} in nerve cells. Other lipid-storage disorders are Gaucher's disease, Niemann-Pick disease, Fabry's disease, generalized gangliosidosis, and fucosidosis. New tests for detecting adult carriers and for prenatal genetic diagnosis now make it possible to control the incidence of these inborn errors of metabolism.

QUESTIONS

1. Distinguish among metabolism, catabolism, and anabolism.
2. Define carbohydrate.
3. Glucose, fructose, and galactose are isomers. What does this mean?
4. Glycogen is often called "animal starch." Why?
5. Glycolysis is often referred to as the anaerobic phase of glucose catabolism. Why?
6. The Krebs cycle is also referred to as the citric acid cycle and the tricarboxylic acid cycle. Why?
7. What is the relationship of the cytochrome system to glycolysis and the Krebs cycle?
8. Fructose is catabolized at a faster rate than glucose. Can you offer an explanation?
9. What is the precise purpose of cellular respiration ("burning of food")?
10. To what molecular feature does ATP owe its energy?
11. What is needed to convert ADP and phosphate into ATP?

12. In the breakdown of glucose, where is the CO_2 produced?
13. Do all metabolites enter the Krebs cycle as pyruvic acid?
14. In which phase of cellular respiration is the great bulk of ATP produced?
15. What is meant by complete glycogenolysis?
16. What is an isomerase?
17. Strictly speaking, glycogenolysis occurs only in the liver. Explain.
18. The portal (blood) system fits in well with the function of glycogenesis. Discuss.
19. An overdose of insulin may be antidoted by an injection of glucagon. Explain.
20. In times of stress, there is an increased output of epinephrine, a hormone that elevates the blood pressure and stimulates glycogenolysis. Explain the common goal of these effects.
21. Glucocorticoids aggravate diabetes mellitus. Why?
22. In times of stress, what effect do ACTH and epinephrine have on carbohydrate metabolism?
23. What are the effects of starvation on glycogenesis and gluconeogenesis?
24. True fats are saturated triglycerides, and true oils are unsaturated triglycerides. Discuss and explain fully.
25. All fats are lipids, but not all lipids are fats. Discuss fully.
26. Carbohydrates and fats contain the same elements. Name them.
27. Even though carbohydrates and fats differ drastically, the intake of excess carbohydrate causes weight gain. How do you account for this?
28. What are the end (waste) products in the complete catabolism of fats and carbohydrates?

29. Before they can enter the Krebs cycle, fatty acids must undergo beta oxidation. Why?
30. β-hydroxybutyric acid is not a ketone, and yet it is referred to as a "ketone body." Why?
31. Acidosis is a possible consequence of a *low*-carbohydrate, *high*-fat diet. Explain.
32. Beta oxidation serves as a pathway for both anabolism and catabolism. Discuss.
33. Account for the fact that triglycerides yield glycerol.
34. What role do phosphatidic acids play in lipid anabolism?
35. What are sterols?
36. Explain the loss of weight that accompanies diabetes mellitus.
37. Insulin, growth hormone, and glucocorticoids regulate fat metabolism in such a way that the rate of fat utilization is related inversely to the rate of carbohydrate metabolism. Discuss.
38. In starvation, tissue proteins, lipids, and glycogen must be catabolized to meet the "energy crises" occasioned by the lack of exogenous glucose. In what order are these substances oxidized, and why?
39. The breakdown of protein into amino acids is essentially depolymerization. Explain.
40. At any given time, blood serum contains a certain concentration of amino acids. What are the sources of these compounds?
41. What is the precise function of a decarboxylase?
42. Abnormally high levels of SGOT (serum glutamic oxaloacetic transaminase) is a well-established finding in cirrhosis and other instances of liver damage. Can you account for this?
43. Where does the CO_2 come from in reactions involving decarboxylation?

44. Transamination results in the formation of a new amino acid and new keto acid. How do you account for this?
45. What purines and pyrimidines are needed in the synthesis of RNA? DNA?
46. "Protoporphyrin IX serves as the nucleus for the heme group in hemoglobin." Illustrate this with a structural formula.
47. That the male has proportionately more muscle tissue than the female illustrates dramatically the effect of androgens on protein metabolism. Precisely, what does it illustrate?
48. Is it any surprise that the growth hormone stimulates protein anabolism? Why?
49. Corticosteroid drugs are *contraindicated* in peptic ulcer. Why?
50. What do glycogenolysis and gluconeogenesis have in common?
51. What is calorimetry?
52. Compare the small calorie and large Calorie.
53. What is the relationship between oxygen consumption and the BMR?
54. Discuss the major factors affecting the BMR.
55. Compute the clinical BMR for a female subject who, under basal conditions, consumes 2 l of oxygen in 10 minutes. She is 25 years old, weighs 125 pounds, and is 5'7" tall.
56. Can you explain why 95% of gouty patients are male?
57. Discuss the meaning of an "inborn error of metabolism."
58. Well over 90% of gouty patients are hyperuricemic. Does this fit in with what we know about the disease?
59. Standard therapy in the management of chronic gout is giving uricosuric drugs. Explain how these drugs improve the condition.

60. The tendency in gout to form kidney stones can be diminished by high fluid intake and alkalinization of the urine. Explain how this helps.
61. Can you account for the hemolytic anemia of erythropoietic porphyria?
62. Barbiturates are true poisons in the face of hepatic porphyria. Can you offer a rather specific chemical explanation?
63. Even though the blood contains high amounts of phenylalanine as a consequence of phenylalanine hydroxylase deficiency, the condition is known as phenylketonuria (PKU). Why?
64. In PKU, phenylalanine is not converted to tyrosine. How does this fit with the clinical signs and symptoms?
65. PKU can be spotted at the time of birth by a very simple test performed on urine. Describe this test.
66. There is no treatment of PKU past the age of 3 years. Why?
67. What apparently is the precise cause of the cataracts and brain damage in galactosemia?
68. Milk is poison to infants with galactosemia. Why?
69. PKU and galactosemia are classic examples of inborn errors of metabolism. In the main, what do all these diseases have in common?
70. Distinguish among lipids, phospholipids, lipoproteins.
71. What is the literal meaning of hyperlipemia?
72. Account for the cardiovascular complications of types II, III, and IV hyperlipemia.
73. What is the basic laboratory difference between type I and type II hyperlipemia?
74. Clofibrate is a standard drug administered in cases of types III, IV, and V hyperlipemia. What is its mechanism of action?
75. Account for the term acanthocytosis (the other name for hypolipemia).

CHAPTER 18

NUTRITION

Literally, as well as figuratively, we are what we eat, a fact so obvious that many of us never really stop to think about it until a problem arises (*low* hemoglobin is an all too common example). The materials and energy lost during the day must be replenished by taking in the right kind and amount of food; this is what nutrition is essentially all about. Starvation is increasingly a way of life in many parts of the world, and even here at home, our great affluence notwithstanding, we encounter widespread malnutrition. The present chapter is intended to underscore the major points pertaining to food and health.

PROTEINS

No matter how much fat or carbohydrate the diet may contain, we sustain a daily loss of an ounce or so of tissue through metabolic wear and tear; this loss can be restored only by the eating of protein. Protein is the only complete nutrient. Proper nutrition calls for balance; the amount of nitrogen ingested—as protein—must equal that excreted in the form of nitrogenous waste products. A negative balance exists when the loss of body proteins exceeds their synthesis. By the same token, an excess of synthesized over degraded proteins constitutes a positive nitrogen balance, a situation encountered during growth, pregnancy, and convalescence from some dis-

eases. In the normal adult, however, a positive balance is short lived because excess protein is catabolized. Increased intakes of protein over and above the minimum protein requirement merely lead to nitrogen balance at a higher level. As a rule of thumb, the daily allowance of protein for the average adult is in the vicinity of 1 g/kg of body weight. Young children need about four times this amount, and pregnancy and lactation call for an intake of about 1.5 g/kg.

Complete proteins

Of the 22 amino acids that go into the making of the protein molecule, only eight are indispensable and must be supplied by the diet: isoleucine, leucine, lysine, methionine, phenylalanine, threonine, tryptophan, and valine. These are indispensable or essential because the body cannot exist without them. We need the remaining amino acids, too, but these can be synthesized from the essential amino acids and other sources. A protein that contains and supplies all the essential amino acids is *complete*. Most animal proteins are complete; most vegetable and grain proteins are incomplete.

Protein deficiency

Protein deficiencies are either primary or secondary. Primary deficiencies stem from a quantitative or qualitative lack of protein in the diet or

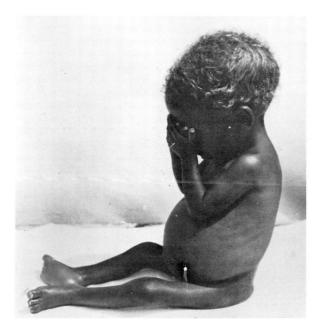

FIG. 18-1

An African child suffering from kwashiorkor. Note uncurled, graying hair, edema, and skin lesions. (FAO photograph by M. Autret.)

from increased utilization of protein (for energy) occasioned by a caloric deficiency. Secondary deficiencies, on the other hand, are "internal matters"—poor digestion, poor absorption, increased catabolism, excessive loss from the body, or increased requirement.

The classic primary deficiencies are marasmus and kwashiorkor (Fig. 18-1). Marasmus is a progressive wasting and emaciation, especially in infants, relating to both protein and caloric deficiencies. Kwashiorkor, which occurs in the face of adequate or even excess caloric intake, relates to a lack of most essential amino acids. It is marked by lethargy, edema, poor skin, decoloration of hair, and enlargement of the liver. A common cause of death is depletion of potassium. Treatment centers on the restoration of

electrolyte balance and the implementation of a diet based on milk.

The treatment of secondary deficiencies involves the giving of proteins (for example, transfusions in hemorrhage) and the correction, if possible, of the underlying cause.

FATS AND CARBOHYDRATES

The bulk of the energy needed by the body is supplied by fats and carbohydrates. Unless these are present in adequate amounts, the tissues must burn protein, and a negative nitrogen balance ensues. This accounts for the well-known nutritional expression, "fats and carbohydrates spare protein." For the most part, fats and carbohydrates can replace each other for this purpose, but neither should be entirely excluded. The nervous system derives its energy almost exclusively from carbohydrates; certain fatty acids are essential for proper growth and cannot be synthesized in the body. Young rats on diets deficient in linoleic, linolenic, or arachidonic acid present a picture of general deterioration and fail to grow. Most authors feel that we should derive about 30% of our energy from fat, 55% from carbohydrate, and 15% from protein.

MINERALS

In a biological and nutritional sense, minerals refer to the inorganic components of food that are essential to life. Those associated with fluid and electrolyte balance (Chapter 8), with the major exception of calcium, are seldom, if ever, deficient in the diet. In addition to calcium, the minerals most likely to be lacking from the diet are iron, phosphorus, iodine, and fluorine.

Iron

Approximately 70% of the iron in the body resides in hemoglobin; the bulk of the remainder is stored in the liver as ferritin and in muscles as myoglobin. The recommended daily allowance is between 10 and 18 mg (Table 18-1). A deficiency

TABLE 18-1

Recommended daily mineral allowance*

Mineral	Daily adult allowance
Calcium	0.8–1.3 g
Phosphorus	0.8–1.3 g
Iodine	80–150 μg
Iron	10–18 mg
Magnesium	300–450 mg

*Food and Nutrition Board (National Research Council).

of dietary iron leads to a lack of hemoglobin and thus to anemia. Iron-deficiency anemia often responds dramatically to such iron salts as ferrous sulfate and ferrous fumarate. Foods rich in iron include liver, raisins, prunes, apricots, and spinach.

Calcium

Calcium is a basic raw material in the making of bone and plays a key role in muscle function. The daily requirement is relatively high and for adults ranges between 0.8 and 1.5 g, depending on age and weight. Pregnancy and lactation call for an increased intake. Milk and milk products are the best dietary sources.

Phosphorus

Phosphorus is a basic raw material in the making of bone and high-energy phosphate compounds. The daily requirement is high and should be equal to the intake of calcium, especially in the diets of children. The daily requirement for adults ranges between 0.8 and 1.5 g, depending on age and weight. Milk, beef, liver, beans, and broccoli are rich sources.

Iodine

Iodine has the special and vital role of serving in the synthesis of the thyroid hormone. A deficiency results in endemic goiter (Chapter 23), a classic condition at one time common in the Alps, the Pyrenees, and the Great Lakes region. Through the use of iodized salt, these so-called goiter belts are becoming a thing of the past. For adults, the recommended daily allowance ranges from 80 to 150 μg, depending on weight and age. Pregnancy and lactation call for increased amounts. Fish and other sea foods are rich in iodine.

Fluorine

Fluorine is essential to proper development and preservation of the teeth; in areas where the water has a low concentration of this element, there is a high incidence of tooth decay. The results of fluoridation are dramatic, and both the American Medical Association and the American Dental Association recommend the practice in areas where the fluoride content is below one part fluoride per 1 billion parts water.

VITAMINS

In 1906, Hopkins proved the existence in normal foods of certain accessory factors essential to the diet, and 5 years later, in 1911, Funk, believing these factors to be chemical amines, coined the term "vitamine." When further work disclosed that the amino group was not characteristic, Drummund saved the day by proposing that the *e* be dropped, giving us our present word "vitamin."

Vitamins sustain life, promote growth, and maintain health. It is hardly an exaggeration to say that the discovery of the vitamin and the various avitaminoses represents the highest achievement in nutrition. Although the precise mechanisms by which most vitamins work are still not completely understood, biochemists apparently have succeeded in unraveling the essential facts. All known vitamins have been isolated in the pure state and synthesized in the laboratory. By convention, they are categorized according to their solubility in water or fat. The water-soluble vitamins include vitamin C and the members of the B complex group, and the fat-soluble vitamins include vitamins A, D, K, and E (Table 18-2).

Vitamin A

Vitamin A (retinol) occurs, as such, only in food of animal origin (chiefly liver, milk, and eggs); most of what we need comes from β-carotene, a pigment found in green and yellow vegetables. Vitamin A content (or "activity") is usually expressed in international units (I.U.); one such unit is equal to 0.3 μg of crystalline vitamin A. The recommended daily allowance for adults is 5,000 units (Table 18-2). Deficiencies result not only from inadequate intake of the vitamin or its precursors but also from poor absorption in such conditions as diarrhea and lack of bile. Vitamin A is needed for the proper development of the epithelium, the bones, and the eyes (Chapter 21), and the chief signs of deficiency relate to these structures. One of the first signs of vitamin A deficiency is nyctalopia, or night blindness. More severe manifestations include keratomalacia, ulceration of the cornea, dental defects, and retarded bone growth.

Vitamin D

The designation *vitamin D* (calciferol) is given to several fat-soluble sterols possessing *antirachitic* properties, the most important being vitamin D_2 (ergocalciferol) and vitamin D_3 (cholecalciferol). Vitamin D_2 results from the action of ultraviolet rays on ergosterol, a vegetable sterol, and vitamin D_3, from the action of ultraviolet rays on 7-dehydrocholesterol, a constituent of the skin. Vitamin D stimulates the absorption of calcium and phosphate from the gastrointestinal tract and thereby promotes the formation of bone and teeth. Disturbances of calcium and phosphorus metabolism caused by vitamin D deficiency result in rickets in infants and children and in osteomalacia in adults. Deficiencies typically arise from inadequate exposure to ultra-

TABLE 18-2

Recommended daily vitamin allowance*

Vitamin	Daily adult allowance
Vitamin A (activity)	5,000 I.U.
Vitamin D	400 I.U.
Vitamin E (activity)	25–30 I.U.
Vitamin C	55–60 mg
Thiamine	1.0–1.5 mg
Riboflavin	1.5–2.0 mg
Niacin	13–20 mg
Pantothenic acid	10 mg
Pyridoxine	1.4–2.5 mg
Folic acid	400 μg
Cyanocobalamin	5–8 mg

*Food and Nutrition Board (National Research Council).

violet rays; it is doubtful whether the normal adult needs the vitamin in food. Infancy, pregnancy, and lactation, however, call for dietary supplements. Like vitamin A, vitamin D activity is expressed in international units. The recommended daily allowance is 400 units for both children and adults (Table 18-2).

Vitamin E

The name *vitamin E* is applied to four closely related substances called tocopherols, of which α-tocopherol is the most abundant and most potent. All are soluble in oil and are absorbed readily from the digestive tract. Vitamin E occurs in wheat-germ oil, fresh vegetables, fruits, liver, and muscle. Deficiency phenomena differ considerably in various species. It has been demonstrated without question that vitamin E is essential to the fertility of male and female rats and mice, but in larger animals and in man this has yet to be proved. Vitamin E deficiency in rats, guinea pigs, and rabbits results in muscular dystrophy and lesions of the spinal cord. The administration of vitamin E to human beings with similar dysfunctions, however, has yielded equivocal results. Vitamin E deficiency in man causes hemolysis and creatinuria. A primary (dietary) deficiency may occur in early infancy but is seldom encountered in the adult; a secondary deficiency may arise as a consequence of malabsorption.

Vitamin K

As used here, vitamin K means vitamin K activity. Vitamins K_1 and K_2 occur naturally and are fat soluble; menadione (fat soluble) and menadione sodium bisulfite (water soluble) are synthetic K vitamins. Vitamin K is essential for prothrombin formation by the liver, and its deficiency (hypoprothrombinemia) results in defective coagulation of the blood and a predisposition to excessive bleeding. Vitamin K is derived in part from the diet but mostly from intestinal bacterial synthesis. Although a deficiency occasionally relates to an inadequate intake, most cases arise from a decrease in bacterial flora. Such a decrease is often encountered during the first few days of life and during therapy with nonabsorbable sulfonamides and antibiotics. Other possible causes of deficiency include malabsorption and excessive use of mineral oil. In medicine, vitamin K preparations are used to prevent and treat bleeding due to a lack of prothrombin, and as antidotes against an overdosage of certain anticoagulants. The vitamin should be given to every newborn to prevent hypoprothrombinemia.

Vitamin C

By unknown mechanisms, vitamin C (ascorbic acid) helps maintain the integrity of intercellular substances throughout the body; these substances include intercellular cement, tissue fibers, bone matrix, and the dentin of teeth. Vitamin C chemically supports the action of the enzyme that converts folic acid to its active form (folinic acid); it facilitates the absorption of iron from food; it is readily oxidized and reversibly reduced, and presumably plays an important role in electron transfer in the cell. A severe deficiency of vitamin C results in scurvy, a disease marked especially by bleeding gums, splotchy hemorrhages beneath the skin, and abnormal formation of bone and teeth. The recommended daily allowance of vitamin C for adults is 55–60 mg. Linus Pauling's belief that large doses of vitamin C reduce the chances of catching cold has stirred much interest and not a little controversy. At the time of this writing, a number of studies tend to support his thesis. At the least, the subject is worthy of further study.

Thiamine

Thiamine, or vitamin B_1, occurs in enriched bread, nuts, pork, whole grains, legumes, and liver. Since it functions as a coenzyme in biooxidation, a deficiency leads to general metabolic disturbance. The major signs of deficiency include polyneuritis and enlargement of the heart. When these derangements occur at the same time, the syndrome is referred to as beriberi. It is wrong to assume, however, that a thiamine avitaminosis and beriberi are synonymous, because lesser deficiencies lead to a variety of vague and "unimportant" symptoms. Indeed, the deficiencies the physician usually sees—and this applies to all the vitamins—are not those depicted by the typical high school biology text. It is uncommon to come across a patient with a single avitaminosis; a poor diet, the usual etiology, leads to multiple deficiencies. The recommended daily allowance of thiamine for adults is 1.0–1.5 mg (Table 18-2).

Riboflavin

Also known as vitamin B_2, riboflavin is present in milk, liver, yeast, kidneys, eggs, nuts, seafoods, meats, cheese, and green leafy vegetables. Like thiamine, riboflavin is a coenzyme in oxidative metabolism. The chief signs of deficiency include a purplish red tongue (glossitis), reddening of the lips, fissuring of the angles of the mouth (cheilosis), and cornification of the skin. The repercussions of a riboflavin deficiency (ariboflavinosis) in the rat are evidenced dra-

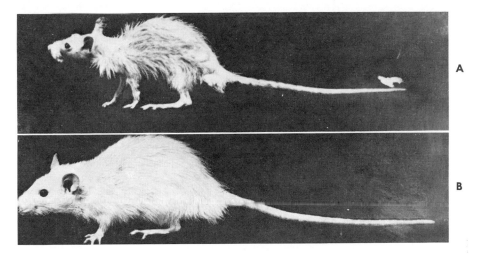

FIG. 18-2

A, White rat with ariboflavinosis. **B,** Same rat 6 weeks later after being placed on diet rich in riboflavin. (United States Department of Agriculture photograph.)

matically in Fig. 18-2. The recommended daily allowance for adults is 1.5–2.0 mg.

Niacin

Niacin (nicotinic acid) or its amide (nicotinamide) is a B-complex vitamin abundantly present in rice, bran, liver, yeasts, meat, peanuts, and fish. It is needed in oxidation-reduction reactions and in carbohydrate and tryptophan metabolism. The major clinical signs and symptoms of niacin deficiency are referable to the skin, gastrointestinal tract, and nervous system. Pellagra, the classic syndrome, is characterized by muscular weakness, discoloration of the skin, and mental derangement; it is actually a multiple deficiency involving thiamine and riboflavin as well as niacin. The recommended daily allowance of niacin for adults is 13–20 mg (Table 18-2).

Vitamin B₁₂ and folic acid

Two members of the B-complex group, vitamin B_{12} (cyanocobalamin) and folic acid, are required for the formation of new protoplasm. This is borne out by the fact that areas of rapid turnover, such as bone marrow, are the first to reflect a deficiency. Lack of vitamin B_{12}, which is usually the result of inadequate intrinsic factor, leads to pernicious anemia (Fig. 18-3). Lack of folic acid leads to the so-called megaloblastic anemias of infancy and pregnancy and generally retards growth and body development, especially in children. Vitamin B_{12} occurs principally in liver, muscle tissue, milk, and cheese; folic acid occurs principally in green leafy vegetables, yeast, soybeans, and wheat. The recommended daily allowances for both vitamins are given in Table 18-2.

Pyridoxine

Pyridoxine (vitamin B_6) acts as a coenzyme in the metabolism of certain amino acids. It is present in egg yolk, nuts, whole grains, legumes, kidneys, muscle, liver, and fish. Though the importance of this B-complex vitamin in animals has long been known, its essential role in human

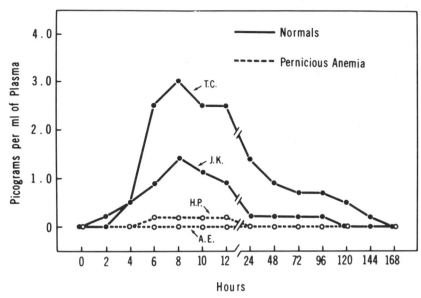

FIG. 18-3

Plasma concentration of radioactive vitamin B_{12} in two normal volunteers and two patients with pernicious anemia after ingestion of 30 g of scrambled egg yolk containing 1.12 μg of radioactive vitamin B_{12}. (Vitamin labeled with cobalt-57 was incorporated in vivo by injecting it into the egg-laying hen.) Note poor showing of patients occasioned by lack of intrinsic factor. (Courtesy Dr. A. Doscherholmen, Veterans Administration Hospital, Minneapolis, Minnesota.)

nutrition was not appreciated until an unfortunate but nevertheless excusable and instructive incident occurred some years ago. This related to widely scattered reports of a nervous syndrome (increased excitability, convulsions, and the like) of unknown origin among infants receiving a commercial milk substitute. An investigation soon demonstrated that the condition resulted from the absence of pyridoxine in the formula. Other deficiency disorders relating to the vitamin are seborrheic dermatitis, impaired growth in infants, and vomiting in pregnancy. The recommended daily allowance of pyridoxine for adults is about 2 mg (Table 18-2).

Pantothenic acid

Pantothenic acid is a B-complex vitamin used by the body in the synthesis of coenzyme A, a

TABLE 18-3

Energy expenditures for various activities

Activity	Cal/kg/hr
Sleeping	0.9
Sitting	1.4
Walking	2
Bicycling	2.5
Sawing wood	6.8
Swimming	7.1
Running	8
Climbing stairs	16

TABLE 18-4

Desirable weights for men and women age 25 and over*

Height (with shoes)		Small frame	Medium frame	Large frame
Feet	Inches			
Men†				
5	2	112–120	118–129	126–141
5	3	115–123	121–133	129–144
5	4	118–126	124–136	132–148
5	5	121–129	127–139	135–152
5	6	124–133	130–143	138–156
5	7	128–137	134–147	142–161
5	8	132–141	138–152	147–166
5	9	136–145	142–156	151–170
5	10	140–150	146–160	155–174
5	11	144–154	150–165	159–179
6	0	148–158	154–170	164–184
6	1	152–162	158–175	168–189
6	2	156–167	162–180	173–194
6	3	160–171	167–185	178–199
6	4	164–175	172–190	182–204
Women‡				
4	10	92– 98	96–107	104–119
4	11	94–101	98–110	106–122
5	0	96–104	101–113	109–125
5	1	99–107	104–116	112–128
5	2	102–110	107–119	115–131
5	3	105–113	110–122	118–134
5	4	108–116	113–126	121–138
5	5	111–119	116–130	125–142
5	6	114–123	120–135	129–146
5	7	118–127	124–139	133–150
5	8	122–131	128–143	137–154
5	9	126–135	132–147	141–158
5	10	130–140	136–151	145–163
5	11	134–144	140–155	149–168
6	0	138–148	144–159	153–173

*Courtesy of the Metropolitan Life Insurance Company.
†Weights in pounds (in indoor clothing; for nude weight deduct 5 to 7 pounds).
‡Weights in pounds (in indoor clothing; for nude weight deduct 2 to 4 pounds).

cellular agent vital to normal metabolism. Since the vitamin is present in such a wide variety of foods, a deficiency in man has not as yet been described except in experimental situations. The avitaminosis is characterized by fatigue, malaise, headache, sleep disturbances, colic, vomiting, and impaired coordination. The daily need is thought to be about 10 mg for adults.

OBESITY

Obesity is a major nutritional problem. It is unsightly, and it shortens life. Certain brain lesions (Chapter 20) and certain endocrine disorders (Chapter 23) result in obesity, but the ultimate cause in most instances remains a mystery. But whatever the ultimate etiology, the immediate cause is an excess caloric intake over caloric output (Table 18-3). We put on weight by overeating or underexercising, and all the "diet revolutions" in the world are not going to alter the situation. The worst diet books are potentially dangerous, and the best merely confirm our common sense. The answer to weight control is self-control. Desirable weights are given in Table 18-4.

QUESTIONS

1. We sustain a daily loss of about an ounce of protein. How many grams is this?
2. Why is protein the only complete nutrient.
3. What elements do fats, carbohydrates, and proteins have in common?
4. Why does nitrogen balance relate specifically to proteins?
5. How do you explain a positive nitrogen balance?
6. A positive nitrogen balance can be sustained in an adolescent but not in an adult. Why?
7. How do you explain a negative nitrogen balance?
8. What is the daily protein requirement (in ounces) for a man weighing 154 pounds?
9. Proteins are polymers of amino acids. What does this mean?
10. Even though 22 amino acids exist in the body, only eight must be supplied by the diet. Explain.

11. Other things being equal, vegetable protein is inferior to animal protein. Explain.
12. A protein deficiency can develop in an individual even in the face of an adequate intake of protein. Explain.
13. What is the etiological distinction between marasmus and kwashiorkor?
14. Eating more protein than the daily allowance calls for is neither biologically nor economically wise. Discuss.
15. Distinguish between an amino acid and a fatty acid.
16. At the molecular level, what do the three essential fatty acids cited in the text (p. 283) have in common?
17. Many minerals occur in food as inorganic salts. What is the best example?
18. The red-cell count may very well be normal in iron-deficiency anemia. Explain.
19. Iron-deficiency anemia is often treated with ferrous sulfate. What is the formula for this compound?
20. Can you explain why liver is rich in iron?
21. On a basis of weight, what is the major metal in the body?
22. The daily requirements for calcium and phosphorus are about the same. What is the chemical reason for this?
23. The normal adult needs about 10 mg of iron per day and about 1 g of calcium. The latter is a much larger "dose." How much larger?
24. Are mineral elements metals or nonmetals?
25. Why are some regions of the world goiter belts?
26. One brand of iodized salt contains 0.01% potassium iodide (KI). How many milligrams of KI would an ounce of this salt supply?
27. Of the amount calculated in question 26, how much is elemental iodine?
28. Assuming a daily iodine requirement of 100 μg, how many days' supply does 1 ounce of iodized salt (question 26) contain?
29. Fluoridation continues to stir up much controversy both here and abroad. The springboard of the antifluoridationists, apparently, is the fact that fluorides are poisonous. (Sodium fluoride is used as a rat poison.) What are your views?
30. The daily requirement of iodine is fantastically low compared to, say, that of calcium. Precisely why?
31. What is the origin of the expression "B complex"?
32. The expression "vitamin A activity" is generally preferred to "vitamin A content." Why?
33. How is β-carotene transformed into vitamin A?
34. Explain why a lack of bile can result in a vitamin A deficiency.
35. How can the excessive use of mineral oil result in a vitamin K deficiency?
36. "Vitamin D activity" is a more accurate expression than "vitamin D content." Why?
37. The role of vitamin D in calcification is actually more indirect than direct. Explain.
38. Rickets may well have been the first disease to be linked to air pollution. What is the story?
39. Strictly speaking, vitamin D is a hormone and not a vitamin. Support this statement.
40. Vitamin E deficiency in humans results in hemolysis and creatinuria. What are these conditions?
41. Almost all cases of vitamin K deficiencies in the adult are secondary. Explain.

42. Vitamin K is a specific antidote against overdoses of certain anticoagulants. Explain its action.
43. Why is it especially important to give the newborn vitamin K?
44. Chemically, what type of compound is vitamin C?
45. Consult the literature and see if you can find hard evidence that vitamin C is effective in the prevention of the common cold.
46. According to the text, thiamine functions as a coenzyme. Does this help explain why the daily requirement is only about 1 mg?
47. What are avitaminoses?
48. Distinguish among niacin, nicotinic acid, and nicotinamide.
49. Tryptophan is converted by the brain into the neurohormone serotonin. Does this fact fit in with what we know about pellagra?
50. In the treatment of pernicious anemia, vitamin B_{12} is much more effective when administered parenterally than when administered orally. Why?
51. Certain drugs used in the treatment of cancer are folic acid antagonists. How does this explain their action?
52. For the most part, what role do B-complex vitamins play in body chemistry?
53. What is hypervitaminosis?
54. "We put on weight by overeating or underexercising." This is both common sense and the unanimous opinion of science. How, then, do you account for the tremendous popularity of "calories-don't-count" books?
55. What are the recommended caloric contributions of fat, carbohydrate, and protein for an intake of 2,500 calories?

CHAPTER 19

THE URINARY SYSTEM

The urinary system is the prime guardian of homeostasis. By excreting unwanted materials and retaining others, it rids the internal environment of wastes and at the same time maintains acid-base balance and fluid and electrolyte balance (Chapter 8). The system consists of the kidneys, the ureters, the bladder, and the urethra (Fig. 19-1).

KIDNEYS

The kidneys are reddish brown, bean-shaped organs embedded in the posterior abdominal wall, one on either side of the vertebral column. Since they are located behind the peritoneum, the kidneys are said to be retroperitoneally situated. Technically, therefore, they lie outside the peritoneal cavity. An average-sized kidney is about 4 inches long, 3 inches wide, and 1 inch thick and weighs between ¼ and ½ pound. Characteristically, a tough, glistening, translucent capsule encases each kidney. Through the hilum on the medial aspect pass the ureter, blood vessels, nerves, and lymphatics.

When cut longitudinally, the kidney is seen to be composed of two regions, an outer cortex and an inner medulla (Fig. 19-2). The ureter dilates into the pelvis, and the pelvis branches into a number of cup-shaped structures called calyces. Leading into the calyces are the apices, or papil-lae, of triangular structures called pyramids. Each pyramid is actually a bundle of collecting tubules that open at the papilla and serve to carry urine into the calyx.

Nephron

The functional unit of the kidney is the nephron, a minute tube about 12 mm long. There are some one million nephrons in each kidney. The basic anatomy of the unit is shown in somewhat idealized form in Fig. 19-3. Its chief features are the renal corpuscle (or malpighian body) and the renal tubule. The corpuscle is composed of a tuft of capillaries, called the glomerulus, enclosed within a double-walled membrane called Bowman's capsule. The latter is actually the dilated and invaginated blind end of the proximal convoluted tubule; the distal convoluted tubule enters a collecting duct. Blood makes its way to the nephron from the renal artery via the interlobar artery, arcuate artery, interlobular artery, and afferent arteriole (vas afferens), in that order. The efferent arteriole (vas efferens) leads the blood from the glomerulus to the capillary networks surrounding the tubules. After working its way through these networks, the blood returns to the renal vein via the interlobular vein, the arcuate vein, and the interlobar vein, in that order (Fig. 19-3).

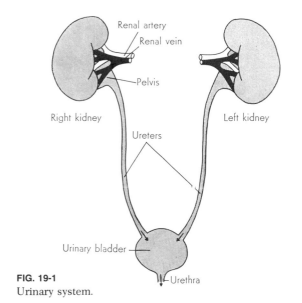

FIG. 19-1
Urinary system.

Formation of urine

The formation of urine by the nephron entails three basic processes: filtration, reabsorption, and secretion. The disparity in diameter between the glomerular arterioles, the afferent being the larger, produces sufficient pressure (somewhere in the vicinity of 70 mm Hg) within the glomerular capillaries to force fluid out of the glomerulus. Some of this fluid, called the glomerular filtrate, eventually becomes urine. Counterpressures, that is, the osmotic pressure of the blood (about 30 mm Hg) and the pressure in Bowman's capsule (about 20 mm Hg), however, tend to keep fluid in the glomerulus. The actual filtration pressure, then, is the difference between the glomerular blood pressure and the osmotic and capsular pressures. Taking the figures just cited, the filtration pressure is 70 −

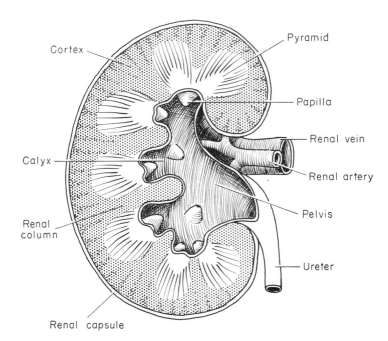

FIG. 19-2
The kidney. (Modified from Anthony, C. P., and Kolthoff, N. J.: Textbook of anatomy and physiology, ed. 8, St. Louis, 1971, The C. V. Mosby Co.)

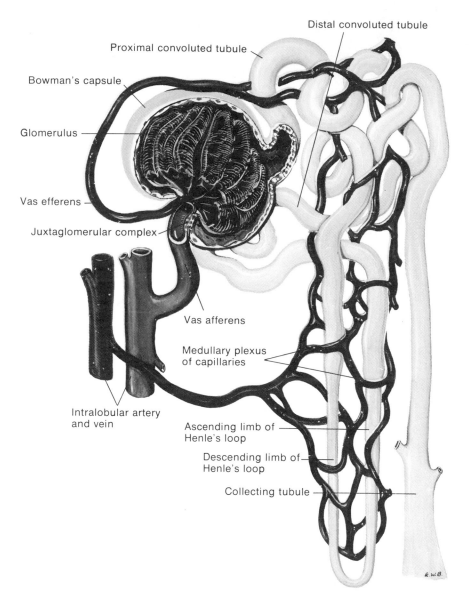

Proximal convoluted tubule

Distal convoluted tubule

Bowman's capsule

Glomerulus

Vas efferens

Juxtaglomerular complex

Vas afferens

Medullary plexus
of capillaries

Intralobular artery
and vein

Ascending limb of
Henle's loop

Descending limb of
Henle's loop

Collecting tubule

FIG. 19-3

Nephron and associated blood supply. Wall of Bowman's capsule has been cut away to reveal detail of glomerulus. (From Schottelius, B. A., and Schottelius, D. D.: Textbook of physiology, ed. 17, St. Louis, 1973, The C. V. Mosby Co.)

(30 + 20) or 20 mm Hg. Clearly, a significant drop in glomerular blood pressure curbs the output of glomerular filtrate and urine. Constriction of the afferent arteriole (via stimulation of sympathetic nerves), for instance, drastically lowers the pressure by impeding the blood flow into the glomerulus. Conversely, dilatation of the afferent arteriole enhances the flow of blood, thereby elevating the pressure and augmenting the output of glomerular filtrate and urine. The systemic arterial pressure, too, plays a key role; when it rises, the increased amount of blood flowing into the glomeruli increases the pressure and thereby the output of urine. Conversely, a decrease in systemic pressure diminishes the output of urine.

Except for the absence of protein and other large molecules, the glomerular filtrate passing into Bowman's capsule is similar in composition to plasma; the walls of the glomerulus are normally permeable only to water and other small molecules. The filtrate is formed at a rate of about 125 ml per minute; during a 24-hour period, this amounts to close to 200 l! Obviously, the great bulk of this volume must be returned to the blood, and therein lies the major role of the tubules. As the filtrate moves along, almost all of the water and most of the dissolved substances are reabsorbed, thereby reducing the 200 l to about 1.5 l of urine. Reabsorbed substances actively or passively leave the tubules by passing through the tubular cells into the interstitial fluid of the kidney; from here they enter the blood by diffusion through the walls of the capillaries surrounding the tubules.

Because the blood associated with the proximal tubules is more concentrated than the newly formed glomerular filtrate, water diffuses back into the blood via osmosis. This commonly is referred to as "obligatory reabsorption." At the same time, specialized cells in the proximal tubules are engaged in the active reabsorption of electrolytes, glucose, amino acids, and other solutes. The loss of these materials also tends to lower the osmotic pressure of the tubular urine, so that even more water diffuses back into the blood. In the distal tubules and collecting ducts, however, water is reabsorbed actively, the extent of the reabsorption varying according to the overall fluid status of the body. At the same time reabsorption is taking place along the tubule, other compounds, including creatinine and potassium, are being secreted actively into the tubular urine by the tubular cells. In sum, then, a blood constituent undergoes filtration (1) with subsequent reabsorption, (2) without subsequent reabsorption or secretion, or (3) with secretion and with or without reabsorption. Finally, there is a point beyond which reabsorbable substances cannot be reabsorbed by the tubules. This is called the *threshold value*. For example, if the amount of glucose in the blood is above the threshold of 150 mg/100 ml, "sugar" will appear in the urine (glycosuria).

Diuresis

An increase in the volume of urine excreted (diuresis) accompanies an increased intake of water. This commonly is referred to as water diuresis. There is also what is called solute or osmotic diuresis, the essential feature of which relates to the inability or decreased ability of the nephron tubules to reabsorb certain solutes from the glomerular filtrate. The sugar mannitol, for example, cannot be reabsorbed, which means that the osmotic pressure of tubular urine rises following ingestion of the sugar and thereby inhibits the obligatory reabsorption of water. The most dramatic example of solute or osmotic diuresis involves the use of specific drugs called diuretics, the more potent of which act by inhibiting the reabsorption of sodium in the proximal convoluted tubule. The upshot, once again, is an increase in the osmotic pressure of tubular urine and an inhibition of the obligatory reabsorption of water. Diuretics are valuable drugs in the management of edema and hypertension (p. 198).

Hormonal control

Water diuresis is readily understood, but the physiological details of the underlying mechanism are far from simple. The accepted explanation runs as follows. When the intake of water is in excess of body needs, the osmotic pressure of the extracellular fluid decreases relative to the intracellular fluid, and additional water diffuses into the cells. Specialized cells called osmoreceptors, located in the hypothalamus of the brain, are activated by this influx of water; they release inhibitory stimuli to the posterior pituitary (Chapter 23) that decrease the gland's output of antidiuretic hormone (ADH), a hormone that in some unexplained way promotes the active reabsorption of water in the distal convoluted tubules. Conversely, if water loss exceeds intake, the osmotic pressure of the extracellular fluid rises, and water diffuses from the cells. This curbs the release of inhibitory stimuli by the osmoreceptors and thereby stimulates the output of ADH and promotes the conservation of water. The active reabsorption of water in the distal convoluted tubules is called *facultative* reabsorption, as opposed to the *passive* or *obligatory* reabsorption in the proximal convoluted tubules.

Another hormone having pronounced influence on kidney function and body fluid is aldosterone, the major mineralocorticoid secreted by the adrenal cortex (Chapter 23). Aldosterone promotes the reabsorption of sodium by the distal tubules, partly at the expense of a less complete reabsorption of potassium. Aldosterone secretion quickly increases if the blood volume and blood pressure decrease markedly or if the sodium concentration in the blood drops below normal. One mechanism postulated for the release of aldosterone is as follows. A decreased blood volume—and therefore a decreased arterial pressure—stimulates the juxtaglomerular apparatus (a group of cells situated near the glomerulus) to release an enzyme, called renin, that converts the plasma protein angiotensinogen to angiotensin I. Another plasma enzyme converts angiotensin I to angiotensin II, and angiotensin II stimulates the adrenal cortex to increase the output of aldosterone. This results in an increased reabsorption of sodium, followed by enhanced obligatory reabsorption of water. Angiotensin II also elevates blood pressure via vasoconstriction.

Acid-base balance

Because sudden death can result if blood pH drops below 6 or rises above 8, the topic of acid-base balance is of great interest and importance. Stripped to the essentials, the maintenance of acid-base balance in the body concerns the hydrogen ion and the way the body gets rid of it to prevent acidosis and retains it to prevent alkalosis. The body does this via three mechanisms working in concert: respiratory regulation, chemical buffer regulation, and renal (kidney) regulation.

The kidneys regulate acid-base balance by three avenues: secretion of hydrogen ions, formation of ammonia, and excretion of bicarbonate ions. Hydrogen-ion secretion is effected principally in the distal tubules, where the free ions, secreted by the tubular epithelium, react with disodium phosphate (Na_2HPO_4) and dipotassium phosphate (K_2HPO_4) in the glomerular filtrate to form monosodium phosphate (NaH_2PO_4) and monopotassium phosphate (KH_2PO_4), respectively. The released sodium and potassium ions are reabsorbed by the tubular epithelium. In effect, hydrogen ions are exchanged for sodium and potassium ions.

In order to spare sodium or to compensate if the above mechanism becomes overtaxed, the tubular cells synthesize ammonia (NH_3), chiefly from glutamine. Ammonia combines with hydrogen ions to form ammonium ions (NH_4^+), and the ammonium ions are excreted in the urine as ammonium salts.

The above mechanisms effect the removal of hydrogen ions and operate to prevent acidosis. A decrease in pH is a more usual tendency than an increase in pH because metabolic wastes are mainly acids (namely, phosphoric, sulfuric, uric, carbonic, and keto acids). Alkalosis, however, is always a possibility, particularly as a result of taking alkaline drugs. To prevent alkalosis, hydrogen ions are retained by the tubules, and bicarbonate ions (basic ions) are excreted.

Kidney function tests

A number of tests and procedures are employed to measure kidney function. Among the more routine are those measuring the levels of *b*lood *u*rea *n*itrogen (BUN), *n*onprotein *n*itrogen (NPN), and creatinine, and those measuring *g*lomerular *f*iltration *r*ate (GFR) and *p*henol*s*ulfon*p*hthalein (PSP) excretion. BUN and NPN are determined by chemical analysis of blood serum and have normal values of 15 mg/100 ml and 25 mg/100 ml, respectively. Creatinine levels (normal, 0.7–1.5 mg/100 ml) are informative to the same degree as BUN and NPN and are actually subject to less day-to-day variation. If the kidneys are not functioning normally, these values typically will rise. The GFR, which is determined clinically by the rate at which creatinine or inulin is removed ("cleared") from the blood, shows a normal value of between 97 and 140 ml/min for men, and between 85 and 125 ml/min for women. Values below these clearly indicate kidney failure. The PSP test is performed by injecting the dye phenolsulfonphthalein into a vein and clocking the time that it takes to be excreted into the urine. A normal kidney clears most of the dye in 1 hour and all of it within 2 hours. If it takes significantly longer for the dye to be cleared

from the blood, one or both kidneys are probably damaged. In this event, more definitive tests are employed to determine whether the impairment is glomerular or tubular.

URETERS

The ureters are cylindrical fibromuscular tubules that convey urine from the renal pelvis to the bladder (Fig. 19-1). They average about 11 inches in length and ¼ inch in diameter and are composed of an inner mucous, a middle muscular, and an outer fibrous layer. Peristaltic waves begin at the kidney and force the urine along into the bladder, where it is held until voided.

BLADDER

The urinary bladder is a tough, muscular storage sac located in the pelvic cavity, immediately before the vagina in the female and immediately before the rectum in the male. Though held in place by folds of peritoneum and fascia, it is subject to considerable movement. The bladder wall is composed of four coats. Beginning with the inside, these include the mucous, the submucous, the muscular, and the serous coats. The muscular coat has an inner longitudinal, a middle circular, and an outer longitudinal layer. The mucous lining consists of transitional epithelium that is thrown into folds, or rugae, when the bladder is empty. The bladder has three openings at the neck, two through which the ureters deliver urine and one through which urine enters the urethra. The ureters open into the posterior angles of a triangular depression called the trigone, and the urethral opening is at the anterior and lower angle. The bladder serves as a reservoir for urine. Its capacity varies widely, but in most adults the bladder can accommodate about 1 l. The urge to urinate, however, is experienced long before this volume is attained, usually when the bladder contains about 300 ml of urine.

URETHRA

The female urethra is a membranous tube about 1½ inches long that extends from the bladder to the urinary meatus, or external orifice, situated between the clitoris and the vaginal opening (Chapter 24). Its walls are composed of three coats: an inner mucous coat, a middle spongy coat containing a plexus of veins, and an outer muscular coat.

The male urethra is about 8 inches long and is divided into three parts: the prostatic, which runs through the prostate gland; the membranous, which pierces the body wall; and the cavernous, which extends the length of the penis (Chapter 24). Its walls have two coats, an inner mucous lining and an outer coat composed of connective tissue. The outer coat serves as a means of attachment to the structures through which the urethra passes.

MICTURITION

The urge to urinate, or micturate, comes when there is about 300 ml of urine in the bladder, at which volume the intrabladder pressure becomes of sufficient intensity to excite the stretch receptors in the bladder wall. As a result, sensory stimuli are sent to the spinal cord. The parasympathetic nerves supplying the musculature are triggered, and the bladder wall starts to contract. Simultaneously, the internal sphincter guarding the opening between the bladder and the urethra is opened. However, if it is not convenient to urinate, the external sphincter around the urethra is voluntarily closed and held closed until it is convenient, at which time the sphincter is relaxed. Micturition, then, is both voluntary and involuntary.

URINE

Freshly voided, normal urine is a clear, pale yellow liquid with a characteristic odor. Upon standing, it becomes cloudy and develops a penetrating odor as a result of the conversion of urea

TABLE 19-1

Average values for normal urine (24-hour sample)*

Volume: 1,200 ml		
Color: pale yellow		
Transparency: clear		
Odor: characteristic		
Acidity: pH 6 (4.7–8.0)		
Specific gravity: 1.020		
Total solids	60	g
Ammonia, as salts	0.7	g
Calcium	0.2	g
Chlorides, as NaCl	12	g
Creatine	0.03	g
Creatinine	1.4	g
Magnesium	0.1	g
Potassium	2	g
Sodium	4	g
Urea	30	g
Other	9.6	g

*From Brooks, S. M.: Basic facts of body water and ions, ed. 3, New York, 1973, Springer Publishing Co., Inc.

into ammonia. Its pH varies from about 4.6 to 8, and its specific gravity ranges from 1.003 to 1.030. In a 24-hour period the total volume of urine expelled from the body averages around 1,200 ml, with a range anywhere from 600 to 2,500 ml, depending on fluid intake, perspiration, and other factors.

Urine is a dilute, aqueous solution containing about 60 g of solids in a 24-hour sample. Approximately one half of this amount is urea, the chief waste product, and the remainder is creatinine, electrolytes, and an assortment of other organic and inorganic constituents (Table 19-1). The characteristic color of urine is due to urochrome, a pigment derived from bile.

Abnormal constituents

Among the abnormal urine constituents of diagnostic importance are blood (hematuria), glucose (glycosuria or glucosuria), albumin (albuminuria), pus (pyuria), and casts. Hematuria

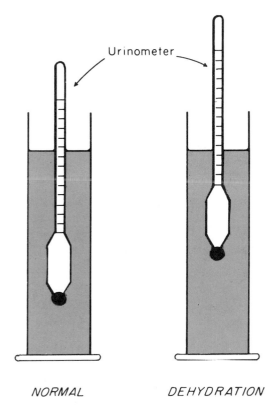

Urinometer

NORMAL   DEHYDRATION

FIG. 19-4
Effect of dehydration on specific gravity of urine. (From Brooks, S. M.: Basic facts of body water and ions, ed. 3, New York, 1973, Springer Publishing Co., Inc.)

indicates bleeding somewhere along the urinary tract, glycosuria generally points to diabetes mellitus, albuminuria is a feature of nephrosis, and pus indicates infection. Urine normally contains a trace of glucose. The complete absence of glucose in urine indicates the presence of bacteria (bacteriuria) because bacteria metabolize and use up glucose. Special tests have been devised for the detection of trace glucose and are now being used in the diagnosis of urinary tract infections.

The physical attributes of urine are also help-ful in diagnosis since a great many pathologic conditions cause a change in its color, volume, and specific gravity (Fig. 19-4). In dehydration, for example, the volume decreases and the specific gravity increases.

Urinary sediment

According to many authorities, examination of the urinary sediment is the single most valuable diagnostic procedure in kidney disease, particularly in regard to differential diagnosis. In acute glomerulonephritis, for example, the sediment discloses casts containing red blood cells and hemoglobin, whereas in chronic glomerulonephritis broad waxy casts are seen. The nephrotic syndrome is characterized by increased numbers of hyaline and granular casts and oval flat bodies.

URINARY INFECTIONS

Infection is the most common disease of the urinary system. Gram-negative bacteria are the usual pathogens, and the species of major concern are *Escherichia coli, Enterobacter aerogenes, Pseudomonas aeruginosa,* and *Proteus* spp. (Gonorrhea, a prime example of genitourinary infection, is discussed in Chapter 24.) Occasionally, gram-positive bacteria, namely, *Staphylococcus epidermidis* and *Streptococcus faecalis* are involved in urinary infections. Infections of the kidney, ureter, bladder, and urethra are referred to as pyelonephritis, ureteritis, cystitis, and urethritis, respectively. Because bacteria are often present in the distal portion of both the male and female urethra, urethritis is the most common genitourinary infection. Urethritis often ascends and leads to cystitis, ureteritis, and pyelonephritis.

GLOMERULONEPHRITIS

Glomerulonephritis is a serious involvement of the kidney marked by changes and damage in the glomeruli. The clinical features include albuminuria, hematuria, edema, oliguria, hypertension, and, in advanced cases, uremia (azotemia). The acute form usually relates to a previous infection with *Streptococcus pyogenes*. This suggests that the disease is an immunologic response either to the bacteria themselves or to their metabolic activities. Chronic glomerulonephritis occurs without apparent relation to strep infections or any other disease process. No specific treatment for glomerulonephritis is known, but general medical management is most important. Many patients with the acute form recover completely within a period of time ranging from a few weeks to 2 years. Beyond 2 years, remission is unlikely, and the disease may be considered chronic. In the event of uremia the patient must be placed on a low-protein diet, and if this and other conservative measures fail, the only recourse is dialysis or transplantation. Kidney transplants involving identical twins yield excellent results in comparison to "nonidentical" transplants. The best hope lies in the eventual development and perfection of an artificial kidney that can be implanted in the body.

NEPHROTIC SYNDROME

Nephrotic syndrome is a disorder characterized by pronounced albuminuria, edema, and hypertension. The underlying causes are many, the most common being the so-called lipoid nephrosis of childhood. Treatment centers on control of the diet and on the use of corticosteroids and diuretics.

ACUTE RENAL FAILURE

Acute renal failure is a severe reduction of kidney function. The most frequent and important cause is damaged nephron tubules ("tubular necrosis"), a potentially reversible condition that occurs in the wake of transfusion reactions, shock, trauma, burns, and nephrotoxic poisonings (for example, poisonings caused by mercury, carbon tetrachloride, and ethylene glycol).

TABLE 19-2

Results of dialysis in a patient with renal failure*†

Blood constituent	Initial	Final
K (mEq/l)	7.3	5.1
Na (mEq/l)	108	130
Cl (mEq/l)	76	106
Ca (mEq/l)	3.7	5.1
NPN (mg/100 ml)‡	162	106
BUN (mg/100 ml)‡	72	48

*After Merrill; from Brooks, S. M.: Basic facts of body water and ions, ed. 3, New York, 1973, Springer Publishing Co., Inc.

†Results obtained after 6 hours of dialysis in a 35-year-old patient with renal insufficiency.

‡NPN, Nonprotein nitrogen; BUN, blood urea nitrogen.

The chief signs are oliguria (anuria is uncommon) and a buildup of potassium (hyperkalemia) and nitrogenous urinary wastes (azotemia). Fluid intake must be restricted to a volume equal to urine output plus an allowance of about 400 ml per day to cover insensible losses. The diet should be free of potassium, protein, and sodium. Hyperkalemia, a major threat to life (Fig. 19-5), can be controlled by oral administration of cation exchange resins (agents that remove potassium from intestinal juice). When these measures fail, dialysis often proves life-saving (Table 19-2).

URINARY CALCULI

Calculi (stones) are formed in the urinary tract by the precipitation of chemical salts in the urine. There are two types of causes: increased urinary concentration of crystalloids that compose stones, and physical changes in the urine or urinary tract that favor precipitation. Specific causes include excessive ingestion of milk, previous infection, dehydration, gout, hyperadrenocorticism, and hyperparathyroidism (Chapter 23). Typical signs and symptoms include excruciating intermittent pain in the kidney area (renal colic), chills, fever, and frequent urination. Many cases of urinary calculi are uncom-

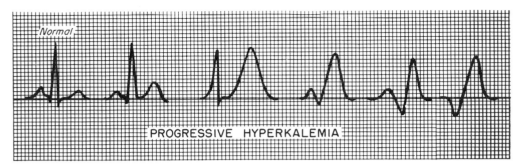

FIG. 19-5
Effect of hyperkalemia upon electrocardiogram.

plicated by obstruction or infection and require no treatment. Obstructing calculi are treated expectantly or surgically.

RENAL HYPERTENSION

Any disorder or disease of the kidney causing low renal blood pressure or inhibited blood flow (renal ischemia) may result in hypertension. As noted previously, low pressure (or flow) stimulates the juxtaglomerular apparatus to produce renin. This, in turn, leads to the formation of aldosterone and causes widespread vasoconstriction. Both effects increase blood pressure. Underlying disorders and diseases include narrowing of the renal artery (stenosis), aneurysms, and blood clots. Surgery, such as removal of the kidney in advanced cases, often restores the pressure to normal.

CANCER

Many types of tumors may occur in the kidney, but only malignant ones are a serious threat to life. With one major exception (Wilms' tumor), kidney cancers usually strike older individuals and afflict twice as many males as females. No definitive causes have been discovered, but speculation points to certain exogenous carcinogens that the kidney presumably concentrates and stores. Lead compounds and hexachlorobenzene have been suggested. Kidney stones, which commonly accompany renal cancer, may play a role. Both suggestions fit in well with the "irritation" theory of cancer development. Some cancers smolder along for years without producing signs and symptoms, while others spread throughout the body in a matter of months. Metastases may appear just about anywhere, but the most common sites are the lungs and bones. Radical removal of the kidney in conjunction with irradiation provides the best treatment in adult involvements. About half of the patients undergoing such treatment can be expected to live 5 years or more. In children, the use of actinomycin D in concert with irradiation and immediate surgery has produced spectacular results, effecting cures in perhaps half the cases.

QUESTIONS

1. Discuss homeostasis.
2. Are the kidneys located in the abdominal cavity or the peritoneal cavity? Explain your answer.
3. Are the renal pyramids associated with the cortex or the medulla?
4. What is the relationship of the calyces to the renal pelvis?
5. What is a translucent capsule?
6. Account for the expression malpighian body.
7. Compare the literal meanings of efferent and afferent.

8. The diameter of the afferent arteriole is greater than that of the efferent arteriole. What would be the effect if the reverse were true?

9. Compare the situational relationship of the proximal and distal convoluted tubules to Bowman's capsule.

10. Trace the route of a red blood cell from the renal artery to the renal vein, naming in order all the vessels along the way.

11. Why is the osmotic pressure of the blood referred to as a "counter pressure" (p. 293)?

12. Calculate the filtration pressure when the glomerular pressure is 60 mm Hg, the osmotic pressure of the blood is 25 mm Hg, and the pressure in Bowman's capsule is 20 mm Hg.

13. Why does the glomerular filtrate contain large amounts of crystalloids, but only traces of colloids?

14. Why does a drop in blood pressure curb the production of glomerular filtrate?

15. Shock causes oliguria. Explain.

16. Distinguish between active and passive reabsorption.

17. Why is the reabsorption of water in the proximal convoluted tubules said to be obligatory?

18. Why is the reabsorption of water in the distal convoluted tubules said to be facultative?

19. A drug called probenecid is often given in conjunction with penicillin to prolong the sojourn of the antibiotic in the blood. By consulting a textbook of pharmacology, describe this drug's mechanism of action?

20. A constituent in the urine underwent reabsorption, excretion, filtration, and secretion. What is the most logical order in which these functions occurred?

21. The disease diabetes mellitus brings to mind the term "threshold." Why?

22. Explain why mannitol inhibits the obligatory reabsorption of water.

23. Explain why drinking water induces diuresis.

24. Chlorothiazide (Diuril) is a potent diuretic. What is its mechanism of action?

25. Certain drugs cause contraction of the bladder and are useful in treating "urinary retention" (which often occurs following surgery). Are such drugs considered diuretics? Discuss.

26. What would be the consequence of an inability of the posterior pituitary to produce ADH?

27. Is there such a disease as suggested in the above question?

28. ADH plays a key role in water balance. Discuss.

29. Hemorrhage powerfully stimulates the output of aldosterone. Cite the chain of events thought to be involved.

30. What effect would a drug that inhibits the action of aldosterone have on urine output?

31. What effect does aldosterone have on blood levels of sodium and potassium?

32. Angiotensin II stimulates the release of aldosterone and produces widespread vasoconstriction. Are these effects complimentary? Discuss.

33. Angiotensin II is marketed under the brand name Hypertensin. Many would consider this a clever name. Why?

34. In alkalosis, would the kidney excrete Na_2HPO_4 or NaH_2PO_4? Explain your answer.

35. In alkalosis, does the kidney excrete more or less NH_4^+ than normal?

36. In acidosis, does the kidney excrete KH_2PO_4 or K_2HPO_4?

37. In acidosis, does the kidney excrete more or less $NaHCO_3$ than normal?

38. The kidney spares sodium by its ability to produce ammonia. Explain.

39. What is renal diabetes?

40. How do the respiratory system and the urinary system respond to metabolic alkalosis?

41. How do the respiratory system and the urinary system respond to metabolic acidosis?

42. What effect does acute renal failure have on respiration?

43. Osmoreceptors in the hypothalamus or elsewhere are probably involved in thirst. Offer an explanation.

44. Specifically, what do BUN values tell us?

45. Name the major blood constituents encompassed by the expression "NPN."

46. In uremic poisoning (uremia) BUN, NPH, and creatinine values all rise. Why?

47. Why is uremia also called azotemia?

48. What does a GFR of 110 ml/min mean?

49. In the event of an abnormal PSP value, other

kidney tests would have to be performed for a definitive diagnosis. Explain and discuss.

50. What is pyelography?
51. The ureters are about 11 inches long. How many centimeters is this?
52. What is transitional epithelium?
53. What is the literal meaning of trigone?
54. The urge to urinate corresponds to a bladder content of about 300 ml. What is this volume in ounces?
55. What is nocturnal enuresis?
56. Account for the difference in length between the male and female urethra.
57. Prostatic hypertrophy results in urinary retention and commonly in uremia. Why?
58. Spinal injuries commonly result in urinary incontinence. Why?
59. Why does urine become alkaline on standing?
60. Other things being equal, urine with a low specific gravity is very light in color. Explain.
61. In dehydration, would you expect the specific gravity of urine to rise or fall?
62. Name some possible specific causes of hematuria.
63. Name some possible specific causes of glycosuria.
64. Name some possible specific causes of pyuria.
65. Albuminuria and proteinuria are commonly used interchangeably. Why?
66. What are casts?
67. A urinometer is a hydrometer. Explain.
68. Urinary retention commonly results in infection. Why?
69. Why is cystitis more common in the female than in the male?
70. Account for the fact that *Escherichia coli* is responsible for most infections of the urinary tract.
71. Sulfisoxazole and nitrofurantoin are often used in urinary tract infections. What types of drugs are these and against what pathogens are they effective?
72. *Pseudomonas aeruginosa* is a gram-negative bacillus. What does this mean?
73. Distinguish between the morphologies of *Staphylococcus epidermidis* and *Streptococcus faecalis*.
74. Most cases of pyelonephritis "start from below." What does this mean?
75. Many authorities believe that acute glomerulonephritis is an immunological disorder, that antibodies are produced that cause injury to the glomeruli. Discuss more fully.
76. Account for the hematuria and albuminuria seen in glomerulonephritis.
77. Why is a low-protein diet essential in the management of glomerulonephritis?
78. Explain the operation of the artificial kidney.
79. Why are transplants involving identical twins the most successful?
80. In "nonidentical" transplants, success depends upon the judicious use of immunosuppressive agents. Explain fully.
81. Why are patients with bleeding tendencies poor candidates for dialysis?
82. Explain the edemas of glomerulonephritis and the nephrotic syndrome.
83. Distinguish between acute and chronic glomerulonephritis.
84. Distinguish between glomerulonephritis and the nephrotic syndrome.
85. How do you explain the low output of urine in acute renal failure?
86. Distinguish between anuria and oliguria.
87. How could a transfusion reaction result in acute renal failure?
88. How can shock lead to acute renal failure?
89. For the most part, acute renal failure is potentially reversible. Discuss.
90. Explain the use of cation exchange resins in acute renal failure.
91. If a patient with acute renal failure has a daily output of 300 ml of urine, about how much fluid should he receive?
92. In acute renal failure, what is the danger of giving more fluid than the body can handle?
93. Why are stones called calculi?
94. What is nephrolithiasis?
95. What is the chemical composition of renal stones?
96. How could dehydration result in urinary calculi?
97. How could hyperparathyroidism result in urinary calculi?
98. Stenosis of the renal artery could very well result in severe hypertension. Explain.
99. Can hypertension ever be cured? Discuss.
100. What are exogenous carcinogens?

CHAPTER 20

THE NERVOUS SYSTEM

The nervous system is an extensive and complicated coordinating mechanism that regulates internal body functions and responses to external stimuli. For convenience the system may be considered to fall into two major divisions, the central and the peripheral. The central division includes the brain and the spinal cord; the peripheral includes the nerves, the ganglia, and the receptors (Fig. 20-1).

NEURONS AND NERVOUS TISSUES

Nervous tissue essentially is made up of two kinds of cells: neurons, the functional units; and neuroglia, the special connective tissue that serves to support the neurons. The cells composing the neuroglia include, among others, astrocytes, oligodendrocytes, microglia, and Schwann cells. Astrocytes (so named because of their shape) form supportive networks within the brain and cord; oligodendrocytes produce myelin sheaths; microglia are phagocytic cells that destroy microbes and remove disintegrating elements within the brain and cord; and Schwann cells form the neurilemma (Fig. 20-2).

The neuron is especially characterized by its intricately branched extensions, or processes, that radiate from the cell body, or soma (Fig. 20-2). The processes that conduct nervous impulses toward the soma are called dendrites; those that carry impulses away from it are called axons. While the dendrites of most neurons are short and extensively branched, axons are often single. Axons may, however, send out one or two side branches called collateral axons.

Returning to the soma, we note, as shown in Fig. 20-2, delicate fibers, or neurofibrils, and chromatophilic granules called Nissl bodies. The Nissl body can be visualized with staining techniques that employ basic dyes; it is actually endoplasmic reticulum. Following injury, Nissl bodies undergo chromatolysis, a pathologic process in which they break up and disappear. This phenomenon has enabled the research physiologist to trace nerve pathways in the brain and cord; different nerves are cut, and the area or areas of chromotolysis are noted.

Nerve fibers

A nerve fiber is a dendrite or axon that extends from the central nervous system or ganglion out into the body, sometimes for several feet. Except for certain autonomic fibers, most nerve fibers are myelinated; that is, they are enveloped by myelin, a whitish material composed of cell-membrane components. The regions along the fiber where myelin is absent are called the nodes of Ranvier (Fig. 20-2). The myelin sheath has a thin, multinucleated outer covering

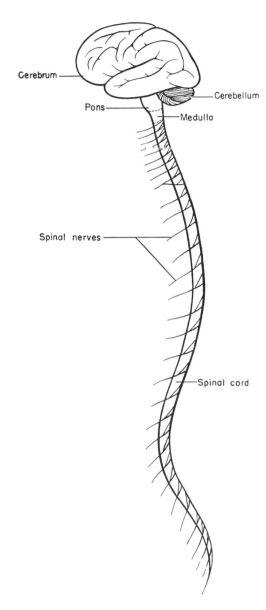

FIG. 20-1

The brain and spinal cord. (From Brooks, S. M.: Basic facts of pharmacology, ed. 2, Philadelphia, 1963, W. B. Saunders Co.)

Labels in figure: Cerebrum, Pons, Cerebellum, Medulla, Spinal nerves, Spinal cord

called the neurilemma or sheath of Schwann. Presumably one Schwann cell provides the sheath for each internode region. The neurilemma is more than just another coat, however, for it serves in the regeneration of the fiber. When a fiber is cut, the distal end degenerates; after a few days the neurilemma extends itself to the distal stump and thereby initiates the regeneration. If the body of the nerve cell is uninjured, the severed fiber or nerve may become operational again, and the affected part returns to normal.

Classification of neurons

Neurons are classified according to both architecture and the direction of transmission of information. The architectural scheme recognizes three types of cells: unipolar, bipolar, and multipolar, depending upon the number of processes projecting from the soma. With respect to function, there are also three types of neurons: afferent, efferent, and internuncial. Afferent or sensory neurons conduct impulses toward the central nervous system; efferent or motor neurons conduct impulses away from the central nervous system; and internuncial or connective neurons conduct impulses from sensory to motor neurons. Afferent neurons are usually unipolar, with a long dendrite and a relatively short axon (Fig. 20-2). In contrast, a motor neuron may have a long axon and short dendrites. Internuncial neurons, situated as they are within the central nervous system, generally have relatively short dendrites and axons and are often multipolar.

Gray and white matter

The gray areas of the central nervous system essentially are composed of cell bodies and unmyelinated fibers. White matter, on the other hand, is composed of myelinated fibers. The outside of the brain is gray, and the inside is white with islands of gray. In contrast, the spinal

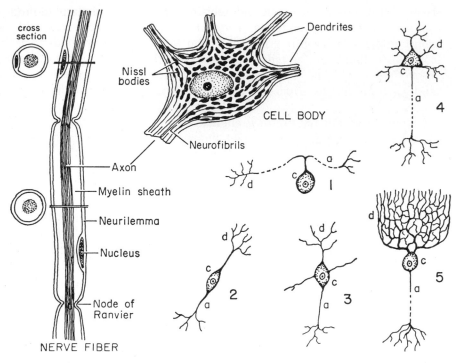

FIG. 20-2

The neuron. Basic types and microscopic details of cell body and nerve fiber. **1,** Unipolar cell. **2,** Bipolar cell. **3,** Cell of cerebral cortex. **4,** Motor cell. **5,** Purkinje cell of cerebellar cortex. *a,* Axon; *c,* cell body; *d,* dendrites.

cord is white on the outside and gray on the inside.

Nerves and tracts

A nerve is an anatomic cable composed of nerve fibers, blood vessels, and lymphatics wrapped together by connective tissue (Fig. 20-3). Afferent fibers carry impulses from the receptors to the cord and brain, and efferent fibers carry impulses to the effectors (muscles and glands). Most nerves are mixed; they contain both afferent and efferent fibers. Most autonomic nerves, however, are considered efferent (or motor); the olfactory, optic, and acoustic nerves are purely afferent (or sensory).

A bundle of fibers within the central nervous system is referred to as a tract. We shall have more to say about tracts later in the present chapter.

NERVE IMPULSE

A nerve impulse is a self-propagating electrical impulse that travels along the surface membrane of a nerve fiber. As presently understood, the membrane actively transports sodium ions out and potassium ions in. In the resting condition there is a slight outward diffusion of potassium ions, which leaves a large organic anion inside and causes a *resting potential* to occur across the membrane; the membrane is positive on its

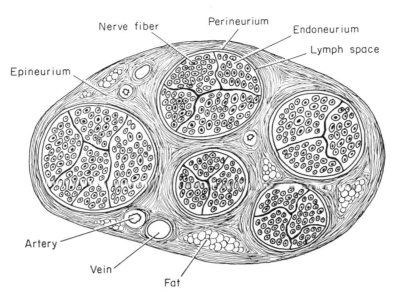

FIG. 20-3

Cross section of nerve. This particular nerve trunk is made up of five bundles (funiculi) of nerve fibers. (After Harris; modified from Schottelius, B. A., and Schottelius, D. D.: Textbook of physiology, ed. 17, St. Louis, 1973, The C. V. Mosby Co.)

outer surface and negative on its inner surface (Fig. 20-4). A stimulus greatly increases the membrane's permeability to sodium ions at the point of stimulation, causing a rapid inrush of sodium ions. This inward current of positive ions causes the membrane to depolarize and actually become repolarized slightly, with the inside about 20 mv positive relative to the outside. The sodium gate closes, and the cell returns to the resting condition, with the inside about 70 mv negative relative to the outside. It is brought to this condition, as indicated, by a brief increase in permeability to potassium ions that occurs just as the sodium gate closes. The transitory increase in sodium permeability followed by the brief increase in potassium permeability produces a voltage change that can be recorded. This is called the *action potential*—the signal or impulse transmitted along the nerve. The action potential at one point on the membrane

causes some current flow into the surrounding medium and within the axon itself. This *electrotonic current* acts as a stimulus to the next, adjacent region of membrane, bringing it to threshold and causing an action potential at that site.

The stimulus

Although an electric shock, because of its convenience, is the stimulus commonly used in the laboratory, nerve fibers can be excited by other types of physical or chemical stimuli. The impulse that travels down the fiber is always electrical, and since the resting potential is the same all along the fiber, the action potential does not diminish as it passes away from the stimulus. In a crude but striking analogy, these features can be dramatized by laying down a line of gunpowder on the sidewalk and igniting one end. "Stimulated" by a spark, match, blowtorch, friction, oxidizer, or whatever, the gunpowder catches fire and burns with undiminished fury along its entire length.

The weakest stimulus that can trigger an action potential is called the *threshold* or *liminal stimulus*. A nerve fiber reacts maximally to such a stimulus. It always produces an action potential of exactly the same amplitude, once it has been brought to threshold. Thus, the nerve fiber obeys the all-or-none law, just as the action potential in the muscle fiber does. This law is in no way invalidated by the fact that a nerve, which is a bundle of fibers, shows a graded response to an increase in the strength of a stimulus, for this simply means that more fibers are being called into play. After all fibers have been engaged, the response cannot be intensified by increasing the strength of the stimulus.

The effect of a stimulus depends upon the irritability of the nerve fiber. This probably explains why a series of subthreshold stimuli delivered at intervals of about 0.001 sec can provoke a nerve impulse. Apparently, the first few stimuli of the series excite the membrane to a point where it can be triggered by subthreshold strength. This phenomenon is called

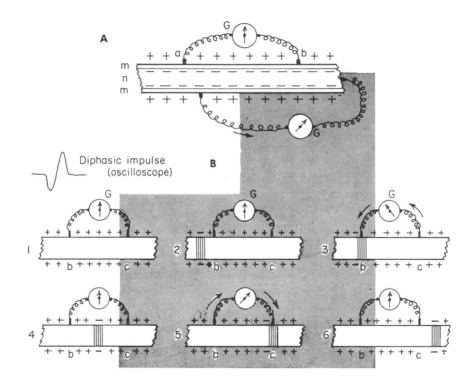

FIG. 20-4

A, Electrical character of resting nerve fiber. Outside membrane, *m*, bears no charge between points *a* and *b* on surface but is charged relative to the interior of the fiber, *n*. **B,** Transmission of nerve impulse. When resting fiber is stimulated, *1*, a negative wave of excitation (the impulse) is propagated, *2* to *6*. This is evidenced by swing of galvanometer needle, *G*, first one way and then the other; hence, the expression *diphasic impulse.*

summation or, more formally, *subliminal summation*. Further, though it is true that all stimuli above the threshold level provoke the same response, it is also true that a stronger stimulus usually provokes the response in a shorter period of time. In other words, the length of time an impulse must be applied to be effective is related inversely to its strength. The time involved, however, is always very short (generally about 0.001 sec).

Refractory period

After the action potential has passed along the nerve fiber, the fiber requires about 0.001 sec to repolarize itself. During this restorative intermission, called the refractory period, the potassium gate is open in order to bring the membrane back to its resting level. Since the membrane must be polarized in order to propagate a new impulse, stimuli received during the refractory period cannot be acted upon. This means that the rate at which stimuli are delivered to a nerve fiber will enhance the response only up to a point. Because the region behind the action potential is refractory, an impulse can only travel forward.

Impulse velocity

The velocity or speed at which an impulse travels is of no little concern in matters involving split-second decisions. The latest figures show that the velocity varies from 1 to 300 miles per hour. Velocity depends upon the diameter of the

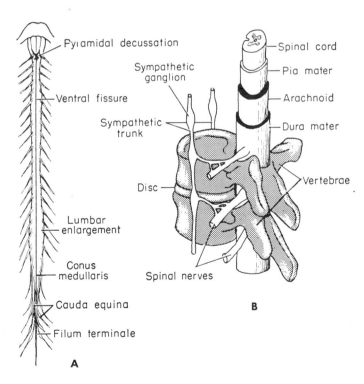

FIG. 20-5
A, Spinal cord. **B,** Its relation to meninges and vertebral canal. (Redrawn from Anthony, C. P., and Kolthoff, N. J.: Textbook of anatomy and physiology, ed. 8, St. Louis, 1971, The C. V. Mosby Co.)

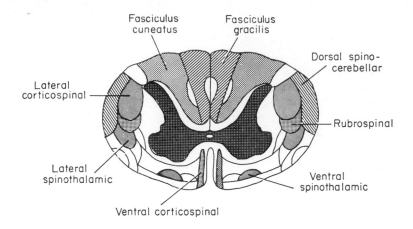

FIG. 20-6

General location of some major spinal tracts. Note that ventral median fissure (not labeled) lies between ventral corticospinal tracts and that dorsal median sulcus lies between right fasciculus gracilis and left fasciculus gracilis. Also, note central canal in central gray, and protrusion of lateral horns.

fiber and upon whether the fiber is myelinated. Fibers of large diameter usually transmit an impulse faster than fibers of small diameter, and myelinated fibers transmit an impulse faster than unmyelinated ones of the same diameter. Generally speaking, voluntary (somatic) nerve fibers are larger than autonomic fibers. Also, voluntary fibers are myelinated, whereas autonomic fibers usually are not. Sensations and skeletal muscle responses therefore are usually transmitted much faster than autonomic responses.

SPINAL CORD

The spinal cord is a fairly soft, white, ovoid structure, about 17 inches long and ½ inch in diameter, occupying the spinal canal from the foramen magnum to the first lumbar vertebra. Aside from two bulges in the cervical and lumbar regions, it tapers slightly, finally terminating as the conus medullaris (Fig. 20-5). The filum terminale, a nonfunctional filament that brings the cord to a point, is not nervous tissue but a

continuation of the pia mater. In cross section the spinal cord has an H-shaped area of gray matter (Fig. 20-6) surrounded by white matter. The forward projections of the H are called ventral (anterior) horns and the backward projections are called dorsal (posterior) horns. The slight protrusions on either end of the cross of the H, or the central gray, are called the lateral horns. The ventral horns are composed of motor neurons; the lateral and dorsal horns are composed of internuncial neurons.

Spinal tracts

The white matter of the cord is divided by the ventral median fissure, the dorsal median sulcus, and the gray matter into three columns (in each half): the ventral, dorsal, and lateral. The fibers that compose these columns are grouped together into tracts, or fasciculi, which are named according to their origin and destination. Those carrying impulses upward to the brain are called ascending tracts, and those carrying impulses downward from the brain are

TABLE 20-1

Major tracts of the spinal cord

Tract	General nature of function
ASCENDING	
Dorsal spinocerebellar	Subconscious kinesthesia
Ventral spinocerebellar	Subconscious kinesthesia
Lateral spinothalamic	Pain and temperature
Ventral spinothalamic	Touch
Fasciculus gracilis	Conscious kinesthesia
Fasciculus cuneatus	Conscious kinesthesia
DESCENDING	
Lateral corticospinal	Skeletal muscle contraction
Ventral corticospinal	Skeletal muscle contraction
Rubrospinal	Unconscious muscle coordination
Vestibulospinal	Muscle tone and equilibrium

called descending tracts. Obviously, the ascending tracts transmit sensory information while the descending tracts transmit motor information. The principal tracts and the nature of the impulses they transmit are presented in Fig. 20-6 and Table 20-1. The location and function of the spinal tracts have been ascertained by experimental procedures in animals and by the study of clinical conditions in human beings.

Meninges

The spinal cord does not lie naked in its bony vault, but rather is ensheathed by three membranes, called meninges, that are continuous with those of the same names that cover the brain. From without inward, they include the dura mater, the arachnoid, and the pia mater (Fig. 20-5). The dura mater, a relatively thick fibrous membrane lining the spinal canal, is slightly separated from the arachnoid, a more delicate structure composed of fibrous and elastic tissue. Between the arachnoid and the pia mater, the most delicate membrane of the three, there is a considerable interval called the subarachnoid space. Since the pia mater, which adheres closely to the surface of the cord, terminates as the filum terminale while the other two membranes continue beyond the cord, a needle can be inserted into the subarachnoid space at the level of the fourth or fifth lumbar vertebra without damaging the cord. This anatomic feature makes lumbar punctures and spinal anesthesia feasible.

Cerebrospinal fluid

The subarachnoid space, the ventricles (spaces within the brain, p. 328), and the central canal are filled with a watery liquid called the cerebrospinal fluid. This fluid is produced by networks of capillaries, called choroid plexuses, in the ventricles and reaches a total volume in a full-grown man of about 130 ml. Qualitatively, but not quantitatively, cerebrospinal fluid resembles plasma and tissue fluid. Cerebrospinal fluid has fewer cells and less protein, calcium, potassium, and glucose, for example, than tissue fluid, but the concentration of sodium chloride in it is appreciably greater. Cerebrospinal fluid, removed by lumbar puncture, has considerable diagnostic value in diseases of the central nervous system, especially meningitis, tumors, and syphilis. In these conditions the protein and cell counts are elevated. A bacteriologic examination of the fluid may disclose pathogenic microbes.

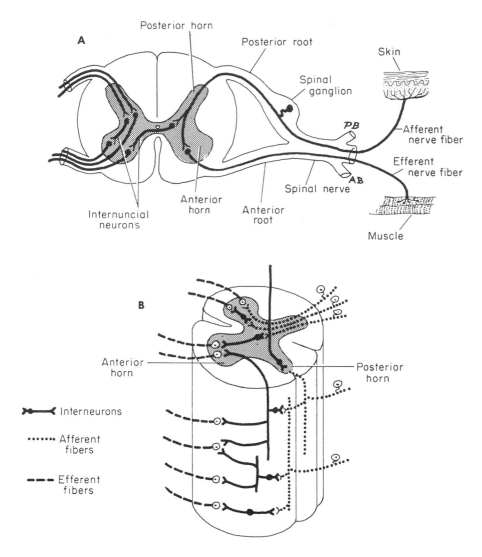

FIG. 20-7

Idealized diagram showing basic features of afferent and efferent fibers as they relate to each other, the cord, and the reflex. **A,** Details of spinal nerve. **B,** A few simple ramifications of fibers passing to higher and lower levels within cord. *PB,* Posterior branch, or ramus, of spinal nerve; *AB,* anterior branch, or ramus, of spinal nerve.

SPINAL NERVES

Along its length the spinal cord gives rise to thirty-one pairs of nerves (Fig. 20-5) that emerge from the spinal canal via the intervertebral foramina. The nerves are named according to the vertebrae from which they exit and include eight pairs of cervical, twelve pairs of thoracic, five pairs of lumbar, five pairs of sacral, and one pair of coccygeal nerves. The cervical and thoracic nerves emerge immediately from the vertebral canal, but the others pass down the canal before they enter the foramina. Thus, the lower spinal nerves bring to mind a horse's tail and are appropriately named the cauda equina.

Fig. 20-7, A, shows an idealized drawing of a spinal nerve and its association with the spinal cord. The dorsal (posterior) and ventral (anterior) roots of the nerve, composed of sensory and motor fibers, respectively, arise from the gray matter of the respective horns. Just before passing through the intervertebral foramina, the roots join to form the mixed spinal nerve trunk, which then splits into two mixed branches called the dorsal and ventral primary rami. The dorsal rami run uninterrupted to the muscles and skin of the back, but most of the ventral rami fuse into complex networks called plexuses (Fig. 20-8).

Plexuses

At a plexus there is a rearrangement of fibers so that the nerves or branches that finally emerge possess fiber groupings different from those in the primary rami. The four main plexuses include the cervical, brachial, lumbar, and sacral.

The cervical plexus, formed by the ventral rami of the cervical spinal nerves, lies deep in the neck in the area of the upper four cervical vertebrae. Its branches supply the diaphragm (phrenic nerve), the muscles and skin of the neck, and the back portion of the skull.

The brachial plexus, formed by the ventral rami of the lower four cervical nerves and first thoracic nerve, lies in the lower neck and the axillary region. Its branches supply the skin and the majority of muscles in the neck, chest, and upper extremities.

The thoracic ventral rami do not form a plexus. They run along the lower borders of the ribs and are called the intercostal nerves. Their terminal endings supply the intercostal muscles of the abdominal wall.

The lumbar plexus, which lies within the psoas major muscle, is formed by the ventral primary rami of the first, second, and third lumbar nerves and most of the fibers of the ventral primary ramus of the fourth lumbar nerve. Its branches supply the skin over the ventral part of the buttocks, the external genitals, and the thighs.

The sacral plexus lies in the vicinity of the piriformis muscle and the internal iliac vessels. It is formed by the fusion of the lumbosacral trunk, the first three sacral nerves, and part of the fibers of the fourth sacral nerve. The lumbosacral trunk is made up of part of the fibers of the fourth lumbar ventral ramus and all the fibers of the fifth ventral lumbar ramus. The branches emerging from the sacral plexus innervate mainly the skin over the gluteal and perianal regions and the muscles of the legs. The sciatic nerve of the sacral plexus is the largest nerve in the body.

Dermatomes

The nerves entering or leaving the spinal cord at each level innervate segments of the body corresponding to that level of the spinal cord; these segments are related to the dermatomes. A dermatome is the area of skin supplied with afferent nerve fibers of a single dorsal spinal root. The loss of feeling in a given dermatomic segment is of considerable diagnostic significance because it points to trouble in the corresponding spinal nerve or cord segment.

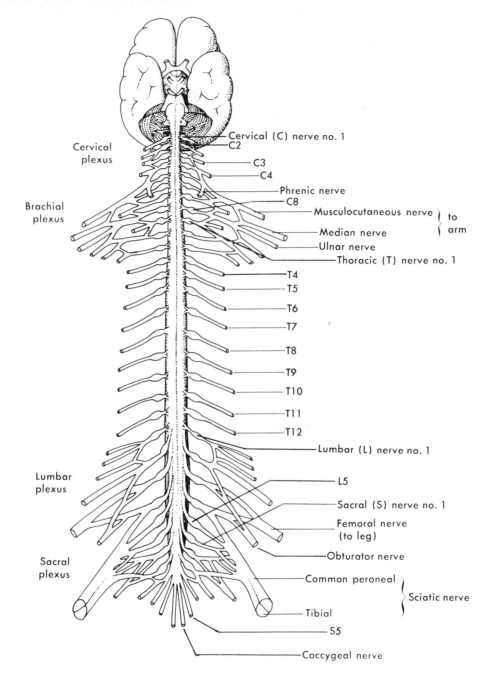

FIG. 20-8
Ventral view of cord, showing spinal nerves, plexuses, and major nerves supplying arm (median, ulnar, etc.) and leg (femoral, sciatic, etc.).

REFLEXES

A reflex is an automatic, involuntary, and typically unconscious response to a stimulus. In its essential form a reflex calls into play an afferent neuron (with or without a receptor), an internuncial neuron, an efferent neuron, and an effector. (In the simplest reflexes internuncial neurons are not involved.)

Stimulation initiates an impulse in the afferent neuron. This impulse is conducted along the afferent dendrite, through the cell body, and along the afferent axon to the terminal endings at the synapse, the region between the processes of two adjacent neurons. Here, via a transmitter substance (neurotransmitter), the impulse is relayed to the dendrites of the internuncial neuron. In a similar fashion the impulse is relayed through the internuncial neuron and across the synapse between the internuncial terminal endings and the efferent dendrites. At the neuro-effector junction, the place of contact between the efferent terminal endings and the effector (muscle or gland), the impulse causes the release of a transmitter substance. This transmitter substance causes the effector to respond.

The transmitter substances released at the synapses in the brain are thought to be acetylcholine, norepinephrine, dopamine, and serotonin. At peripheral skeletal and central autonomic synapses, the transmitter substance is acetylcholine. The transmitter released at the neuroeffector junction is thought to be acetylcholine in the instances of somatic and parasympathetic nerves, and norepinephrine in the instance of sympathetic nerves. It is likely that transmitter substances are stored in the neurovesicles present in the synaptic knobs at the axon terminals. The nerve impulse apparently causes these structures to release their contents.

Reflex arc

The complete pathway over which the impulse must travel from an afferent dendrite ending (or receptor) to an effector is called the reflex arc (Fig. 20-7, A). Though few reflexes in man are as simple as the essential form illustrated (most arcs involve hundreds or even thousands of neurons), the general idea is always the same. Afferent dendrites are found in the spinal nerves and generally run long distances before they reach the cell bodies in the spinal ganglia on the dorsal (posterior) roots. From here the afferent axons extend into the dorsal gray horns of the cord, where they synapse with one or more internuncial neurons and sometimes directly with efferent neurons. The dendrites and cell bodies of the efferent neurons are confined to the ventral (anterior) gray horns, and their axons extend into the spinal nerve via the ventral roots. A spinal nerve proper is mixed; it carries both sensory and motor fibers. The dorsal roots of the spinal nerves carry only sensory fibers, and the ventral roots carry only motor fibers.

Fig. 20-7, B, depicts other significant features. An afferent neuron may trigger an efferent neuron on the same (ipsilateral) side of the cord, on the opposite (contralateral) side of the cord, or on both sides. An afferent axon may trigger internuncial and efferent neurons at one level of the cord, producing a *segmental reflex*, or at more than one level, producing an *intersegmental reflex* arc. Thus, a single sensory neuron can call scores of motor neurons into action. Conversely, a number of sensory neurons may synapse, or focus, upon a single motor neuron. Regardless of the complexity of the reflex arc, the basic mechanism and purpose are always the same: sensory impulses, through the agency of synaptic relays, induce motor impulses, which in turn effect the response.

Inhibition and facilitation

In our discussion so far, we have implied that an incoming impulse always triggers an internuncial or motor neuron. Actually, the cord and brain have some influence on what will and will not pass across synapses. If this were not true, the enormous number of incoming impulses would keep the central nervous system in continual electrical frenzy. Just how this central inhibitory state is effected still remains largely unknown, but it is generally thought to center largely about the synapse. A dramatic example of reflex inhibition is shown when one stems a sneeze by applying pressure to the upper lip.

Sensory impulses may be reinforced, or facilitated, via a corollary phenomenon known as the central excitatory state. In some way, as yet poorly understood, synaptic resistance is decreased, thereby permitting a weak stimulus to evoke a strong response. For example, the knee-jerk reflex that accompanies a tap to the patellar tendon can be greatly intensified by clasping the hands and pulling at the time of the tap.

Types of reflexes

Reflexes are named in various ways. Those triggered by stimulated exteroceptors (receptors located in the skin and the surface membranes) are called exteroceptive reflexes. Coughing, winking, tearing, sneezing, and response to pain, touch, and heat applied to the surface are all exteroceptive reflexes. Interoceptive (visceroceptive) reflexes, on the other hand, result from the stimulation of interoceptive receptors (interoceptors) located in the viscera. The stimulation of respiration by carbon dioxide, the cardiac reflexes, and micturition are examples of interoceptive reflexes.

Special interoceptors, called proprioceptors, are located in the muscles, tendons, and semicircular canals. Proprioceptors are sensitive to stretch, pressure, and movement and generally maintain muscle tone and body balance.

Most reflexes classed according to response are flexor or extensor. A flexor reflex is marked by the flexion of a muscle. Since such reflexes usually indicate that the body is protecting itself against a harmful stimulus (for example, flexing at the knee in response to stepping on a nail or at the elbow when the finger is burned), they are sometimes referred to as nociceptive reflexes. Extensor reflexes are responses effected by contraction of the extensor muscles. Because they oppose the force of gravity, they operate to keep the body balanced. When one steps on a nail, for instance, the extensors in the opposite leg are convulsed to enhance that leg's support.

In clinical work the physician generally refers to a reflex according to the most obvious manifestation or to the part affected. Some of the more classic reflexes of this type include the knee jerk, the corneal reflex (winking), the pupil reflex (constricting of the pupil in response to light), and the Babinski big toe reflex. The latter reflex, which involves extension of the big toe and fanning of the others, is elicited during the first few months of infancy by tickling the sole of the foot. Thereafter, however, this reflex is abnormal and indicative of lesions of the upper motor neurons. The other reflexes mentioned also have diagnostic value.

At birth we have certain natural reflexes. When the nipple is placed in the baby's mouth, he sucks without hesitancy; when his foot is tickled, his toes move; when his eye is touched, he winks. As the baby grows, however, he begins to acquire learned or conditioned reflexes and continues to do so throughout life. Whereas a natural or unconditioned reflex is mediated through the spinal cord or the lower levels of the brain, a conditioned reflex involves the higher centers of the brain as well. For example, when an infant is shown a piece of strawberry shortcake, he probably will make no response other than to put his finger into the whipped cream; an older child, on the other hand, may feel his mouth water upon just hearing the words "strawberry shortcake." Similarly, a gourmet may experience a flow of saliva just from thinking about a juicy steak, let alone smelling it.

In the conditioned reflex, stimuli that do not

at first evoke a response are conditioned into action by cerebral association gained through experience and training. In the case of the juicy steak, the cortical areas of sight and smell have established pathways with the salivary centers at the lower levels. At maturity the brain has a vast repertoire of conditioned reflexes relating to many processes important to the welfare of the body. Since these functions stem from experience, they are of cardinal interest to the psychologist. In experimental psychology, for example, the conditioned reflex affords a valuable tool for studying animal behavior.

BRAIN

The brain, or encephalon, is a 3-pound mass of nervous tissue composed of approximately 12 billion neurons complexly organized and interconnected. The embryonic brain is divided into three primordial parts: the prosencephalon (forebrain), the mesencephalon (midbrain), and the rhombencephalon (hindbrain). The prosencephalon develops into the telencephalon (cerebrum) and diencephalon ('tweenbrain), and the rhombencephalon develops into the metencephalon and the myelencephalon (Fig. 20-9).

Cerebrum

The cerebrum is the largest part of the brain. Its outer gray surface, or cortex, is thrown into convolutions called gyri, and it is divided lengthwise into hemispheres by the longitudinal fissure. Two other fissures, the fissure of Rolando (central fissure) and the fissure of Sylvius (lateral

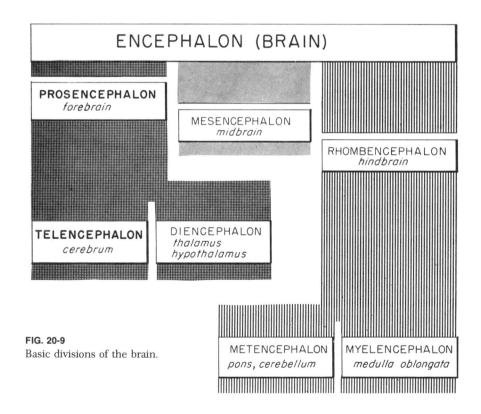

FIG. 20-9
Basic divisions of the brain.

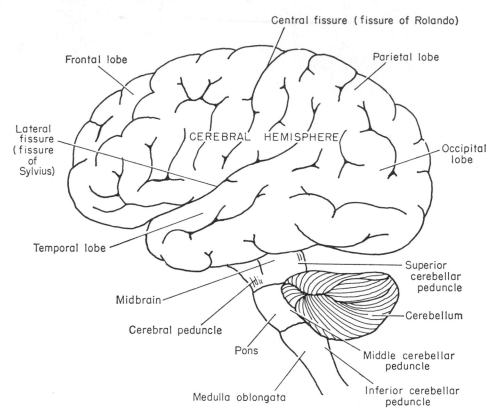

FIG. 20-10
General lateral view of brain showing lobes, fissures, and major divisions. (After Quain; from Schottelius, B. A. and Schottelius, D. D.: Textbook of physiology, ed. 17, St. Louis, 1973, The C. V. Mosby Co.)

fissure), divide each hemisphere into lobes. The deep grooves between the gyri are the *fissures*; the shallow grooves are called *sulci*.

Cerebral lobes

Each of the two cerebral hemispheres has five lobes: the frontal lobe, the parietal lobe, the temporal lobe, the occipital lobe, and the insula, or island of Reil. As shown in Fig. 20-10, the frontal lobe is before the fissure of Rolando, and the parietal lobe, immediately behind it; the fissure of Sylvius serves to separate these two

lobes from the temporal lobe below. The occipital lobe is the pyramidal portion of the hemisphere lying behind the parietal and temporal lobes. The insula is located below the lateral fissure and cannot be seen unless the adjacent portions of the frontal and temporal lobes are raised.

Cerebral tracts

Whereas the cerebral cortex is made up of gray matter, the interior contains both gray and white matter. The white matter is composed of trillions of nerve fibers running in tracts (Fig.

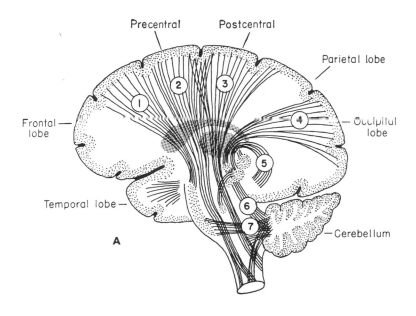

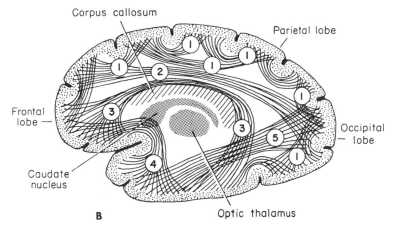

FIG. 20-11

A, Projection tracts of the brain. *1,* Tract connecting frontal convolutions with cells lying in pons; *2,* tract descending into cord after decussating in medulla; *3,* sensory tract carrying impulses to postcentral gyri; *4,* visual tract; *5,* auditory tract; *6,* superior cerebellar peduncle; *7,* middle cerebellar peduncle. **B,** Association tracts of brain. *1,* Tracts between adjacent convolutions; *2,* tract between frontal and occipital areas; *3* and *4,* tracts between frontal and temporal areas; *5,* tract between occipital and temporal areas. Note corpus callosum, through which pass commissural tracts between hemispheres.

20-11) that may be grouped conveniently into three categories: association tracts, commissural tracts, and projection tracts. Association tracts connect different areas of the cortex of the same hemisphere; commissural tracts run from one hemisphere to the other, principally via the corpus callosum; projection tracts, the longest of the three types, extend from the cerebrum to other parts of the brain and the spinal cord. Depending upon the character of the impulses carried, the tracts are designated as ascending or descending; ascending tracts convey impulses to the brain, and descending tracts carry impulses from the cortex to the lower portions of the brain and to the spinal cord. The projection tracts that extend into the cord become spinal tracts. For example, the corticospinal, a descending tract, extends from the cerebral cortex through the brainstem and into the spinal cord, where it synapses with neurons in the ventral horns; the spinothalamic, an ascending tract, orginates in a dorsal horn, runs up the spinal cord, and synapses at the thalamus.

Cerebral nuclei

The large subcortical mass of gray and white matter in front of the thalamus (p. 328) in each cerebral hemisphere is called the corpus striatum. The gray matter is arranged in two principal nuclei ("basal ganglia") called the caudate nucleus and the lentiform nucleus,* which lie medial and lateral to the internal capsule (Fig. 20-12), respectively. The internal capsule is a strip of white matter that passes between the lentiform nucleus and the thalamus. Because the tracts of the internal capsule are compressed

---

*The lentiform, or lenticular, nucleus, is actually made up of two other nuclei—the putamen and the smaller (and more medial) globus pallidus.

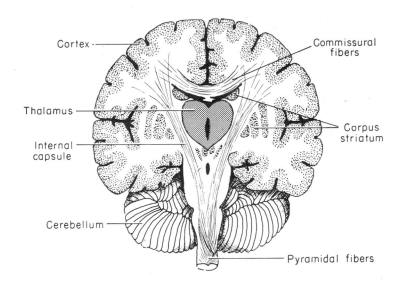

FIG. 20-12

Coronal section through brain showing internal structure. Corpus striatum is composed of caudate nucleus and lentiform nucleus, which are medial and lateral to internal capsule, respectively. (From The human body: its anatomy and physiology, third edition, by C. H. Best and N. B. Taylor. Copyright © 1932, 1948, 1956, by Holt, Rinehart and Winston, Inc. Adapted and reprinted by permission of Holt, Rinehart and Winston, Inc.)

into such a narrow space, a lesion in this region can have pronounced effects on the impulses going to and from the brain.

Cerebral function

There can be little doubt that the cerebrum is the seat of that evasive thing we call the mind. Speech, intelligence, memory, reason, and all other mental attributes are involved here. Although earlier views held that these attributes were peculiar to the frontal lobes, today most physiologists believe that no one region can be singled out as the most significant; mental performance is essentially the highest sum total of cerebral function.

Animal experiments, especially those on apes and monkeys, clinical findings in cases of brain damage, and instrumentation of the exposed

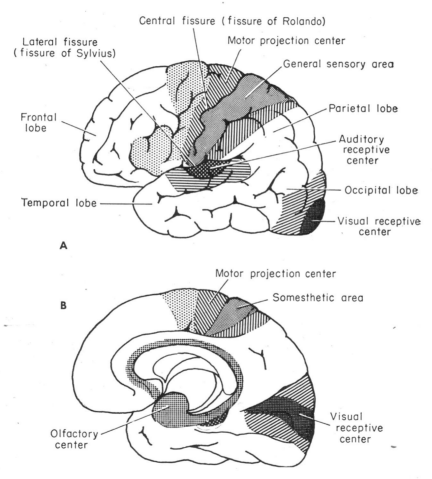

FIG. 20-13

Diagram of brain showing various functional areas and centers. **A**, Lateral view. **B**, Medial view.

cortex in patients under local anesthesia have enabled the researcher to define certain areas of the brain associated with certain functions (Fig. 20-13). Among others, these include the motor area, the somesthetic area, and the centers of taste, hearing, smell, vision, and speech.

The motor area, situated in the gyrus immediately anterior to the fissure of Rolando, contains neurons, or pyramidal cells (Fig. 20-2), that give rise to the corticospinal (or pyramidal) tracts. These are descending tracts that transmit impulses to the muscles and make possible voluntary movement. Whereas the traditional view held the motor area to be the "highest" level of motor impulse integration, it now appears, according to researcher Edward Evarts, that this distinction belongs to the basal ganglia. When the functioning of the basal ganglia is impaired, faulty signals pass to the motor cortex via the thalamus and cause postural disturbances, muscular tremor, and difficulty in initiating movement.

The somesthetic, or general sensory, area resides in the gyrus immediately posterior to the fissure of Rolando. Here the sensations of touch, pain, heat, cold, and body movement (kinesthetic sense) are received and experienced.

At the lower end of this same gyrus is situated the gustatory (taste) center. The auditory (hearing) center lies in the upper part of the temporal lobe, and the visual area is located on the medial aspect of the occipital lobe. The olfactory (smell)

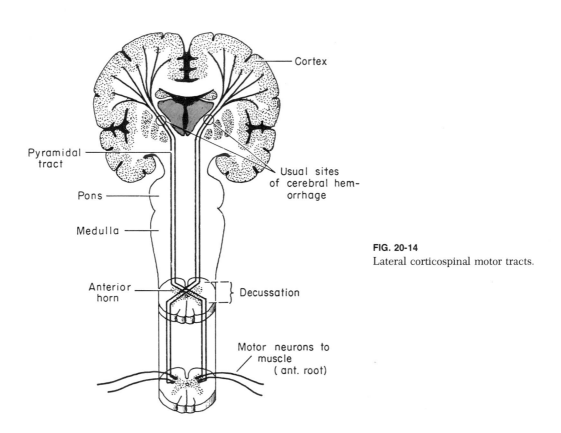

FIG. 20-14
Lateral corticospinal motor tracts.

area lies deeper within the brain, in a region roughly adjacent to the anterior pole of the temporal lobe. Neither the olfactory area nor the gustatory area have been precisely located. A very special area, now well defined, is Broca's motor speech area situated in the third (inferior) frontal gyrus. Damage to this area, especially to the left cerebral hemisphere, results in motor aphasia, a condition in which the individual becomes unable to express his ideas in spoken words, even though he is not actually unable to speak.

Motor pathways

Motor impulses are conveyed to the skeletal muscles via the corticospinal tracts and the peripheral nerves. These impulses leave the cerebral cortex via the axons of the pyramidal cells. Such fibers make up the corticospinal tracts of the internal capsule, the cerebral peduncles, the pons, and the medulla oblongata, in that order (Fig. 20-14). At the level of the medulla oblongata these tracts appear on the anterior surface as two well-defined columns called pyramids (Fig. 20-12), each of which divides into the anterior corticospinal tract and the lateral corticospinal tract. Whereas the anterior tracts directly descend the cord, the lateral tracts cross over (decussate). The corticospinal fibers synapse with neurons in the ventral horns, which in turn innervate the skeletal muscles via the spinal nerves. In summary, then, an impulse travels from the cortex to muscle via two neurons, one whose cell body resides in the cortex and the other whose cell body resides in the ventral horn of the cord. The former is sometimes referred to as an upper motor neuron and the latter, as a lower motor neuron.

From the foregoing it should be evident that injury or damage along the motor pathways will result in paralysis. There are four etiologic possibilities: cortical damage, tract damage, ventral horn damage, and peripheral nerve damage. In upper motor neuron damage (cortical or tract damage) the muscles are stiff, or spastic, and the knee jerk or tendon reflex is exaggerated. On the other hand, lower neuron damage (ventral horn or peripheral nerve damage) leads to a flaccid paralysis in which the muscles are flabby and the limbs are uncontrollably loose; the tendon reflexes cannot be demonstrated, and there is extreme wasting of muscle tissue.

Sensory pathways

Sensory axons that travel up the spinal cord via the dorsal columns synapse in the lower medulla with dendrites in the nucleus gracilis and the nucleus cuneatus (Fig. 20-15). The nucleus gracilis receives the axons of the fasciculus gracilis, and the nucleus cuneatus receives the axons of the fasciculus cuneatus. The axons that emerge from the nuclear neurons then decussate and ascend to the thalamus through the pons and the midbrain as a tract called the lateral lemniscus. At the thalamus there is another synaptic connection, and the axons from the thalamic neurons terminate in the cortical somesthetic area. In brief, then, the impulse is carried from receptor to cortex via three sensory neurons, the first with its cell body located in the dorsal root ganglion, the second with its cell body in the medullary nucleus, and the third with its cell body in the thalamus. Sensory impulses carried up the cord via the dorsal and lateral spinothalamic tracts pass through the medulla without interruption. They travel through the sensory lemniscus and reach the cortex via synaptic connections in the thalamus.

Thalamus

The diencephalon or 'tween-brain, the part of the prosencephalon situated between the cerebrum and the midbrain, surrounds the third ventricle (p. 328). Its main divisions are the thalamus and the hypothalamus (Fig. 20-16). The thalamus is a large area of gray matter situated on either side of the third ventricle. The right thalamus bulges into the right lateral wall of the ventricle, and the smaller left thalamus bulges into the left lateral wall. Both thalami consist of scores of nuclei of known and unknown functions. The posterior ventricular nuclei, for example, serve as relay centers for sensory impulses on their way to the cerebral cortex. Similar nuclei serve as relay centers between the cerebral cortex and basal ganglia. Other functions relate to complex reflex movements, arousal mechanisms, and emotional responses.

Hypothalamus

The hypothalamus is a small region of the diencephalon lying immediately behind the optic chiasm (p. 329) and beneath the floor of the

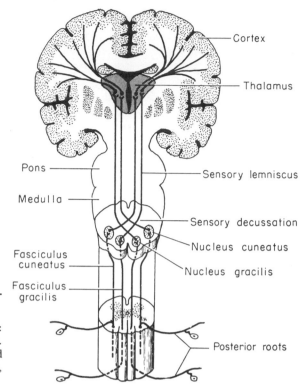

FIG. 20-15
Sensory pathways from cord to brain. (From The human body: its anatomy and physiology, third edition, by C. H. Best and N. B. Taylor. Copyright © 1932, 1948, 1956, by Holt, Rinehart and Winston, Inc. Adapted and reprinted by permission of Holt, Rinehart and Winston, Inc.)

FIG. 20-16
Sagittal view of right side of brain. (Modified from Rogers, T. A.: Elementary human physiology, New York, 1961, John Wiley & Sons, Inc.)

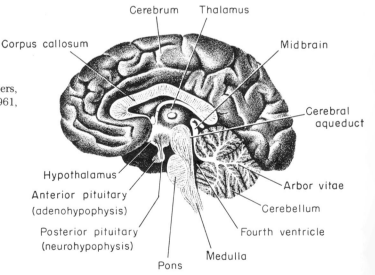

third ventricle. It is an area that plays a key role in directing the autonomic activities of the body. The array of nuclei spread throughout its substance apparently is in one way or another associated with just about every major impulse tract in the brain. The hypothalamus is considered the control center for autonomic activity and temperature and appetite regulation, and also is thought to play an essential role in maintaining the waking state via some arousal or alerting mechanism. The neurons of its supraoptic and paraventricular nuclei manufacture the hormones that the posterior pituitary stores and secretes.

Midbrain

The midbrain, or mesencephalon, is situated posterior to the diencephalon (Fig. 20-16). Its ventral position is formed by the two cerebral peduncles, and its dorsal portion, or tectum, by the corpora quadrigemina. The cerebral aqueduct (or aqueduct of Sylvius), which connects the third and fourth ventricles, marks off the ventral and dorsal aspects. Although the bulk of the midbrain is composed of white matter, vital areas of gray matter are present in its interior around the cerebral aqueduct and in the region ventral to the superior corpora quadrigemina.

The peduncles, which are composed of both ascending and descending tracts, plunge into the undersurface of the cerebral hemispheres. The midbrain, then, is the main functional as well as anatomical link between the higher and lower parts of the brain. The corpora quadrigemina (the four hemispherical eminences) are reflex centers for muscular movements (for example, turning the head and eyes) provoked by visual and auditory stimuli. The gray area about the cerebral aqueduct gives rise to the third and fourth cranial nerves, and the two gray areas (red nuclei) in the tectum give rise to the rubrospinal tracts. Because the red nuclei receive information from the cerebellum, these tracts carry impulses effecting muscle coordination.

Cerebellum

The cerebellum, or "little brain," is a large mass of nervous tissue that lies beneath the back part of the cerebrum (Fig. 20-16). Its two lateral halves, or hemispheres, are joined together centrally by a wormlike bridge aptly called the vermis. The surface of the cerebellum (cerebellar cortex) is composed of gray matter fashioned into shallow, parallel grooves and ridges, a pattern differing markedly from the large irregular convolutions that characterize the cerebral cortex. Within the cerebellum we find white matter and islands of gray matter, the most important of which are the dentate nuclei. These nuclei give rise to axons that travel to the cerebrum, the red nucleus, and the midbrain. The only paths of communication between the cerebellum and the rest of the brain reside in three pairs of tracts called cerebellar peduncles. The superior peduncles, composed chiefly of axons that make connections with neurons in the midbrain, relay impulses upward to the cerebrum and downward to the pontine nuclei (gray areas within the pons); the middle peduncles are made up of transverse fibers connecting the cerebellum to the pons; the inferior peduncles, which connect the cerebellum with the spinal cord, carry impulses from the muscles, tendons, joints, and semicircular canals (Chapter 22).

The cerebellum performs its duties below the level of consciousness. These functions include maintenance of proper muscle tension and reinforcement and refinement of the impulses transmitted by the motor area of the cerebral cortex. The cerebellum ensures the strength, precision, and smoothness of muscular movements through the coordination of reflex activities. It receives much input from the cerebral cortex and the peripheral nerves. Damage to the cerebellum results in faulty signals to the motor area (via the thalamus) and pronounced muscular dysfunction, including loss of muscle tone, tremors, difficulty in walking, and poor equilibrium.

Pons

The pons, or pons varolii, is an egg-shaped mass immediately above the medulla (Fig. 20-16) and in front of the fourth ventricle. It consists of a bridge of fibers that run to the cerebral hemispheres and join the midbrain above with the medulla below. The pons forms an important connecting link in the corticopontocerebellar path by which the cerebellum and cerebrum are united. The several small internal nuclei of gray matter in the pons give rise to the fifth, sixth, seventh, and eighth cranial nerves (p. 329), and a very special area (the pneumotaxic center) plays a role in respiratory rhythmicity.

Medulla oblongata

The medulla oblongata is actually an enlargement of the spinal cord just below the pons (Fig. 20-16). It contains the centers for the various reflexes regulating heart rate, blood pressure, respiration, vomiting, sneezing, coughing, micturation, and defecation. These centers reside in small gray nuclei distributed throughout the interior among the projection tracts. The pyramidal tracts decussate on the anterior aspect of the medulla, and the nucleus gracilis and the nucleus cuneatus serve as relay centers for sensory impulses on their way up to the thalamus. The medulla gives rise to the ninth, tenth, eleventh, and twelfth cranial nerves.

Reticular formation

In recent years an area of the medulla called the reticular or bulboreticular formation has aroused considerable neurophysiologic interest. Composed of a maze of neurons and interlacing fibers that run in all directions from the level of the upper spinal cord to the diencephalon, its chief function seems to be to coordinate muscular activity and to arouse the cerebral cortex via an hypothesized "wake center" situated in the hypothalamus. One view of wakefulness and sleep holds that the wake center stimulates the reticular formation, which in turn causes an increase in muscle tone throughout the body. The proprioceptors in the muscle respond to the increased tone and transmit sensory impulses (via the thalamus) to the wake center, thereby causing it to become re-excited. If this oscillatory effect continues, we remain awake by virtue of the high degree of neural activity. Conversely, when the oscillatory system becomes sluggish, as a result of fatigue, for example, we fall asleep. Even in the absence of fatigue we fall asleep if the body is relaxed, if there are fewer sensory impulses feeding into the reticular formation.

Certain other facts about sleep may be explained on the basis of the foregoing theory. For instance, recent work has shown that certain depressants, such as hypnotics and tranquilizers, owe their pharmacologic action to their ability to inhibit the reticular formation. On the other hand, certain stimulants, such as the amphetamines, stimulate the reticular formation.

Limbic system

If we take the liberty of regarding the brain as three concentric circles situated at the upper end of the spinal cord, the central ring, the ring between the core ring (brainstem) and the outer ring (cerebral cortex), represents what is called the limbic system. This system, often likened to a primitive cortex, is constituted by certain nuclei of the thalamus and hypothalamus, basal nuclei (including the amygdaloid nucleus), and the hippocampus. The amygdaloid nucleus, or amygdala, is situated beneath the tip of the temporal lobe; the hippocampus is a curved structure on the floor of the middle horn of the lateral ventricle. Sensations are projected by various pathways to the limbic system, where they are perceived in a manner presumably analogous to, but "cruder" than, the perception by the cerebral cortex. Presumably, basic sensations—hunger, fear, rage, sexual drive, and the like—are developed here and then converted to modified conscious forms in the cerebral cortex. The limbic system is apparently the seat of our raw emotions and has proved of tremendous psychiatric interest. A number of tranquilizers and certain other behavioral drugs are known to alter its function.

Brain metabolism

Even though the brain makes up only about 2% of body weight, it contributes up to as much as 10% of the basal metabolism. The brain derives its energy almost exclusively from glucose, a fact dramatically underscored by the mental confusion following a decrease in blood sugar. A pronounced decrease, as in an overdose of insulin, can cause convulsions and even unconsciousness. Especially critical, too, is the supply of oxygen. A complete shutoff of blood to the brain may cause permanent damage in 5 minutes.

Electroencephalogram

The constant electrical activity of the cerebral cortex can be detected and recorded by means of an oscillograph connected to electrodes applied to various regions of the scalp. Such a record of electrical potentials ("brain waves") is called an electroencephalogram (EEG). In a normal subject the form is quite characteristic of the region and of the level of wakefulness and mental activity. Alpha waves are fairly regular waves with a frequency of 8 to 12 per second; beta waves are less regular and have a frequency of 18 to 32 per second. Many disorders of the brain give rise to abnormal patterns, so that in skilled hands the EEG is a valuable diagnostic tool. For example, it is possible to distinguish among the various kinds of epilepsy by this means (Fig. 20-17).

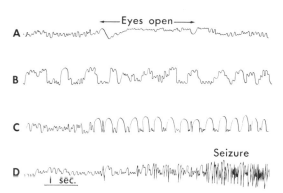

FIG. 20-17

Electroencephalogram tracings. **A,** Normal, showing diminution of waves upon opening of eyes. **B,** Effect of sleep. **C,** Petit mal epilepsy. **D,** Grand mal epilepsy. (From Main, R. J.: Synopsis of physiology, St. Louis, 1946, The C. V. Mosby Co.)

MENINGES

The three membranes, or meninges, that cover the spinal cord continue upward and envelop the brain. The dura mater, the outermost membrane, is a double-layered membrane that serves both as a covering for the brain and as a periosteal lining for the skull. Its inner layer extends between the parts of the brain, forming septa. Two of these, the falx cerebri and the tentorium cerebelli, are especially prominent. The falx cerebri is a sickle-shaped fold that runs vertically between the cerebral hemispheres; the tentorium cerebelli runs between the cerebellum and the occipital lobes above. The dura mater also forms blood sinuses that drain into the jugular veins. The larger of these, the superior sagittal sinus, forms the upper border of the falx cerebri, and the inferior sagittal sinus forms the lower border.

The arachnoid membrane forms a rather loose investment about the brain and is separated from the dura mater by the subdural space, which is crisscrossed by a fine network of connective tissue fibers. The arachnoid membrane

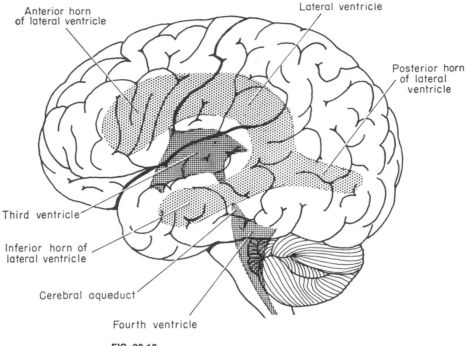

FIG. 20-18

Ventricles of the brain.

VENTRICLES

is separated from the pia mater by a greater interval called the subarachnoid space. In the areas where this space is expanded, the term "subarachnoid cisterna" is used. The largest of these spaces, the cisterna magna, is located between the undersurface of the cerebellum and the posterior surface of the medulla oblongata.

VENTRICLES

Within the brain itself are four interconnecting spaces, or ventricles, that communicate with the central canal of the spinal cord and with the subarachnoid space about both the brain and the cord. The largest ventricles are the two lateral ventricles, one in each cerebral hemisphere. As shown in Fig. 20-18, the central portion of each

lateral ventricle extends into three horns—anterior, inferior, and posterior—that invade the frontal, temporal, and occipital lobes, respectively. The third ventricle is a single, irregular space lying below the anterior horns and between the two thalami. The fourth ventricle is a single, diamond-shaped space immediately before the cerebellum.

Cerebrospinal fluid (p. 311) seeps from each lateral ventricle into the third ventricle via the interventricular foramen, or foramen of Monro, and from the third it enters the fourth ventricle via the aqueduct of Sylvius. Through the three openings in the fourth ventricle—the foramina of Luschka and the foramen of Magendie—the fluid escapes into the subarachnoid space. From there it enters the arachnoid villi (fingerlike

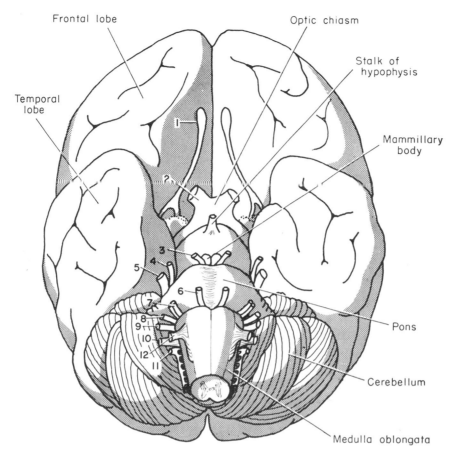

FIG. 20-19

Undersurface of brain showing origin of cranial nerves. *1*, Olfactory; *2*, optic; *3*, oculomotor; *4*, trochlear; *5*, trigeminal; *6*, abducens; *7*, facial; *8*, acoustic; *9*, glossopharyngeal; *10*, vagus; *11*, accessory; *12*, hypoglossal.

processes projecting into the venous sinuses) and then the systemic circulation.

CRANIAL NERVES

The twelve pairs of cranial nerves that emerge from the undersurface of the brain (Fig. 20-19) are numbered from front to back according to their function or according to the structure or structures that they innervate (Table 20-2). The first, second, and eighth are purely sensory. The others are mixed; they contain both afferent and efferent fibers. Whereas the cell bodies of the efferent fibers are located within the brainstem (midbrain, pons, and medulla), most of the afferent cell bodies are situated in ganglia outside the brain.

AUTONOMIC NERVOUS SYSTEM

The autonomic nervous system (Fig. 20-20) concerns those efferent nerve fibers of the pe-

TABLE 20-2

Cranial nerves

Nerve	Function
I. Olfactory	Smell
II. Optic	Vision
III. Oculomotor	Eye movements; regulation of size of pupil
IV. Trochlear	Eye movements; proprioception
V. Trigeminal	Mastication; sensations of head and face
VI. Abducens	Abduction of eye; proprioception
VII. Facial	Facial expressions; salivary secretion; taste
VIII. Acoustic	
A. Cochlear or auditory branch	Hearing
B. Vestibular branch	Equilibrium
IX. Glossopharyngeal	Swallowing; secretion of saliva; taste
X. Vagus	Sensory and parasympathetic fibers to major abdominal viscera
XI. Spinal accessory	Motor fibers to shoulder and head
XII. Hypoglossal	Tongue movements

ripheral nervous system that transmit efferent impulses to glands and smooth muscle. In contrast to the nerves that innervate skeletal muscle, autonomic nerves are not under conscious control and act involuntarily. The autonomic nervous system governs the internal environment of the body. It is not an independent unit, for its centers, all situated in the hypothalamus, are directly in the path of "impulse traffic" to and from the cerebral cortex and the cord. For example, while it is quite true that the presence of food in the mouth automatically provokes salivation, it is also true that merely thinking about food—LEMON JUICE—often has the same effect. Peptic ulcer supposedly results from excessive secretion of gastric juice, which in turn stems from excessive cortical stimulation (from overwork, worry, and the like) of the autonomic centers. The same is true of nervous hypertension (p. 344) and a variety of other psychosomatic disorders. Thus, the nervous system is interdependent; it would be wrong to consider one part as functioning independently of another. Provided that we keep this in mind, it is acceptable, convenient, and helpful to discuss the autonomic system as a separate entity.

The autonomic nervous system may be divided anatomically and physiologically into the parasympathetic division and the sympathetic division. In general, the viscera receive impulses from both divisions, and with few exceptions their effects are antagonistic. For example, sympathetic stimulation speeds the heart, whereas parasympathetic stimulation slows it. (Throughout the discussion that follows the student should refer to Fig. 20-20.)

Parasympathetic division

Typically an autonomic impulse originates within the central nervous system and travels to the periphery via two neurons. The parasympathetic division is characterized anatomically by the location of its "central" cell bodies in the midbrain, pons, medulla, and sacral portions of the spinal cord. It is sometimes referred to as the craniosacral division. The axons of the central neurons follow certain cranial and pelvic nerves and finally synapse with peripheral neu-

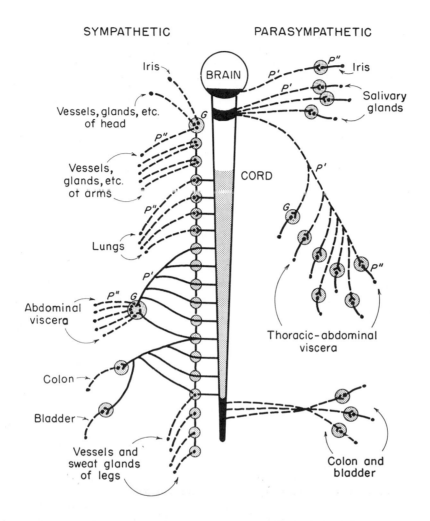

SYMPATHETIC   PARASYMPATHETIC

Iris

BRAIN

Vessels, glands, etc.
of head

Vessels,
glands, etc.
of arms

CORD

Lungs

Abdominal
viscera

Colon

Bladder

Vessels and
sweat glands
of legs

Iris

Salivary
glands

Thoracic-abdominal
viscera

Colon and
bladder

FIG. 20-20

Autonomic nervous system. Gray portion of spinal cord represents thoracolumbar region; dark portion at bottom represents sacral region. Dark areas of brain represent midbrain (upper) and medulla (lower). *P′*, Preganglionic fiber; *P″*, postganglionic fiber; *G*, ganglion. (From Brooks, S. M.: Basic facts of pharmacology, ed. 2, Philadelphia, 1963, W. B. Saunders Co.)

rons whose cell bodies are located in ganglia close to or upon the structures innervated. The comparatively short axons of the latter neurons terminate at neuroeffector junctions. By convention, the axon of the central neuron is called the *preganglionic fiber*, and the axon of the peripheral neuron, the *postganglionic fiber*.

Sympathetic division

In contrast to the parasympathetic division, the cell bodies of the preganglionic sympathetic fibers originate in the lateral gray horns of the thoracic and first four lumbar segments of the spinal cord—hence the expression thoracolumbar division. From these sites the fibers issue

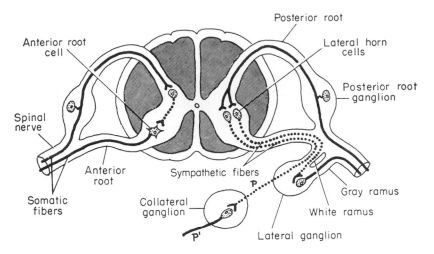

FIG. 20-21
Relationship between spinal and sympathetic nerves. *P*, Preganglionic fiber; *P'*, postganglionic fiber. (After Gaskell; modified from Mountcastle, V. B., editor: Medical physiology, ed. 13, St. Louis, 1974, The C. V. Mosby Co.)

through the ventral roots of the corresponding spinal nerves and via side branches, called white rami, enter the two sympathetic chains of ganglia that parallel the spinal cord (Fig. 20-21). Here most of them synapse with the postganglionic fibers at the same level or pass up or down the chain and then synapse. Fibers innervating the gut extend through the sympathetic chain and become splanchnic nerves (that is, pure autonomic nerves) that synapse with postganglionic fibers at the collateral ganglia. Regardless of the path a preganglionic fiber takes, it generally synapses with several postganglionic fibers, thereby provoking a widespread response. This contrasts with the parasympathetic system in which a single preganglionic fiber typically makes but a single synapse.

Autonomic action

Though the parasympathetic system, like the sympathetic, is always in operation, it prevails in times of quietude. For example, it decreases the heart rate (Fig. 20-22). A summary of its

effects upon specific organs and structures is presented in Table 20-3. In looking over these effects, it is to be remembered that a given response is due either to stimulation or inhibition of glandular secretion or to the contraction or relaxation of smooth muscle. For instance, parasympathetic stimulation causes the bronchioles to constrict because the smooth muscle in their walls is made to contract. On the other hand, parasympathetic stimulation causes certain

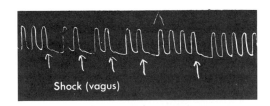

FIG. 20-22
Effect on frog heart of electrical stimulation of vagus nerve. (From Brooks, S. M.: Basic facts of pharmacology, ed. 2, Philadelphia, 1963, W. B. Saunders Co.)

TABLE 20-3

Autonomic function

Visceral effector	Sympathetic effects	Parasympathetic effects
Adrenal medulla	Stimulates epinephrine secretion	—
Anal sphincter	Closes	Opens
Blood vessels of brain and viscera	Constricts	Dilates
Blood vessels of skeletal muscles	Dilates	—
Blood vessels of skin	Constricts	—
Bronchi	Dilates	Constricts
Digestive glands	Decreases secretion	Stimulates secretion
Digestive tract musculature	Slows peristalsis	Accelerates peristalsis
Eye, ciliary body	Accommodates for far vision	Accommodates for near vision
Eye, iris	Dilates pupil	Constricts pupil
Heart	Accelerates rate	Slows rate
Liver	Stimulates glycogenolysis	—
Pancreas	Decreases secretion	Stimulates secretion
Pilomotor muscles	Causes "goose pimples"	—
Sweat glands	Stimulates secretion	—
Urinary bladder	Relaxes	Contracts
Urinary sphincter	Closes	Opens

blood vessels to dilate because the smooth muscle in their walls is made to relax. These effects relate to the nature of the properties of the cells at the neuromuscular junction.

The sympathetic division prevails in times of stress and in emergencies. It prepares the body for fight or flight. Among other effects, it dilates the pupils, facilitates breathing, increases the blood pressure, and stimulates the release of epinephrine from the medulla of the adrenal gland (Table 20-3). These responses, at least in modern man, are not essential to life. In malignant hypertension, sympathetic fibers are sometimes cut (sympathectomy) to relieve the pressure. Completely sympathectomized animals, provided they are cared for properly, have a normal span of life. On the other hand, the body cannot get along without its parasympathetic division.

Chemical transmission

One of the most ingenious experiments known to physiology was first performed in 1921 by Otto Loewi. Using a very simple setup (shown diagrammatically in Fig. 20-23), Loewi demonstrated that some chemical substance was the mediator of nerve impulses at the neuroeffector junction. Not knowing its identity, he called the agent "Vagusstoff." Eventually, Loewi's Vagusstoff turned out to be a comparatively simple chemical called acetylcholine. It is now known that acetylcholine transmits the nerve impulse across the ganglionic synapses of both autonomic divisions and across the neuroeffector junctions

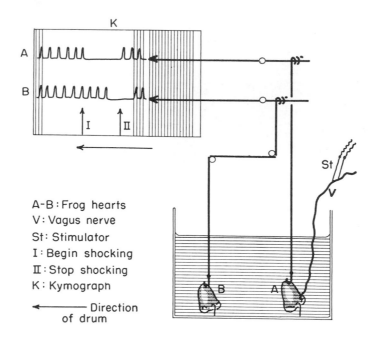

FIG. 20-23

Schematic representation of Loewi's original experiment demonstrating that acetylcholine (called *Vagusstoff* by Loewi) is released at ends of parasympathetic nerve fibers when these fibers are stimualted. Note that stopping of heart *A*, caused by shocking vagus nerve, is followed shortly by stopping of heart *B*. Thus, acetylcholine formed at myoneural junctions of heart *A* is carried via solution to heart *B*. (From Brooks, S. M.: Basic facts of pharmacology, ed. 2, Philadelphia, 1963, W. B. Saunders Co.)

A-B : Frog hearts
V : Vagus nerve
St : Stimulator
I : Begin shocking
II : Stop shocking
K : Kymograph

⟵ Direction of drum

of all nerves except the sympathetic (Fig. 20-24). The neurotransmitter in the latter case has now been proved to be norepinephrine.* Once formed, acetylcholine and norepinephrine are deactivated immediately by the enzymes cholinesterase and catechol-o-methyl transferase, respectively. The terms cholinergic and adrenergic are commonly used. According to this scheme, all voluntary motor fibers, all preganglionic autonomic fibers, and all postganglionic parasympathetic fibers are said to be cholinergic because they release acetylcholine at their endings. Postganglionic sympathetic fibers are said to be adrenergic because they release norepinephrine (noradrenaline).

---

*Norepinephrine is also believed to serve as a neurotransmitter in certain brain synapses.

Autonomic drugs

When injected into the body, acetylcholine mimics parasympathetic stimulation (for example, constricts pupils, decreases heart rate, and stimulates secretion of saliva), and norepinephrine mimics sympathetic stimulation (for example, dilates pupils, increases heart rate, and decreases secretion of saliva). The site of action in both cases is believed to be the neuroeffector junction. Though acetylcholine is not used therapeutically because of its fleeting action,* chemically related agents of similar action are employed in a variety of disorders in which parasympathetic stimulation yields salutary effects. In

---

*Acetylcholine is quickly metabolized by the many esterases present in the blood.

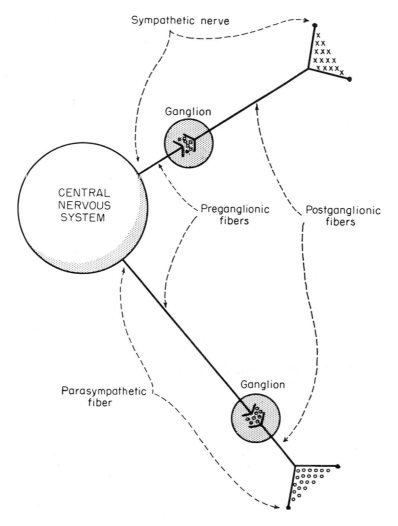

FIG. 20-24

Neurohumoral transmission of autonomic nerve impulses. *x*, Norepinephrine; *o*, acetylcholine. (From Brooks, S. M.: Basic facts of general chemistry, Philadelphia, 1956, W. B. Saunders Co.)

urinary retention following surgery, for example, bethanechol (Urecholine) stimulates the atonic bladder. Norepinephrine is used in shock to raise the blood pressure; epinephrine is used in asthma to dilate the bronchioles; and so on. Autonomic drugs that mimic the parasympa-

thetic system are aptly called parasympathomimetics or *cholinergics*. Their sympathetic counterparts are called sympathomimetics or *adrenergics*. There are also autonomic blocking agents, that is, drugs which inhibit the autonomic system. Drugs that block parasympathetic action are called parasympatholytics, and their sympathetic counterparts are called sympatholytics.

DISEASES OF THE NERVOUS SYSTEM

A functional or organic disorder of nervous tissue, especially one involving the brain and cord, is often serious. Many times little can be done other than treating the symptoms and hoping for the best. Such is usually the case in pathologic situations in which nerve cells are destroyed. Since these cells are not replaced, their loss is irreversible. The more common manifestations of neurologic disorders include pain, headache, dizziness, tics, hiccups, recurrent attacks of sleep (narcolepsy), tremors, convulsions, and unconsciousness. Major highlights of some specific disorders follow.

Unconsciousness

Unconsciousness is a state of insensibility that is normal only during sleep. It may range in depth from stupor or semiconsciousness to coma, a profound unconsciousness from which the victim cannot be aroused even by powerful stimulants. Because unconsciousness is a manifestation of an underlying injury or disease, its duration and intensity naturally depend upon the basic cause. Among the many and varied causes are simple fainting, acute alcoholism, head injury, cerebrovascular accidents, depressant poisoning, epilepsy, diabetic acidosis, infections, heart failure, severe anemia, shock, uremia, hepatic failure, hysteria, heat stroke, exposure to extreme cold, eclampsia, and Addison's disease.

Unconsciousness, of course, is an emergency situation. Pending diagnosis of the underlying disorder, the physician must direct his skills toward maintenance of respiration and circulation. This includes such measures as controlling hemorrhage, giving oxygen, and applying artificial respiration. Stimulants are contraindicated unless the etiologic factor is depressant poisoning (opiates, barbiturates, and the like).

Epilepsy

Epilepsy is a chronic disorder of cerebral function characterized by recurrent attacks of altered consciousness, often accompanied by convulsions in which the patient falls involuntarily. Although in most cases no significant underlying cause is revealed, in other instances a tumor or a cerebral lesion of some magnitude may be present. These may have resulted from injury or from other morbid processes.

There are five types of epileptic seizure, each with a specific pattern: grand mal, petit mal, psychomotor, infantile spasms, and epileptic equivalents. In approximately 70% of the patients only one type of seizure occurs; the remaining 30% have two or more types. It is most important to understand that most epileptics are normal between attacks; 75% of the noninstitutionalized epileptic patients are mentally competent, and when there is mental deterioration, it generally is related to accompanying brain damage.

Treatment is multidimensional. It includes correction of the causative or precipitating factor (for example, surgery to remove a tumor or relieve pressure on brain), physical and mental hygiene, and drug therapy. Drugs that are effective in controlling and preventing one type of seizure often fail to control or prevent the others. Diphenylhydantoin (Dilantin), for instance, yields good results in grand mal epilepsy but fails to control the petit mal form. This indicates, of course, that these forms of epilepsy are of diverse etiology.

Cerebral palsy

Cerebral palsy is a nonprogressive disturbance of the motor system resulting from injury at birth or from intrauterine cerebral degeneration. The most common manifestation is spastic weakness of the extremities, characterized by scissor gait and exaggerated tendon reflexes. In the most severe cases the extremities are stiff, and there is great difficulty in swallowing. Though most patients afflicted with cerebral palsy exhibit some degree of mental retardation, this poor showing may well result from difficulty in self-expression. Drugs to relieve spasticity and control convulsions, if and when they occur, special courses, and vocational guidance are essential if the patient is to live out his years in a meaningful way, especially in the milder cases, in which the potential for a normal way of life exists. As indicated, many cerebral palsy victims are much brighter than they appear.

Multiple sclerosis

Multiple sclerosis is marked histologically by disseminated, sclerosed or demyelinated patches in the brain and spinal cord; clinically, it is characterized by progressive weakness, incoordination, jerking movements, abnormal mental exhilaration, and disturbances of speech and vision. The cause remains unknown. Postulated etiologies include autoimmune mechanisms, "slow viruses" (agents producing symptoms after a prolonged incubation period), certain poisons, and trauma. The duration of the disease averages about 12 years following onset, but the course is highly variable. In some cases attacks are frequent; in others there may be remissions of as long as 20 years or more. At present, there is no specific therapy. The patient is made as comfortable as possible, and the physician may recommend such measures as physiotherapy and psychotherapy. Moving to a warmer climate may help. Multiple sclerosis is relatively uncommon in the tropics.

Parkinsonism

Parkinsonism, also referred to as Parkinson's syndrome, paralysis agitans, and shaking palsy, is a chronic disorder marked by tremor, muscular rigidity, and slowness of movement. Most cases are of unknown etiology ("idiopathic"), but the disorder may be produced by various agents, including carbon monoxide, manganese (poisoning), and certain tranquilizers. Postencephalitic parkinsonism commonly occurred following attacks of epidemic encephalitis (p. 340) between 1919 and 1924.

The discovery that parkinsonism is accompanied by a decrease in the level of dopamine, a neurotransmitter concentrated in some of the basal ganglia, led to the therapeutic use of levodopa (Larodopa), the metabolic precursor of dopamine. Although this drug is not always effective and may provoke serious side effects, it does, in balance, represent a definite breakthrough in treatment. Other drugs in use include piperidines, antihistamines, and atropine-like agents. Surgical destruction of certain areas of the thalamus by cryotherapy has effected cures in selected cases.

Stroke

A stroke (cerebral apoplexy) is a "cerebrovascular accident" (CVA) resulting from hemorrhage, thrombosis, embolism, or other vascular insufficiencies. There is injury to the substance of the brain, usually from inability of specific regions to obtain sufficient oxygen. Common symptoms, which are typically immediate, include headache, vomiting, convulsions, and coma. The specific symptoms will naturally depend upon the site of the lesion. In cerebral hemorrhage, for example, the lesion is usually well within the brain substance, and the characteristic manifestations are hemiplegia and hemianesthesia (paralysis and loss of sensation, respectively, of one side of the body).

The specific treatment depends upon the cause. If the diagnosis is thrombosis or embolism, anticoagulants are indicated in the hope of preventing further enlargement of the clot. General measures include skillful nursing care,

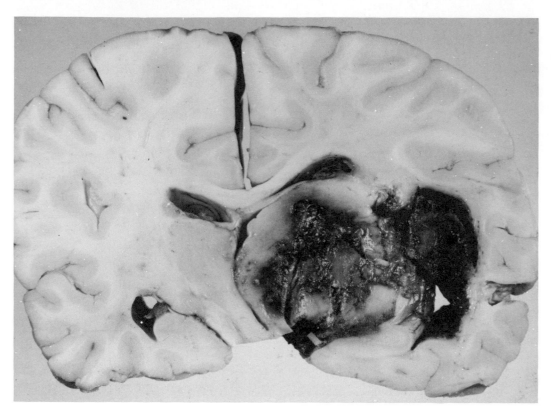

FIG. 20-25

Massive hemorrhage within substance of brain tumor. Note marked displacement of midline structures. (From Ackerman, L. V., and Rosai, J.: Surgical pathology, ed. 5, St. Louis, 1973, The C. V. Mosby Co.)

lumbar puncture, physiotherapy, and operative removal of the blood clot in the face of increased intracranial pressure. If only a small vessel is involved, almost complete recovery can be expected sometime within a year. Massive brain damage, however, generally results in death a week or so following the attack.

Tumors

Tumors of the brain and spinal cord kill about 8,000 Americans annually. Of chief concern are the intracranial growths, perhaps a third of which are metastatic. Primary intracranial tumors arise from the brain itself or from lesser structures (namely, blood vessels, cranial nerves, meninges, and the pituitary gland). Nearly half of all primary tumors are gliomas, or growths involving the neuroglia (Fig. 20-25). In the adult the most common glioma—and the most common brain cancer—is the rapidly expansive, invasive, and deadly glioblastoma multiforme of the cerebral hemispheres. Intracranial tumors in children occur almost exclusively in the cerebellum. The most common are the benign astrocytoma and the highly malignant medulloblastoma.

Actual manifestations are best thought of in

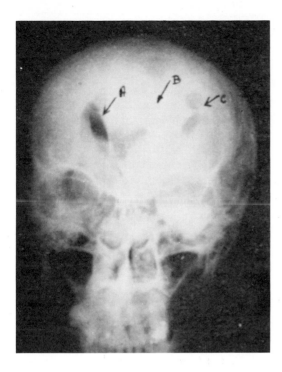

FIG. 20-26
Ventriculogram (face forward) showing normal ventricle *A* of left cerebral hemisphere and displaced ventricle *C* of the left cerebral hemisphere. Displacement is caused by "invisible tumor" at *B*. (From Ray, B.: The nervous system. In Davis, L. (ed.): Christopher's textbook of surgery, ed. 8, Philadelphia, 1964, W. B. Saunders Co.)

terms of pressure and location. The pressure arises because the cranium does not give and because the tumor generally interferes with the drainage of cerebrospinal fluid. Symptoms caused by pressure include, among others, headache, vomiting, and failing vision. The signs and symptoms are referable to the function of the brain region involved. Frontal lobe tumors cause a change in personality; occipital lobe tumors cause visual hallucinations; and so on. Diagnosis involves a number of radiological techniques, including radioscanning, photoscanning, angiography, and ventriculography

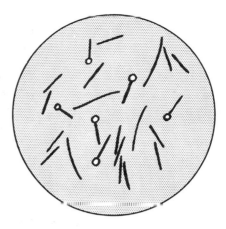

FIG. 20-27
Clostridium tetani. Note terminal spores. (From Brooks, S. M.: Basic facts of medical microbiology, ed. 2, Philadelphia, 1962, W. B. Saunders Co.)

(Fig. 20-26). Treatment centers on surgery; in the curable case, surgical removal remains the one and only cure.

Tetanus

Tetanus, or lockjaw, is a vicious infection caused by the exotoxin released by *Clostridium tetani,* a slender, motile, gram-positive, anaerobic bacillus with terminal spores that are considerably wider than the body of the bacillus and make the organism look like microscopic tennis rackets (Fig. 20-27). The exotoxin is believed to act by blocking inhibitory synapses and thereby causing great excitation of many cells and pathways. Generalized hypertonus with intermittent convulsions results. Stiffness of the jaw, the most common symptom, is of utmost diagnostic significance. Death may result from respiratory failure. As in many conditions, the heart continues to beat for several minutes after breathing has stopped.

The ubiquitous *Clostridium tetani* is introduced into the tissues via contaminated wounds, particularly wounds of the puncture type (for example, stab wounds, gunshot wounds, bites, wounds sustained in stepping on a nail, and the like). The deeper the wound, the greater the likelihood of tetanus because *Clostridium tetani,* being an anaerobe, thrives in the absence of air. Soon after the bacilli have established a foothold in the underlying tissues, they start to multiply and turn out the deadly exotoxin. Since it takes but a fantastically small amount of the toxin to destroy life, time is of the essence. Statistics show that when treatment is delayed 1 day after the onset of symptoms, the mortality rate is about 50%! On the other hand, if the victim survives, he usually is free of all ill effects in about a month.

World War II demonstrated that tetanus toxoid affords 100% protection, which means that no one need die at the hands of this disease. The recommended procedure is to give a primary injection, followed by booster shots 1 year later and every 3 to 5 years thereafter. Because one third to one half of tetanus cases occur with no history of injury, such precaution is highly desirable.

Tetanus prophylaxis in injured patients centers upon whether such patients have been previously immunized. The recommended procedure is as follows:

Patients previously immunized
1. Adequate cleansing and debridement of wound
2. Toxoid booster
3. Antitoxin*—may be needed rarely (along with toxoid) for very extensive wounds, for persons seen more than 48 hours after injury, and for patients whose last toxoid booster had been given more than 20 years previously
4. Antibotics—given if danger of clostridial infection is very great or if indicated for other bacterial infection

Patients not previously immunized
1. Adequate cleansing and debridement of wound
2. Antitoxin—may be omitted for trivial wounds; indicated for all tetanus-prone wounds
3. Antibotics—given if clostridial infection is likely or if indicated for other bacterial infection

---

*Whenever possible antitoxin of human origin ("tetanus immune globulin") should be used. Horse serum antitoxin is not without danger.

4. Toxoid—first dose gives no protection for wound being treated but provides opportunity to start active immunization

Meningitis

Meningitis, or inflammation of the meninges, is caused by a great many organisms, including *Neisseria meningitidis, Haemophilus influenzae, Diplococcus pneumoniae, Streptococcus pyogenes, Escherichia coli,* and *Staphylococcus aureus.* The classic signs and symptoms include severe headache, fever, stiffness in the neck and back, vomiting, malaise, irritability, and finally drowsiness and coma. At autopsy there is often a purulent exudate in the subarachnoid space over the brain as well as a thin layer of exudation over the cord when the disease extends down from the brain.

The most common form of the infection, called meningococcal, or epidemic, cerebrospinal meningitis, is caused by the gram-negative diplococcus *N. meningitidis.* Spread via droplet spray from the nose and mouth of carriers and persons with upper respiratory meningococcal infection, this form of the disease has struck one fourth of the population during an epidemic. Diagnosis is based upon the clinical picture and the presence of the pathogen in the cerebrospinal fluid. However, it may be fatal to await a definitive diagnosis and more often than not antibiotic therapy must be started before the pathogen is identified. Ampicillin is considered by many to be the drug of choice pending identification. Thereafter, the appropriate antibiotic is selected. Penicillin G, for example, is the drug of choice in meningococcal meningitis.

Encephalitis

The various forms of encephalitis, or inflammation of the brain, almost defy enumeration. Fortunately, they are relatively uncommon in the United States. The chief forms of the disease include eastern equine encephalomyelitis (EEE), western equine encephalomyelitis (WEE), Venezuelan equine encephalomyelitis (VEE), Japanese B encephalitis, and St. Louis encephalitis. Each is caused by a virus, and the mosquito serves as the vector; for the most part, wild birds serve as the reservoir. The specific signs and symptoms of encephalitis depend upon the particular type. In general, however, there is usually some degree of headache, fever, muscle pain, stiffness, and delirium. In severe cases there may be convulsions and coma. At this time there is no specific treatment.

Poliomyelitis

Poliomyelitis, also called acute anterior poliomyelitis and infantile paralysis, is an acute viral infection (Fig. 20-28) characterized by fever, headache, stiffness, and flaccid paralysis. However, these signs and symptoms apply only to the full-blown infection. In most instances, an attack goes unnoticed, the patient merely complaining of slight malaise, headache, sore throat, and gastrointestinal upset. Contrary to popular belief, less than one fourth of the patients who develop paralysis sustain permanent disability. Although poliomyelitis still strikes chiefly at the younger segments of the population, over the years it has definitely broadened its attack. Whereas 90% of the victims in an epidemic that hit New York City in 1916 were children under 5 years old, 35% of the victims in a more recent out-

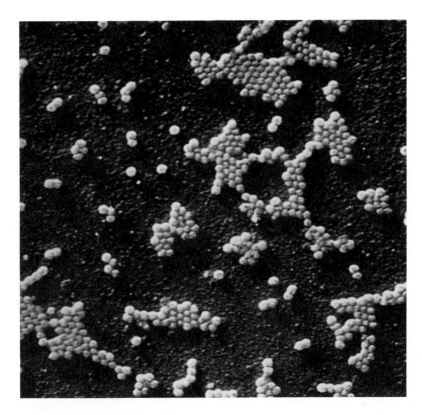

FIG. 20-28
Type 2 poliovirus. ($\times$ 79,000.) (Courtesy Parke, Davis & Co.)

break in New England were over 15 years old.

The polio pathogen is an enterovirus that exists in three immunologic types: type I (Brunhilde), type II (Lansing), and type III (Leon). Authorities now believe that the virus enters the body via the oral route and, having multiplied in the gastrointestinal mucosa, invades the blood and ultimately the central nervous system, where it attacks the cord or medulla or both. It now seems reasonable to assume that in a mild case of poliomyelitis the virus either does not pass beyond the mucosa or, if it does, fails to invade nervous tissue. The introduction of an effective vaccine against the disease represents one of the great medical conquests of the present century. The original Salk vaccine, a formalin-killed preparation containing a strain of each of the three types of virus, proved from 70% to 90% effective. The oral attenuated (Sabin) vaccine, however, is superior and has largely supplanted the Salk preparation.

Rabies

Rabies, or hydrophobia, is perhaps the most vicious disease of the nervous system. The virus, present in the saliva of rabid animals, is introduced into the tissues as the result of a bite or scratch. Dogs cause most cases of rabies among human beings, but cats, foxes, wolves, skunks, squirrels, bears, and bats are also occasionally responsible. From the portal of entry the virus travels along the nerves to the spinal cord and then on to the brain, where it multiplies and invades the vital areas. In "unprotected" persons, death can be expected. As of this writing only two or three victims have survived once the signs and symptoms were in evidence.

Clinical highlights of rabies include an early period of mental depression, fever, and restlessness, followed by fanatic excitement and painful spasms in the larynx and throat. Since an attempt to drink precipitates these spasms, the victim shuns fluid (hence the term "hydro-

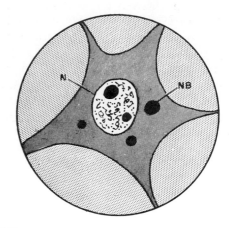

FIG. 20-29
Neuron from brain of rabid animal, showing Negri bodies, *NB. N* is the nucleus.

phobia"). At autopsy, engorged vessels, tiny diffuse areas of hemorrhage, and nerve cell destruction are seen. Characteristically, the neurons in certain areas show cytoplasmic structures called Negri bodies (Fig. 20-29), the presence of which represents a definitive diagnosis.

Any bite or scratch, particularly about the head or on unprotected surfaces (clothing affords some protection), demands immediate attention, for, as indicated, there is no specific treatment for rabies once the disease is in progress. Licks of abraded skin should be considered in the same light as actual bites. The wound should be laid open and washed repeatedly with soap and water. This simple but vital procedure enhances bleeding and kills the exposed virus. Immediate vaccination is indicated (1) when the animal is known to be rapid, (2) if the animal is not available for observation, (3) following all wild animal bites, and (4) in all cases of severe bites, especially of the head and neck. Otherwise, the animal is caged and observed by a veterinarian for at least 10 days. If during this time the animal shows signs of the infection, immunization is started by giving both hyperimmune serum and rabies vaccine.

Herpes zoster

Herpes zoster, or shingles, is an infection of the central nervous system with cutaneous manifestations. The details of this disease were presented earlier (p. 119).

Psychiatric disorders

Psychiatry is the branch of medicine that deals with disorders of the psyche, or our conscious and unconscious mental life. Basically, such disorders fall into two major categories: those caused by or associated with damage of brain tissue, and those of psychogenic origin without demonstrable physical cause. Mental disorders of the first category can be ameliorated or sometimes even corrected completely by treatment of the underlying organic cause; mental disorders of the second category, however, require the special talents of psychiatry in addition to those of physical medicine.

Mental disorders of psychogenic origin are classified, according to one scheme, into four subcategories: psychoneurotic disorders, psychotic disorders, psychosomatic disorders, and personality disorders.

Psychoneurotic disorders, or neuroses, are relatively mild abnormalities in which the symptoms play some compensatory or protective role in the patient's mental life. The neurotic's behavior, however, remains more or less intact. Also, his emotions, thoughts, and impulses seem strange and foreign even to himself. Characteristic of all forms of neurosis is anxiety. Indeed, the various defense mechanisms employed by the neurotic patient in coping with his anxiety are the basis of the various recognized types of the illness. Hysterical blindness, for example, may occur at the moment a person witnesses an overwhelming act of violence, as when a mother sees her child run over by a car.

Psychotic disorders, or psychoses, are more profound and far-reaching, and the aberrations are more prolonged than in neuroses. There is disintegration of the personality, and the conscious ego is unable to distinguish effectively between the real and the unreal. The two major psychotic categories are schizophrenic reaction and manic-depressive reaction.

The schizophrenic reaction, or schizophrenia (formerly, dementia praecox), represents cleavage of the mental functions. Characteristically, there are progressive withdrawal from the environment and deterioration of emotional response. In simple terms we may consider the disorder to arise from a personality structure that is unable to meet the demands of adjustment. Increasing evidence points to hereditary factors and some sort of biochemical defect as the underlying etiology.

In the manic-depressive reaction the patient experiences alternating periods of mania and depression. Some patients, however, experience only one phase of the illness and are either elated or depressed. The manic phase is characterized by verbosity, singing, shouting, frenzied dancing, and the like. The depressive phase may range from downheartedness to frank stupor in which the patient fails to respond to external stimuli. Studies indicate that both hereditary and environmental factors are etiologic possibilities.

Psychosomatic disorders are personality situations in which the defense against anxiety is expressed through the viscera. Since the viscera are innervated by the autonomic nervous system, and since the functions of the central and autonomic nervous systems are integrated, it is not difficult to understand how emotional conflict in the former can physiologically trigger the latter. The upshot is excessive autonomic stimulation and malfunctioning internal organs.

Perhaps the most common psychosomatic disorder is the proverbial nervous breakdown. In this condition the body becomes overstimulated and extremely tense because of excessive sympathetic bombardment. Another potentially serious variety is peptic ulcer. Whereas the normal gastric mucosa releases just enough juice to deal with digestion, the "psychosomatic mucosa" (as a result of too much parasympathetic stimulation) is thought to pour out so much juice that a hole is eaten in the stomach or duodenal wall. Other commonly encountered psychosomatic complaints include palpitation of the heart, hypertension, constipation, and diarrhea. In each instance, the symptoms can be explained by autonomic activity.

Personality disorders* include developmental defects in the personality structure. Since there are no acute mental or emotional symptoms, a person with such a condition may pass through life without realizing his basic problem. Ineptness, poor judgment, social incompatability, self-consciousness, and sexual deviation are all examples of this category.

DRUG ABUSE

In general, drugs affecting psychological functions or behavior fall into three categories based on their fundamental action. Psycholeptics depress mental activity and include alcohol, narcotics, barbiturates, and tranquilizers; psychoanaleptics stimulate mental activity and include imipramine, methylphenidate, and the amphetamines; and the now well known psychodysleptics, also called psychedelics and hallucinogens, create a delusional disturbance of judgment with a distortion of reality and include psilocybin, psilocin, peyote, mescaline, lysergic acid diethylamide (LSD), and marijuana (Fig. 20-30). These agents cause a state of psychic or physical dependence and, taken as a whole, present a

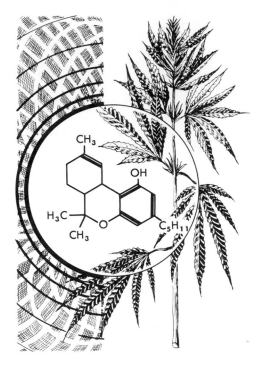

FIG. 20-30
Structural formula for tetrahydrocannabinol, chief active constituent of marijuana.

very serious medical problem, not to mention their social ramifications.

Alcoholism is now a well-established disease (2 million cases in this country alone), and a recent study disclosed the startling statistic that about 10% of first-admission patients to mental hospitals are alcoholics.

The problem of drug abuse is very complex, and its solution does not appear on the horizon. Clearly, we need to know much more than we do and to explore all possible avenues.

QUESTIONS

1. Distinguish between neuron and neuroglia.
2. In regard to the nerve cell, what are processes?
3. The distinction between a dendrite and an axon is basically functional. Explain.

---

*Personality disorders should be distinguished from the value judgments of one person or group about the behavior of another person or group.

4. Distinguish between neurofibrils and Nissl bodies.
5. Distinguish between a nerve and a nerve fiber.
6. Of what value is chromatolysis in neurological research?
7. What is the relationship between myelin and the nodes of Ranvier?
8. Discuss the classification of neurons.
9. Distinguish between nerves and tracts.
10. Most nerves are mixed. Explain.
11. Distinguish between resting potential and action potential.
12. The surface membrane of a nerve fiber is often characterized as a sodium pump. Why?
13. Why does a stimulus applied to a nerve fiber cause depolarization?
14. A nerve impulse is undiminished as it travels down a nerve fiber. How is this possible?
15. Account for the fact that repolarization occurs immediately in the wake of depolarization.
16. A nerve trunk shows a graded response to an increase in strength of the stimulus. In a way this proves the all-or-none law. Explain.
17. Discuss subliminal summation.
18. A nerve fiber is not responsive during the refractory period. Why?
19. What factors determine the speed of a nerve impulse?
20. Distinguish between synapse and neuroeffector junction.
21. Are neuroeffector junction and neuromuscular junction synonymous? Discuss.
22. In regard to the reflex, distinguish between receptor and effector.
23. What is the role of neurotransmitters at synapses?
24. What is the role of neurotransmitters at neuroeffector junctions?
25. What would be the result of a severed dorsal root?
26. Where are the cell bodies of afferent neurons of the spinal nerves located?
27. What is meant by a contralateral intersegmental reflex?
28. Distinguish between convergence and divergence.
29. What role does the central inhibitory state play in reflex action?
30. What role does the central excitatory state play in reflex action?
31. Distinguish between exteroceptors and interoceptors.
32. Suggest some possible reasons for a given reflex, for example, the knee jerk, being abnormal.
33. The corneal reflex and certain others are useful in delineating the stages and depth of anesthesia. Discuss.
34. Distinguish between natural and conditioned reflexes.
35. What was Pavlov's contribution to physiology and psychology?
36. In regard to the spinal cord, where are internuncial and efferent cell bodies located?
37. What makes the spinal cord gray on the inside and white on the outside?
38. Distinguish between the horns and columns of the spinal cord.
39. Ascending tracts are said to be sensory, while descending tracts are "motor." Why?
40. One major spinal tract is the ventral spinocerebellar. What, specifically, does this name actually tell us?
41. What would be the result following injury to the fasciculus gracilis?
42. What would be the result following injury to the lateral corticospinal tract?
43. Distinguish between vertebral foramen and intervertebral foramen.
44. What is spinal anesthesia?
45. Locate the subarachnoid space.
46. What is the relationship of the pia mater to the filum terminale?
47. What is the approximate volume of cerebrospinal fluid, in ounces?
48. Is cerebrospinal fluid clear or cloudy?
49. What prevents the volume of cerebrospinal fluid from going above a certain value?
50. In essence, spinal nerves arise from the horns. Elaborate.
51. How many pairs of nerves comprise the cauda equina?
52. What is the relationship between ventral rami and plexuses?
53. Distinguish between roots and primary rami.
54. A loss of feeling in a given dermatomic segment

is pathognomonic of the corresponding spinal nerve or cord segment. Elaborate.

55. What is the difference between a sulcus and a fissure?

56. How are cerebral lobes named?

57. The cerebral cortex is gray. What does this tell us?

58. We associate peppermint with a certain taste, a certain smell, and a certain color. What type of cerebral tract does this bring to mind?

59. What type of cerebral tract do the terms sensory and motor bring to mind?

60. What is the purpose of commissural tracts?

61. Relative to nervous tissue, what are nuclei and ganglia?

62. Discuss the structure and function of the corpus striatum.

63. Discuss how clinical findings have contributed to our understanding of brain function.

64. Locate the cerebral areas associated with vision, audition, olfaction, and gustation.

65. Support the view that mental performance is an interplay of cerebral function.

66. What would happen following injury to the posterior central gyrus?

67. Why are the motor neurons of the anterior central gyrus called pyramidal cells?

68. What is meant by upper neuron and lower neuron?

69. Distinguish between spastic paralysis and flaccid paralysis.

70. The knee jerk is exaggerated in upper motor neuron damage and diminished or absent in lower motor neuron damage. Can you account for this?

71. A cerebral hemorrhage involving the left hemisphere results in paralysis of the right side of the body. Explain.

72. What is the lateral lemniscus?

73. Trace a nerve impulse from a receptor to the somesthetic area, identifying all axons, dendrites, and synapses along the way.

74. At one time it was thought that aspirin acted on the thalamus. Why was this an attractive theory?

75. The thalamus is often characterized as a relay center. Why?

76. Discuss some possible repercussions following injury to the hypothalamus.

77. Some authors have referred to the hypothalamus as the bridge between the nervous and endocrine systems. Why?

78. Are the cerebral peduncles composed of white matter or gray matter?

79. What would happen following injury to the midbrain's red muclei?

80. What is the literal meaning of pons varolii?

81. What would happen following damage to the pons nuclei associated with the acoustic nerve?

82. Compare the functions of the cerebral peduncles and the cerebellar peduncles.

83. That the dentate nuclei give rise to axons that travel to the red nucleus tells us much about the purpose of the cerebellum. Discuss.

84. If the medulla oblongata were not the part of the brain least sensitive to depressants, anesthesia would not be feasible. Why?

85. Codeine is an effective drug in checking cough. What is its mechanism of action?

86. Certain tranquilizers induce sleep and muscle relaxation. How do these effects fit in with the hypothesized function of the reticular formation?

87. Chlordiazepoxide (Librium) is thought to act upon the limbic system. Does this fit in with its clinical effects?

88. Some cranial nerves are sensory, and others are mixed. What does this mean?

89. What would happen if the foramen of Monro were to be obstructed?

90. To what feature does the arachnoid membrane owe its name?

91. The success of resuscitation in respiratory or cardiac arrest is directly related to how much time elapses before the resuscitation begins. Precisely why?

92. What does a cessation of brain waves ("flat EEG") indicate?

93. What is the relationship of the autonomic nervous system to homeostasis?

94. Distinguish between a dorsal root ganglion and a lateral ganglion.

95. Even though there are autonomic reflexes, the autonomic system by convention is restricted to efferent neurons. What about the afferent neurons?

96. In a sense the gray rami serve as a bridge between the autonomic system and the spinal nerves. Explain.

97. Where are the cell bodies that give rise to postganglionic sympathetic fibers located?

98. Cholinergic fibers are associated with both the parasympathetic system and the sympathetic system. Elaborate.

99. Cholinergic fibers derive their name from their release of acetylcholine. Why are the fibers that release norepinephrine called adrenergic?

100. What would be the effect of a vagotomy on gastric function?

101. What would be the effect of a lumbar sympathectomy on blood pressure?

102. Atropine blocks the action of acetylcholine. What, then, is its action on the heart?

103. Why are cholinergics contraindicated in patients with asthma?

104. Nasal decongestants contain sympathomimetics. What is the pharmacologic rationale?

105. Both sympathomimetics and parasympatholytics dilate the pupil. Explain.

106. What type of autonomic drug can be used to antidote poisoning by a cholinergic drug?

107. How could severe anemia cause unconsciousness?

108. How is a definitive diagnosis of epilepsy made?

109. Typically, an anticonvulsant is effective in one form of epilepsy and not another. What does this tell us about the etiology of the disease?

110. What part of the brain seems to be particularly involved in cerebral palsy?

111. How do the sclerosed or demyelinated patches of multiple sclerosis fit in with the signs and symptoms?

112. Assuming that low levels of dopamine result in parkinsonism, why isn't it used in treatment instead of its precursor levodopa?

113. The hemiplegia of a stroke easily brings to mind two anatomical considerations: internal capsule and decussation. Elaborate.

114. What would be a major clinical feature associated with cancer of the cerebellum?

115. What is the literal meaning of eastern equine encephalomyelitis?

116. Even before the introduction of the polio vaccine most persons past the age of fifty years were immune to the disease. How do you account for this?

117. When first introduced, the Salk vaccine caused a number of cases of paralytic polio. What happened?

118. Why is the oral vaccine more effective against polio than the Salk vaccine?

119. Suspected cases of meningitis are treated with ampicillin until a definitive diagnosis can be reached. Why isn't penicillin given instead?

120. What is meant by aseptic meningitis?

121. In stark contrast to vaccination against other infections, vaccination against rabies is performed following exposure. Vaccination following exposure to *Clostridium tetani,* for example, affords no protection unless the victim has been previously vaccinated. Discuss.

122. A dog has severely bitten a small child on the leg. Discuss in detail the handling of the case from start to finish, pretending you are the doctor.

123. In regard to tetanus, the term vaccination has the correct connotation, but is literally inaccurate. Why?

124. That *Clostridium tetani* is a spore-forming anaerobe tells us much about the circumstances surrounding the contraction of the infection. Elaborate.

125. Antidepressants are drugs used to cause mood elevation in depressed patients. Some of these agents are known to inhibit monoamine oxidase, the enzyme responsible for the deactivation of serotonin (p. 315). Elaborate on any possible connection.

126. What is the distinction between habituation and addiction?

CHAPTER 21

THE EYE

The eyes are the organs of vision. They are spherical, about 1 inch in diameter, and have a clear circular window, the cornea, in front to permit the entrance of light. The optic nerve, which issues from the posterior pole of the eyeball, carries impulses from the retina, the light-sensitive tissue of the eye, to the brain. Except for the small area that is exposed anteriorly, the eye is enclosed in a bony socket of the skull, the orbital cavity or orbit, and, as further protection, is embedded in the loose fat that lines the orbital cavity.

Inserted about the circumference of the eyeball are six small muscles that originate at the back part of the orbit. By their contractions they permit the eye to turn and roll about in its fatty bed. Other structures associated with the eye include the eyelids, the lacrimal glands, and the lacrimal ducts.

COATS AND CAVITIES

The eyeball is composed of three coats: outer, middle, and inner (Fig. 21-1). The outer or sclerotic coat (sclera) is a tough white supporting tunic composed of dense fibrous tissue that covers the entire eyeball except in front, where it becomes the cornea. Part of the sclera is seen as the white of the eye. The middle or choroid coat (the choroid) is a dark brown vas-cular layer that nourishes the retina and the lens. Anteriorly, in the area behind the cornea, this coat becomes the iris and the ciliary body.

Retina

In actuality the retina, the innermost coat, is the receptor of the eye, for it is here that light is converted into nerve impulses that make their way to the brain via the optic nerve. This highly specialized structure consists of ten microscopic layers, of which the pigment layer is the outermost, and the layer of nerve fibers, the innermost. The layer of rod and cone cells adjacent to the pigment layer is now known to be the light-sensitive layer. The biochemistry of this particular layer has been explored in great detail and has yielded much to our knowledge of vision.

Macula lutea. The macula lutea (yellow spot) is an oval depression on the retina temporal to the entrance of the optic nerve and the retinal blood vessels (Fig. 21-2). In the center of this area is a tiny depression (about 0.5 mm. across) called the fovea centralis, where the retina is the most sensitive to daylight and color. The fovea centralis is made up exclusively of cone cells; in moving away from the fovea centralis, the number of cones progressively decreases and the number of rods progressively increases.

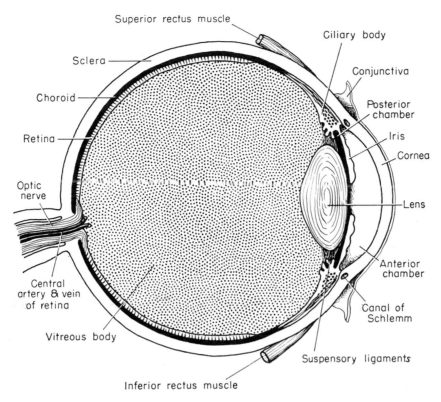

FIG. 21-1
Section through the eye showing major anatomic features.

These facts can be vividly demonstrated by two simple experiments. If we focus our eyes on a single object, say the tip of a pencil, other objects not in the line of vision become vague and colorless because only the object that we are concentrating on is striking dead center at the fovea centralis. All other objects in the visual field are forming images peripheral to the "retinal bull's-eye." The second experiment is best performed out-of-doors at night. Looking up at the heavens, one discovers that more stars come into view when the eyes are moved to the side than when they are fixed straight ahead. This shows that the rods, which are more sensitive than the cones to dim light (night light), are concentrated in the retinal periphery, away from the fovea centralis.

Optic disc. The optic disc, a white spot in the fundus medial to the macula lutea, corresponds to the entrance of the optic nerve (Fig. 21-2). Since there is no retina in this area, it is actually a blind spot (Fig. 21-3).

Cornea

The cornea, the transparent structure of the anterior part of the eye, may be considered a derivative of the sclerotic coat. Although it is as clear as glass, microscopic examination discloses a heterogeneous structure. Four layers are recognized: the outer corneal epithelium (continuous

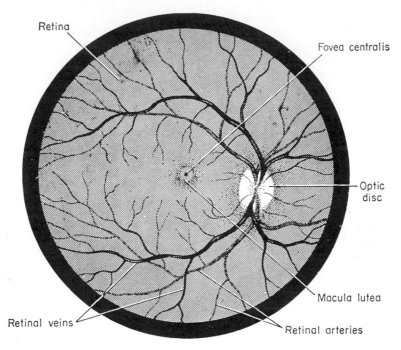

FIG. 21-2
Fundus (eyegrounds) of right eye as seen with ophthalmoscope.

FIG. 21-3
Demonstrating the blind spot. With left eye closed, hold figure about 12 inches in front of right eye and, focusing on white disc, slowly move book toward eye until cross disappears. When this occurs, image of cross falls upon optic disc, or blind spot.

with the conjunctiva), the substantia propria, the elastic propria, and the endothelium of the anterior chamber, in that order.

Iris

The iris, the colored doughnut-shaped structure behind the cornea, derives from the choroid coat. It is composed of smooth muscle arranged so that the pupil, the hole in the "doughnut," narrows or widens, depending upon whether the circular fibers or the radial fibers are contracted. Two delicate nerves, the constrictor and the dilator, innervate the circular and radial muscles, respectively. In strong light the constrictor nerve (a parasympathetic nerve) predominates, producing constriction of the pupil (*miosis*); in dim light the dilator nerve (a sympathetic nerve) predominates, producing dilatation of the pupil (*mydriasis*). Thus, the iris controls the amount of light that enters the eye, much as the diaphragm controls the amount of light entering a camera. When looking at near objects, the pupils narrow to sharpen the image.

Ciliary body

The ciliary body, also derived from the choroid coat, is situated about the interior periphery of the iris (Fig. 21-1). It is composed of the ciliary muscle and the ciliary process, the latter giving rise to the suspensory ligament that inserts into the capsule enclosing the lens. The engineering is such that contraction of the ciliary muscle pulls the ciliary process slightly forward, thereby relieving the tension on the suspensory ligament and the squeeze on the lens. The lens, being elastic, then assumes a more convex shape. Conversely, when the ciliary muscle relaxes, the ciliary process falls back, increasing the tension on the capsule and compressing the lens. This thinning and thickening of the lens permits the eye to focus upon far and near objects, respectively.

Lens

The lens is a glass-clear biconvex elastic disc suspended behind the iris (Fig. 21-1). It is enclosed in a transparent capsule and held in position by the suspensory ligament. The lens serves as the major element in focusing.

Cavities

The eyeball is divided into the anterior and posterior cavities. The posterior cavity is the larger and is filled with a clear jellylike substance called the vitreous humor (Fig. 21-1). The anterior cavity, which lies before the lens, is subdivided into the anterior and posterior chambers. The former is the space between the cornea and the iris; the latter, the space behind the iris. As shown in the illustration, these chambers are interconnected via the pupil and are filled with a water fluid called aqueous humor.

Canal of Schlemm. The canal of Schlemm is a tiny channel that circumscribes the cornea (Fig. 21-1). The aqueous humor, produced by filtration from the capillaries in the ciliary body and iris, enters the canal via the trabecular meshwork (Fig. 21-10). The importance of this drainage system in glaucoma will be discussed later in this chapter.

MUSCLES

The intrinsic muscles of the eye, the smooth muscles in the ciliary body and iris, have been discussed previously.

The *extrinsic* muscles are of the skeletal type and are under voluntary control. There are six such muscles for each eye. They originate in the bone of the orbit and insert into the sclera and include the superior rectus, the inferior rectus, the lateral rectus, the medial rectus, the superior oblique, and the inferior oblique (Fig. 21-4). The desired movement of the eye is accomplished by the reciprocating action of opposing muscles. For example, if one wishes to look up, the superior rectus muscles are contracted, and the inferior rectus muscles are relaxed; the opposite occurs when one looks down.

Whereas the intrinsic muscles are innervated by autonomic fibers from the third and fourth cranial nerves, the extrinsic muscles receive somatic fibers from the third, fourth, and sixth cranial nerves.

ACCESSORY STRUCTURES

Aside from the eyeball proper and its associated extrinsic muscles, there are certain accessory structures that have to do with the workings of the eyes. These include the eyelids, the lacrimal apparatus, and the eyebrows.

Eyelids

The eyes are protected in front by the eyelids, or palpebrae, two movable folds of skin con-

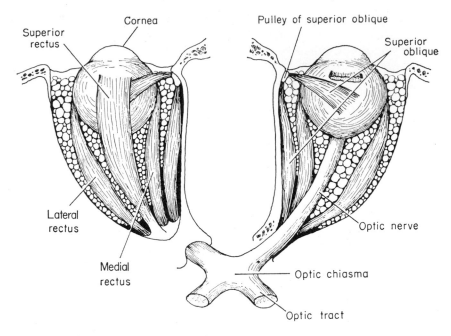

FIG. 21-4

Four of the six extrinsic eye muscles. Inferior rectus and inferior oblique muscles are situated below eyeball. (Redrawn from Schottelius, B. A., and Schottelius, D. D.: Textbook of physiology, ed. 17, St. Louis, 1973, The C. V. Mosby Co.)

taining skeletal muscle and a border of thick connective tissue, the tarsal plate, from which project the eyelashes. There are a number of small sweat and sebaceous (meibomian) glands in the lids that empty by minute openings along the free margin. Though commonly considered cosmetic elements, the eyelashes serve as valuable devices against the entrance of foreign bodies.

The inside surfaces of the eyelids are lined with a transparent mucous membrane, the *conjuctiva*, that continues over the surface of the eyeball. Any inflammation of this membrane, a fairly common occurrence, is referred to as conjunctivitis.

The upper and lower eyelids join at the corners to form angles, called the inner or medial canthus and the outer or lateral canthus. The opening between the lids, the palpebral fissure, extends between the canthi. The width of the fissure ordinarily determines the apparent size of the eye. "Those great big beautiful eyes" are due not to the size of the eyeballs but to the eyelids being naturally held wide apart; conversely, the person with small eyes owes this feature to a narrow palpebral fissure.

Lacrimal apparatus

The lacrimal apparatus is composed of the lacrimal glands, the lacrimal ducts, the lacrimal canaliculi, the lacrimal sacs, and the nasolacrimal ducts (Fig. 21-5). The glands reside in a depression of the frontal bone at the upper, outer margin of the orbit. Several short lacrimal ducts lead the tears from these glands onto the surface of the conjunctiva. In addition to moistening

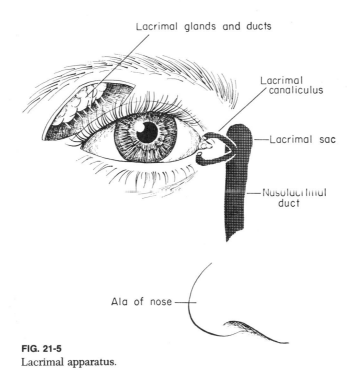

Lacrimal glands and ducts

Lacrimal canaliculus

Lacrimal sac

Nasolacrimal duct

Ala of nose

FIG. 21-5
Lacrimal apparatus.

the exterior part of the eye and lubricating the eyelids, the tear secretions wash down any debris upon the surface into the conjunctival trough, or sac, between the lower lid and the eyeball. As soon as the secretions start to accumulate, they are drained away into the inferior nasal meatus of the nose via the lacrimal canaliculus, the lacrimal sac, and the nasolacrimal duct, in that order.

We are unhappily reminded of this unique drainage system during a head cold. As a result of nasal inflammation, the nasolacrimal ducts become plugged and squeezed, thereby obstructing the flow of tears into the nose and producing watering of the eyes.

Crying, of course, is something else. We can only speculate concerning its purpose. Perhaps crying occurs because sorrowful experiences are often allied with danger and the body is mus-

tering all its defenses; that is, the tears are called forth to wash out and refresh the eye.

FOCUSING

Vision, the act of seeing, results from the reception of nerve impulses at the visual areas of the cerebral cortex. These impulses are sent out in response to light by the rods and cones of the retina and are carried to the brain via the optic nerve. In order for this to be meaningful, however, light rays must be focused upon the retina. Focusing essentially entails three separate physical acts: refraction, constriction of the pupil, and convergence.

Refraction

Refraction, the bending of light rays, occurs when light passes from one medium into

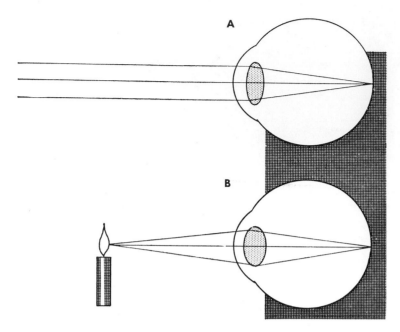

FIG. 21-6
Adjustment of lens of eye to **A,** far and **B,** near, objects. Note change in convexity.

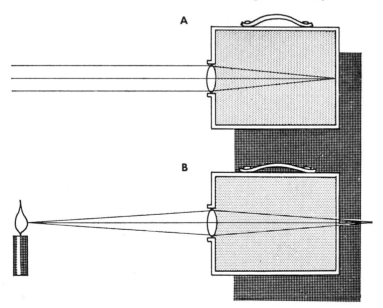

FIG. 21-7
Lens of fixed convexity (such as that of a camera) focusing on **A,** far, and **B,** near, objects. (Compare with Fig. 21-6.)

another. In the case of the eye, light rays are refracted three times: at the anterior surface of the cornea, at the anterior surface of the lens, and at the interface of the posterior surface of the lens and the vitreous humor. In the normal eye these refracting mechanisms are such that an object 20 feet or more away forms a clear image upon the retina; the relaxed eye is set for distant vision.

In order to form clear images of objects at a distance of less than 20 feet, the refractive power of the eye must be increased to bend the parallel and divergent rays that impinge upon the cornea. (Light rays from distant objects are parallel.) Since the degree of divergence increases rapidly as the object comes closer to the eye, the degree of refraction must increase to bring the rays into focus. If an object is too close to the eye, it cannot be focused at all.

The eye (Fig. 21-6), but not the camera (Fig. 21-7), can increase its refractive power for near vision by accommodation, that is, by increasing the convexity of the lens. As explained earlier, this is accomplished by contraction of the ciliary muscle. With age the lens becomes less elastic and therefore less able to focus near objects, an occurrence referred to as presbyopia.

Constriction of pupil

The iris serves the eye in two ways: It protects the retina against bright light by constricting the pupil, and it aids in the focusing of near objects, also by constricting the pupil. The former response can be demonstrated by shining a light into the eyes, and the latter, by watching the pupils of a subject who is focusing on the tip of a pencil as it is brought toward the eyes. Conversely, in dim light and in focusing on distant objects the pupils enlarge.

The reason that the size of the pupil varies with the intensity of light is quite obvious. It is equally obvious in near vision, too, if we remember that the more peripheral divergent rays cannot be focused; by cutting out these rays the sharpness of the focus is enhanced.

Convergence

Although we see with both eyes (binocular vision), we do not see double images (diplopia) because when we train our eyes on an object, the two images fall on corresponding parts of the retina. This is true whether we are looking at a distant object or at a near object. Obviously, the nearer the object, the more the eyes must converge. Convergence is effected by contraction of the medial rectus muscles and by relaxation of the lateral rectus muscles. Proper focusing and vision, therefore, depend upon keen functional balance between these antagonistic extrinsic muscles.

A simple experiment will vividly demonstrate what has been said. Gently pressing in on one of the eyeballs while focusing on a single object causes two images to appear because the image in the disturbed eyeball falls on a different or noncorresponding part of the retina.

VISUAL PATH

The fibers from the rods and cones gather together at the optic disc and leave each eyeball as the optic or second cranial nerve. The two nerves then curve medially and meet each other just above the pituitary gland at the optic chiasma (Fig. 21-4). Here there is a 50% *decussation* of fibers; those from the medial half of the retina cross over to the optic nerve on the opposite side, while the fibers from the lateral portion pass directly into the optic tract on the same side. The fibers continue on to the visual areas of the two occipital lobes via the lateral geniculate bodies (of the thalamus) and the superior colliculi (of the midbrain). Thus, each occipital lobe "sees with both eyes," or, turned around, each eye "sees with both lobes." This unique feature explains the rather peculiar visual disturbances resulting from injury or damage to the brain.

CHEMISTRY OF VISION

The emission of nerve impulses by the retinal rods and cones results from photochemical reactions within these microscopic structures. It has now been established that dim-light vision results from the extreme photosensitivity of a substance in the rods called rhodopsin, or visual purple, a material derived from the protein scotopsin and the carotenoid retinal, the latter being an aldehyde derived from vitamin A. In bright light, rhodopsin is bleached into these two components and therefore is unresponsive; in dim light the components are resynthesized into the active molecule. Thus, vitamin A serves as a raw material for vision, for without it the rods cannot synthesize retinal, and without retinal there can be no rhodopsin. The earliest sign of vitamin A deficiency is night blindness, or nyctalopia.

The classic Young-Hemholtz theory of color vision holds that there are three types of cones, each containing a photosensitive pigment responsive to a primary color (red, green, or blue). This theory is supported by current research. Thus, we experience red, green, and blue when the red, green, and blue cones, respectively, are stimulated by the corresponding frequency. Other color experiences result from the summation of outputs from two or three respective types of cone cells. Yellow, for example, is due to stimulation of equal numbers of green and red cones and very few blue cones. The sensation of white apparently results when the three types of cones are stimulated equally.

VISUAL ACUITY AND OPTICAL DISORDERS

If a person can barely read letters 20 feet away that are barely readable at that distance to all persons with normal vision, his visual acuity, or eyesight, is said to be 20/20; if he can barely read letters at 20 feet that are barely readable at 40 feet to all persons with normal vision, his visual acuity is said to be 20/40; if he can barely read letters at 20 feet that are barely readable at 100 feet to all persons with normal vision, his visual acuity is 20/100; and so on. The more common optical defects that relate to visual acuity are discussed below.

Astigmatism

Astigmatism is due to an unevenness in the surface of either the cornea or the lens or both. In geometric terms this means that the radius of curvature in one plane is longer or shorter than the radius at right angles to it. The rays passing through the plane of greater curvature are refracted more than the rays passing through the other plane, resulting in two focal points instead of one and in the blurring of vision (Fig. 21-8).

Because the cornea and lens are never optically perfect, most of us experience some degree of astigmatism. If this were not true, stars would appear as small bright dots instead of star-shaped bodies (that is, bright centers with

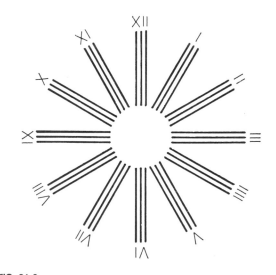

FIG. 21-8
To normal eye, all black lines on chart have same intensity. To astigmatic eye, some appear darker than others.

radiating short lines). Stars twinkle because the constant rapid movement of the eyeballs makes the radiating lines shift upon the retina. The natural imperfection of the refractive apparatus of the eye also explains why a light in the darkness appears to come to the eye in radiating beams.

To correct astigmatism, the oculist prescribes glasses that correct the unevenness of the corneal and lens surfaces; the convexity of each lens in the glasses is increased or decreased to compensate for the defect in the corresponding diameter, or meridian, of the eye.

Myopia

Myopia, or nearsightedness, is an eye condition in which the focus falls in front of the retina (Fig. 21-9). To put the image on the

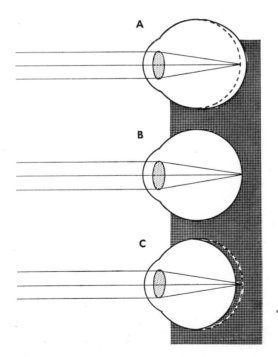

FIG. 21-9
A, Near sightedness. **B,** Normal sight. **C,** Farsightedness.

retina where it belongs, the myopic person must look at an object at closer range than the emmetropic person. To remedy the defect, the oculist prescribes glasses with concave lenses that diverge the light rays hitting the cornea just enough to force the focus back upon the retina. The cause of myopia may be either that the refractive power of the eye is too great or that the eyeball is too long.

Hyperopia

In contrast to the myopic eye, the hyperopic eye sees objects most clearly at far range; hyperopia, or hypermetropia, is farsightedness. Instead of the focus falling before the retina, as in myopia, it falls behind it (Fig. 21-9). The usual cause of hyperopia is a flattened lens or cornea, with a consequent loss of refractive power, or an eyeball that is too short. To correct the condition, convex lenses are prescribed to concentrate the light rays and cause the focus to retreat to the retina.*

Color blindness

Color blindness is the inability to distinguish differences in color; it is seldom, if ever, complete. Usually hereditary and rarely seen in women, the condition is said to affect about 8% of all men. In dichromatism, the most common variety, the person does not see red or green but only yellow and blue or combinations thereof. Apparently, the retinas of such persons lack the pigments that are sensitive to these fundamental colors. Since color blindness may lead to serious accidents in driving and in all occupa-

---

*The refractive power, or strength, of a lens is expressed in diopters (D). A lens that can focus parallel light rays to a point 100 cm behind it is arbitrarily given a strength of 1 D. Lenses with shorter focal distances—that is, those with greater refractive power—have strengths greater than 1 D. For example, a lens with a focal distance of 50 cm has a refractive power of 2 D (100/50 = 2); a lens with a focal distance of 20 cm is 5 D, and so on.

tions in which colored signals are used, it is of the greatest importance that the condition be detected at an early age.

Glaucoma

Glaucoma, the leading cause of blindness in this country, results from faulty drainage of aqueous humor. Because the inflow and outflow of the fluid are normally balanced, faulty drainage results in an increase in the intraocular pressure, sometimes four times the average range (14 to 25 mm Hg). The affected eye may feel as hard as a marble.

In the acute condition pain is intense, vision is fogged, and lights appear ringed with halos. In the chronic condition, in which the pressure rises over long periods of time, the symptoms are transitory and mild. In either circumstance, however, the prolonged elevation of pressure ultimately destroys the optic nerve.

The drainage difficulty resides at the periphery of the anterior chamber in the angle between the cornea and the iris. Here, as shown in Fig. 21-10, a spongy network of fibers, the trabecular meshwork, affords a series of microscopic channels that extend from the anterior chamber to the canal of Schlemm. The aqueous humor drains through this meshwork into the canal of Schlemm and returns to the blood via the veins in the sclera.

In *angle-closure glaucoma* the space between the cornea and the iris is less than average, so that when the pupil dilates somewhat, the thickened iris closes off the narrowed angle (Fig. 21-10). The outflow of fluid stops, and the pressure begins to rise.

In *open-angle glaucoma* the primary trouble is in the trabecular meshwork itself. Although the angle is normal ("open"), the tiny channels for some reason become narrowed or clogged. The end result is poor drainage and increased intraocular pressure, just as in angle-closure glaucoma.

The standard approach to acute glaucoma is the use of parasympathomimetic drugs (for ex-

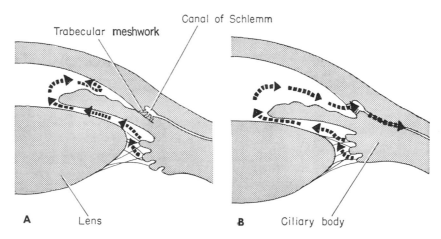

FIG. 21-10

Glaucoma. **A,** In angle-closure glaucoma, base of iris prevents flow (arrows) of aqueous humor into trabecular meshwork because of narrow angle. **B,** Normal flow into canal of Schlemm via trabecular meshwork. In open-angle glaucoma, angle is normal (as in **B**) but trabecular meshwork is occluded.

ample, pilocarpine) to reduce the size of the pupil. This procedure thins out the iris at the periphery, thereby enlarging the angle and facilitating drainage into the trabecular meshwork. Acetazolamide (Diamox) improves the condition by inhibiting the production of aqueous humor.

Drugs, of course, do not correct the underlying factors, and because of the nature of the disease, a cure must come by way of surgical intervention. Great hope in this area has arisen, chiefly from an operation called peripheral iridectomy, a procedure whereby a tiny piece is snipped from the edge of the iris. This permits the aqueous humor to pass directly from the posterior chamber to the anterior chamber instead of going through the pupil. Since the normal, roundabout way through the pupil causes the pressure in the posterior chamber to be slightly higher than in the anterior chamber, the iris normally is always pushed a little forward, narrowing the angle. The operation equalizes the pressure in the two chambers, straightens the iris, widens the angle, and facilitates drainage, in that order. In early pure angle-closure glaucoma, peripheral iridectomy can sometimes effect a complete cure; in the open-angle form a modified operation is helpful. However, in the latter variety of glaucoma, ophthalmologists usually rely upon drugs alone, particularly if the patient responds and there is no further visual loss.

Cataract

A cataract is an opacity of the lens or its capsule. The developmental cataracts seen in young persons result from infection of the mother in the first trimester of pregnancy, or they may be hereditary. Degenerative cataracts, on the other hand, occur in elderly persons (senile cataracts) and in persons exposed to excessive radiation and heat. The senile variety makes up the bulk of all cases. When there is considerable impairment of vision, operative removal of the lens is essential.

Tumors

The main tumors of the eye are the retinoblastoma in children and the malignant melanoma in persons past middle age.

The retinoblastoma is a gray, soft tumor of the retina that may invade the vitreous humor and often the entire eyeball. In the final stages it works its way along the optic nerve and extends to the cranial cavity. In addition to mechanical damage, there is necrosis and hemorrhage.

The melanoma is a black tumor that starts in the choroid, the iris, or the ciliary body and later projects into the eyeball and the tissue surrounding the eye. As in other malignant growths, metastases may occur, particularly in the liver.

Less common tumors include the myoma, the neurofibroma, and the angioma.

Infections of the eye

The tissues and various structures of the eye are fertile ground for a variety of pathogens. The more common infections include hordeolum, or sty, an involvement of one or more sebaceous glands of the eyelids; blepharitis, an involvement of the margin of the eyelids; conjunctivitis, an involvement of the conjunctiva; keratitis, an involvement of the cornea; and iritis, an involvement of the iris. Gonorrheal conjunctivitis in the newborn infant, commonly referred to as *ophthalmia neonatorum*, is usually preventable by the instillation of 2 drops of 1% silver nitrate in each eye at the time of delivery.

QUESTIONS

1. Describe the general features of the eyeball.
2. Describe the structure of the retina.
3. Describe the workings of the iris.
4. Explain how the ciliary body controls the convexity of the crystalline lens.

5. Compare the anterior and posterior chambers of the eye.
6. Discuss the extrinsic muscles and how they move the eyeball.
7. Discuss the functions of the eyelids and the meibomian glands.
8. Describe the structure and function of the lacrimal apparatus.
9. Why do the eyes water during a severe head cold?
10. Discuss in detail how the eye refracts and focuses light rays.
11. Explain why we see only a single image even though we have two eyes.
12. Even though one eye is closed, we see with both visual areas. Explain.
13. Explain why nyctalopia is an early sign of vitamin A deficiency.
14. Discuss color vision.
15. Why does a blue shirt appear black when viewed in red light?
16. Compare the rods and the cones.
17. We "see" with the brain. Explain.
18. Why do most of us have some degree of astigmatism?
19. What is the cause of hypermetropia? How is it corrected?
20. What is dichromatism?
21. In examining the eye, the ophthalmologist uses mydriatic drugs. Explain the action of such agents.
22. Why do the pupils enlarge during excitement?
23. Discuss the cause and treatment of glaucoma.
24. Why must glasses be worn following a cataract operation?
25. Discuss the prophylaxis of ophthalmia neonatorum.
26. Explain why our night vision is temporarily impaired if we are dazzled by the headlights of an oncoming car.
27. What does it mean for a person to have 20/15 vision?

CHAPTER 22

THE EAR

The ear is concerned with hearing and equilibrium. On the basis of structure it is divided into three parts: outer, middle, and inner. The first two parts conduct sound waves to the receptor in the inner ear.

OUTER EAR

The outer, or external, ear comprises the auricle, or pinna, and the auditory canal (Fig. 22-1). The auricle is what is commonly referred to as the ear. It is composed of an irregular plate of elastic cartilage covered with skin. Its size and shape vary considerably from individual to individual, usually being larger in the male than in the female. Though a considerable number of muscle fibers are attached to the auricle, modern man has all but lost the ability to wiggle his ears.

The auditory canal is a bent tube, a little more than 1 inch long, which leads from the auricle to the eardrum. There is cartilage in its outer third, and bone constitutes the remainder. The skin over the cartilaginous portion characteristically contains hairs and special glands that secrete *cerumen*, a waxlike material commonly called earwax. The auditory canal is also supplied with a large number of sensory fibers from certain cranial nerves.

MIDDLE EAR

The middle ear (Fig. 22-1) is a small chamber located within the temporal bone and separated from the external ear by the eardrum, or tympanum (less correctly, the tympanic membrane), a thin, translucent disc composed of fibrous tissue. The back of the chamber communicates with the mastoid air cells through a tiny opening, and the front of the chamber communicates with the eustachian tube, a passageway that leads to the nasopharynx. The inner wall is composed of bone, except for the oval and round windows, which communicate with the inner ear. The footplate of the stapes fits over the oval window, and a fibrous disc fits over the round window (Fig. 22-4). The entire chamber is lined with a mucous membrane that is continuous with that which lines the mastoid air cells and nasopharynx. Tiny muscles are attached to the ear bones from below, and all the soft structures mentioned receive sensory fibers.

Ossicles

The three ear bones, or ossicles, in each ear are aptly named the malleus (hammer), the incus (anvil), and the stapes (stirrup). These tiny irregular bones extend across the chamber

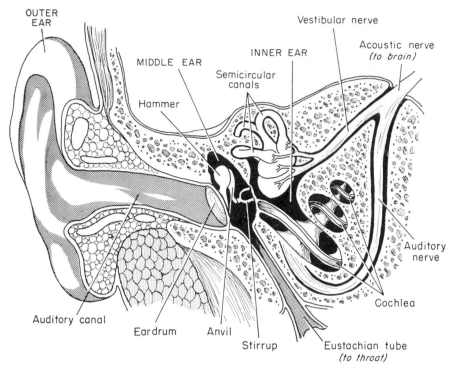

FIG. 22-1
The ear. (Auditory nerve also called cochlear nerve.)

from the tympanum to the oval window; the malleus articulates with the incus, the incus, with the stapes. In this manner, tympanic vibrations, produced by sound waves, are transmitted across the middle ear to the oval window and on to the inner ear.

INNER EAR

The inner ear, or labyrinth, consists of the osseous portion and the membranous portion. The latter is a soft structure composed of two small sacs—the utricle and the saccule—the three semicircular canals, and the membranous cochlea (Fig. 22-2).

A watery fluid, the endolymph, fills the membranous labyrinth, and a similar fluid, the perilymph, fills the space between the membranous labyrinth and the osseous labyrinth. These fluids serve the dual purpose of cushioning the soft structures and conducting sound waves from the middle ear to the organ of Corti, the actual receptor of sound.

Cochlea

The osseous cochlea is divided by the vestibular membrane (Reissner's membrane) and the basilar membrane into three spiral tubes called scalae: the scala vestibuli, the scala media, and the scala tympani (Fig. 22-3). The scala vestibuli communicates with the middle ear via the oval window, which is secured by the footplate of the stapes. The scala tympani is secured by the

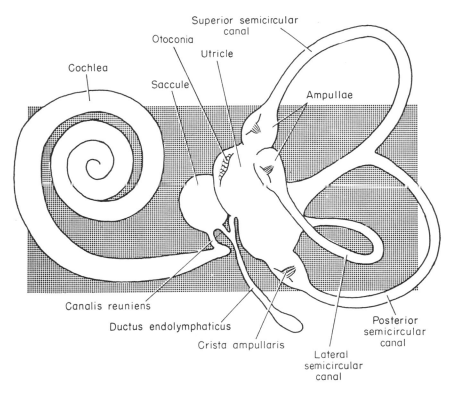

FIG. 22-2
The membranous labyrinth. (Ductus endolymphaticus shown pulled down.) (From The human body: its anatomy and physiology, third edition, by C. H. Best and N. B. Taylor. Copyright © 1932, 1948, 1956, by Holt, Rinehart and Winston, Inc. Adapted and reprinted by permission of Holt, Rinehart and Winston, Inc.)

round window, and the scala media, which is actually the membranous cochlea, is sealed off from the middle ear by bone (Fig. 22-4).

Organ of Corti. Upon the delicate basilar membrane rest receptor cells supplied with nerve fibers from the auditory or cochlear nerve (Fig. 22-3). These long, slender cells, which stand on end and in most instances are capped by a row of hairlike processes, are known collectively as the organ of Corti. Overlying this structure and touching the hairs is a veillike tissue called the tectorial membrane, whose motion in the endolymph somewhat resembles a long string of seaweed waving in the water.

Semicircular canals

The semicircular canals include the three sickle-shaped, fluid-filled, membranous tubes of the labyrinth that communicate with the utricle (Fig. 22-2). These canals are positioned at right angles to each other and form a swelling, called the ampulla, just before entering the utricle.

Within the ampullae are special rotary receptors, called cristae, that are groups of hair cells with their hairlike processes embedded in a gelatinous covering, the cupula. The cristae are supplied with sensory nerve fibers that convey impulses to the brain via the vestibular portion

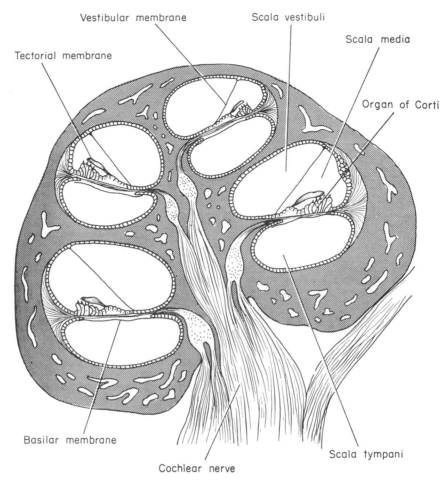

FIG. 22-3

Cross section of cochlea, showing scalae and organ of Corti. (Modified from Goss, C. M., editor: Gray's anatomy of the human body, ed. 29, Philadelphia, 1973, Lea & Febiger.)

of the acoustic nerve. The function of the semicircular canals will be discussed later in the chapter.

Utricle

The utricle is the fluid-filled membranous sac, situated within the vestibule of the osseous labyrinth, that communicates with the three semicircular canals. Inside are special gravity receptors constructed of hair cells and calcareous bodies (the otoliths or otoconia). Because of their weight, the otoliths pull upon the hairs with a force that depends upon the sideward position of the head. Sensations from these cells are transmitted to the brain via the vestibular nerve.

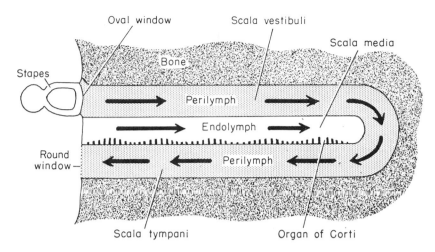

FIG. 22-4

Basic relationship between osseous and membranous cochlea. (After Williams; modified from Anthony, C. P.: Textbook of anatomy and physiology, ed. 8, St. Louis, 1971, The C. V. Mosby Co.)

Saccule

The saccule, a much smaller sac than the utricle, resides within the vestibule and communicates with the utricle and the membranous cochlea. Much like the utricle the saccule is a gravity receptor, but here activation results from moving the head forward or backward.

HEARING

The sense of hearing is likely to be the musician's most precious sense; for others, it is at least next to vision in importance. The mechanics of the ear and the cerebral mechanisms responsible for the experience of sound are one of nature's greatest engineering feats. The more one knows about the mechanics of sound, the more phenomenal the sense of hearing becomes. A sound wave with a pressure of only one tenbillionth of 1 atmosphere* can provoke an audible sensation!

To understand the salient features of hearing,

———————

*One atmosphere is the pressure of the air at sea level.

we must begin with the sound waves funneled into the auditory canal by the pinna. (Here, of course, we do not do as well as the dog because of our inability to prick up our ears and aim them in the direction of the sound.) When sound waves reach the end of the canal, the tympanum is set in vibration and, in turn, so are the hammer, anvil, and stirrup. Thus, across the middle ear sound is conducted as vibrating bone. At the footplate of the stapes (at the oval window) the vibrations are conveyed to the perilymph in the scala vestibuli and then through the fluid in the scala media and scala tympani. The fluid vibrations cause the basilar membrane to vibrate, resulting in a shearing movement of the hairs against the tectorial membrane. This movement of the hairs initiates a graded potential in the hair cells that in turn induces an action potential in the nerve fibers associated with them. From here the impulses pass along nerve fibers to the auditory area via relay nuclei situated in the medulla, pons, midbrain, and thalamus, in that order.

Pitch

Several theories have been proposed to explain pitch, but the idea of sympathetic resonance is now generally accepted. According to this theory, first proposed by Helmholtz, the 24,000-odd delicate cross fibers of the basilar membrane, instead of vibrating en masse, vibrate in groups. The longer fibers near the top of the cochlea vibrate, or resonate, in sympathy with the lower-pitched sounds, while the shorter fibers at the bottom resonate with the higher-pitched sounds. (If the 1½-inch basilar membrane were to be removed and laid out flat, we might liken its arrangement to a microscopic stringed instrument.) The theory holds that a sound of a given pitch sets off certain basilar fibers, which in turn cause certain hair cells to move against the tectorial membrane and produce an impulse as described above.

The human ear is not sensitive to sound waves with a frequency of less than 16 or more than 20,000 cycles per second. It is most sensitive to sound vibrations between 500 and 5,000 cycles per second. Cats and dogs pick up sound waves far above this range. The human voice ranges between 100 and 800 cycles per second, and musical instruments, between 30 and 4,000 cycles per second.

Eustachian tube

Were it not for the eustachian tube, which runs from the nasopharynx to the middle ear and allows the air pressure on both sides of the eardrum to be equalized, our hearing would be impaired. If air were not allowed to enter and leave the middle ear chamber, the eardrum would be pushed in and out and would not vibrate the way it should.

We become acutely aware of the function of the tube when we go from one air pressure to another, as in a sudden ascent or descent in an elevator or airplane, because the pressure in the middle ear does not adjust instantly to external pressure. The condition is easily remedied by swallowing several times in quick succession, for it is only when we swallow that the eustachian tubes open.

SENSE OF BALANCE

Because the semicircular canals lie in three different planes at right angles to one another, the muscles are immediately activated to maintain proper balance, or equilibrium, should the entire body or the head be quickly rotated about any axis. If the body is spun about a vertical axis, as in a swivel chair, the semicircular canals in the horizontal plane will be particularly stimulated; receptors, or cristae, in the ampullae of these two structures will be bent because of the inertia of the endolymph. At the instant the motion begins, the receptors turn with the body while the endolymph stands still. The bent hair cells are activated and induce impulses in the vestibular nerve fibers associated with the hair cells. These impulses stimulate the balance centers in the brain, and thereby we experience the sensation of motion; also, reflexes are initiated that effect the appropriate movement of the head and limbs.

The sensation of rotation, however, is experienced only in the first few seconds. Prolonged rotation, if smooth, does not provoke the sensation because once the endolymph loses its resting inertia, it moves with and does not bend or stimulate the receptors. When rotation stops, however, the endolymph continues to move, once again bending and stimulating the receptors. This time the sensation of rotation is in the opposite direction because the hair cells are bent in the opposite direction.

Working hand in hand with the semicircular canals in maintenance of equilibrium and position are the receptors in the utricle and saccule. However, while the semicircular receptors are triggered by rotation, the utricle and saccule receptors are triggered by the position of the head with respect to gravity. That is, when the head is placed in various positions, the otoliths resting upon the hair cells, because of their weight, pull in different directions with varying degrees of tension. This triggers nerve impulses that upon arrival in the brain effect the appropriate muscular movements to right the body. The coordinating mechanism of this righting reflex resides within the cerebellum. The efficiency of this reflex is dramatically illustrated by holding a kitten about two feet above a soft surface and allowing it to fall back downward. It will quickly turn in midair and land upon all four feet in perfect balance.

DISEASES OF THE EAR

The ear is subject to a number and variety of injuries, obstructions, infections, and abnormalities. The more common of these will be discussed.

Conduction deafness

Deafness is complete or partial lack or loss of the sense of hearing. Conduction deafness is due to some defect in the sound-conducting system, that is, some defect relating to the auditory canal, eardrum, ear bones, or eustachian tube. Most cases of conduction deafness in children result from the presence of excessive lymphoid tissue in and about the eustachian tube opening in the nasopharynx. This interferes with proper ventilation of the middle ear, thereby upsetting the normally equalized pressures on either side of the eardrum. As a result, tympanic vibrations are inhibited and hearing is impaired. The treatment usually recommended is surgical removal of the lymphoid obstruction.

The usual causes of conduction deafness in adults are impacted cerumen (earwax) and otosclerosis. Otosclerosis is a chronic disease characterized by the formation of spongy bone in the osseous labyrinth and the ossification of the annular ligament by which the stapes footplate is attached to the circumference of the oval window. The result, once again, is poor vibration and impaired hearing. A dramatic and usually successful surgical procedure is the stapes mobilization, or stapes substitution, operation.

Perceptive deafness

Perceptive deafness results from disorders of the auditory nerve, the cerebral pathways, or the auditory center. The etiologic factors may be infection, tumors, psychogenic disturbances, injuries, and certain drugs (for example, quinine and streptomycin), to name a few. Treatment is directed at eradicating the underlying cause. If the condition is irreversible, careful rehabilitation (hearing aid, lip reading, and the like) is essential.

Tinnitus

Tinnitus is an annoying symptom characterized by hissing, ringing, buzzing, thumping, whistling, or roaring in the ears. The underlying cause can be almost anything, including an infection, drug intoxication, obstruction of the external auditory canal, obstruction of the eustachian tube, or a dental disorder. When this symptom is present, therefore, the physician must consider every possibility in searching for the answer.

Ménière's syndrome

Of the various labyrinthine (inner ear) disturbances, Ménière's syndrome is perhaps the most common. Possibly of an allergic nature, the condition is characterized by bouts of deafness, tinnitus, vertigo, and nausea and vomiting. Antihistamines and vasodilators help in some cases, but not in others. When they fail, a variety of surgical procedures to destroy or bypass the equilibrium mechanism are available.

Infection

From the standpoint of a general threat to health, infection, particularly the acute form, is probably the most serious ear condition. Infection of the outer ear, *otitis externa*, involves the auricle or auditory canal or both. Since the area is relatively easy to reach, administration of appropriate anti-infective drugs usually effects a swift cure.

Otitis media, or middle ear infection, however, is always potentially dangerous. Although almost any pathogen can incite an infection in the area, the most commonly encountered include certain hemolytic streptococci, pneumococci, staphylococci, *Corynebacterium diphtheriae*, *Pseudomonas aeruginosa*, *Mycobacterium tuberculosis*, and *Escherichia coli*. Often the infection is secondary and follows in the wake of scarlet fever, measles, mumps, pneumonia, or influenza. The principal danger is mastoiditis, or infection of the mastoid antrum and cells, which not infrequently complicates acute purulent otitis media. Mastoiditis may lead to infection of the brain and sudden death.

QUESTIONS

1. Distinguish among eardrum, tympanum, and tympanic membrane.
2. The middle ear has two openings and two windows. Elaborate.
3. What is the origin of the word eustachian?
4. What is the literal meaning of ossicles?
5. What is the Latin meaning of cerumen?
6. Are the ossicles fused together?
7. What is the literal meaning of labyrinth?
8. What are the two membranes that divide the cochlea into the osseous portion and the membranous portion?
9. Why is the membranous cochlea also called the scala media?
10. What is the Latin meaning of scala?
11. The distinction between perilymph and endolymph is essentially one of location. Explain.
12. Distinguish among vestibular nerve, cochlear nerve, acoustic nerve, and auditory nerve.
13. What is the Latin meaning of cochlea?
14. What is the specific relationship of the organ of Corti to the acoustic nerve?
15. What is the specific relationship of the crista ampullaris to the eighth cranial nerve?
16. The transmission of sound from the source to the organ of Corti is mechanical; from the organ of Corti to the brain, it is electrical. Elaborate.
17. What is the relationship between pitch and frequency?
18. Relative to the tympanum, what is the difference between the response to a loud sound and the response to a soft sound?
19. Swallowing relieves pressure imbalances between the outer and middle ear. Does swallowing cause air to enter or leave the middle ear?
20. Which semicircular canal is especially called into play in a somersault?
21. Dizziness, or vertigo, relates etiologically to the labyrinth, acoustic nerve, or brain. Can you be more specific with regard to each of these possibilities?
22. Occlusion of the eustachian tube results in conduction deafness. Why, precisely, is this form of deafness called "conduction"?
23. Of all types of conduction deafness, which is usually the simplest to correct?
24. In high dosage, streptomycin can cause perceptive deafness. Is the damage to the cochlea, acoustic nerve, or brain?
25. Explain how a "strep" throat can lead to mastoiditis.

CHAPTER 23

THE ENDOCRINE SYSTEM

In conjunction with the nervous system, the endocrine system (Fig. 23-1) regulates countless physiological processes and maintains the constancy of the body's internal environment. The endocrine system refers to the many and sundry glands of internal secretion and their hormones. In contradistinction to the glands of external secretion, such as sweat and salivary glands, those of internal secretion are without ducts and discharge their secretions into the blood or lymph. Appropriately, then, we call the glands that secrete externally duct, or *exocrine*, glands and the glands that secrete internally ductless, or *endocrine*, glands.

Hormones are phenomenally potent regulators of cellular functions. Their mechanism of action relates to RNA transcription, enzyme activity, and the like. Hormones vary greatly in chemical composition, ranging all the way from complex proteins to simple amino acids. Those of major concern are presented in this chapter.

PITUITARY GLAND

The pituitary gland, or hypophysis, a tiny gland that lies in the sella turcica of the sphenoid bone, is divided into two completely separate parts, the anterior lobe, or adenohypophysis, and the posterior lobe, or neurohypophysis (see Fig. 20-16). The hypophysis is connected to the hypothalamus by a stalk that permits neurohormonal control of pituitary function.

Hormones of anterior lobe

The insignificant-looking anterior lobe secretes at least six known hormones. Five of these potent regulators, the tropic hormones, act upon endocrine glands situated elsewhere; further, their output is stimulated by *releasing factors* from the hypothalamus.* As one authority put it, the hypothalamus and pituitary are, respectively, the orchestra conductor and band master of the endocrine system. Anterior lobe hormones are secreted by two kinds of cells: acidophilic, which take an acid stain, and basophilic, which take a basic stain. The acidophilic cells secrete the growth hormone and prolactin, and the basophilic cells secrete thyrotropic hormone, follicle-stimulating hormone, luteinizing hormone, and adrenocorticotropic hormone.

Growth hormone. By increasing the synthesis of cellular elements, the growth hormone (GH, somatotropin) stimulates the development and

---

*The relationship among the hypothalamus, pituitary, and target gland (the gland stimulated by the tropic hormone) is very complex, involving both negative and positive feedback. For example, thyroid hormones increase the output of thyrotropin-releasing factor (TRF) but interfere with its stimulating action on the pituitary.

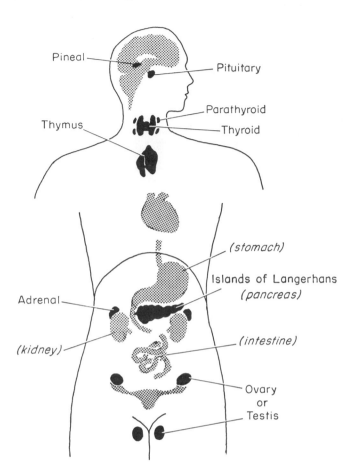

FIG. 23-1
Endocrine glands. (Hormonal function of thymus poorly understood.)

Labels in figure: Pineal, Pituitary, Parathyroid, Thyroid, Thymus, (stomach), Islands of Langerhans (pancreas), Adrenal, (kidney), (intestine), Ovary or Testis

enlargement of all the tissues. Although it is released in highest concentrations during preadolescence, the anterior lobe never completely curtails its manufacture.

One of the causes of dwarfism is an insufficient output of the growth hormone. The degree of underdevelopment, of course, will depend upon the extent of the insufficiency. Gigantism, on the other hand, usually results from a tumor of the acidophilic cells of the anterior lobe during preadolescence, resulting in an increased output of the growth hormone and a tremendous enlargement of all parts of the body (Fig. 23-2).

If a tumor occurs after adolescence, when most of the bones have fused, the skeleton grows disproportionately, in thickness instead of length; the result is not gigantism but acromegaly. Characteristically, the lower jaw, nose, lips, hands, and feet become tremendously enlarged.

Prolactin. Prolactin, or the lactogenic hormone, is one of several hormones involved in the production of milk by the mammary glands. Prolactin stimulates milk secretion once the glands have been primed by sex hormones during pregnancy. At one time prolactin was known as the luteotropic hormone (LTH).

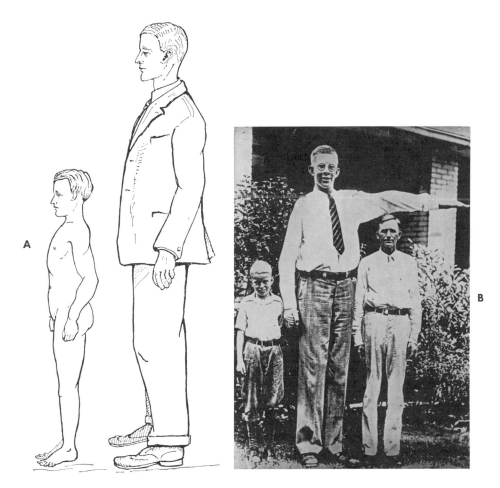

FIG. 23-2

A, Midget type of dwarfism due to pituitary deficiency. Dwarf is 21 years of age. Body proportions are those of 12-year-old boy, with relatively large head. Sexual and mental development are normal. Man on the right is about average height. **B,** Example of gigantism. Boy in the center, 13 years of age, is 7 feet 5 inches tall and weighs 290 pounds. He is shown with his father and brother, who are of normal size. (From The human body: its anatomy and physiology, fourth edition, by C. H. Best and N. B. Taylor. Copyright © 1932, 1948, 1956, 1963, by Holt, Rinehart and Winston, Inc. Reprinted by permission of Holt, Rinehart and Winston, Inc.)

Thyrotropic hormone. The thyroid-stimulating hormone (TSH), or thyrotropin, promotes growth and multiplication of the cells in the thyroid gland and the secretion of thyroxine and triiodothyronine. We shall have much to say about both hormones later in the chapter in the discussion of the thyroid gland.

Adrenocorticotropic hormone. In much the same fashion as the thyrotropic hormone acts upon the thyroid, the adrenocorticotropic hormone, or ACTH, acts upon the adrenal cortex to stimulate release of the adrenocortical hormones; directly it tends to produce increased pigmentation by acting on the melanocytes in the skin.

Gonadotropic hormones. The gonadotropic hormones are concerned with sexuality. Those of prime concern are the follicle-stimulating hormone (FSH) and the luteinizing hormone (LH). In the female, FSH stimulates development of the follicles in the ovaries, bringing about maturation of the ova and the release of estrogens, one of the two types of female hormones. In the male, FSH stimulates development of the testes and promotes the manufacture of sperm.

In the female the luteinizing hormone causes rupture and release of the ovum from the follicle (Chapter 24); in the male, where it is called the interstitial cell–stimulating hormone, or ICSH, it stimulates the interstitial cells of the testes to secrete testosterone, the male hormone.

Hormones of posterior lobe

The posterior lobe releases two hormones: vasopressin and oxytocin. The actual synthesis of these hormones takes place in the hypothalamus. Once formed, they migrate as bead-like droplets down the connecting fibers leading to the neurohypophysis.

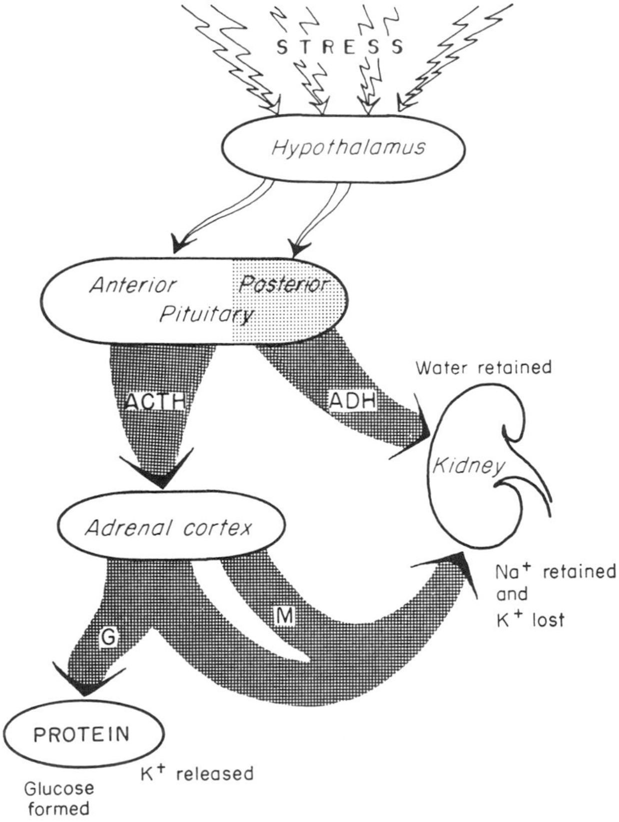

FIG. 23-3

Effect of stress upon body water and ions. *G* and *M* are glucocorticoids and mineralocorticoids, respectively. Aldosterone (a mineralocorticoid) is also released independently of ACTH. (From Brooks, S. M.: Basic facts of body water and ions, ed. 2, New York, 1968, Springer Publishing Co., Inc.)

Vasopressin. Vasopressin, or antidiuretic hormone (ADH), has two actions. It contracts smooth muscle, particularly in the blood vessels, thereby elevating blood pressure, and, most important, it stimulates the reabsorption of water from the distal tubules in the kidney (Fig. 23-3). This renal mechanism is the principal means that the body has of conserving water and regulating the concentration of the extracellular compartment. One theory holds that ADH is released when the osmoreceptors of the anterior hypothalamus are stimulated by an increased osmotic pressure of the plasma and interstitial fluid. As a result of such stimulation these receptors transmit impulses to the posterior pituitary, bringing about the release of ADH, and the latter, by cutting down the production of urine, conserves water and reduces the concentration of the extracellular solutes. This mechanism operates until the osmotic pressure is reduced to normal. A drop in osmotic pressure causes the pituitary (via osmoreceptors) to curtail output of ADH.

When posterior pituitary function is inhibited, the body starts losing great amounts of water via the urine. This derangement, called diabetes insipidus, is also marked by voracious appetite, weakness, and emaciation.

Oxytocin. Oxytocin acts upon the uterine musculature, causing forceful contractions. Even so, most indications are that the hormone does not play a significant physiological role in the initiation of labor. Therapeutically, however, the physician uses Pitocin, a commercial form of oxytocin, to induce labor and control postpartum hemorrhage (Chapter 24).

Oxytocin also causes the alveoli and milk ducts of the breasts to contract and release milk. Experiments show that the stimulus for this release arises in response to sucking.

THYROID GLAND

The thyroid is a comparatively large endocrine gland situated beneath the muscles of the neck at the anterior juncture of the larynx and the trachea. It consists of two lateral masses, or lobes, connected at the midline by a bar of tissue called the isthmus. Histologically, the gland is seen as a mass of follicles, each of which is lined with a single layer of epithelium and filled with an amorphous material called colloid (Fig. 23-4). Colloid is composed of thyroglobulin, a conjugated protein from which the hormones thyroxine and triiodothyronine are released in response to the thyrotropic hormone of the anterior pituitary. The follicle cells synthesize thyroxine, generally considered the chief hormone, according to the reaction summarized at the bottom of the page. A more recently discovered thyroid hormone is thyrocalcitonin (calcitonin), the action of which was presented earlier (p. 129).

$$2\left[HO-\!\!\left\langle\!\!\!\bigcirc\!\!\!\right\rangle\!\!-CH_2-CHNH_2-COOH \right] + 4I \longrightarrow$$

Tyrosine

$$HO-\!\!\left\langle\!\!\!\bigcirc\!\!\!\right\rangle\!\!-O-\!\!\left\langle\!\!\!\bigcirc\!\!\!\right\rangle\!\!-CH_2-CHNH_2-COOH + \text{Alanine}$$

Thyroxine

FIG. 23-4

A, Parathyroid gland and part of thyroid of monkey, low magnification. **B,** Follicles of human thyroid, showing colloid secretions in follicles. (× 640.) (**A** from Bevelander, G.: Outline of histology, ed. 7, St. Louis, 1971, The C. V. Mosby Co.; **B** from Bevelander, G.: Essentials of histology, ed. 6, St. Louis, 1970, The C. V. Mosby Co.)

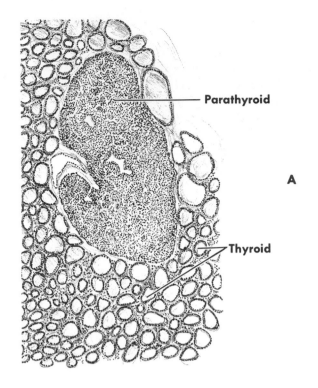

Parathyroid

Thyroid

A

Colloid

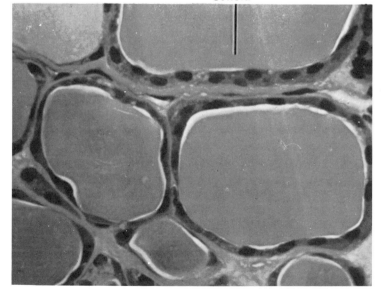

B

Thyroxine is gradually released as needed to the blood, where it forms a loose complex with the circulating plasma proteins. From these proteins it enters the interstitial fluids and finally the cells, where it increases the rate of metabolism and thereby enhances the activities of all tissues, organs, and systems. Some idea of the potency of the hormone can be gained by considering that lack of it can drop the metabolic rate to as low as −50! An excess, on the other hand, can elevate the rate to about three times the normal value. The particulars of the mechanism of the effects of thyroxine remain something of a mystery.

Hyperthyroidism

Hyperthyroidism refers to the condition resulting from overproduction of thyroid hormones. The cardinal signs and symptoms include an enlarged thyroid (goiter), loss of weight, nervousness, and tachycardia. Because the tissues are literally being burned out, there are degenerative changes, particularly of heart muscle. Hyperthyroidism commonly is accompanied by exophthalmos (protrusion of the eyes), a condition resulting from hypertrophy and edema of the orbital tissue. The basal metabolic rate (BMR) is always elevated, and the level of protein-bound iodine (PBI) is above normal.

The usual methods employed in the treatment of hyperthyroidism include surgical removal of part of the gland and administration of drugs (for example, propylthiouracil) to inhibit synthesis of the hormones. More recently, radioactive iodine (^{131}I) has been used with striking success. Since iodine is almost completely absorbed by the gland, a high dose of radiation is delivered precisely at the spot where it is needed.

Hypothyroidism

Since hypothyroidism results from lack of thyroid hormones, the pathologic picture is physiologically opposite that of hyperthyroidism.

The usual features include a low BMR and PBI, mental sluggishness, myxedema, cold dry skin, coarse hair, deep voice, and generalized retardation. Occasionally, there is a goiter, which represents an attempt by the gland to compensate for its deficiency. The goiter mechanism apparently relates to PBI: low PBI calls forth an increased output of thyrotropic hormone, which causes the thyroid gland to enlarge. Even though the gland continues to grow larger, however, the growth is to no avail because the gland has lost its ability to produce thyroxine in sufficient amounts to meet the needs of the body.

When hypothyroidism occurs congenitally, the body and mind fail to develop properly, and the result is a pathetic creature called a cretin (cretinism). Hypothyroidism, when acquired later in life (the adult form), is referred to as myxedema. It should be appreciated, however, that there are degrees of hypothyroidism and therefore instances in which the signs of a deficiency are subtle (for instance, poor skin and hair, poor academic performance, and the like).

Treatment of hypothyroidism is highly specific, effective, and dramatic. It involves the use of thyroid extract, which is prepared by drying and powdering the glands removed from cattle, sheep, and hogs. The yellowish powder, usually called "thyroid," contains a large amount of the hormone and is readily absorbed by the intestinal mucosa. Preparations of the pure hormones, such as Synthroid and Cytomel, are also available.

Endemic goiter

In areas of the world where the soil and food are deficient in iodine, the population at large will be afflicted with endemic goiter unless the element is added to the diet. Because of the wide use of iodized salt (regular salt containing a trace of some iodide), this disorder is now rare in the United States. The mechanism of the enlargement concerns elaboration of excessive amounts of the thyrotropic hormone, which, as explained previously, is secreted in response to a deficiency of thyroxine, in this instance occasioned by lack of iodine. As before, this enlargement represents a mode of compensation, but here the compensation is usually good enough to maintain a normal BMR.

PARATHYROID GLANDS

The parathyroids, four tiny glands embedded in the posterior aspect of the thyroid, secrete a hormone called parathormone (parathyroid hormone) that stimulates the release of calcium from bone into the blood and inhibits reabsorption of phosphate by the kidney in order to maintain a steady concentration of these ions in

the extracellular fluid. A severe decrease in calcium ion concentration can cause tetany and sudden death. Fortunately, almost any factor that serves to decrease calcium ion concentration stimulates production of parathormone.

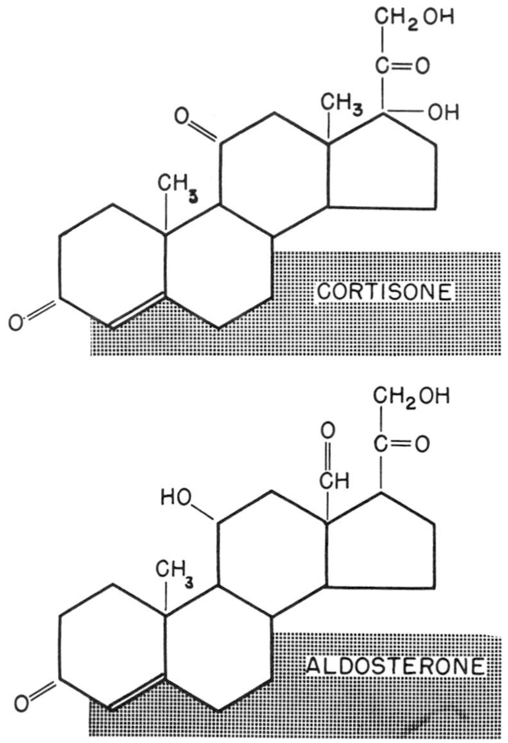

Hypoparathyroidism

Although it occasionally develops otherwise, hypoparathyroidism most commonly results from the inadvertent removal of the parathyroids during thyroidectomy. When this happens, tetany develops in a day or two. Calcium gluconate or other calcium salts are used in emergency treatment, and vitamin D is effective in maintenance therapy.

Hyperparathyroidism

Quite rarely, a parathyroid gland may develop a tumor and start pouring out excessive amounts of hormone, resulting in decalcification, soft bones, and kidney stones. Sometimes the bones become so weak and brittle that they break under the slightest amount of pressure or tension. Treatment is surgical removal of the tumor.

ADRENAL GLANDS

The two adrenal glands are elongated, flattened bodies situated at the top of each kidney. The outer portion of the gland, the cortex, secretes an array of hormones collectively called the adrenocortical hormones; the inside of the gland, the medulla, secretes the catecholamines, epinephrine and norepinephrine. These vital hormones produce a variety of physiological effects.

Adrenocortical hormones

The adrenocortical hormones* are similar chemically (Fig. 23-5) but different physiologi-

FIG. 23-5
Structural formulas for two corticosteroids. Cortisone is a glucocorticoid; aldosterone is a mineralocorticoid.

---

*Also called adrenocorticoids, adrenocorticosteroids, or simply corticosteroids.

cally. They are divided into three categories: mineralocorticoids, glucocorticoids, and androgens. Since the androgens are produced in insignificant amounts in the healthy gland, they are normally of little physiological concern.

Mineralocorticoids. The principal mineralocorticoid is aldosterone. These hormones are so named because they control the extracellular concentrations of sodium and potassium by their action upon the distal renal tubes: they stimulate the reabsorption of sodium and inhibit the reabsorption of potassium. In the process, large amounts of water are returned to the blood along with the sodium ions. A major effect of all this, aside from conservation of sodium, which the body needs in appreciable amounts, and the excretion of potassium, which is needed only in minute amounts, is the maintenance of blood volume.

The body increases the output of the mineralocorticoids in three instances: when the extracellular concentration of sodium starts to fall; during periods of physical stress when an increased blood pressure is generally desirable; and when there is a decrease in blood pressure. Just how the increased output is mediated is not known for certain. It may be that a low sodium ion concentration triggers the adrenal cortices directly or indirectly (via ACTH) or both. In the case of stress, perhaps ACTH may be the important avenue (Fig. 23-3). In regard to the decrease in blood pressure, the general feeling is that the mechanism centers about the kidney enzyme renin, which converts a plasma globulin to angiotensin I, which in turn is converted by an unknown enzyme into angiotensin II, an agent that stimulates the adrenal cortices to produce aldosterone as well as causing vasoconstriction.

Glucocorticoids. The adrenal cortices secrete a number of glucocorticoids that have a pronounced effect upon the metabolism of protein, fat, and carbohydrate. To a lesser extent they act upon inorganic metabolism in much the same manner as the mineralocorticoids. From the data currently available, hydrocortisone (cortisol) appears to be the chief hormone in this category.

In some fashion not yet completely understood, the glucocorticoids cause the breakdown of protein into amino acids and assist in the transport of these acids across the cellular membrane into the extracellular compartment. In a word, then, these hormones "mobilize" protein. The purpose of such drastic chemical measures is not difficult to understand in light of the greatly increased output of the glucocorticoids during periods of physical stress (infection, exposure, trauma, and the like); having amino acids immediately available allows distressed areas to use them to repair protoplasmic damage, to synthesize new cytoplasmic elements, and to provide energy. Also, the increased concentration of amino acids in the blood provides the liver with the raw materials to make glucose (gluconeogenesis). Because nerve cells cannot survive without glucose, gluconeogenesis is especially important during stress. The ability of the glucocorticoids to elevate the concentration of blood glucose was their first discovered action, hence, their name.

Fats are handled in much the same manner, being split into fatty acids for energy and other emergency needs. Unless these acids are used up right away, acidosis may result.

Stress response. The increased output of the adrenocortical hormones in response to stress, first emphasized by Selye, represents the body's call to arms. These hormones provide the tissues, organs, and systems with emergency materials: amino acids, fatty acids, glucose, sodium, and water. Very generally, we can say that the mineralocorticoids attend to fluid and electrolyte balance, and the glucocorticoids, to energy and tissue resistance.

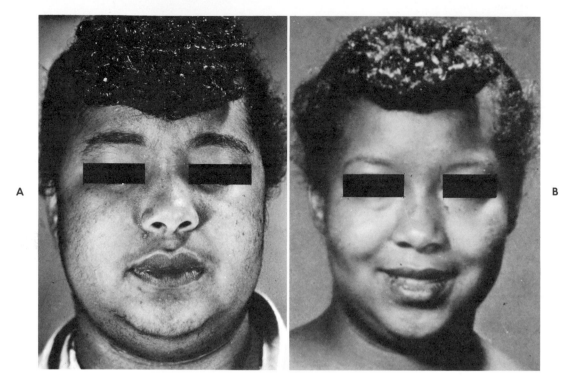

FIG. 23-6
Cushing's syndrome, an endocrine derangement caused by excessive output of hormones by adrenal cortex. **A,** Preoperative. **B,** Six months after removal of tumor from adrenal cortex. (Courtesy Dr. William McKendree Jefferies, Highland View Hospital, Cleveland, Ohio.)

Hypoadrenocorticism

If the adrenal cortices atrophy or are damaged by disease, hyposecretion of adrenocortical hormones (hypoadrenocorticism) results. Commonly known as Addison's disease, hypoadrenocorticism leads to death in a very short time unless treated. The immediate threat to life is the loss of sodium and the retention of potassium. Excess potassium is toxic, and the loss of sodium, and the water it carries away with it, reduces the extracellular compartment in general and the blood volume in particular. Simply taking extra amounts of salt will prolong life somewhat by replacing the lost sodium.

The full treatment of hypoadrenocorticism includes, in addition to taking salt, the administration of fludrocortisone (a synthetic mineralocorticoid) and hydrocortisone (or another glucocorticoid). Fludrocortisone corrects the electrolyte and fluid disturbance, and the glucocorticoid corrects the low state of metabolism and bolsters resistance.

Hyperadrenocorticism

Excessive secretion of the glucocorticoids by the adrenal cortices, one form of hyperadrenocorticism, usually occurs as a result of a tumor of the adrenal cortex; in rare cases it is due to a tumor of the pituitary gland. In the former instance, the overactivity results from an increase in cortical tissue (Cushing's syndrome); in the latter, the cortices are overstimulated by excess ACTH (Cushing's disease).

The major repercussions include edema, due to retention of salt and water; hypertension, due to the increased blood volume; loss of weight and weakness, due to gluconeogenesis; masculinization, due to excess androgens; and psychiatric disturbances. Typically, there is truncal obesity, plethoric appearance, and rounded ("moon") facies (Fig. 23-6).

Corticosteroids

The term corticosteroids is applied to adrenocortical hormones and their synthetic substitutes; the latter differ from the natural chiefly in relation to electrolyte balance and potency. Cortisone (17-hydroxy-11-dehydrocorticosterone) was the first glucocorticoid to be used clinically. Although it proved to be a "wonder drug," its considerable side effects stimulated the search for drugs that did not have such drawbacks. Newer drugs possess no new properties but afford better therapeutic adjustment.

Considering the wide and profound metabolic effects that the corticosteroids produce, it is not difficult to appreciate their ability to ameliorate a great many diverse disorders. In addition to their effectiveness in the treatment of hypoadrenocorticism, the corticosteroids frequently produce dramatic results in the treatment of rheumatoid arthritis, rheumatic fever, allergies, inflammatory eye conditions, acute leukemias, and many heretofore fatal skin diseases.

In addition to the corticosteroids, the physician sometimes employs the adrenocorticotropic hormone (ACTH). As we might expect, its clinical effects closely parallel those of the corticosteroids.

Epinephrine

In emergency situations the adrenal medulla secretes epinephrine (commonly called adrenaline) and the closely related hormone norepinephrine (noradrenaline). Since norepinephrine is also released at sympathetic nerve endings, we can readily see that the adrenal medulla functions to intensify sympathetic bombardment. Apparently, then, the body employs these two hormones together for defense and protection.

An injection of epinephrine does many things, all uniquely designed to prepare the body for fight or flight. It elevates the blood pressure, widens the pupils, dilates the bronchioles, decreases peristalsis, stimulates glycogenolysis,* and constricts all vessels except those in the coronary system and muscles.

The effects of norepinephrine are similar but not identical. For example, this hormone exerts a much more pronounced action upon blood pressure. Whereas epinephrine elevates pressure chiefly by increasing cardiac output, norepinephrine does so via vigorous vasoconstriction. Norepinephrine is now considered a drug of choice in the treatment of shock. On the other hand, epinephrine is the drug of choice in the treatment of acute asthma.

Pheochromocytoma. Pheochromocytoma is a chromaffin cell tumor of the adrenal medulla that results in the release of excessive amounts of epinephrine and norepinephrine. Hypertension and its associated manifestations are the main symptoms. Characteristically, the rise in blood pressure is intermittent, an "attack" ranging anywhere from a few minutes to several hours. The condition is remedied by surgical removal of the tumor.

---

*The mechanism by which epinephrine promotes the production of glucose by liver cells (glycogenolysis) has recently been deciphered. Attached to the cell membrane is the enzyme adenyl cyclase, which in the presence of the hormone (acting as an activator) changes adenosine triphosphate (ATP) into cyclic adenosine monophosphate (cyclic AMP), a messenger substance that activates a second enzyme that brings into play a five-step sequence leading to glucose.

PANCREAS AND INSULIN

Because of the millions of diabetic persons whose lives depend upon insulin, it is the most important hormone used in medical practice. Before its isolation by Banting and Best in 1923 there was little that the diabetic could do but starve and wait for the inevitable. Today, the outlook for patients with diabetes is almost as good as for persons with normal metabolism.

Like other hormones, insulin enters the blood directly. It is a relatively simple protein produced in the pancreas by the beta cells of the islands of Langerhans (Fig. 23-7). The release of insulin by these cells probably results from the direct action of excess glucose (hyperglycemia). Insulin lowers blood glucose by stimulating the conversion of glucose to glycogen (glycogenesis) and by stimulating the cellular uptake of glucose.

The glucose tolerance test (Fig. 23-8) is an important tool for determining the functional ability of the pancreas. The test is performed by giving an oral dose of glucose and then taking blood samples at specified intervals. Whereas the blood glucose level in a nondiabetic person will rise to around 145 mg/100 ml and return to the normal value of about 90 mg/100 ml in about 2 hours, that of the diabetic patient may run well over 300 mg/100 ml and not return to normal for several hours.

Diabetes

Though diabetes mellitus is most conveniently viewed as a disease caused by lazy or incapacitated islands of Langerhans, recent findings, including the fact that some diabetics have high circulating levels of insulin, disclose a much more elaborate pathology. Other endocrine glands may be involved; insulin may circulate in a bound, inactive form; immunologic factors may be involved; and so on.

There is a great mass of clinical data linking the disease to obesity and heredity. In 25% of

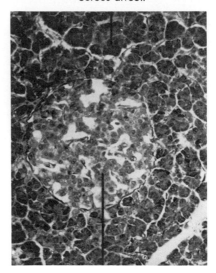

Serous alveoli

Islet of Langerhans

FIG. 23-7
Photomicrograph of human pancreas. (From Bevelander, G.: Essentials of histology, ed. 5, St. Louis, 1965, The C. V. Mosby Co.)

the patients there is a family history of diabetes. Other conditions associated with the development of the disease include certain infections, pancreatic tumors, and even trauma. In animals the feeding of certain chemicals, such as alloxan, incites diabetes by a demonstrable destruction of beta cells.

The signs and symptoms of diabetes mellitus are mainly referable to the failure of the body to utilize glucose and include hyperglycemia, glycosuria, polydipsia, polyphagia, polyuria, weakness, and loss of weight. From the facts previously presented, the student should have little difficulty figuring out the relationship of excess glucose to these derangements.

Diabetic coma is the immediate threat to life, and once again we go back to glucose. Without glucose, the cells are forced to burn fats, and

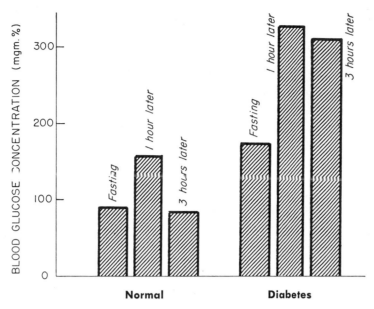

FIG. 23-8
Comparison of blood glucose levels of normal and diabetic subjects following ingestion of 50 g of glucose.

when fats are burned in excess, keto acids (namely, acetoacetic and β-hydroxybutyric acids) are formed. These intermediate products of fat metabolism, which are normally metabolized in the Krebs cycle, accumulate in the blood and produce acidosis. Unless treated with insulin, the patient will lapse into coma and die.

Diagnosis. Persistent glycosuria (sugar in the urine) is presumptive evidence of diabetes. The diagnosis, however, rests largely upon the blood sugar level and the glucose tolerance test.

Testing the urine for glucose is a simple but highly useful procedure. It serves as a screening test for the disease and is a speedy way for the patient to gauge his insulin. Although there are several modifications, the basic laboratory test depends upon reduction of Cu^{++} to Cu^{+} by glucose. The latter ion then couples with oxygen to form cuprous oxide (Cu_2O), a red precipitate, disclosing not only the presence of glucose but

also, depending upon the degree of change (Cu^{++} is blue), its concentration. Test papers that undergo a color change incident to the action of glucose oxidase (for example, Tes-Tape) are available for use by the patient.

Treatment. The treatment of diabetes centers upon the character of the diet and the administration of exogenous insulin, both of which must be tailored to the severity of the disturbance and the needs of the patient.

From the various forms of insulin available, the physician selects the appropriate preparation. These vary with respect to duration of action and presence of foreign protein. Crystalline zinc insulin (CZI), for example, has a quick onset and short duration of action (4 to 6 hours); on the other hand, protamine zinc insulin (PZI) and NPH insulin have a delayed onset and much longer duration of action. CZI is the preparation of choice in an emergency situation; the other agents are better for a mild case of diabetes in which sustained action represents the most desirable feature. Lente insulins have the advantage of being free of foreign protein and therefore less likely to cause allergic reactions.

The need to administer insulin by injection long ago sparked the search for an oral preparation. To some degree, this search has borne fruit, for several drugs (for example, Orinase) are now available that make it possible to control certain types of diabetes without using insulin. Although these drugs have their shortcomings and are not infrequently ineffectual, an ideal drug may someday be developed. These agents, which bear no chemical relationship to the hormone, act by stimulating the beta cells. In cases in which the drugs are not effective, the beta cells have probably lost their ability to make insulin.

Hyperinsulinism

Hyperinsulinism may result from an overdose of injected hormone or from oversecretion as caused by a pancreatic tumor involving the beta cells. In either instance the upshot is *hypoglycemia*. Since nerve cells must receive a constant supply of glucose for proper metabolism, the repercussions of a low blood sugar are ominous. The initial effects are characterized by excitement and, if the deprivation is severe, convulsions. Later these subside, and the patient lapses into coma. Paradoxically—and of prime clinical significance—is the fact that the coma of hyperinsulinism mimics diabetic coma. Not infrequently the physician must tax his diagnostic skills to determine whether the diabetic patient in coma has taken too much or too little insulin. Diabetics, of course, are always on guard against hyperinsulinism and carry a piece of candy or other sweet to take at the very first sign of intoxication.

The advent of insulin shock in the treatment of certain forms of mental illness was a milestone in psychotherapy. There are some authorities who still believe that this mode of therapy yields better results than electric shock and even more recently developed procedures. Insulin shock is produced by injecting just enough of the hormone to produce a few moments of unconsciousness. Just exactly how it benefits the higher mental processes is a good but moot question. Perhaps it erases the fresh neural paths forged by impulses that have gone astray.

Glucagon

Whereas the beta cells in the island tissue elaborate insulin, the alpha cells secrete a polypeptide hormone called glucagon. By increasing the activity of phosphorylase, the enzyme that initiates the first step in glycogenolysis, glucagon causes the liver to release glucose to the blood. The hormone apparently is released by the pancreas during those times when the tissues are in immediate need of extra glucose.

Recently, glucagon was introduced into medical practice as an antidote in hyperinsulinism. Since only a very small dose of the hormone is required, and since it can be given simply by an injection under the skin (subcutaneously), it is vastly superior to intravenous glucose for use outside the hospital.

OVARY

In girls at the age of about 12 years the anterior pituitary starts to release the follicle-stimulating hormone (FSH) and the luteinizing hormone (LH). As will be explained in some detail in the next chapter, these agents stimulate the development of the ovaries and cause them to release appreciable amounts of estrogens and progesterone. As soon as the effects of these hormones become manifest, puberty has commenced.

Estrogens

The estrogens are steroid substances secreted by the cells of the graafian follicle. Estrone and estradiol, particularly the latter, are the principal members of the group. Their chemical structure differs little from that of progesterone and the male hormone.

Besides stimulating growth of the sex organs, the estrogens are responsible for the development and maintenance of the secondary sex characteristics, that is, distribution of body fat and hair, breasts, texture of skin, and character of voice. The role that they play in the menstrual cycle is discussed in Chapter 24.

Undersecretion of estrogens (hypogonadism) results in retardation of the sex organs and secondary characteristics. This may be due to either a primary deficiency or lack of the follicle-stimulating hormone. Excessive secretion of hormones (hypergonadism), on the other hand, is characterized by sexual precocity. This may result from a tumor of either the ovary or the anterior pituitary. Hypogonadism is treated with estrogens (natural or synthetic) and hypergonadism by surgical intervention.

Progesterone

Progesterone is a steroid released by the corpus luteum (Chapter 24) in response to the LH hormone. Its known effects include inhibition of uterine muscle, preparation and maintenance of the lining of the uterus, development of the breasts, and suppression of ovulation in the latter half of the menstrual cycle. These features are taken up in more detail in Chapter 24.

As with estrogens, there is a lack of progesterone in hypogonadism and an excess in hypergonadism. Therefore, the manifestations of these two derangements stem from the combined effects of both hormones. Hypogonadal disorders (for example, amenorrhea) are treated with synthetic progesterone or its derivatives, often in combination with estrogens.

TESTIS

The interstitial tissue of the testis secretes a variety of male hormones collectively called androgens. Testosterone, the chief androgen, is closely related chemically to progesterone. Androgens are required for the development of the male sex organs and secondary sex characteristics and, in conjunction with FSH, the maintenance of spermatogenesis. In addition, they inhibit the pituitary gland, stimulate the synthesis of protein, and cause the retention of potassium and phosphate. When injected into the female, androgens suppress menstruation and lactation and tranquilize uterine activity.

The output of male hormone in the fetus is stimulated by chorionic gonadotropin secreted by the placenta. At birth, however, the testes become dormant and remain so until puberty, at which time the interstitial cell–stimulating hormone (ICSH) of the adenohypophysis stimulates them to produce androgens. ICSH is identical to the LH of the female. In the event of hyposecretion, the male characteristics fail to develop. Castration, for example, causes atrophy of the penis and accessory structures.

PINEAL GLAND

The pineal gland, a tiny white structure weighing about 0.1 g and shaped somewhat like a pine cone, is buried nearly in the center of the brain. Like the thymus, it is slowly yielding its secrets to the techniques of present-day biochemistry and neurophysiology. In the rat at least, the pineal now appears to be a neuroendocrine *transducer*, that is, a gland that converts nervous input into hormonal output. The pineal manufactures a potent hormone called melatonin that slows the estrus or sex cycle. Sympathetic bombardment of the gland, as a consequence of impulses relayed from the retina upon stimulation by light, inhibits the output of the hormone and thereby accelerates estrus. Thus, the current view is that the pineal is a true biological clock that regulates sexuality in mammals, and perhaps the timing of menstrual cycles in human beings. Future developments are enthusiastically awaited.

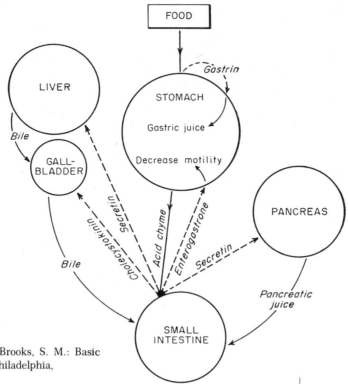

FIG. 23-9
Digestive hormones. (From Brooks, S. M.: Basic facts of general chemistry, Philadelphia, 1956, W. B. Saunders Co.)

DIGESTIVE HORMONES

A number of hormones are called into play in the digestive process (Fig. 23-9). The most important of these are briefly described.

Gastrin

Certain protein factors in food stimulate the stomach mucosa to secrete gastrin, which in turn provokes the release of gastric juice. This hormone is chemically and physiologically related to the ubiquitous histamine, and at one time they were thought to be the same. Among other pronounced effects, both agents provoke a copious flow of gastric juice. An injection of histamine is sometimes given in order to secure a sample of the juice for gastric analysis.

Secretin

When acid chyme enters the duodenum, it causes certain cells in the mucosa to release a hormone called secretin. Once in the blood, secretin stimulates the pancreas to form pancreatic juice and the liver to form bile. This is one of the ingenious ways in which these two organs are apprised of the presence of food in the intestine and of the urgent need for digestive juices.

Pancreozymin

Pancreozymin is a duodenal enzyme released into the blood in response to acid chyme. As indicated by its name, it induces the secretion of pancreatic juice. However, in contrast to secretin, which provokes chiefly water and sodium bicarbonate, pancreozymin stimulates chiefly the elaboration of digestive enzymes.

Cholecystokinin

Bile is forced into the duodenum by contraction of the gallbladder, an action initiated by a hormone called cholecystokinin that is released by the duodenal mucosa in response to food, especially those rich in fat, such as cream and egg yolk. This response is one of direct action of the hormone upon the musculature within the wall of the gallbladder.

Enterocrinin

Intestinal juice is called forth by a number of factors, one of which is hormonal. Several hormones, collectively called enterocrinin, are released by the mucosa and are carried to the intestinal glands via the bloodstream. It is believed that these agents are especially important in determining the character of the juice; protein provokes peptidases, and carbohydrate provokes carbohydrases.

Enterogastrone

When fat-bearing food enters the duodenum, the mucosa releases enterogastrone. This hormone travels via the bloodstream to the stomach, where it decreases gastric secretion and motility.

PROSTAGLANDINS

The prostaglandins, a family of hormonelike fatty acids present throughout the tissues of the body, are among the most potent of all known biologic materials. Their various effects are numerous: lowering and increasing blood pressure, contracting the uterus, decreasing gastric acidity, relaxing the bronchial tubes, slowing the heart, contraception, and so on. Recent work points to the cell membrane as the site of the formation of these agents and also as the site of their basic action. The action of aspirin apparently relates to its ability to block the synthesis of certain prostaglandins.

QUESTIONS

1. Locate the hypophysis and describe its structure.
2. Why is the pituitary called the master gland?
3. Compare acromegaly and gigantism.
4. Discuss the action of the gonadotropic hormones.
5. Compare the action of vasopressin and oxytocin.
6. What is Pitocin?
7. What is the cause and treatment of diabetes insipidus?
8. What are osmoreceptors?
9. What is thyroglobulin?
10. Distinguish between myxedema and cretinism.
11. Discuss the function of thyroxine.
12. Discuss the treatment of hyperthyroidism.
13. Distinguish between simple goiter and toxic goiter.
14. What is endemic goiter?
15. Discuss the use of thyroid extract.
16. How does the parathyroid hormone control the $[Ca^{++}]$ of blood?
17. What is the immediate cause of tetany? What is the treatment?
18. Compare the glucocorticoids and the mineralocorticoids.
19. What is the relationship of stress to adrenal function?
20. Why are drugs such as cortisone often contraindicated in diabetes mellitus?
21. What is the relationship of ACTH to the adrenal cortices?
22. What is the effect of stress upon fluid balance?
23. A characteristic finding in Addison's disease is hyperkalemia. Explain.
24. Discuss the etiology, signs, symptoms, and treatment of Cushing's disease.
25. Compare the effects of epinephrine and norepinephrine.
26. What is a pheochromocytoma?
27. What is the mechanism of action of insulin?
28. Why does insulin have to be injected?
29. Describe and explain the glucose tolerance test.
30. What is the mechanism of action of Orinase and the other "oral insulins"?

31. Explain the etiology of diabetic acidosis.
32. Explain the color change of Benedict's solution occasioned by the presence of glucose.
33. Compare the action of protamine zinc insulin and crystalline zinc insulin.
34. State the action of estradiol, progesterone, and testosterone.
35. Excess histamine might cause a gastric ulcer. Explain.
36. What is the derivation of the term pituitary?
37. What does the suffix -tropic denote?
38. What are acidophilic cells?
39. How do melanocytes darken the skin?
40. Popular writings speak of the ovaries producing "estrogen." What is your reaction to the use of the singular?
41. Discuss negative feedback.
42. Elaborate on the idea of the hypothalamus being an "orchestra conductor" (p. 369).
43. In a sense the neurohypophysis is not a true endocrine gland. Elaborate.
44. What is postpartum hemorrhage?
45. Tyrosine is not an essential amino acid, and yet it is needed in the synthesis of thyroxine. Isn't this contradictory?
46. In preparation for thyroidectomy the patient is heavily dosed with antithyroid drugs. What is the reason for doing this?

47. A high level of PBI does not necessarily mean hyperthyroidism. Why?
48. Why, specifically, does a lack of iodine result in a lack of thyroid hormone?
49. How do you account for the correction of a lack of parathyroid hormone by the administration of calcium salts?
50. How do you explain the kidney stones in hyperparathyroidism?
51. What are catecholamines?
52. Discuss the literal meaning of glucocorticoid.
53. Account for the masculinization that may result from a tumor of the adrenal cortex.
54. Explain the polydipsia and polyuria of diabetes mellitus.
55. What advantage does glucagon have over intravenous glucose in the management of hyperinsulinism?
56. What effect does the luteinizing hormone (LH) have on the output of LH-releasing factor?
57. Why, specifically, is the "male LH" called interstitital cell-stimulating hormone?
58. Why is it that fat stimulates the output of cholecystokinin?
59. What is the specific chemical difference between ATP and cyclic AMP? (See footnote, p. 379.)
60. What is the origin of the term postaglandin?

CHAPTER 24

REPRODUCTION

Sex cells, or gametes, are produced by reproductive organs or gonads—the testes in the male and the ovaries in the female. There are a number of secondary or accessory reproductive organs. The actual reproductive process encompasses fertilization, gestation, and parturition.

THE MALE

The reproductive apparatus in the male includes the testes, the seminal ducts, the seminal vesicles, certain glands, the urethra, and the penis (Fig. 24-1). These structures will be discussed in the order named, and in studying them the student should make full use of the illustrations.

Testes

The testes (testicles) are ovoid bodies enclosed in the scrotum, a cutaneous pouch suspended from the pubic and perineal regions. In the fetus, however, they lie within the lower abdominal cavity until about 2 months before birth, at which time they descend into the scrotum. In the event the testes do not descend, sterility results, for human spermatozoa cannot properly develop or thrive unless their environment is below body temperature, as it is in the scrotum.

The interior of the testis is divided by fibrous partitions into a number of wedge-shaped lobes, each containing one to three convoluted seminiferous tubules (Fig. 24-2). These tortuous structures unite to form a series of straight ducts that immediately unite in plexiform fashion to form the rete testis. From this plexus emerge other ducts that ultimately unite into a long, single, convoluted duct called the epididymis. This structure, situated on the posterior aspect of the testis, gives rise to the vas deferens (also called the ductus deferens and the seminal duct), which ascends the posterior border of the testis and enters the abdominal cavity where it travels several inches before joining the seminal vesicle.

Seminal vesicles

The seminal vesicles are two coiled tubes with sacculated walls situated just behind the lower portion of the bladder. Each joins a vas deferens and in so doing gives rise to the ejaculatory duct, a short tube that passes through the prostate gland to join the prostatic urethra.

Glands

The sex glands include the prostate and the bulbourethral (Cowper's gland). The prostate is about the size of a walnut and surrounds the neck of the bladder and the urethra. Its median

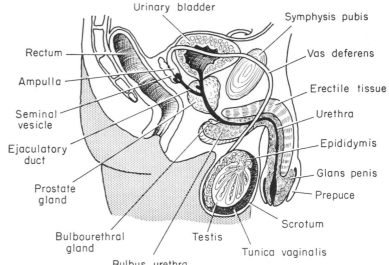

FIG. 24-1
Sagittal view of male reproductive system.

Urinary bladder

Symphysis pubis

Rectum

Vas deferens

Ampulla

Erectile tissue

Seminal vesicle

Urethra

Ejaculatory duct

Epididymis

Prostate gland

Glans penis

Prepuce

Bulbourethral gland

Testis

Scrotum

Bulbus urethra

Tunica vaginalis

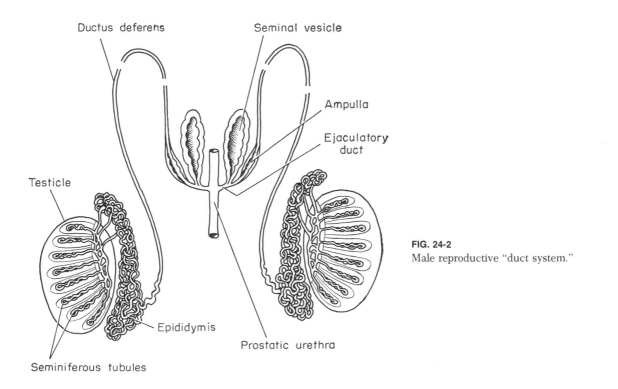

Ductus deferens

Seminal vesicle

Ampulla

Ejaculatory duct

Testicle

FIG. 24-2
Male reproductive "duct system."

Epididymis

Prostatic urethra

Seminiferous tubules

and two lateral lobes are composed partly of glandular matter and partly of muscular fibers, the latter encircling the urethra. The two bulbourethral glands lie near the bulb of the corpus cavernosum urethrae and send out ducts into the posterior section of the cavernous urethra. Both glands secrete a thin fluid that enters into the formation of semen.

Urethra

The male urethra is a membranous tube that conveys urine and semen to the surface; it extends from the neck of the bladder to the urinary meatus and measures about 8 inches long. On the basis of the structures through which it passes, the urethra is divided into three parts (Fig. 24-1): the prostatic, the membranous, and the cavernous (or spongy) portions. About the membranous portion is a band of circular striated muscle fibers, the external sphincter, that remains contracted except during micturition.

Penis

The penis, the male organ of copulation, has three divisions: the root, the body, and the extremity, or glans penis. The root is attached to the descending portions of the pubic bone by the crura, or extremities, of the corpora cavernosa. The body, the major portion of the structure, consists of the two parallel corpora cavernosa and the corpus cavernosum urethrae (Fig. 24-3). Through the latter passes the urethra. The glans penis is covered with mucous membrane and is ensheathed by the prepuce or foreskin.

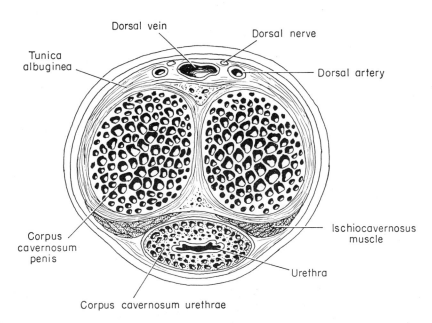

FIG. 24-3
Cross section of penis. (Modified from Francis, C. C: Introduction to human anatomy, ed. 6, St. Louis, 1973, The C. V. Mosby Co.)

Semen

The collective purpose of the structures just described is to produce and deliver semen to the female. Semen, the whitish fluid ejaculated in copulation, is composed of male sex cells, or spermatozoa, suspended in the nutrient secretions contributed by the prostate, the seminal vesicles, and Cowper's glands. The spermatozoa arise in the tubules and are conveyed to the epididymis through the complex system of channels mentioned earlier. Here they assemble, mature, and await discharge.

Coitus

Sexual union between male and female (coitus, copulation, intercourse) amounts to insertion of the erect penis into the vagina, followed by ejaculation. Erection results from parasympathetic nerve stimuli causing the spaces in the spongy or erectile penile tissue (corpora cavernosa penis and corpus cavernosum urethrae) to fill with blood and the venous outlets simultaneously to contract. In this fashion the tissue is expanded by increased blood pressure. Ejaculation is the convulsive contraction of the epididymis, vasa deferentia, and seminal vesicles, propelling semen through the urethra. The amount of semen varies from 3 to 7 ml.

Male hormone

In order for the gonads, accessory structures, and secondary sex characteristics to develop, the male hormone testosterone must be present. The particulars about this key androgen were presented earlier (p. 383).

THE FEMALE

The female reproductive system consists of the ovaries, the uterine tubes, the uterus, the vagina, and the vulva (Fig. 24-4).

Ovaries

The ovaries, or female gonads, each about the size and shape of an almond, are located in a shallow depression on the lateral wall of the pelvis, one on each side, and are connected with the posterior surface of the broad ligament (Fig. 24-5). The infundibulopelvic ligament, a fold of

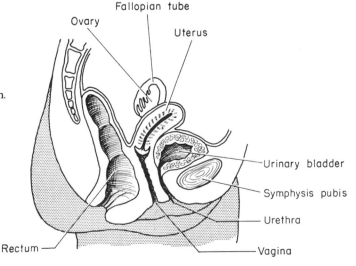

FIG. 24-4
General outline of female reproductive system.

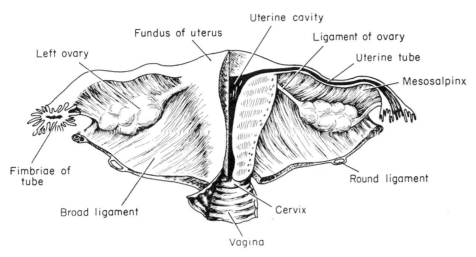

FIG. 24-5
Posterior view of uterus and allied structures. (Modified from Francis, C. C: Introduction to human anatomy, ed. 6, St. Louis, 1973, The C. V. Mosby Co.)

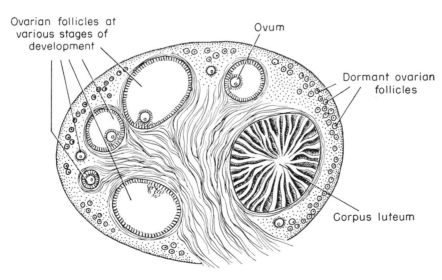

FIG. 24-6
Section through cat ovary. (After Schrön; modified from Tuttle, W. W., and Schottelius, B. A.: Textbook of physiology, ed. 15, St. Louis, 1965, The C. V. Mosby Co.)

peritoneum that passes from the pelvic wall to the ovary, carries blood vessels and nerves.

Histologically, the ovary is made up of thousands of ova-containing follicles embedded in connective tissue called stoma. Whereas the fetal ovary and ovaries of children consist almost entirely of immature follicles, in the sexually mature female some of these develop into the mature graafian follicle (Fig. 24-6). This does not occur until the time of puberty because the follicle-stimulating hormone (FSH) of the hypophysis is not available until then. Only the mature follicle can produce hormones and discharge its ovum.

Uterine tubes

The uterine, or fallopian, tubes are the two, 4-inch long, slender tubes that arise from the upper lateral angles of the uterus, each running to the ovary of the same side (Fig. 24-5). They are attached to the broad ligament by the mesosalpinx and enlarge into a funnel-shaped mouth called the infundibulum, the rim of which is formed into fringelike extensions called fimbriae. To propel along the ova passed into the infundibula, the mucosa of the tubes is equipped with cilia, and the walls, with smooth muscle.

Uterus

The uterus is the hollow, muscular, pear-shaped organ that houses the embryo and fetus. It is about 3 inches long, 2 inches wide, 1 inch thick, and has a broad, flattened body above and a narrow, cylindrical part called the cervix below. The rounded portion that passes above the openings of the fallopian tubes is referred to as the fundus. The organ is anchored to the pelvic walls, rectum, and bladder by the broad ligaments, the round ligaments, the uterosacral ligaments, the anterior ligaments, and the posterior ligament. The posterior ligament, a fold of peritoneum extending from the posterior surface of the uterus to the rectum, forms a

deep pouch between the uterus and rectum known as the rectouterine pouch or cul-de-sac of Douglas. As shown in Fig. 24-5, the uterine cavity opens into the vagina below through the mouth or external cervical os. The walls of the uterus are formed of smooth muscle (the myometrium), and its lining is made of a special mucous membrane, called the endometrium, that plays a key role in the reproductive process.

Vagina

The vagina, a curved, musculomembranous canal leading from the vulva to the cervix, receives the erect penis in copulation. It is lined with stratified squamous epithelium, and its walls are composed of an inner circular and an outer longitudinal layer of smooth muscle. In the virgin, the vaginal orifice is partly closed by a fold of mucous membrane called the hymen. As a consequence of intercourse and parturition, this structure is only fragmentary in the non-virgin. Often the condition of the hymen upon vaginal examination has medicolegal significance in cases of rape.

Vulva

The vulva, the external female genitals, consists of the mons pubis, the labia majora, the labia minora, the vestibule, and the clitoris (Fig. 24-7). The mons pubis, or mons veneris ("mount of Venus"), is a cushionlike, rounded prominence overlying the symphysis pubis. After the age of puberty, this area becomes covered with hair.

The labia majora (sing., labium majus) are the two large folds of skin and fatty tissue that extend backward and downward from the mons pubis to within about 1 inch of the anal opening. The skin of the labia majora contains hair follicles and sebaceous glands. Medial to and lying under the cover of these structures are the labia minora (sing., labium minus), two smaller folds of mucous membrane extending backward from

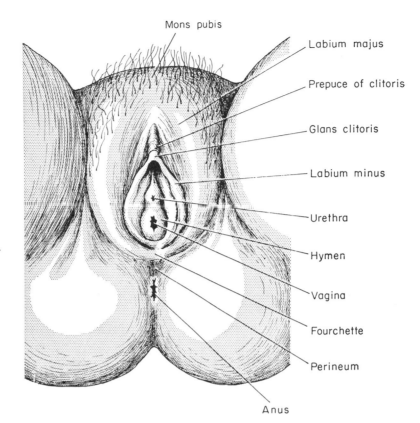

FIG. 24-7
Vulva. Labia majora shown parted.

Mons pubis

Labium majus

Prepuce of clitoris

Glans clitoris

Labium minus

Urethra

Hymen

Vagina

Fourchette

Perineum

Anus

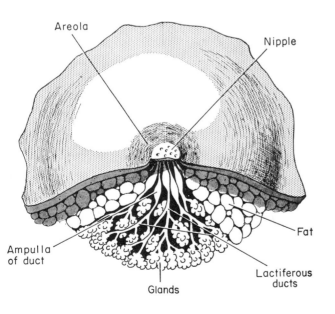

Areola

Nipple

Ampulla
of duct

Glands

Lactiferous
ducts

Fat

FIG. 24-8
Structure of the breast.

the clitoris. The labia minora do not contain hair follicles but have many glands and blood vessels. The cleft between them leads into the vestibule, the slight recess containing the vaginal and urethral orifices. Opening into the vestibule are two ducts from Bartholin's glands, which secrete a lubricating fluid, and two ducts from Skene's glands.

The clitoris, the small elongated body situated at the anterior angle of the vulva, corresponds to the penis in the male. It is composed of erectile tissue and becomes hard and erect upon sexual stimulation.

Mammary glands

The mammary glands, or breasts, are the milk-secreting organs in the female (Fig. 24-8). They are composed of glandular tissue organized into some twenty lobes, which in turn are organized into lobules. The lobes are partitioned by connective tissue, and the whole mass is embedded in a variable but large amount of fatty tissue.

At its center the breast is surmounted by the nipple, a small, dark, conical structure composed of erectile tissue. The lactiferous ducts, one from each lobe, meet here and open to the exterior upon its surface. About the nipple is a circular area of pigmented skin known as the areola.

Before the age of puberty the mammary glands are composed mostly of connective tissue, but with the onset of puberty the ducts and glandular tissue undergo rapid development. This hyperplastic activity stems from the activity of the female hormones. Estrogens stimulate development of the duct system, and progesterone, the alveoli, the basic milk-secreting units of the gland. This is especially true during pregnancy, when these hormones reach high levels.

Lactation. The secretion of milk by the breasts (lactation) is provoked chiefly by prolactin, the lactogenic hormone of the anterior pituitary. It is believed that the sudden drop in progesterone concentration at the end of pregnancy brings about the release of prolactin, since the former hormone is known to inhibit production of the latter. The act of sucking plays a role via oxytocin. This hormone does not increase the production of milk, but it does enhance its release by contracting the smooth muscle in the alveoli and milk ducts. Sucking is probably not the only stimulus, however. Music in the stable is said to facilitate the milking of cows.

Milk. For the first 6 months of life, human milk provides all the materials necessary for proper growth. Although human milk is noticeably low in iron, the newborn infant has absorbed enough of this mineral in utero to meet the needs of the first year or so. Like cow's milk, human milk is low in vitamin D, and for this reason the infant's diet should be supplemented. Human milk and cow's milk are very much alike except for the concentrations of protein and sugar (Table 24-1).

An often overlooked fact is that, aside from nutrition, milk also provides the offspring with immunologic protection; the glandular tissue in the breasts extracts antibodies as well as nutrients from the mother's blood. This is a form of passive immunity that will carry the infant through the first few months in his new microbe-infested world.

Menstrual cycle

The periodic discharge of blood from the vagina (menstruation) is only one phase of

TABLE 24-1

Human milk versus cow's milk (percent composition)

	Human	*Cow*
Water	88.5	87.2
Fat	3.3	3.5
Lactose	6.5	4.9
Protein	1.4	3.5
Electrolytes	0.3	0.9

a tremendously interesting "hormonal clock" whose mainspring is situated in the anterior pituitary.

Somewhere between the ages of 10 and 14 years, the anterior pituitary begins to secrete the follicle-stimulating hormone (FSH) and the luteinizing hormone (LH). The follicle-stimulating hormone causes a few of the immature follicles to grow and release estrogens (Fig. 24-9). A few days following the release of FSH, the pituitary starts putting out LH, which increases the rate of follicular growth and secretion even more. Finally, one of the follicles becomes so large that it ruptures, expelling its

ovum into the abdominal cavity. When this happens, the follicular cells, still under the influence of LH, increase in size, become fatty and yellow, and thereby become a structure called the corpus luteum (Fig. 24-6).

The corpus luteum soon secretes large quantities of both estrogens and progesterone. The former hormones cause the endometrium to grow in thickness (Fig. 24-9), and the latter enhances endometrial blood flow and nutrient secretion. The purpose of the body in doing this, of course, is to provide a suitable environment in which the fertilized ovum can grow.

If fertilization occurs, the developing ovum

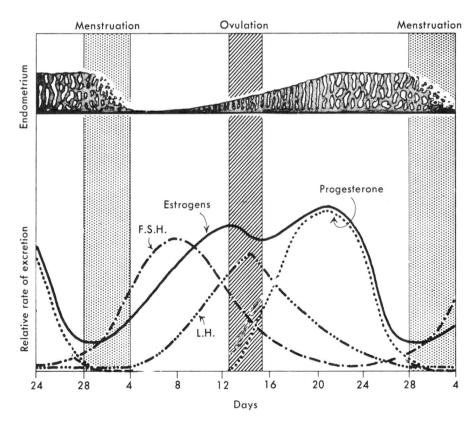

FIG. 24-9
Menstrual cycle in graphic form. *FSH,* Follicle-stimulating hormone; *LH,* luteinizing hormone.

itself becomes an endocrine gland and releases a gonadotropic hormone that sustains the corpus luteum throughout pregnancy. Otherwise, the corpus luteum undergoes involution, and the output of estrogens and progesterone is brought to a halt. As a consequence, the endometrium degenerates and sloughs off into the uterine cavity, the necrotic debris measuring about 50 ml and containing blood, serous exudate, and dead endometrial cells. Concomitantly, the low levels of estrogen and progesterone stimulate the output of FSH and thereby set the stage for a new cycle (negative feedback). High levels of estrogen and progesterone inhibit the ouput of FSH and thereby prevent ovulation; hence, their effectiveness in the contraceptive pill. Statistically, a cycle is 28 days, menstruation lasting 4 days, and ovulation occurring on the fourteenth day. By convention, the cycle begins and ends on the first day that the menses appears.

Period of fertility. Since the ovum can be fertilized by a sperm cell only during the 24-hour period following ovulation, and since the sperm cell usually can live in the vaginal canal for no more than 24 hours, it stands to reason that fertilization cannot ordinarily occur unless there is intercourse either shortly before, during, or shortly after ovulation. Because ovulation in most instances occurs on or about the fourteenth day of a 28-day cycle, the meeting of the ovum and sperm is statistically most likely to occur somewhere around the thirteenth, fourteenth, or fifteenth day. Although many women have cycles as short as 20 days or as long as 40 days, it appears that ovulation still follows a rather definite schedule—14 days before menstruation. For example, if a woman has regular cycles, let us say of 33 days' duration, ovulation will occur on about the nineteenth day ($33 - 14 = 19$).

In the event that man and wife desire to prevent conception, these facts and figures are useful and usually dependable in the presence of regular cycles. Obviously, when the cycles are erratic (for example, 28 days one time, 20 the next, 35 the next, and so on), ovulation cannot be predicted with any degree of accuracy. In the rhythm method of birth control the average of several successive cycles is computed and, as explained above, the number 14 is subtracted from this average to derive the day of ovulation. If abstinence is observed for the 5 days preceding and the 5 days following the calculated day of ovulation, fertilization will almost never occur. This method is not applicable to erratic menstrual cycles.

At this time there is no absolute method of birth control short of total abstinence and sterilization.

Menopause

In the average woman, at about the age of 45 years the follicles no longer secrete their hormones in full amounts, and the monthly cycles cease. This is termed the menopause. Aside from the not always pleasant fact that this period of one's life signals advancing years, the menopause is not infrequently accompanied by distressing signs and symptoms: nervousness, headache, "hot flashes," and the like. Since the syndrome is usually benefited by estrogenic therapy, there is good reason to believe that the sudden drop in hormone output is the underlying cause of the menopausal symptoms. Psychosomatic factors undoubtedly play a role, perhaps the major role in some instances.

Coitus

The sexual act is enhanced by the swelling of the erectile tissue about the vaginal opening and the secretion of large quantities of fluid and mucus by Bartholin's and Skene's glands. The tight but distensible and lubricated opening thus provided intensifies the stimulation resulting from the to-and-fro movement of the penis in the vaginal canal. In part the stimuli leading to the female climax stem from the movement of the penis against the clitoris. The climax itself is characterized by an exotic sensation and rhythmic, peristalsislike contraction of the uterus, fallopian tubes, and vaginal walls. These effects may serve to hasten along the ejaculated semen into the upper reaches of the fallopian tubes. Some say the act reaches its highest emotional development when the male and female experience the climax at the same time, but this is by no means a prerequisite for successful fertilization. At the proper time in the menstrual cycle, fertilization may be accomplished via artificial insemination.

FERTILIZATION AND GESTATION

The normal charge of semen deposited in the vagina generally contains from 150 to 300 million spermatozoa per milliliter. Traveling at a speed of about 1 foot per hour, most will arrive in the upper reaches of the fallopian tubes in a little over half an hour. In order for fertilization to occur, sperm must encounter the fresh ovum either here or in the abdominal cavity before the ovum enters the tube. An encounter elsewhere is generally of no avail, for by the time the ovum reaches the uterus, it has acquired an armor of mucus.

A sperm count below fifty million generally results in infertility, although many pregnancies have occurred with sperm counts of this magnitude. There is some evidence for believing that a multitude of spermatozoa are needed to secrete sufficient amounts of the enzyme hyaluronidase to remove the corona radiata, a relatively thick barrier of cells surrounding the ovum. The enzyme does this by digesting away the protein that links these cells together. Once the barrier is removed, a single spermatozoon enters the ovum to effect fertilization (Fig. 24-10). Shortly thereafter its head swells into the male pronucleus, the original nucleus of the ovum being called the female pronucleus. The two nuclei, each containing the haploid number of chromo-

somes, then fuse into the 46-chromosome zygote, the cell destined to be the new individual. In about 24 hours the zygote undergoes the first cleavage (division) and continues to do so every 12 to 15 hours thereafter. By the time (about 1 week) the fertilized structure reaches the uterus, the segmented mass, now called the morula, contains in the vicinity of 25 cells. It is still barely visible to the unaided eye.

As growth continues, the new cells arrange themselves in such a way that a cavity forms within the mass, with a cluster of cells (the inner cell mass) projecting into the cavity. This hollow-ball structure is called the blastocyst, and the cavity within, the blastocoele. The cells forming the outer layer of the blastocyst are referred to en masse as the trophoblast, or trophectoderm. The trophoblast secretes proteolytic enzymes that digest away a tiny bit of the endometrium, and its cells phagocytize the digested products. In this fashion the blastocyst implants itself into the uterine wall and in the process derives its sustenance.

By the end of 2 weeks the blastocyst is completely embedded within the endometrium, and the trophoblastic cells are rapidly growing and dividing. Soon they and the adjacent cells start to form the fetal membranes. The chorion, the outer membrane, sends out thousands of microscopic projections, called villi, that invade the surrounding mucosa and lay the groundwork for the placenta, the cakelike mass within the uterus that will establish communication between the mother and the child via the umbilical cord.

Embryonic disc

While the process just described is going on outside, drastic changes are taking place in the inner cell mass. Two cavities have appeared in the mass, and a new layer of cells (the mesoderm) has grown over the original lining of the blastocoele, passing between the two new cavi-

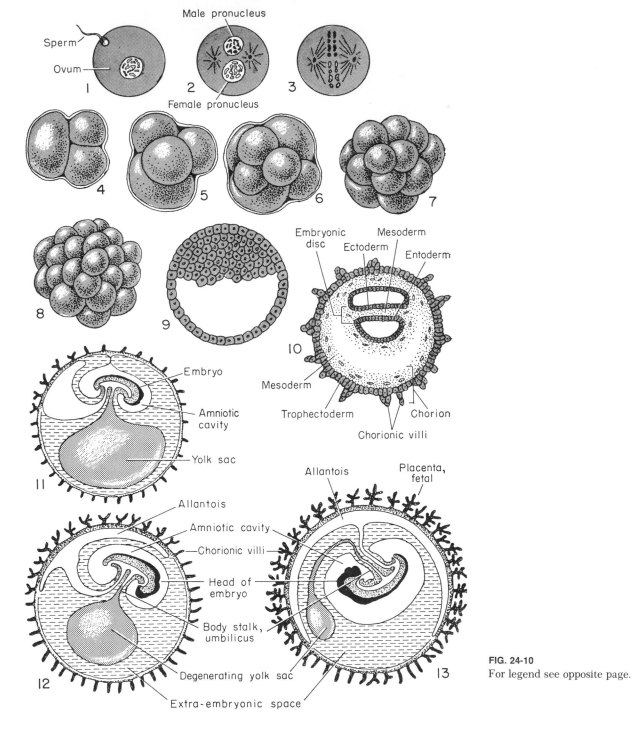

FIG. 24-10
For legend see opposite page.

ties. The cavity closest to the trophoblast, called the amniotic cavity, is destined to house the embryo, which has not yet appeared; the outer cavity, called the yolk sac, serves no purpose in man and will ultimately degenerate and disappear. The three-layered plate of cells running between the amniotic cavity and the yolk sac, aptly referred to as the embryonic disc, now becomes the crucial area of development, for this is where the embryo is formed.

The ectoderm, the outermost of the three primary germ layers of the disc, will evolve into the skin, the nervous system, the external sense organs, and the mucous membrane of the mouth and anus; the mesoderm, the middle layer, will evolve into the connective tissues, muscles, blood vessels, sex organs, and epithelium of the pleura, pericardium, peritoneum, and kidney; and the endoderm (or entoderm), the innermost layer, will evolve into the epithelium of the pharynx, respiratory tract, gastrointestinal tract bladder, and urethra. Some idea of how these transformations take place in the early stages can be obtained by a close study of the diagrammatic sections shown in Fig. 24-10.

Sex. In each human somatic cell, two of the 46 chromosomes are concerned with sex. When two X chromosomes appear together following gametic union, the result is a female individual; when an X and Y chromosome are together, the result is a male. It stands to reason, then, that all ova will carry an X chromosome, whereas half the sperm cells carry an X chromosome and the other half a Y chromosome. Thus, when an X sperm cell fertilizes an ovum, the offspring is a girl; conversely, when a Y sperm cell fertilizes an ovum, the offspring is a boy. Since it is the father who carries the odd chromosome, he should by no means blame the mother in the event the sex of the offspring is not to his liking.

Twinning. If two or more ova, instead of the customary one, are released and are fertilized simultaneously, fraternal twins result. Identical twins, on the other hand, result from a single fertilized ovum that has split one or more times into cell masses that develop into separate but identical offspring. In the instance of quintuplets there are four such divisions prior to implantation.

The placenta

The placenta makes intrauterine life feasible. As shown in Fig. 24-11, this structure is essentially a mass of blood sinuses formed by the placental septa. Into these sinuses extend chorionic projections from the fetal portion of the placenta, each covered with an enormous number of microscopic villi containing blood capillaries.

Maternal blood flows into and out of these sinuses by means of a well-channeled system of vessels derived from the uterine wall. Fetal blood is led into the villi via the two umbilical arteries and then led back via the umbilical vein. As the blood courses through the villi, nutrients are absorbed and waste products are excreted. This is largely effected through simple diffusion; since the concentration of oxygen and nutrients is greater on the maternal side of the placental barrier, these materials flow from the mother to the offspring. By the same token, fetal wastes (carbon dioxide, urea, and the like) diffuse from the villi into the maternal blood, whence they are removed by the kidneys.

FIG 24-10

Embryologic highlights of mammalian gestation from time of fertilization, *1,* to development of the fetal placenta, *13.* Stages *8* and *9* are called the morula and the blastocyst, respectively. (*11* to *13* adapted from Woodruff, L. L.: Animal biology, New York, 1961, The Macmillan Co.)

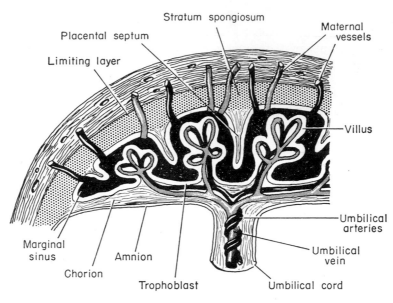

FIG. 24-11

Gross structure of the placenta. (After Gray; modified from Guyton, A. C.: Function of the human body, Philadelphia, 1959, W. B. Saunders Co.)

Placental hormones. Aside from serving as a food source and purifier, the placenta also secretes hormones, without which pregnancy cannot continue. Earlier it was pointed out that when fertilization occurs the corpus luteum, which normally degenerates at the end of each menstrual cycle, is maintained by gonadotropin secreted by the developing ovum, at first by the trophoblast and later by the chorionic membrane. However, after about the fourth month of pregnancy, the chorionic gonadotropin concentration drops to low levels, and the corpus luteum ceases to be stimulated sufficiently to produce the necessary high levels of estrogens and progesterone. At this time the placenta takes over the job and pushes the concentration of these hormones to well over 50 times their peak value during nonpregnancy.

Estrogens and progesterone are especially vital during pregnancy. In brief, the estrogens thicken the uterine musculature, greatly enhance the uterine blood supply, enlarge the breasts, and facilitate embryonic development; progesterone relaxes the uterine musculature until the time of birth, aids the development of the endometrium, prevents ovulation, and produces an alveolar arrangement of cells in the breast to make ready for milk production and secretion.

Fetal membranes

There are two membranes surrounding the fetus: the amnion and the chorion (Fig. 24-12).

The amnion, the thin, transparent, innermost membrane, encloses the amniotic cavity. This cavity is filled with a watery liquid, called amniotic fluid, that equalizes and cushions the pressures bearing upon the fetus.

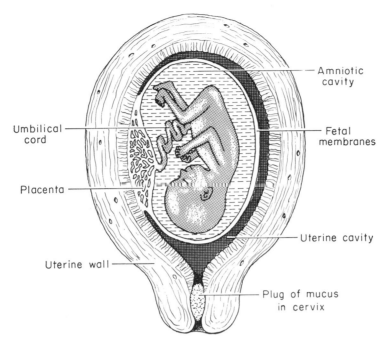

FIG. 24-12
Fetus in utero. (Modified from Guyton, A. C.: Function of the human body, Philadelphia, 1959, W. B. Saunders Co.)

The chorion, the thick outermost membrane, is actually composed of two layers, an outer ectoderm and an inner mesoderm. The portion of the chorionic surface that gives rise to the villi and forms the embryonic and fetal placenta is called the chorion frondosum. The remaining surface, called the chorion laeve, is membranous and smooth.

Fetal circulation

Because the fetus has no need for its liver or lungs, the blood flow through these organs is cut to a minimum via three shunts: the ductus venosus, the foramen ovale, and the ductus arteriosus (Fig. 24-13). The ductus venosus runs straight through the liver and carries blood from the umbilical vein and portal vein directly into the inferior vena cava, thereby shutting off most

of the blood supply to the liver. Upon reaching the heart, the blood bypasses the nonaerated lungs by flowing through the foramen ovale (the opening between the atria) and the ductus arteriosus. The former shunts part of the blood returning to the right atrium into the left atrium and thereby cuts down on the amount passing into the right ventricle and pulmonary artery. Considerable blood, however, does pass into the artery, and it is for this reason that a second shunt, the ductus arteriosus, directs most of the right ventricular blood directly into the aorta.

Other major features of the fetal circulatory system are the three vessels composing the umbilical cord: the umbilical vein and the two umbilical arteries. The vein carries fresh blood from the placenta to the fetus, and the arteries return blood from the fetus to the placenta. A substance

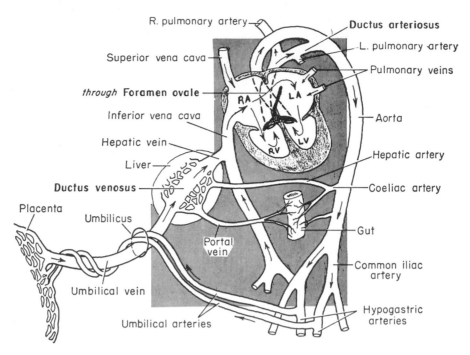

FIG. 24-13
Circulation of blood in fetus. *RA*, Right atrium; *RV*, right ventricle; *LA*, left atrium; *LV*, left ventricle.

called Wharton's jelly, a soft pulpy type of connective tissue, constitutes the protective matrix about these vessels as they pass through the cord.

Fetal growth

The fetus grows in an explosive fashion. Metabolizing prodigious amounts of nutrients supplied by the mother's blood, the fetus evolves from an average weight of 1 ounce at 3 months to an average weight of 7 pounds at the time of birth. The bulk of this terrific increase occurs during the last 3 months of pregnancy (Table 24-2).

Maternal physiology

Pregnancy is by no means confined to the anabolic happenings within the amniotic cavity.

The mother, too, undergoes tremendous change. Her metabolism is accelerated, her tissues retain excess fluid (about 3 quarts), and she gains an average of about 20 pounds (fetus, 7 pounds; uterus, 2 pounds; placenta and membranes, 2½ pounds; breasts, 2 pounds; fat and extra fluid, 6½ pounds). If all goes well throughout pregnancy and during delivery, these anatomical and physiological alterations reverse themselves in an amazingly short period of time. There is good reason to believe that a normal pregnancy strengthens the body.

Parturition

The duration of pregnancy averages 280 days from the beginning of the last menstrual period; about 90% of all births occur within a week before or after this figure. Although there is still no

TABLE 24-2
Fetal growth

Age (weeks)	Length (inches)	Weight	
8	1	$^1/_{10}$	ounce
12	4	1	ounce
20	10	12	ounces
28	14	2	pounds
36	18	5	pounds
40	20	7	pounds

unanimous agreement on the mechanism or mechanisms behind such precise timing, there is good reason to believe that the placental hormones play the principal role. The support of this view stems from the changes in their concentrations that occur a month or so prior to birth. The concentration of progesterone, which inhibits uterine motility, begins to decrease, whereas that of estrogens, which enhance motility, begins to increase. The myometrium becomes progressively more active as the fetus approaches the end of its intrauterine life. Also, the fetus grows larger and larger, and this in itself aids in stimulating uterine contraction.

Thus, uterine activity steadily increases during the last 2 months of pregnancy, finally terminating in intense rhythmic contractions a few hours before birth. The period from the onset of these terminal contractions until delivery is referred to as labor, or parturition.

The duration of labor usually runs between 8 and 18 hours. In the first stage the cervix is dilated and, generally, the amniotic membrane is torn, causing expulsion (the "show") of the amniotic fluid. During the second stage the powerful contractions push the baby through the birth canal and out into the world. The third stage is an anticlimax—the expulsion of the afterbirth (fetal membranes and placenta). As soon as the baby is born, that is, at the end of the second stage, the umbilical cord is tied and cut.

The pulling away of the placenta from the uterine wall leaves a wide area of bare, oozing vessels and accounts for the considerable bleeding that accompanies parturition. Though the loss of a pint or so of blood (the usual amount) is well tolerated and is actually provided for by the mother's enhanced supply, larger losses are obviously dangerous. This is why the physician often administers oxytocic drugs. Such agents cause the uterus to "clamp shut," thereby closing dilated blood vessels and facilitating clotting.

Pregnancy tests

Chorionic gonadotropin, the hormone produced by the trophoblast shortly after implantation, appears in the urine and may be detected by its action on the mammalian ovary and on the gonads of various amphibia. The original test, the Aschheim-Zondek (A-Z) test, depends upon development of hemorrhagic follicles in the ovaries of immature female white mice. Six injections of urine must be made over a period of 2 days. In the Friedman test the mature female rabbit is used.

The simplest and most accurate biological pregnancy test utilizes the South African toad (*Xenopus laevis*). One milliliter of concentrated urine is injected into the dorsal lymph sac, and if a sufficient concentration of chorionic gonadotropin is present, ovulation occurs and myriads of eggs are extruded in 8 to 16 hours. The toad test is 96% to 100% accurate.

Since the South African toad is expensive and occasionally difficult to procure, many laboratories now use the American male frog *Rana pipiens* as the test animal. In this version of the test, 5 ml of urine is injected into the dorsal or lateral lymph sac. The presence of spermatozoa in the urine of the frog after 2 to 4 hours is interpreted to indicate a positive test. This test has proved about 95% to 96% accurate and is entirely satisfactory in most laboratories.

Recently a number of in vitro tests have been developed and are available commercially as

Gravindex, Pregnosticon, and UCG Test, to name a few. While each differs in procedure, all are based on the use of antihuman chorionic gonadotropin serum derived from rabbits that have been caused to produce antibodies against human chorionic gonadotropins.

Complications of pregnancy

Pregnancy not uncommonly assumes pathological proportions—nausea and vomiting, abortion, eclampsia, and the like. While some degree of nausea and vomiting is to be expected and usually proves anything but serious, pernicious vomiting (hyperemesis gravidarum) is a real problem and often demands hospitalization. In some cases of severe vomiting, liver damage and hemorrhagic retinitis may occur. Treatment entails the administration of intravenous fluid to remedy dehydration and supply calories and the use of such antiemetic drugs as pyridoxine and prochlorperazine (Compazine).

Abortion, or the interruption of pregnancy prior to the period of fetal viability, may be spontaneous or induced. An induced (or therapeutic) abortion is done for reasons of health or because of the likelihood of a defective baby. Spontaneous abortion may be due to embryonal abnormalities, acute infectious diseases, uterine abnormalities, severe injury, or dysfunction of the thyroid, ovary, or pituitary gland.

Eclampsia is a disorder of unknown etiology, usually occurring during the first pregnancy or with twins, and is characterized by hypertension, edema, convulsions, and albuminuria. Usually it may be prevented by adequate prenatal care, with careful attention to diet and salt restriction. When the full-blown disease process occurs, it may cause death of the fetus, of the mother, or of both. Treatment includes measures to lower blood pressure, to increase blood flow and oxygenation of vital tissues, and to secure an adequate urinary output. Eclampsia is cured only by delivery.

TERATOLOGY

Teratology is the division of embryology and pathology that deals with abnormal development and congenital malformations. The various possible etiological factors fall into three categories: environmental, genetic, and multifactorial (environmental plus genetic). There are also abnormalities that cannot be traced to any environmental or genetic factor, but instead appear to be simply the result of a certain small fraction of embryos failing to develop properly.

Currently much research concerns teratogens, the specific environmental agents causing congenital malformations. Known teratogens against the human embryo include irradiation, thalidomide, carbon monoxide, rubella virus, and the protozoan *Toxoplasma gondii*.

Inborn errors of metabolism were discussed earlier (p. 276), and chromosomal diseases are discussed in Chapter 25.

DISEASES OF THE MALE

Disorders and diseases peculiar to the male reproductive system are many and various. They mainly relate to infection, congenital anomalies, trauma, impotence, infertility, and tumors. Common problems of major concern are undescended testes (cryptorchidism), enlargement of the prostate (benign prostatic hypertrophy), and cancer of the prostate. In 1972 cancer of the prostate struck some 36,000 and killed an estimated 18,000, making it the third most common male cancer. Radical prostatectomy, the removal of the prostate and associated lymph nodes, affords the only cure, but is effective only in those cases where the cancer has not spread beyond the gland. All other forms of therapy are palliative, providing the metastatic victim with a few more months or years of life.

DISEASES OF THE FEMALE

The principal diseases of the female reproductive system include infectious and noninfectious disorders involving the vagina, uterus, ovaries, and breasts. An average of 24% of 30-year-old women will develop some kind of cancer during their lives. About 25% of these victims will have breast cancer, the most common female cancer, and about 17% will have cancer of the uterus, the second most common female cancer. In 1972 the estimated number of new cases of breast cancer was 71,000, and the estimated number of deaths from this disease was 32,000; the estimated number of new cases of uterine cancer stood at 43,000, and the death toll, at 13,000. As in all cancer, a favorable prognosis relates to early diagnosis and appropriate surgery.

SYPHILIS

The incidence (number of new cases occurring per year) of syphilis rose to an all time high of 106,539 in 1947 and then, thanks to penicillin, dropped to an all time low of 6,500 in 1955. It is now up again to about 20,000. The prevalence (total number of new and old cases) continues to decline and is estimated currently to be about 500,000. The number of cases of congenital syphilis continues to decline also. In 1941 the official figure stood at 17,600, and in 1970 it was 1,903.

Though much less common than gonorrhea, syphilis is much more dangerous. Untreated, it results in the degradation of mind and body and, ultimately, in death. The causative organism is the spirochete *Treponema pallidum* (Fig. 24-14). In acquired syphilis treponemes typically are passed along via the sexual act, but any form of intimate body contact suffices if it involves the transfer of liquid infectious material. Kissing and abnormal sex practices are recognized modes of transfer, and initial syphilitic lesions upon occasion involve the lips, tongue, tonsils, eyelids, breasts, or fingers. Congenital syphilis results when the spirochete is passed from mother to fetus.

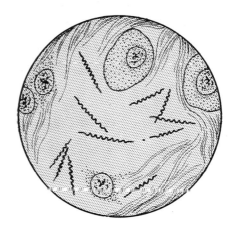

FIG. 24-14
Treponema pallidum in tissue.

Acquired syphilis

Following sexual intercourse with an infected person, the chancre, the hallmark of primary syphilis (Fig. 24-15), appears in about 3 weeks at the portal of entry; the usual sites are the penis, scrotum, vulva, vagina, and cervix. Usually single rather than multiple, the chancre has a firm base and a raised border; its size varies from that of a pinhead to that of the end of the thumb. The surface appears eroded, and gentle pressure calls forth a watery discharge rather than pus. Another characteristic feature of the primary stage is the development of swollen regional lymph nodes (buboes).

In a month or so the chancre heals, with or without treatment, leaving a pale scar; the buboes, if they occur, may or may not persist. The passing of the chancre heralds the end of the primary stage and the beginning of the secondary incubation period, during which time the treponemes are swimming about in increasing numbers and setting up foci of infection throughout the body.

This goes on for weeks or months (about 6 weeks on the average) until one day the patient

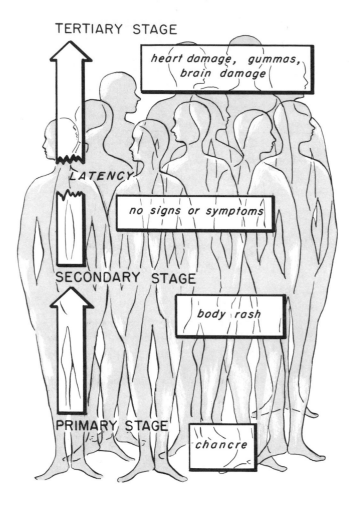

TERTIARY STAGE

heart damage, gummas, brain damage

LATENCY

no signs or symptoms

SECONDARY STAGE

body rash

PRIMARY STAGE

chancre

FIG. 24-15
Stages of syphilis. (From Brooks, S. M.:
The V. D. story, New York, 1973,
A. S. Barnes & Co., Inc.)

arrives at the secondary stage, the characteristic features of which are a generalized skin eruption and mucous patches. The rash is highly varied and may simulate almost any skin lesion—measles, for instance. The mucous patches are circular, multiple areas of erosion on the membranes of the mouth, throat, genitalia, and rectum. They teem with treponemes. The rash, patches, and other signs and symptoms last from 6 weeks to 6 months and then disappear, with or without treatment.

The untreated patient then enters the asymptomatic latent period, which lasts from a year to a lifetime, depending on the outcome of the continuing battle between the treponemes and the forces of immunity.

Some one third of afflicted individuals pass into the tertiary, or late, stage of the disease, the cardinal lesion of which is an inflammatory scarring and weakening of the aorta. Commonly, the semilunar valve is damaged, resulting in regurgitation and heart failure. Other cardio-

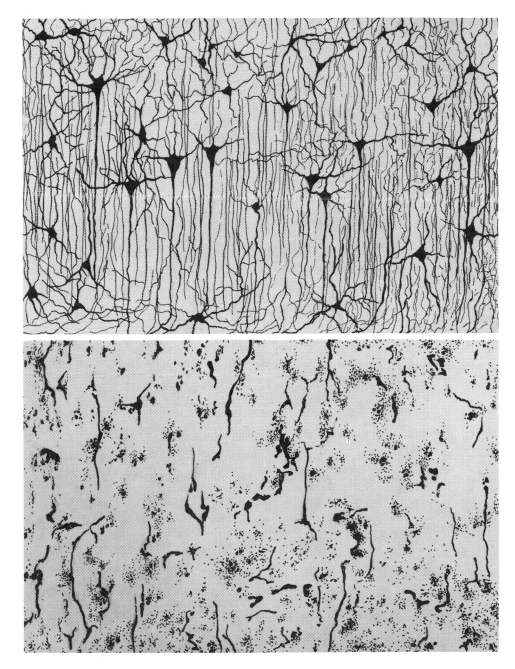

FIG. 24-16
Normal cerebral cortex (above) and cortex of neurosyphilis (below). (From Brooks, S. M.: The V. D. story, New York, 1973, A. S. Barnes & Co., Inc.)

vascular possibilities include an outpouching (aneurysm) of the aortic arch and a direct treponemal attack against the muscle proper.

Some patients develop neurosyphilis, the most severe form of which is a general paresis arising from a treponemal invasion of the brain. Histologically, there is considerable dishevelment of the cerebral cortex and widespread loss of nerve cells (Fig. 24-16). The clinical features are changes in the intellect, memory, mood, and behavior; terminally, there is a dementing psychosis.

About 10% of the patients in the third stage develop soft rubbery tumors called gummas. The location, number, and effects of these lesions are extremely varied and, along with the cardio-vascular and nervous ramifications, make syphilis the great imitator that it is.

Congenital syphilis

The fetus "acquires" syphilis sometime after the fifth month of pregnancy. In early syphilis, pregnancy always results in miscarriage, stillbirth, or diseased babies. Quite to the contrary, women who become pregnant many years after infection often give birth to normal babies.

Babies born with congenital syphilis are not likely to show evidence of the disease for 3 or 4 weeks, at which time some sort of skin eruption almost always appears. Other distinguishing marks in the early days are cracking of the lips, enlarged spleen, snuffles, and peculiar cry. In

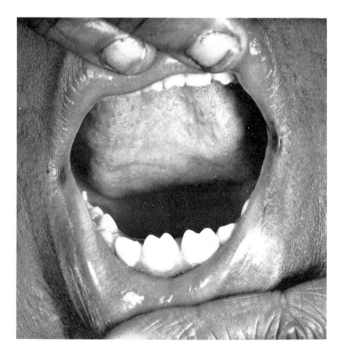

FIG. 24-17
Congenital syphilis. Hutchinson's teeth. (From Top, F. H., and Wehrle, P. F.: Communicable and infectious diseases, ed. 7, St. Louis, 1972, The C. V. Mosby Co.)

overall appearance, the syphilitic infant is puny, withered, and shriveled and has the face of a little old man.

Juvenile lesions appear at an average age of 10 years. The classic feature here is Hutchinson's triad: inflammation of the cornea, often resulting in blindness; deafness, due to auditory nerve damage; and notching of the upper incisors (Fig. 24-17). Additionally, there are considerable and widespread skeletal changes and, not uncommonly, neurosyphilis

Diagnosis and treatment

To establish whether a chancre is really a chancre or a mucous patch is really a mucous patch, all one need do is examine a drop or two of exudate for the presence of treponemes. Unfortunately, the diagnosis of syphilis is seldom this straightforward because well over three quarters of syphilitic patients seen by doctors are in latency and thus are asymptomatic. In this situation diagnosis must be made on a basis of telltale antibodies and serological tests. By

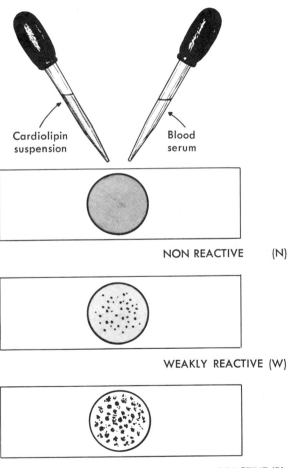

FIG. 24-18
VDRL test for syphilis. Presence of antibody in syphilitic serum causes extremely tiny suspended particles of cardiolipin to clump together into larger particles (flocculi). Degree of flocculation depends on concentration of antibody. Because antibody (reagin) is associated with certain other infections and conditions, the VDRL serves only as a diagnostic screening test, and not as a confirmatory test. (From Brooks, S. M.: The V. D. story, New York, 1971, A. S. Barnes & Co., Inc.)

far the most useful screening test is the VDRL (Fig. 24-18); the most accurate confirmatory test is the FTA-ABS.

The management of syphilis centers upon the use of penicillin to kill the treponemes and follow-up serology to monitor the response. Generally speaking, cures are effected in about 90% of early syphilitics receiving one course of treatment and in a substantial proportion of those who must be retreated. The prognosis in late syphilis is from good to excellent when the cardiovascular and nervous systems are not involved. In asymptomatic neurosyphilis, where the only sign is a "positive" cerebrospinal fluid, immediate therapy can prevent the development of the various symptomatic forms. Congenital syphilis responds dramatically to penicillin, and complete cures are the rule when the infection is caught early. Afflicted youngsters under 2 years old usually are cured in 1 year, and even in late congenital syphilis (12 years or older) the prognosis is good provided there has been no real damage before the commencement of treatment.

GONORRHEA

Gonorrhea is the most common venereal disease of universal stature; its prevalence in the United States is presently estimated to be at least 2 million cases! The cause is *Neisseria gonorrhoeae,* a gram-negative diplococcus (Fig. 24-19) that cultures only on special media in an atmosphere of 10% carbon dioxide. The infected partner infects the noninfected partner by introducing the gonococci into the fluid products of the sexual act. The gonococci are massaged and washed into the urethra in a most efficient manner, and in a matter of days infection is established. The usual incubation period runs between 3 and 9 days.

In the typical male case the first signs are burning on urination and the appearance of a pussy discharge. The urethral orifice is usually

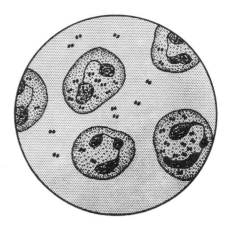

FIG. 24-19
Neisseria gonorrhoeae in pus. Note that many organisms have been ingested by white cells.

inflamed and puffy. The infection works its way upward and backward through the urethra and shortly incites real trouble in the prostate gland and the seminal vesicles, the former (prostatitis) squeezing the urethra and causing urinary retention and the latter causing fever and pain. Further advancement of the infection into the seminal duct leads to inflammation of the epididymis and testis. Complications include sterility and inflammation of the joints and heart valves.

In the typical female case the early stages usually do not appear; on good authority, perhaps nine out of ten victims may be completely without initial signs and symptoms. The female becomes a healthy carrier par excellence. When the early phase is in evidence, however, it begins with painful urination and vaginal discharge. The cervix is involved almost immediately, and from this location the infection spreads to the fallopian tubes. In untreated cases the tubes get larger and larger and fill with pus; the entire pelvic area eventually gives way to fibrosis, abscesses, and adhesions. Systemic complications are the same as in the male: involvement of the joints and heart valves.

Diagnosis centers on the isolation of *Neisseria gonorrhoeae*, and treatment amounts to the administration of large doses of penicillin or other antibiotics. The response is dramatic; "single shot" cures are almost the rule. Untreated gonorrhea is communicable for months and often, especially in women, for years. For the sake of precious time, all contacts should be considered infected and given penicillin or tetracycline at the earliest hour. Expectant mothers known to have gonorrhea should be given appropriate therapy during pregnancy, and if this is not possible, the infant should be given penicillin immediately after delivery.

The conjunctiva of the eye affords rich soil for *Neisseria gonorrhoeae*. If unchecked, gonorrheal conjunctivitis can easily result in blindness. The adult patient may infect his own eyes or someone else's via contaminated fingers; newborn babies pick up the eye involvement (ophthalmia neonatorum) as a consequence of having passed through a birth canal laden with gonococci. Instilling 2 drops of 1% silver nitrate solution into each eye immediately after birth has reduced tremendously the incidence of blindness traceable to this cause.

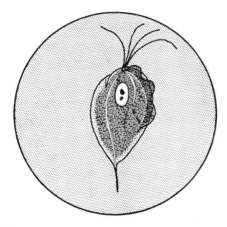

FIG. 24-20
Trichomonas vaginalis.

TRICHOMONAS VAGINITIS

The pear-shaped protozoan *Trichomonas vaginalis* (Fig. 24-20) is believed to inhabit the vagina or the male urethra of a quarter of the population. It never seems to cause trouble in the male, but in the female it is responsible for a severe vaginitis. Untreated, the vagina and vulva itch beyond belief, and a profuse, foamy, foul-smelling, yellowish discharge makes its way to the surface. Moreover, once the infection is established, sexual intercourse keeps it going. In the view of many authorities, "man and wife" trichomoniasis is our most common venereal disease. The diagnosis is made by demonstrating the protozoan in the vaginal discharge, and treatment centers on the use of metronidazole (Flagyl). This drug is just about 100% effective.

QUESTIONS

1. Naming all the major structures in order, trace the path of a spermatozoon from its place of origin to the exterior.
2. What is the literal meaning of epididymis?
3. What is the function of the seminal vesicles and Cowper's glands?
4. Distinguish between "corpus cavernosum penis" and "corpora cavernosa penis."
5. Distinguish between "corpus cavernosum penis" and "corpus cavernosum urethrae."
6. What is the longest part of the male urethra?
7. In regard to the penis, what are the crura?
8. Discuss the composition of semen.
9. Erection of the penis is essentially a matter of blood pressure. Explain.
10. What are androgens?
11. What is a follicle?
12. Why, specifically, are prepuberty ovarian follicles immature?
13. The size of the uterus can be memorized easily as "1-2-3." Explain.
14. Who was Fallopio?
15. What is the literal meaning of infundibulum?
16. Locate the cul-de-sac of Douglas.
17. Distinguish between myometrium and endometrium.

18. The uterus is anchored to the pelvic walls and the rectum by eight ligaments. Which are they?
19. What is the purpose of the muscle in the walls of the vagina?
20. Distinguish between labia minora and labium minus.
21. What is the function of the clitoris?
22. What is the relationship of the labia majora to the mons pubis?
23. Discuss the role of estrogens, progesterone, prolactin, and oxytocin in lactation.
24. What does lactiferous mean?
25. What is "progestin"?
26. What is "estrogen"?
27. At about the twenty-second day of the menstrual cycle, estrogens and progesterone are being excreted at a high rate, and FSH and LH, at a low rate. Explain.
28. Discuss the composition and mechanism of action of the contraceptive pill.
29. A dozen or more methods of contraception are in use in addition to the rhythm method and the contraceptive pill. Discuss some of the methods in detail.
30. If, hypothetically, ovulation occurs on the fifteenth day of a cycle, can fertilization occur as a consequence of intercourse on the fourteenth day? on the eighteenth day? Discuss.
31. Distinguish between menses and menstruation.
32. Compare mitosis and meiosis.
33. Compare spermatogenesis and oogenesis.
34. Unless the ovum encounters a sperm cell in the upper part of the uterine tube, fertilization does not occur. Can you suggest a reason why?
35. Hyaluronic acid is sometimes spoken of as the "intercellular cement." Why?
36. Distinguish between the zona pellucida and the corona radiata.
37. Who determines the sex of the offspring, the mother or father?
38. Precisely, why are identical twins identical?
39. In a sextuplet birth, two might be fraternal and four identical. How do you explain this?
40. If the eye cup of one amphibian embryo is transplanted to an abnormal location in another amphibian, say the belly ectoderm, the cells of the belly ectoderm overlying the eye cup develop into a lens, and eventually a structurally perfect eye develops. What does this experiment tell us about development and differentiation?
41. Describe the events leading from the zygote to the blastocyst.
42. If one cell of a two-cell frog embryo is destroyed, the other develops into one half of an embryo. Would you say this indicates chemical or mechanical influence?
43. What does the amniotic cavity refer to?
44. Describe the placenta.
45. There is no mixing of blood between fetus and mother. Why not?
46. Oxygen diffuses into fetal villi, and wastes diffuse into maternal blood sinuses. Why not the other way around?
47. A patent ductus arteriosus and failure of the foramen ovale to close over both result in a "blue baby." Explain.
48. Habitual abortion may be due to a hormonal deficiency. Which hormone comes to mind first, and why?
49. Why are the placenta and fetal membranes referred to as afterbirth?
50. What does the doctor do to control postpartum hemorrhage?
51. Is there any way to safely stimulate labor?
52. What is Wharton's jelly?
53. Discuss the origin of the word teratology.
54. What was the "thalidomide disaster"?
55. What is toxoplasmosis?
56. Discuss the risks and precautions in the use of the rubella vaccine.

57. Distinguish between hereditary and congenital.
58. What is an inborn error of metabolism?
59. Prostatic hypertrophy leads to urinary retention and infection. Discuss fully.
60. Distinguish between impotence and infertility.
61. What is the origin of the word cryptorchidism?
62. Cite a congenital anomaly relative to the male reproductive system.
63. What is the purpose of removing lymph nodes in a prostatectomy?
64. Describe the Pap test.
65. What is cervical cancer?
66. What is a radical mastectomy?
67. Distinguish between incidence and prevalence of a disease.
68. Distinguish between acquired syphilis and congenital syphilis.
69. In syphilis, what is meant by the primary incubation period?
70. Why is the interval between secondary and tertiary syphilis called latent?
71. In regard to the VDRL test for syphilis, what is meant by a false-positive?
72. What is the FTA-ABS test for syphilis?
73. *Neisseria gonorrhoeae* is a gram-negative diplococcus. What does this mean?
74. The organisms responsible for syphilis and gonorrhea are extremely fragile and easily destroyed. How, then, do you account for the high incidence of these infections?
75. The fact that penicillin is so effective in gonorrhea and syphilis is probably one reason why these infections are once again a problem. How do you reconcile this paradox?
76. In the treatment of trichomonas vaginitis, metronidazole must be given to both husband and wife. Why?

CHAPTER 25

BASIC GENETICS

When red-eyed fruit flies of pure stock (that is, flies that produce only red-eyed offspring) are mated with brown-eyed fruit flies of pure stock, all the offspring are red-eyed. If these red-eyed hybrids are mated with one another, about three quarters of the offspring are red-eyed and the rest brown-eyed. And if these brown-eyed flies are mated among themselves, only brown-eyed offspring are produced.

In a nutshell this is the essence of mendelian genetics, a concept that stands next to Darwin's theory of evolution in biological significance. Prior to the time of Gregor Mendel, and for almost half a century after his historic paper in 1866, it was thought that heredity was a blending process, much like the mixing of different kinds of paint. Using garden peas, Mendel showed that heredity is of a particulate nature. Today we call these particles genes, and their mode of operation, the genetic code (Chapter 6).

GENES AND CHROMOSOMES

Chromosomes are composed of hundreds and hundreds of genes, and each gene carries a genetic or hereditary message. Such messages come in pairs, as do the chromosomes that contain them. Referring to the fruit fly, when a sex cell carrying a red gene unites with a sex cell carrying a brown gene, the resulting zygote gives rise to an organism carrying a red gene and a brown gene in every one of its body cells. Only in the eyes, however, is the influence of these genes manifest. Moreover, only the red gene has a noticeable effect; hence, the red gene is said to be *dominant* and the brown gene *recessive*. Brown eyes, a recessive trait, occurs only in the absence of the dominant gene; both recessive genes must be present to produce the recessive trait.

Sometimes a gene is only partially dominant. For example, in some flowers a cross between a red and a white produces pink hybrids. In this particular instance a hybrid can be distingushed just by inspection, whereas in fruit flies, hybrid red eyes are just as red as those of the pure stock. The inheritance of eye color in fruit flies is an example of complete dominance.

Alleles

Two genes that alter a characteristic in contrasting ways are said to be alleles of each other. In the fruit fly the red gene is an allele of the brown gene, and vice versa. Alleles occupy the same position, or locus, on corresponding or homologous chromosomes. Whereas body cells carry alleles in pairs, sex cells carry only one of the partner alleles. For example, in the fruit fly the body cells carry two eye color genes: two red, two brown, or one brown and one red. The sex cell carries either a red or a brown gene.

Genotype versus phenotype

The particular set of genes present in the cells of an organism is called its genotype, and the actual physical appearance of an organism is called its phenotype. The phenotype is an expression of the genotype. Individuals with the same phenotype do not necessarily have the same genotype. For example, a red-eyed fruit fly may have either two red genes per cell or one red gene and its brown allele. On the other hand, all brown-eyed fruit flies have the same genotype relative to this trait.

Homozygous versus heterozygous

More or less hand in hand with the expressions genotype and phenotype go the expressions homozygous and heterozygous. Homozygous refers to an organism possessing a pair of alleles for a given trait that are of the same kind; a heterozygous organism possesses a pair of unlike alleles for a given trait. Thus, red-eyed fruit flies may be either homozygous or heterozygous, whereas brown-eyed fruit flies are always homozygous.

Multiple alleles

A given locus on a chromosome may be occupied by more than two kinds of alleles. For example, the four major blood groups—A, B, AB, and O—relate to three kinds of alleles (A, B, and O). In any given individual, however, only one pair of alleles is involved. Type A blood stems from the genotype AA or AO (O is recessive); type B blood stems from genotype BB or BO (O is recessive here too); type O blood stems from genotype OO; and type AB blood stems from genotype AB. Note that while the A and B alleles are both dominant to the O allele, they display no dominant relationship to each other. This

situation is also known to obtain among alleles associated with other characteristics.

LINKAGE AND CROSSING OVER

Mendel noted that certain traits tend to be inherited together, and today we know why. In brief, these linked traits arise from a linear series of genes linked together on the same chromosome; as the chromosome is passed on and on in the reproductive process, so too are its constituent linked genes. Sometimes, however, such genes become unlinked, due to what is called crossing over. This occurs during gametogenesis when paired chromosomes twist around each other and break, the broken pieces then fusing back together the wrong way; crossing over amounts to a mutual exchange of genes between homologous chromosomes. For example, if a chromosome containing genes A, B, C, and D, in that sequence, crosses over with its partner chromosome, with genes a, b, c, and d, the result might be the new chromosomes ABcd and abCD.

DEFECTIVE GENES

Some genes carry the wrong chemical information and result in structural or metabolic defects. Those that kill the organism in utero are known as lethal genes. Some lethal genes are dominant and some are recessive. So-called dominant recessive lethals are recessive lethal genes that produce noticeable deleterious effects even in the heterozygous condition. A true recessive lethal produces no effects whatsoever in the heterozygous condition. Diseases and disorders stemming from defective genes include, among many others, brachydactyly, Huntington's chorea, Tay-Sachs disease, hemophilia, red-green color blindness, and inborn errors of metabolism (p. 276).

SEX (AUTOSOMES AND HETEROSOMES)

As pointed out earlier, the determination of sex involves the sex, or X and Y, chromosomes.

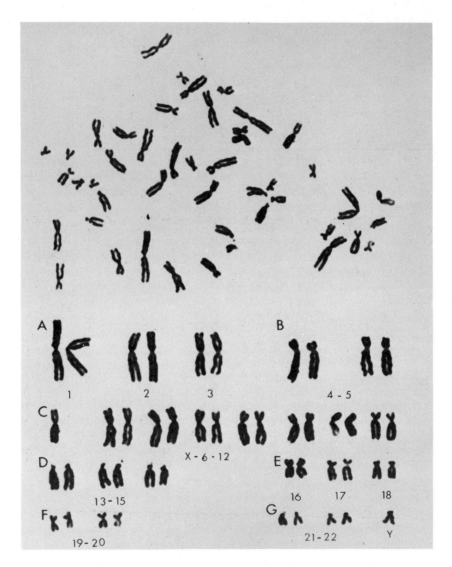

FIG. 25-1

Karyotype of normal human chromosomes. Note arrangement according to size and shape into seven groups of homologous pairs (*A* through *G*). Note, too, that X and Y chromosomes belong to different groups—*C* and *G,* respectively. (From Rowley, J. D.: J.A.M.A. **207:**914, 1969.)

These two chromosomes are formally known as heterosomes; all the other chromosomes are known as autosomes. Each human somatic, or body, cell contains 22 pairs of autosomes and one pair of heterosomes, for a total of 46 (the *diploid* number) chromosomes. A formal arrangement of these chromosomes, called a karyotype, is shown in Fig. 25-1. Each sex cell contains 23 (the *haploid* number) chromosomes. The X chromosome carries "female genes," and the Y chromosome carries "male genes"; two X's produce a female, an X and Y produce a male. Apparently, then, the Y chromosome is a more potent genetic force than the X chromosome inasmuch as it takes only the single chromosome to induce maleness. When there is doubt about an individual's sex, a simple test is now available to establish genetic sex based on the presence or absence of so-called Barr bodies in the cells of buccal mucosal smears or other suitable specimens (Fig. 25-2).

In the process of gametogenesis the diploid number of chromosomes is reduced by meiosis to the haploid number. Each human ovum carries 22 autosomes and one X heterosome; each human sperm carries 22 autosomes and either one X or one Y heterosome. On the average, therefore, half the sperm cells carry the X heterosome and half carry the Y heterosome, which means in practice that the chances for a baby being a boy or being a girl are equal.

CHROMOSOMAL DISORDERS

Sometimes during gametogenesis things do not happen the way they should, and the sex cells end up with too few or too many chromosomes. Down's syndrome (mongolism) results when a normal sperm cell fertilizes an ovum carrying 24 chromosomes. The body cells of the ensuing individual carry a triple dose of one of the kinds of autosomes, a condition called *trisomy* (as contrasted to the normal condition of the double dose, or *diploidy*). Chromosomal disorders involving heterosomes include the metafemale (XXX), Turner's syndrome (XO; that is, only one heterosome), and Klinefelter's syndrome (XXY). In conditions such as Turner's syndrome, where there is a single dose of one of the kinds of chromosomes in the body cells, the term *monosomy* is used.

SEX-LINKED TRAITS

The Y heterosome in man apparently carries no genes other than those relating to maleness; the X heterosome, however, carries an array of genes other than those relating to femaleness. Understandably, then, certain traits are sex linked, a phenomenon underscored in red-green color blindness (the inability to distinguish between red and green). The gene involved here (let us call it "c") is recessive, meaning that it does not cause color blindness in association with the normal gene ("C"). This means that although a Cc female does not exhibit color blindness, she nonetheless is a *carrier*. The female is color blind only when two c's occur together. In the male, however, it takes but one c to cause color blindness because the Y heterosome carries no overriding allele.

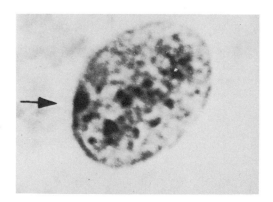

FIG. 25-2

Dark body in nucleus (indicated by arrow) is the now-famous Barr body. Since it represents two X chromosomes, its presence is proof of femaleness. (From Rowley, J. D.: J.A.M.A. **207**:914, 1969.)

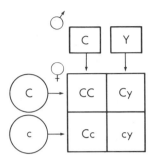

FIG. 25-3

Possibilities of color blindness in children of normal father, CY, and carrier mother, *Cc*.

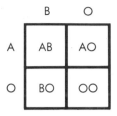

FIG. 25-4

Possible blood types in children of parents heterozygous for blood types A (AO) and B (BO).

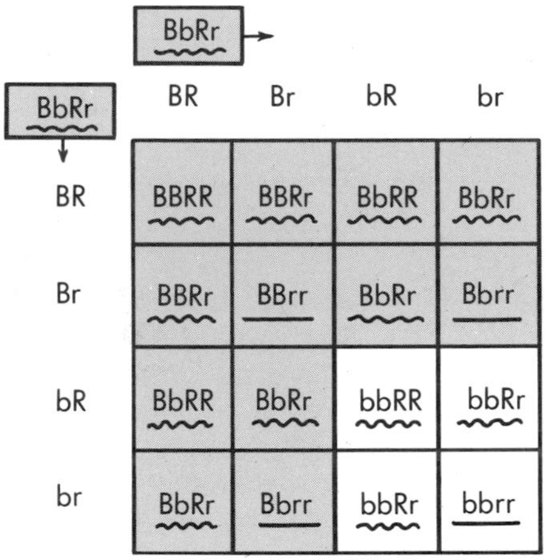

By way of example, let us suppose a normal man (CY)—that is, one who has the C on the X chromosome—marries a carrier (Cc). Half the sperm would carry C and half (the "Y sperm") would carry no genes relating to the trait. In contrast, half the ova would carry C and half would carry the c. As shown in Fig. 25-3, there are four possible offspring combinations: normal female (CC), carrier female (Cc), normal male (CY), and color-blind male (cY). Among the boys resulting from such a marriage there would be a fifty-fifty chance of color blindness; among the girls there would be no color blindness, but there would be a fifty-fifty chance of the carrier state.

Exactly the same situation obtains in the sex-linked disease of hemophilia, where the abnormal gene (call it "h") is recessive to the normal gene ("H"). Thus, HY and HH stand for the normal male and female, respectively; hY and hh stand for the male and female hemophiliac, respectively; and Hh stands for the female carrier.

THE PUNNETT SQUARE

To predict the genetic character of the offspring of parents of known genotypes, we may use algebra or, thanks to the geneticist R. C.

FIG. 25-5

Possible offspring in matings between guinea pigs heterozygous for black rough coat, *BbRr. BR, Br, bR,* and *br* are male and female gametes; wavy lines represent rough coats; straight lines represent smooth coats; shading represents black coats.

Punnett, the so-called Punnett square. Actually, we have already used the square in the solution of color blindness (Fig. 25-3); all one need do is draw a block of squares and "interact" the sex cells (indicated at the top and at the side) so as to produce a zygote in each square.

As another example, if a man heterozygous for

type A blood marries a woman heterozygous for type B blood, then the possible blood types of the offspring would be as shown in Fig. 25-4.

A more involved situation is shown in Fig. 25-5. Here we are dealing with the interaction of sex cells carrying two kinds of traits—color (black or white) and texture (rough or smooth) of the coats of guinea pigs. The square predicts the offspring in the matings of a pair of guinea pigs hybrid for black and rough coats, that is, pigs that have the genotype BbRr and produce the sex cells BR, Br, bR, and br. Black, B, is dominant over white, b; rough, R, is dominant over smooth, r. The square tells us that the offspring will be in the ratio of 9 black-rough to 3 black-smooth to 3 white-rough to 1 white-smooth.

MUTATIONS

Rarely genes may undergo a sudden change in character, a change that will be inherited if it occurs in the sex cells. This is called mutation, and the new genetic individual that exhibits the trait we call a mutant. For the most part, mutations are recessive and undesirable; they are undesirable because in most instances a given species already represents perfection through the natural selection of prior mutations. Evolution employs the mutation mechanism to make the best possible fit between a species and its environment, but once the right fit has been found, any subsequent alteration can only be for the worse, or neutral at best. The big exception, of course, is an environmental change that proves disadvantageous to the previously perfected species and thereby affords room for improvement.

A great jump in genetic thinking came in 1927 when Herman J. Muller showed that exposure to X rays induced mutations in fruit flies. Since that time all sorts of physical and chemical agents have been found to be mutagenic; such agents are called mutagens. Understandably,

man's great genetic problem at the moment is determining exactly how much damage is being caused by the increasing use of ionizing rays and radioactive materials, not to mention radioactive fallout. Of course, we are all now greatly aware of the ominous possibilities, and this in itself affords some degree of hope for our genetic future. Moreover, there is reason to believe that man may one day control his genetic destiny through biological engineering.

HEREDITY AND ENVIRONMENT

One of the most important things for the beginning student to unlearn is that a gene always does exactly what it is supposed to do. This is not true; in actual practice the genotype represents potential, not realization. A given gene does carry the blueprint for a certain trait, but most certainly environmental factors are also involved. By environment we mean every single influence that a gene encounters from within and without. Above all, there is the influence of other genes, those that have the power to turn a gene "on" or "off." In rabbits, for example, gene B (the gene for black) operates only in the presence of gene C.

In regard to the "outside environment," there are examples galore, some of which are of medical concern. It is said that 10% of the general population carry the gene for epilepsy, yet only 5% of these individuals ever develop the condition. Similarly, the individual who carries the gene for diabetes often fails to develop the disease if he avoids sweets and fats. Thus, when such genes are known to exist, the door to possible prevention is clearly indicated.

QUESTIONS

1. What is meant by a hybrid?
2. In a trait produced by a dominant gene, is it possible to tell the difference between hybrids and pure stock?
3. What is meant by partial dominance?

4. Mendel showed that for each trait there are two contrasting factors, one dominant and the other recessive. Discuss this statement in the framework of present-day genetics.

5. What is the essential difference between somatoplasm and germ plasm?

6. An individual that displays a trait's recessive expression is generally of pure stock. Explain.

7. Distinguish between chromatin, chromosomes, and genes.

8. What is the derivation of the term allele?

9. What are homologous chromosomes?

10. Alleles occupy corresponding loci. Explain.

11. Other things being equal, organisms with the same genotype have the same phenotype, but the reverse may or may not be true. Explain.

12. Pure stocks are homozygous; hybrids are heterozygous. Explain.

13. The idea of multiple alleles goes beyond classic mendelism. Explain.

14. Crossing over makes for variety, whereas linkage does not. Explain.

15. The "one gene one enzyme" theory of genetics is demonstrated by a metabolic disorder such as galactosemia. Explain.

16. What do you think might be done in the year 2001 about the problem of defective and lethal genes?

17. In gametogenesis the diploid number becomes reduced to the haploid number. Explain.

18. What is reduction division?

19. A recessive gene on the X chromosome in the male becomes dominant. Explain.

20. How do you account for the genotypes of individuals with Klinefelter's syndrome and Turner's syndrome?

21. Distinguish among trisomy, diploidy, and monosomy.

22. For some species at least, the statement that mutations are usually harmful, or at best neutral, did not apply 100 million years ago. Explain.

23. What, if any, were the genetic effects of the radioactivity released in the bombings of Hiroshima and Nagasaki?

24. What precautions are taken among those who work with ionizing rays and radioactive materials?

25. On a basis of molecular biology, how do mutagens produce mutations?

26. DNA is composed of only four kinds of nucleotides. How then do you account for the very large number of genes?

27. Somatic mutations are not inheritable. Explain.

28. How do you account for a mutated gene's not making its presence known for some time?

29. What are the Rh possibilities in the offspring of heterozygous Rh-positive parents? What are the possibilities when the man is heterozygous Rh positive and the wife Rh negative?

30. What are the possibilities of hemophilia in a match between a normal man and his carrier wife? What are the possibilities in a match between a hemophiliac male and a normal female?

31. If a baby has type O blood and the mother type A blood, then Mr. X with type AB blood could not be the father. Explain.

GLOSSARY

abdomen the portion of the body between the thorax and pelvis.

abduction the act of withdrawing a part from the midline.

abscess a localized collection of pus in a cavity formed by degeneration of tissue.

absolute temperature centigrade temperature added algebraically to 273 degrees.

absorption the taking up of fluid or other substances by the mucous membranes.

acapnia a condition of diminished carbon dioxide in the blood.

acetabulum the large cup-shaped cavity on the lateral surface of the hip.

acetone bodies the ketone compounds (for example, acetone) that appear in the blood in diabetes.

Achilles tendon the tendon that attaches the gastrocnemius muscle to the heel bone.

acid a compound that furnishes hydrogen ions or donates protons.

acid-fast not decolorized when treated with acid-alcohol after having been stained; said of bacteria.

acidosis an acidemia, or condition of lowered blood pH.

acquired immunity any type of immunity that is not inherited.

acromegaly an enlargement of bones of the face and extremities caused by an excess of the growth hormone.

acromion the outermost projection of the spine of the scapula.

actinomycosis an infection caused by *Actinomyces bovis*.

active immunity the production of antibodies by a person's own body.

acute sudden, sharp, and severe.

addition a reaction between an unsaturated compound and some element whereby the element "adds" on at the double bond.

adduction the act of drawing a part toward the midline.

adenohypophysis the anterior pituitary.

adenoids hypertrophy of the pharyngeal tonsil.

adenoma a benign epithelial tumor with a glandlike structure.

adenosine triphosphate a high-energy compound of the cell.

adhesion the attraction of unlike molecules.

adipose fatty; fat.

adrenal gland an endocrine gland located on the top of the kidney.

adrenaline another name for epinephrine, the hormone of the adrenal medulla.

adrenergic pertaining to sympathetic nervous system.

adventitia tunica externa.

Aedes aegypti a species of mosquito that transmits the yellow fever virus.

aerobe a microbe that requires oxygen.

afferent neuron a neuron that carries impulses to the central nervous system.

agar a dried mucilaginous substance derived from seaweed.

agglutination the clumping of cells.

agglutinins antibodies that cause agglutination.

agonist the prime mover (muscle); a muscle opposed to the action of another.

agranulocytosis an acute disease marked by leukopenia and ulcerative lesions of the mucous membranes.

albuminuria the presence of albumin in the urine.

alcohol an organic compound containing one or more hydroxyl (OH) groups.

aldehyde an organic compound containing a — CHO group.

algae thallophytes that contain chlorophyll.

alicyclic compounds cyclic organic compounds that resemble the paraffins.

alkali a strong, soluble base, for example, NaOH; physiologically, any chemical that neutralizes acid.

alkalosis an alkalemia, or condition of increased blood pH.

allantois an extraembryonic membrane.

allele one of two or more contrasting genes that occupy the same locus on homologous chromosomes.

allergen an antigen that provokes an allergy.

allergy a hypersensitive state acquired through exposure to a particular allergen.

allotropism a condition in which an element exists in two or more forms.

alpha particles helium nuclei emitted in radioactive decay.

alveolus a tiny cavity or space; for example, the alveoli of the lung.

ameba a protozoan belonging to the class Rhizopoda.

amebiasis a protozoiasis caused by *Entamoeba histolytica*.

amenorrhea the cessation of menstruation.

amino acids a carboxylic acid containing one or more amino (— NH$_2$) groups.

amnion an extraembryonic membrane that immediately encloses the embryo.

ampere the unit of intensity of an electric current.

amphiarthrosis an articulation permitting little motion.

amphitrichate bacilli equipped with flagella at both ends.

ampulla a flasklike dilatation of a tubular structure.

amylase any starch-digesting enzyme.

amylopsin pancreatic amylase.

anabolism the synthesis of nutrients into protoplasm; constructive metabolism.

anaerobe an organism that does not require oxygen.

analgesic a drug that inhibits pain.

anaphase the stage of mitosis characterized by the movement of chromosomes toward spindle poles.

anaphylaxis an exaggerated allergic reaction (especially in animals) to a foreign substance.

androgen a male hormone.

anemia a condition in which the blood is deficient in quantity or quality.

anesthesia loss of feeling or sensation.

aneurysm a blood-filled sac formed by the dilatation of the wall of an artery or vein.

angina a suffocating pain.

angina pectoris a paroxysmal thoracic pain usually resulting from myocardial anoxia.

angstrom a unit of wavelength equal to 10^{-8} cm.

anhydrous without water.

anion a negative ion.

anode a positive electrode.

Anopheles a genus of mosquito that transmits malaria.

anorexia the lack of appetite.

anoxemia the lack of oxygen in the blood.

anoxia the lack of oxygen.

antagonist a muscle that acts in opposition to another muscle (called the agonist).

anterior situated in front of.

antibiotic an agent produced by one microorganism that inhibits or kills another microorganism.

antibody a specific immune substance present in the blood.

anticoagulant a drug that interferes with the clotting mechanism.

antigen an agent that provokes the production of antibodies in the body.

antiseptic an agent that inhibits microorganisms.

antiserum a serum that contains antibodies.

antitoxin an antibody that neutralizes a toxin.

antrum a cavity or chamber within a bone.

anuria the absence of excretion of urine from the body.

anus the outlet of the rectum.

aorta the largest artery of the body.

apex a pointed extremity.

aphasia the partial or total loss of the ability to articulate ideas in any form as a result of brain damage.

apnea the transient cessation of the breathing impulse.

apoenzyme the protein portion of an enzyme.

aponeurosis a sheet of white fibrous tissue serving mainly as an investment for muscle.

apoplexy a condition caused by acute vascular lesions of the brain, such as hemorrhage, thrombosis, or embolism.

appendectomy the removal of the appendix.

aqueous containing water.

arachnoid the delicate, weblike membrane interposed between pia mater and dura mater.

Archimedes' principle a submerged or floating body is buoyed up by a force equal to its weight.

areola a dark, circular area.

Arnold sterilizer a device that sterilizes by free-flowing steam.

arrhythmia any variation from the normal rhythm of the heartbeat.

arteriole a microscopic artery.

arteriosclerosis the thickening, hardening, and loss of elasticity of the arteries.

artery a vessel that carries blood away from the heart.

arthrosis a joint or articulation.

articulation a joint.

aseptic sterile.

asexual having no sex.

asphyxia suffocation.

asthma recurrent attacks of paroxysmal dyspnea caused by constriction of bronchioles.

astigmatism defective curvature of the refractive surfaces of the eye.

astrocytes star-shaped cells of neuroglia.

ataxia the loss of muscle coordination.

atelectasis the incomplete expansion of the lungs at birth.

atherosclerosis a condition of the arteries marked by deposits of yellowish fatty plaques in the intima.

atlas the first cervical vertebra.

atom the smallest "practical unit" of an element.

atomic number the number of protons in the nucleus of an atom.

atomic theory the theory that all matter is composed of basic units called atoms.

atomic weight the weight of an atom compared to carbon-12.

atrium (pl., atria) the upper chambers of the heart.

atrophy a diminution in size.

attenuated weakened.

auricle the atrium of the heart.

auscultation the act of listening for sounds in the body.

autoclave a device that sterilizes by steam under pressure.

autonomic nervous system the functional division of the nervous system that supplies the glands, heart, and smooth muscles.

autosome any chromosome not related to sex.

avitaminosis a vitamin deficiency.

Avogadro's law under the same conditions of temperature and pressure, equal volumes of gases contain equal numbers of molecules.

Avogadro's number 6.02331×10^{23}; the number of molecules contained in 1 mole of any substance.

axilla the armpit.

axis a line about which a revolving body turns.

axon the efferent fiber of a nerve cell.

bacillary dysentery a severe enteritis caused by bacteria of the genus *Shigella*.

bacillus a rod-shaped bacterium.

bacteriolysin an antibody that causes the dissolution of bacteria.

bacteriophage a virus that destroys bacteria.

Bang's disease a disease of cattle caused by *Brucella abortus*.

barometer an instrument used to measure atmospheric pressure.

baroreceptor a receptor sensitive to changes in pressure.

Barr body the characteristic dark spot present in the nucleus of female somatic cells.

basal ganglia islands of gray matter deep inside the cerebrum.

basal metabolism the lowest level of energy expenditure.

base a chemical that furnishes OH^- ions; a proton acceptor (Brönsted theory).

basophil a granulocyte whose granules and nucleus take a basic stain.

BCG vaccine a vaccine against tuberculosis prepared from attenuated tubercle bacilli.

beta rays high-speed electrons.

biceps two-headed.

bile the greenish yellow secretion of the liver.

bilirubin a red pigment of bile.

biliverdin a green pigment of bile.

binary fission reproduction in which a one celled organism divides (splits) in two.

biotin a member of the vitamin B complex.

biuret test a test for protein.

blastocoele the cavity of the blastocyst.

blastocyst an early stage in the development of the embryo, produced by cleavage.

blastomycosis a chronic systemic mycosis caused by *Blastomyces dermatitidis*.

bleeding time the duration of the bleeding that follows puncture of the skin (about 6 minutes).

blind spot the area of the retina where the optic nerve leaves the eyeball.

boil a furuncle; a circumscribed area of inflammation of the skin and subcutaneous tissue containing a central core.

bond the linkage between atoms.

botulism a severe type of food poisoning caused by *Clostridium botulinum*.

Bowman's capsule the invaginated proximal portion of the nephron tubule.

Boyle's law provided the temperature remains constant, the volume occupied by a gas is inversely proportional to the pressure.

bradycardia an abnormal slowness of the heartbeat.

broad-spectrum antibiotic an antibiotic effective against a wide variety of pathogenic microbes.

bronchioles the smallest air tubes in the lung.

bronchitis an inflammation of the bronchi.

bronchus an air tube in the lung.

Brönsted theory the theory that acids are proton donors and bases are proton acceptors.

brownian movement the dancing motion of particles suspended in a fluid.

brucellosis an infection caused by *Brucella abortus, B. suis*, or *B. melitensis*.

buccal pertaining to the cheek of the mouth.

budding an asexual mode of reproduction in yeasts.

buffer an agent that resists a change in pH.

BUN blood urea nitrogen.

bundle of His the neuromuscular bundle located in the wall of the interventricular septum of the heart.

bursa a fluid-filled sac or pouch, usually lined with a synovial membrane.

buttock the prominence over the gluteal muscle.

calculus a stone.

calorie a small calorie is the amount of heat required to raise the temperature of 1 g of water 1° C; a large Calorie, or kilocalorie, equals 1,000 small calories.

calyx a recess of the pelvis of the kidney.

cancellous spongy; for example, cancellous bone.

canthus the angle formed by the meeting of the upper and lower eyelids.

capillaries the minute vessels that connect the arterioles and the venules.

capsule a gelatinous covering that surrounds a bacterium.

carbohydrates a class of organic compounds containing the starches and sugars.

carbonate a salt of carbonic acid.

carbonyl group the organic radical — $C = O$.

carboxyl group the — COOH group of organic acids.

carboxyhemoglobin a compound formed by union of carbon monoxide and hemoglobin.

carbuncle an inflammation of the subcutaneous tissue terminating in slough and suppuration and accompanied by marked systemic symptoms.

carcinoma a malignant tumor composed of epithelial tissue.

cardia the opening of the esophagus into the stomach.

caries the decay of bone or teeth.

carpus the wrist.

casein the principal protein in milk.

cast effused matter molded to the shape of a hollow organ such as a renal tubule.

castration the removal of ovaries or testes.

catabolism destructive metabolism.

catecholamines epinephrine and norepinephrine.

cathode the negative electrode.

cation a positive ion.

cauda equina the bundle of spinal nerves issuing from the lower portion of the spinal cord.

caudal toward the tail.

cecum the pouch at the proximal end of the large intestine.

celiac pertaining to abdomen.

cell a microscopic circumscribed mass of protoplasm containing a nucleus.

cellulose an indigestible carbohydrate forming the basic framework of most plant structures.

centimeter a unit of length in the metric system; 0.3937 inch.

central nervous system the brain and the spinal cord.

centriole a minute organelle lying within the centrosome.

centromere the structure that joins each pair of chromatids.

centrosome the region of condensed cytoplasm surrounding the centrioles.

cephalic pertaining to head.

cerebellum the division of the brain immediately below the occipital lobes of the cerebrum.

cerebrum the two great hemispheres forming the upper and larger portions of the brain.

cerumen the waxlike material found in the external meatus of the ear; earwax.

cervix the neck or any necklike part.

chancre the primary lesion of syphilis.

Charles' law provided that the pressure remains constant, the volume of a gas is directly proportional to the absolute temperature.

chemoreceptor a receptor adapted to chemical stimulation.

chemotherapy the treatment of disease using chemicals.

Cheyne-Stokes respiration abnormal breathing in grave diseases in which the respirations increase progressively in depth and vigor to a maximum, then diminish again, and finally cease.

chiasma a crossing.

chicken pox a dermotropic virus infection.

chloride a salt of hydrochloric acid.

cholecystectomy the removal of the gallbladder.

cholecystitis an inflammation of the gallbladder.

cholera an acute infectious disease caused by *Vibrio cholerae*.

cholesterol a fatty steroid alcohol present in animal fats and oils.

cholinergic pertaining to the parasympathetic nervous system.

cholinesterase the enzyme that hydrolyzes acetylcholine into acetic acid and choline.

chordae tendineae the cords that attach the heart valves to the papillary muscles.

choroid the eye coat between the sclera and the retina.

chromatid a newly formed chromosome.

chromatin the easily stained network of fibrils within the nucleus.

chromosome dark-staining, rod-shaped bodies that appear in the nucleus at the time of mitosis.

chronic long continued.

chyle lymph containing absorbed fat.

chyme partially digested food as it leaves the stomach.

cilia microscopic hairlike processes.

cisterna chyli an enlargement at the base of the thoracic duct.

class the main division of a phylum.

Clostridium a genus of anaerobic, spore-forming, gram-positive bacilli.

coal tar a product obtained by the destructive distillation of coal.

coccus a spherical bacterium.

cochlea the spiral-shaped structure that forms part of the inner ear.

codon the segment (or nucleotide triplet) of the DNA molecule that specifies a certain amino acid.

coenzyme a heat stable, nonprotein, organic molecule that must be loosely associated with an enzyme for the enzyme to function.

collagen the chief supportive protein of the skin and connective tissue.

colloid a state of matter in which minute particles of one substance are dispersed throughout some other substance.

colon the large intestine exclusive of the cecum and the rectum.

colony an isolated group of bacteria on a solid culture medium.

compartment a theoretical division of body water.

complemental air the volume of air that can be inspired over and above the tidal volume.

compound a substance composed of two or more elements in a definite proportion.

concentrated solution a solution that contains a relatively large amount of solute.

conchae the bony plates of the nasal cavity.

condyle a rounded projection of a bone.

congenital existing at or before birth.

conjunctivitis an inflammation of the conjunctiva.

contracture the incomplete relaxation of muscle.

convex having a rounded or elevated surface.

corium the true skin, or dermis.

cornea the transparent portion of the eyeball.

coronal lying in the direction of the coronal suture; said of a transverse plane parallel to the long axis of the body.

corpus a body.

cortex the outer part of an organ.

costal pertaining to the ribs.

covalence the sharing of electrons.

crenation an osmotic shriveling of a cell; also called plasmolysis.

cretinism a condition caused by a congenital lack of thyroid secretion.

crystalloid a compound that forms a true solution.

cubital pertaining to the forearm.

culture a growth of microorganisms.

cutaneous pertaining to the skin.

cyanosis a bluing of the skin, usually due to lack of oxygen.

cyst any sac, normal or otherwise, filled with a liquid or semisolid.

cystitis an inflammation of the bladder.

cytokinesis the dividing of the cytoplasm.

cytology the study of cells.

cytoplasm the protoplasm of a cell exclusive of the nucleus.

dark-field microscope an optical microscope modified to permit lateral lighting of the specimen.

deamination the removal of the amino group ($-NH_2$) from an amino acid.

deciduous shedding at a certain stage of growth; for example, deciduous teeth.

decussation a crossing over.

defecation the elimination of wastes from the intestine.

deferens carrying away

deglutition the act of swallowing.

dehydration the removal of water.

deltoid the deltoid muscle.

dendrite (dendron) an afferent nerve process.

density mass per unit volume.

dentin the main part of the tooth under the enamel.

depolarization the loss of electrical charge; discharge.

dermatitis an inflammation of the skin.

dermatome an area of skin supplied with afferent nerve fibers by a single dorsal spinal root.

dermatomycosis a fungal infection of the skin.

dermis the true skin (also, corium).

dermotropic having an affinity for the skin.

destructive distillation the heating of a substance out of contact with air.

dextrose glucose.

diabetes insipidus a metabolic disorder caused by lack of antidiuretic hormone.

diabetes mellitus a metabolic disease caused by a deficiency or deactivation of insulin.

dialysis the separation of a crystalloid and a colloid by means of a semipermeable membrane.

diapedesis the passage of blood cells through the wall of the intact capillary.

diaphragm the muscular sheath between the thorax and the abdomen.

diaphysis the shaft of a long bone.

diarthrosis a freely movable joint.

diastole the relaxation of the heart between contractions.

Dick test a skin test to determine susceptibility to scarlet fever.

Dick toxin a toxin produced by the pathogen *Streptococcus pyogenes*.

diencephalon the portion of the brain between the cerebrum and the midbrain.

diffusion the spreading out of particles.

digestion the conversion of food into assimilable molecules.

diopter a unit of refractive power of a lens.

diphtheria a severe infection caused by *Corynebacterium diphtheriae*.

diplococcus cocci that occur in pairs.

diploid number the number of chromosomes in a "body cell" (in man, 46).

diplopia double vision.

direct current a current that flows in only one direction and has polarity.

disaccharide a sugar with the formula $C_{12}H_{22}O_{11}$.

disinfectant an agent that kills pathogens.

disinfection the destruction of pathogenic organisms.

dissociation the release of ions in solution; ionization.

distal farther from some point of reference.

distillation the heating of a liquid and condensation of its vapors.

diuretic a drug that increases the production of urine.

diverticulum an outpouching of the intestine.

DNA deoxyribonucleic acid.

dominant the one of two alternative characteristics that is evident in heterozygous conditions.

dorsal in back of; posterior.

droplet infection the transmission of an infection via minute particles of sputum.

duodenum the first portion ("12 fingerbreadths in length") of the small intestine.

dura mater the outermost layer of the meninges.

dwarfism a condition caused by a deficiency of the growth hormone.

dynamic equilibrium the state of balance reached by a reversible reaction when it proceeds at equa! rates in both directions.

dyne a unit of force.

dysentery an acute enteritis, especially of the colon.

dyspnea labored breathing.

dystrophy faulty nutrition.

ectoderm the outermost germ layer.

ectoparasite a parasite that lives on the surface of the body.

eczema an inflammatory skin disease with vesiculation.

edema the accumulation of fluid in the tissues.

effector a responding organ.

efferent neuron a neuron that carries an impulse away from the central nervous system.

electrocardiography the making of graphic recordings of the heart's action potentials.

electroencephalography the graphic recording of the brain's action potentials.

electrolysis the decomposition of a compound by means of an electric current.

electrolyte a compound that, in solution, conducts an electric current.

electron a negative particle of electricity.

electron theory the theory that the atom is composed of three basic particles: electrons, protons, and neutrons.

electron volt a measure of energy.

electrophoresis the migration of colloidal particles to an electrode.

electrostatics the study of static electricity.

electrovalence the loss or gain of electrons.

element a substance composed of like atoms.

elephantiasis an edematous swelling of the legs or genitals usually caused by the presence of filarial worms in the lymph vessels.

embolism a sudden blocking of an artery or vein by a clot brought by the bloodstream.

embryo an organism in the early stages of development; in man, the developing organism before the third month.

emmetropia the normal condition of the eye as regards refraction.

emphysema the presence of air in the intra-alveolar tissue of the lungs.

emulsion a colloidal system of one liquid dispersed in another liquid.

encephalitis an inflammation of the brain.

endemic prevalent in a particular area.

endemic typhus fever a rickettsial disease transmitted from rat to man by the bite of the rat flea.

endocardium the inner lining of the heart.

endocrine gland a ductless gland.

endoderm the innermost germ layer of the embryo (also entoderm).

endolymph the fluid contained in the membranous labyrinth of the inner ear.

endometrium the inner lining of the uterus.

endoplasmic reticulum the network of fine tubules and vesicles within cytoplasm.

endospore a spore within the bacterial cell.

endosteum the lining of the medullary cavity of long bones.

endothelium the epithelial lining of heart and blood vessels.

energy the ability to do work.

enteric pertaining to the intestine.

enterokinase an enzyme of the small intestine that converts trypsinogen into trypsin.

enzyme an organic catalyst.

eosinophil a granulocyte whose granules are stained readily by eosin.

epidemic attacking many people in a given region at the same time.

epidermis the outer layer of skin.

epididymis a group of coiled tubules continuous with the seminiferous tubules of the testis.

epinephrine adrenaline.

epiphysis the end of a long bone.

epithelium a basic tissue that forms the epidermis and lines the ducts and hollow organs.

equivalent weight the atomic weight of an element divided by its valence.

erepsin a mixture of proteolytic enzymes produced by the intestinal glands.

erythema a pathologic redness of the skin.

erythroblast a marrow cell that gives rise to red cell.

erythrocyte a red blood cell.

esterification the reaction between an alcohol and an acid to form an ester and water.

estrogen a type of female sex hormone.

ethylene series a series of hydrocarbons whose members contain a double bond.

etiology the study of the cause of disease.

eupnea normal respiration.

eustachian tube the channel that connects the middle ear with the nasopharynx.

excretion the elimination of waste products.

exocrine glands glands that have ducts.

exophthalmos an abnormal protrusion of the eyes.

exothermic reaction a reaction in which heat is released.

exotoxin a toxin released by a living bacterium.

extension the act of stretching out.

extracellular compartment fluids located outside the cells.

extrinsic coming from outside.

facultative anaerobe a microbe that lives best in the presence of oxygen but can live in its absence.

family a main division of an order.

fascia the tissue that binds organs or parts of organs together.

fat a glyceryl ester of a saturated fatty acid.

fauces the space bounded by the soft palate, tongue, and tonsils.

feces the material excreted from the rectum.

fermentation anaerobic oxidation.

fertilization the union of male and female sex cells to form a zygote.

fetus the developing human organism after the end of the second month.

fibrillation the spontaneous contraction of muscle fibers no longer under nervous control.

    ventricular f. convulsive movements of the ventricles of the heart.

fibrin protein threads that form the framework of the blood clot.

fibrinogen a protein of the blood plasma.

fimbriae fringe.

fissure a cleft or groove, for example, a fissure of the brain.

flaccid lax and soft.

flagellum a whiplike process of a microbe.

flexion the act of bending.

flexor a muscle that effects flexion.

flutter a condition of cardiac arrhythmia in which atrial contractions are extremely rapid (up to 400 per minute).

follicle a very small sac containing a secretion.

fomite a contaminated object.

fontanel a membranous spot in the infant's skull.

foramen an opening.

fossa a depression; a cavity.

fractional distillation the separation of liquids by distillation.

frambesia an infection caused by *Treponema pertenue*.

frequency the number of cycles or vibrations per second.

FSH follicle-stimulating hormone.

fundus a base.

fungi yeasts and molds.

fungicide an agent that destroys fungi.

furuncle a boil.

gallbladder a muscular sac, contiguous with the liver, that stores and concentrates bile.

galvanometer a sensitive instrument for measuring weak electric currents.

gamete a sex cell.

gametogenesis the development of gametes.

gamma globulin the fraction of the blood that contains the antibodies.

gamma rays one of three types of rays emitted by radioactive substances; similar to but more penetrating than X rays.

ganglion a collection or mass of nerve cells.

gas gangrene a morbid infection caused by several species of the genus *Clostridium*.

gastric pertaining to stomach.

gastritis an inflammation of the stomach.

gastrocnemius the calf muscle.

gastrula an early stage in the embryonic development produced by invagination of the blastula.

gel a semisolid colloid.

gene the hereditary unit of the chromosome.

genitals the reproductive organs.

genotype a cell's genetic constitution.

German measles a dermotropic virus infection characterized by a rash.

germicide an agent that kills germs (microbes).

gestation pregnancy.

gigantism an abnormal growth caused by excessive secretions of the anterior lobe of the hypophysis.

gingivitis an inflammation of the gums.

glaucoma an eye disease characterized by elevated intraocular pressure.

glomerulus a coiled mass of capillaries within Bowman's capsule of the kidney.

glossitis an inflammation of the tongue.

glottis the vocal apparatus of the larynx.

gluconeogenesis the formation of glucose from protein or fat.

glucose dextrose ($C_6H_{12}O_6$).

glycogen a carbohydrate, $(C_6H_{10}O_5)_x$, formed in the liver and muscle from glucose.

glycogenesis the formation of glycogen from glucose.

glycogenolysis the breakdown of glycogen into glucose.

glycolysis the anaerobic breakdown of glucose into pyruvic acid.

glycosuria the presence of sugar in the urine (also, glucosuria).

goiter an enlargement of the thyroid gland.

Golgi body a cytoplasmic organelle involved in the production of secretions.

gonadotropin a hormone of the anterior pituitary that stimulates the sex glands.

gonads sex glands.

gonorrhea a venereal disease caused by *Neisseria gonorrhoeae*.

graafian follicle an ovarian follicle.

gram a unit of weight in the metric system.

gram-negative taking the counterstain when stained according to Gram's method.

gustatory pertaining to the sense of taste.

gyrus a convolution of the cerebral cortex.

half-life the time required for a given mass of radioactive element to lose half of its radioactivity.

halogens the elements fluorine, chlorine, bromine, and iodine.

haploid having a single set of chromosomes.

haustrum the sacculations of the colon.

hemiplegia paralysis of one side of the body.

hemoglobin the oxygen-carrying substance of red blood cells.

hemoglobinuria the presence of hemoglobin in the urine.

hemolysin an agent that destroys red cells.

hemolysis destruction of red cells.

hemophilia a sex-linked hereditary blood disease characterized by a prolonged bleeding time.

hemopoiesis the formation of blood.

hemorrhage the loss of blood from the vessels.

heparin an anticoagulant present in many tissues, especially the liver.

hernia the protrusion of a loop of the intestine or of another organ through an abnormal opening.

herpes simplex cold sore.

herpes zoster shingles.

heterosome a sex chromosome.

heterozygous possessing a pair of unlike alleles for a given trait.

hilus (hilum) a depression where vessels enter an organ.

histamine a chemical agent thought to account for certain allergies.

histology the study of tissues; tissue anatomy.

histoplasmosis a systemic mycosis caused by *Histoplasma capsulatum*.

homeostasis a tendency toward uniformity in the internal environment of the organism.

homozygous possessing a pair of identical alleles for a given trait.

hormone a chemical, produced by an endocrine gland, that influences the activity of other organs.

hyalin glasslike.

hyaluronidase an enzyme that dissolves the intercellular cement of tissue.

hybrid an organism that is heterozygous for one or more traits.

hydrocarbon a compound containing only carbon and hydrogen.

hymen a membranous fold that partially or wholly occludes the external orifice of the vagina.

hypercapnia abnormal levels of CO_2 in the blood.

hyperemia an excess of blood in any part of the body.

hyperglycemia an excess of sugar in the blood.

hyperkalemia an excess of potassium in the blood.

hypermetropia (hyperopia) farsightedness.

hypernatremia an excess of sodium in the blood.

hyperplasia an increase in the number of cells.

hyperpnea abnormally rapid breathing.

hypertension high blood pressure.

hyperthyroidism overactivity of the thyroid gland.

hypertonic having a greater osmotic pressure than solution being compared.

hypertrophy an overgrowth.

hypervolemia an abnormally high blood volume.

hyphae filaments of a mold.

hypochondrium the region under the cartilages of the ribs.

hypogastric under the stomach.

hypokalemia a deficiency of potassium in the blood.

hyponatremia a deficiency of sodium in the blood.

hypoparathyroidism decreased activity of the parathyroid glands.

hypophysis the pituitary gland.

hypothalamus a part of the diencephalon.

hypotonic having a lesser osmotic pressure than solution being compared.

hypovolemia an abnormally decreased blood volume.

hypoxia an oxygen deficiency.

idiogram a diagrammatic representation of a chromosome complement or karyotype.

ileum the distal portion of the small intestine.

immunity resistance to disease.

inclusion bodies intracellular lesions of viral origin.

incubation period the period between the entrance of a pathogen and the first manifestation of infection.

infarct an area of necrosis due to lack of blood.

infectious hepatitis a severe infection of the liver caused by a virus.

infectious mononucleosis a viral infection characterized by an increase in abnormal leukocytes.

inferior lower.

influenza an acute catarrhal inflammation of the respiratory tract caused by a virus.

infundibulum the stalk of the pituitary gland.

inguinal of the groin.

insertion the attachment of a muscle to a bone at the more freely movable end.

insulin the hormone secreted by beta cells of the islands of Langerhans.

intercellular between the cells.

interstitial fluid intercellular fluid; tissue fluid.

intima the innermost of the three coats of a blood vessel.

intracellular within the cell.

intrathoracic within the chest.

intrinsic present within.

involution a rolling or turning inward; a retrograde change.

ion a charged particle.

ipsilateral situated on the same side.

iris the colored circular muscle of the eye behind the cornea.

ischemia a local and temporary deficiency of blood.

islands of Langerhans cells of the pancreas that secrete insulin and glucagon.

isotonic having the same osmotic pressure as a solution being compared.

isotopes elements that have the same atomic number but different atomic weights.

jaundice a disease characterized by yellowing of the skin.

jejunum the second portion of the small intestine.

jugular pertaining to the neck.

juxtaglomerular cells cells in the media of the afferent arterioles of the kidney that secrete renin.

karyokinesis mitosis.

karyotype the chromosomal characteristics of an individual presented as metaphase chromosomes of a single cell arranged in pairs in descending order of size.

keratin the chief protein of the epidermis.

ketones organic compounds containing the carbonyl (CO) group.

ketosis a condition marked by excessive formation of ketones in the body.

kidney the excretory organ that forms urine.

kilocalorie a large calorie; one thousand small calories.

kilogram the metric unit of weight (mass); equal to 2.2 pounds.

kinase an agent that activates an enzyme.

kinesthesia the sense by which motion, position, muscular motion, and so on are perceived.

kinetic pertaining or relating to or producing motion; for example, kinetic energy.

Klebs-Löffler bacillus the common name for *Corynebacterium diphtheriae*.

Kline test a precipitation test for syphilis.

Koch's postulates the criteria by which the causal relationship between an infection and its etiologic agent is established.

Koplik's spots the small reddish spots, each with a bluish white central speck, present on the lining of the mouth during the prodromal stages of measles.

kymograph an instrument used to record physiological activity.

kyphosis hunchback.

labia lips.

lacrimal pertaining to tears.

lactase an enzyme that acts upon lactose.

lactation the secretion of milk.

lacteals lymph vessels of the small intestine.

lactose a disaccharide sugar ($C_{12}H_{22}O_{11}$) present in milk.

lacuna a small space.

lamella a thin layer.

larynx the "voice box."

latent syphilis the asymptomatic stage of syphilis.

lateral of or toward the side.

leprosy an infectious disease caused by *Mycobacterium leprae*.

lesion a pathological change in tissue.

leukemia a fatal disease marked by leukocytosis and by enlargement and proliferation of the spleen, lymph glands, and bone marrow.

leukocyte a white blood cell.

leukocytosis an increase in the number of leukocytes.

leukopenia a reduction in the number of leukocytes.

lichen a symbiotic combination of fungi and algae.

ligament any tough, fibrous band that connects bones or supports viscera.

limbic system the part of the cerebrum associated with primitive emotions.

lipase a fat-splitting enzyme.

lipids fatty substances.

liter the unit of volume in the metric system.

lobar pneumonia an inflammation of one or more lobes of the lung.

loin the part of the back between the thorax and pelvis.

lumbar pertaining to the loin.

lumen the space within a tube.

luteum yellow.

lymph the watery fluid that circulates through the lymphatic system.

lymphocyte a type of white cell.

lysis the dissolution of cells.

lysosome an organelle containing digestive enzymes.

macromolecule a molecule of very high molecular weight; for example, proteins and polysaccharides.

macrophage a large phagocytic cell of the reticulo-endothelial system.

maltase an enzyme that converts maltose into glucose.

mammary pertaining to the breasts.

manometer an instrument used for measuring the pressure of liquids.

manubrium the top part of the sternum.

mastication the chewing of food.

matrix intercellular material.

meatus a passageway.

medial toward the midline.

mediastinum the space in the middle of the thorax.

medulla the central part of a gland or organ.

medulla oblongata the portion of the brain just above the spinal cord.

meiosis cellular division in which the chromosome number is halved; also called *reduction division*.

melanin the pigment of the skin.

membrane a thin layer or sheet.

meninges the membranes covering the brain and spinal cord.

meningitis an inflammation of the meninges.

menopause the cessation of menstruation at the close of the reproductive period.

menstruation the cyclic (monthly) discharge of blood (menses) from the uterus.

mesencephalon the midbrain.

mesenchyme embryonic connective tissue.

mesentery the large expanse of tissue that anchors the intestine to the posterior abdominal wall.

mesoderm the middle germ layer of the embryo.

metabolism the sum total chemical processes of life.

metacarpus the part of the hand between the wrist and the fingers.

metaphase the middle phase of mitosis, when chromosomes are lined up at the equatorial plate.

metastasis the transfer of disease from one organ or part to another not directly connected to it.

metatarsus the part of the foot between the tarsus and the toes.

meter the metric unit of length; equal to 39.371 inches.

micrometer 0.001 mm (1/25,000 inch); formerly, micron.

milliequivalent one thousandth of an equivalent; abbreviated mEq.

minerals inorganic elements or salts.

mitochondria minute bodies in the cytoplasm of the cell charged with respiratory enzymes.

mitosis cell division by means of which the two daughter cells receive the same number of chromosomes as the parent cell.

mitral pertaining to the left atrioventricular heart valve.

mold a filamentous fungus.

mole the quantity of a chemical compound that has a weight in grams equivalent to its molecular weight.

molecule the smallest possible particle of a compound.

monosaccharide a simple sugar ($C_6H_{12}O_6$).

morphology the study of shapes and structure.

morula the ball of cells resulting from the cleavage of a zygote.

motor neuron an efferent neuron.

mucosa a mucous membrane.

mucus a slimy, clear fluid secreted by mucous membranes.

mutation an inheritable alteration of the genes.

mycelium a tuft of hyphae.

myelin the fatlike substance forming the sheath around certain nerve fibers.

myocardium heart muscle.

myofibril a component or unit of a muscle fiber.

myopia nearsightedness.

myxedema a condition of hypothyroidism in the adult.

NADP nicotinamide adenine dinucleotide phosphate.

nares the nostrils.

necrosis the death of a circumscribed portion of tissue.

Negri bodies inclusion bodies that appear in the brain tissue of rabid animals.

nematode a class of roundworms of phylum Aschelminthes.

neoplasm a tumor.

nephritis an inflammation of the kidney.

nephron the physiological unit of the kidney.

nephrosis any disease of the kidney, particularly one marked by degenerative changes in the tubules.

nerve a bundle of nerve fibers.

neurilemma the outermost covering of a nerve fiber.

neuroglia the supporting structure of nervous tissue.

neurohypophysis the posterior pituitary.

neuron a unit of nerve tissue.

neutralization the reaction between an acid and a base to form a salt and water.

neutrophil a granulocyte that stains with neutral dyes.

niacin (nicotinic acid) a B complex vitamin.

normal saline a 0.9% solution of NaCl.

nucleolus the dark roundish body within the nucleus of a cell.

nucleoprotein a conjugated protein whose prosthetic group is nucleic acid.

nucleotide a compound of phosphoric acid, a five-carbon sugar (ribose or deoxyribose), and a nitrogenous base (adenine, cytosine, guanine, thymine, or uracil).

nucleus the dark, spherical structure within a cell containing genetic material; also, a group of neuron cell bodies in the brain.

nutrient a nutritive substance.

nystagmus involuntary oscillatory movements of the eyes.

obesity a condition marked by excessive accumulation of fat in the body.

occiput the back of the head.

olecranon the elbow.

olfactory pertaining to the sense of smell.

omentum a fold of peritoneum attached to the stomach.

ophthalmia neonatorum a gonorrheal conjunctivitis of the newborn infant.

opsonins antibodies that facilitate phagocytosis.

optically active capable of rotating the plane of polarized light.

organ of Corti the receptor of hearing located in the cochlea.

organelles specialized parts of a cell.

organism a living thing.

os a bone; a mouth or orifice.

osmosis the passage of water through a semipermeable membrane separating two solutions, the chief flow being from the less dense to the more dense solution.

osmotic pressure the pressure generated by osmosis.

ossicles little bones; in particular the ear bones.

ossification the formation of bone.

otitis media an inflammation of the middle ear.

ovary the female sex gland.

oviduct the uterine or fallopian tube.

ovulation the release of the ovum from the ovary.

ovum the female sex cell.

oxidation the loss of electrons.

oxide a compound of oxygen and another element.

oxyhemoglobin a compound of oxygen and hemoglobin.

oxytocin a hormone of the posterior pituitary gland that stimulates contraction of the uterus.

palate the roof of the mouth.

pancreas a digestive gland located in the loop of the duodenum.

pandemic a worldwide epidemic.

paralysis the loss of motor function.

parasite an organism that lives upon another living organism.

parasympathetic nervous system the autonomic fibers that originate in the lower part of the brain and the sacral portion of the cord.

parathyroids four tiny glands located on the posterior surface of the thyroid.

parietal pertaining to the wall.

parotitis an inflammation of the parotid glands; mumps.

parturition childbirth.

passive immunity resistance to disease through antibodies produced elsewhere than in one's own body.

pasteurization the destruction of pathogens by the use of moderate heat; milk is pasteurized by heating at 143° F for 30 minutes.

pathogen a disease-producing microbe.

pellagra an avitaminosis caused by lack of nicotinic acid.

pelvis the cavity in the lowest part of the trunk; a basin or basinlike structure.

peptide a compound of two or more amino acids.

peptone an intermediate derivative formed from the digestion of proteins.

pericardium the membranous sac enclosing the heart.

periodic table an arrangement of the elements according to their atomic numbers.

periosteum the membranous covering around bone.

peripheral away from.

peristalsis the wormlike movement by which the alimentary canal propels its contents.

peritoneum the serous membrane that surrounds the organs and lines the abdominal cavity.

peritonitis an inflammation of the peritoneum.

pertussis whooping cough, caused by *Bordetella pertussis*.

Petri dish a shallow, covered, glass dish used for solid culture media.

Peyer's patches the oval elevated areas of lymphoid tissue in the mucosa of the small intestine.

pH a measure of acidity and alkalinity, equal to the logarithm of the reciprocal of the amount of hydrogen ion (in grams) in a liter of solution.

phagocytosis the engulfing and destruction of microbes and other foreign particles by granulocytes, macrophages, and certain other cells of the reticuloendothelial system.

phalanges the bones of the fingers and toes.

pharynx the part of the alimentary canal that connects the mouth and the esophagus.

phenol coefficient a number that expresses the germicidal action of an agent compared to phenol.

phenotype an organism's physical appearance.

phosphate a salt of phosphoric acid.

phosphorylation the metabolic reactions involving addition of phosphate groups to a compound.

phylum the primary division of the plant or animal kingdom or of a subkingdom.

pia mater the innermost membrane of the meninges.

pineal gland a small body situated at base of brain.

pituitary gland the endocrine gland that lies beneath the brain in the sella turcica of the sphenoid bone.

placenta the membranous structure that provides for exchange of nutrients and wastes between the blood of the mother and the fetus.

plantar pertaining to sole of the foot.

plasma the fluid portion of the blood.

plasmolysis the shrinking of the protoplasm away from the cell wall due to the loss of water by osmotic action.

plasmoptysis the escape of protoplasm through a ruptured cell wall.

platelets blood cells involved in clotting; also called thrombocytes.

Platyhelminthes a phylum of the animal kingdom comprised of parasitic flatworms.

pleomorphism indefinite morphology.

pleura the serous membrane that lines the thorax and covers the lungs.

plexus a network of nerves, veins, or lymphatics.

pneumonia an inflammation of the lungs.

pneumothorax an accumulation of air in the thoracic cavity, but outside the lung.

poliomyelitis a neurotropic virus infection characterized by fever, sore throat, headache, vomiting, stiffness of the muscles, and often paralysis.

polymer a large molecule composed of smaller molecules (monomers) linked together.

polymerization the building up of large molecules from smaller molecules.

polysaccharide a carbohydrate having the general formula $(C_6H_{10}O_5)_x$.

polyuria the excretion of large volumes of dilute urine.

pons a bridge.

posterior situated behind or toward the rear.

potential electrical charge.

presbyopia the loss of the eye's power of accommodation with aging.

pressoreceptors receptors stimulated by a change in pressure (also called baroreceptors).

progesterone one of the hormones produced by the corpus luteum.

prone lying with face downward; of the hand, having the palm turned downward.

prophase the first stage of mitosis.

proprioceptive impulses impulses received from muscles, tendons, and joints.

prostate the male gland that surrounds the base of the urethra.

prosthetic group a small molecule or ion attached to a protein.

proteinase an enzyme that acts on proteins.

proteins polymers of amino acids.

proton the positively charged particle of the atomic nucleus.

protoplasm the matter of living cells.

protozoa one-celled animals.

proximal nearest; opposite of distal.

psittacosis a viral infection of birds and man; also called parrot fever.

psychic pertaining to the mind.

psychosomatic pertaining to the influence of the mind on body functions.

pytalin salivary amylase.

puberty the age at which the reproductive organs become functional.

purines a class of nitrogenous bases to which belong adenine and guanine.

purulent pus-forming.

pyelitis an inflammation of the pelvis of the kidney.

pyemia a generalized septicemia with pus-forming lesions.

pylorus the opening of the stomach into the duodenum.

pyramidal tracts the cerebral motor fibers extending into the spinal cord.

pyrimidines a class of nitrogenous bases to which belong cytosine, thymine, and uracil.

Q fever a pneumotropic rickettsial infection.

quadriceps four-headed.

quarantine to isolate because of suspected contagion.

rabies a fatal neurotropic virus infection; hydrophobia.

radical a group of two or more elements that act as a unit.

radioactivity the spontaneous emission of alpha, beta, and gamma rays.

ramus a branch.

receptor the peripheral ending of an afferent (sensory) neuron.

reflection the turning back of a light ray from a surface it does not penetrate.

reflex an involuntary reaction to a stimulus.

refraction the bending of a light ray as it passes from one medium into another.

refractory period the short period of reduced irritability following functional activity of a nerve or muscle.

renal pertaining to the kidney.

rennin a milk-coagulating enzyme of the gastric juice.

resonance the intensification of a sound by means of a cavity.

reticuloendothelial system the tissues of the spleen, lymph nodes, bone marrow, and liver engaged in phagocytosis.

retina the nerve coat of the eye.

rheobase the minimal strength of electrical current needed to produce stimulation.

riboflavin one of the vitamins of the B complex.

ribosome cytoplasmic granules composed of protein and ribonucleic acid.

rickettsiae microscopic rod-shaped obligate parasites.

RNA ribonucleic acid.

roentgen the international unit of intensity of X rays or gamma rays.

rubella German measles.

rubeola measles.

rugae folds or wrinkles on the inside of the stomach.

sagittal pertaining to a plane that divides the body into right and left portions.

saprophyte an organism that lives on dead organic matter.

sarcolemma the delicate sheath that invests striated muscle fibers.

saturated compound a compound without double or triple bonds between carbon atoms.

scarlet fever an infection caused by *Streptococcus pyogenes* and marked by generalized erythema.

Schick test a test to determine susceptibility to diphtheria.

Schultz-Charlton test a diagnostic test for scarlet fever.

sclera the outer, white coat of eye.

scrotum the pouch that contains the testes.

scrub typhus a severe rickettsial infection; tsutsugamushi disease.

scurvy a condition caused by lack of vitamin C.

sebum the secretion of the sebaceous glands.

secretin an intestinal hormone that stimulates the pancreas and liver.

seminal vesicle a secretory gland of the male reproductive system.

septic sore throat a severe pharyngitis caused by *Streptococcus pyogenes*.

septicemia the presence and active multiplication of bacteria in the blood.

serum plasma minus fibrinogen.

sign an objective manifestation of disease.

sigmoid S-shaped.

sinus a cavity or recess.

sinusoid a terminal blood channel having a lining of phagocytic cells.

smallpox a severe dermotropic virus infection.

soap a metallic salt of a fatty acid.

soft chancre a venereal disease caused by *Haemophilus ducreyi*.

solute that which is dissolved in the solvent.

solvent the liquid in which another substance is dissolved.

somatic pertaining to the body.

specific gravity the ratio of a given weight of a substance to the weight of an equal volume of water.

spermatozoa male sex cells, sperm.

sphincter a muscle that closes an orifice.

sphygmomanometer an instrument used for taking blood pressure.

spirometer an instrument for determining the volume of air taken into or exhaled from the lungs.

splanchnic visceral.

squamous scalelike.

staphylococci cocci that occur in bunches.

starch a polysaccharide that forms a purple color with iodine.

stasis a stoppage.

stenosis a constriction or narrowing of a channel or opening.

sterilization the killing or removal of all microorganisms and their spores.

sternum the breastbone.

steroid a group name for compounds resembling sterol; pertaining to such compounds.

sterol a complex four-ringed alcohol.

stratum a layer.

streptococci cocci that occur in chains.

stroma the framework of an organ.

stye an inflammation of a sebaceous gland of the eyelid.

subarachnoid space the space between the arachnoid membrane and the pia mater.

subcutaneous under the skin.

substrate a substance on which an enzyme acts.

sucrase an enzyme that converts sucrose to glucose and fructose.

sudoriferous secreting sweat.

sugar a sweet, white, crystalline, water-soluble carbohydrate.

sulcus a groove or fissure.

supination the act of turning the palm upward.

symbiosis the living together of two dissimilar organisms.

sympathetic nervous system that portion of the autonomic system whose fibers originate in the thoracic and lumbar portions of the spinal cord.

symphysis pubis the place where the pubic bones join together.

synapse the junction between the terminal endings of one neuron and the dendrites of another.

synarthrosis an immovable joint.

syndrome a typical set of conditions that characterize a disease.

synovia the fluid found in joint cavities, bursae, and tendon sheaths.

systole the contracting of the heart chambers.

talus the ankle.

tarsus the instep.

temperature the intensity of heat.

tendon a band or cord of white connective tissue that attaches a muscle to a bone or other structure.

testes the male reproductive glands.

testosterone the chief male sex hormone.

tetanus an acute infectious disease caused by a toxin elaborated by *Clostridium tetani.*

tetany a syndrome marked by muscle twitching, cramps, and convulsions; caused by hypocalcemia.

thalamus the middle portion of the diencephalon and main relay center for sensory impulses to the cerebral cortex.

thorax the chest.

thrombocytes blood cells involved in the clotting of blood; also called platelets.

thrombus a blood clot formed within the heart or the blood vessels.

thrush an infection of the mucous membranes of the mouth caused by *Candida albicans.*

thymus an endocrine gland in the upper chest; present in the infant but absent in the adult.

thyroid the endocrine gland located in the neck just below the thyroid cartilage.

thyroxine a hormone of the thyroid gland.

tibia the shin bone.

tidal air the volume of air inspired or expired during quiet breathing.

tissue a group of similar cells.

tonus the partial contraction of muscle.

toxin a poisonous agent of plant or animal origin.

toxoid a neutralized toxin.

tract a bundle of nerve fibers within the central nervous system.

transaminase an enzyme involved in the exchange of amine groups between keto acids and amino acids.

trauma an injury.

tuberculin test a skin test used in the diagnosis of tuberculosis.

tuberculosis an infection caused by *Mycobacterium tuberculosis.*

tularemia an infection caused by *Pasteurella tularensis.*

tunica a coat.

turbinate spiral-shaped.

tympanum the eardrum.

typhoid fever a severe enteric infection caused by *Salmonella typhosa.*

umbilicus the navel.

undulant fever brucellosis.

unsaturated compounds compounds with double or triple bonds between carbon atoms.

urea the chief nitrogenous urinary waste.

ureter the duct that conveys urine from the kidney to the bladder.

urethra the passageway through which urine leaves the bladder.

urethritis an inflammation of the urethra.

urinometer an instrument for determining the specific gravity of the urine.

urochrome a yellow pigment of urine.

urticaria hives.

uterus the pear-shaped muscular organ of the female in which the fetus develops.

uvula the posterior tip of the soft palate.

vaccine a preparation made from a killed or attenuated pathogen.

vagina the passageway from the uterus to the vulva.

valence the combining power of an element.

varicella chicken pox.

vas deferens the duct that carries sperm from the testis to the ejaculatory duct.

vasopressin a hormone of the posterior pituitary that stimulates the absorption of water from the renal tubules.

vastus wide.

vector a quantity that has magnitude and direction, commonly represented by a directed line segment.

vector (animal) an insect or arachnid that transmits infection.

vein a vessel that conveys blood to the heart.

ventral situated toward the front of the body.

ventricle a small cavity.

venule a small vein.

vermiform worm-shaped.

vertebrae the bones of the spinal column.

vertigo dizziness.

vesicle a circular elevation of the skin containing a clear watery fluid.

villi microscopic fingerlike projections of the intestinal mucosa.

Vincent's angina trench mouth.

virus an ultramicroscopic obligate parasite.

viscera internal organs.

viscus an internal organ.

vitamin an essential nutrient present in foods in small amounts.

volar pertaining to palm of hand.

volt the unit of electrical potential.

vulva the female external genitals.

Wassermann test a complement-fixation test formerly used in the diagnosis of syphilis.

water balance the proper distribution of water between the intracellular and extracellular compartments of the body.

wavelength the distance from the top of one wave to the top of the next.

Weil-Felix test an agglutination reaction used to diagnose rickettsial infections.

white matter nerve tissue composed of myelinated nerve fibers.

whooping cough pertussis.

Widal's test an agglutination test originally used in the diagnosis of typhoid fever.

xanthoproteic test a test for protein; a yellow color is produced when concentrated nitric acid is added if protein is present.

xerophthalmia lesions of the cornea of eye due to lack of vitamin A.

xiphoid sword-shaped.

X rays electromagnetic radiations of great penetrating power.

yeasts fungi characterized by large oval cells.

yellow fever a viral infection of the liver transmitted by the *Aedes aegypti* mosquito.

Ziehl-Neelsen stain a method for doing the acid-fast stain.

zona pellucida the transparent layer surrounding the ovum.

zygoma the malar, or cheek, bone.

zygote the fertilized ovum.

zymase the collective term for the enzymes of yeast that convert glucose into alcohol and carbon dioxide.

zymogen an inactive precursor of an enzyme.

REFERENCES

Ackerman, L. V., and del Regato, J. A.: Cancer: diagnosis, treatment, and prognosis, ed. 4, St. Louis, 1970, The C. V. Mosby Co.

Anderson, D. A.: Introduction to microbiology, St. Louis, 1973, The C. V. Mosby Co.

Anderson, W. A. D., and Scotti, T. M.: Synopsis of pathology, ed. 8, St. Louis, 1972, The C. V. Mosby Co.

Anthony, C. P., and Kolthoff, N. J.: Textbook of anatomy and physiology, ed. 8, St. Louis, 1971, The C. V. Mosby Co.

Arey, L. B.: Developmental anatomy—a textbook and laboratory manual of embryology, ed. 7, Philadelphia, 1965, W. B. Saunders Co.

Baker, J. J. W., and Allen, G. E.: Matter, energy, and life, Reading, Mass., 1965, Addison-Wesley Publishing Co., Inc.

Balinsky, B. I.: Introduction to embryology, Philadelphia, 1962, W. B. Saunders Co.

Beaver, W. C., and Noland, G. B.: General biology, ed. 8, St. Louis, 1970, The C. V. Mosby Co.

Beeson, P. B., and McDermott, W.: Cecil-Loeb's textbook of medicine, ed. 13, Philadelphia, 1971, W. B. Saunders Co.

Best, C. H., and Taylor, N. B.: Physiological basis of medical practice, ed. 8, Baltimore, 1966, The Williams & Wilkins Co.

Best, C. H., and Taylor, N. B.: The human body, ed. 5, Baltimore, 1963, The Williams & Wilkins Co.

Bevelander, G.: Essentials of histology, ed. 6, St. Louis, 1970, The C. V. Mosby Co.

Bevelander, G.: Outline of histology, ed. 7, St. Louis, 1971, The C. V. Mosby Co.

Biddle, H. C., and Floutz, V. W.: Chemistry in health and disease, ed. 7, Philadelphia, 1969, F. A. Davis Co.

Bloom, W., and Fawcett, D. W.: A textbook of histology, ed. 9, Philadelphia, 1968, W. B. Saunders Co.

Boyd, W.: An introduction to the study of disease, ed. 6, Philadelphia, 1971, Lea & Febiger.

Brock, T. D.: Biology of microorganisms, ed. 2, Englewood Cliffs, N. J., 1974, Prentice-Hall, Inc.

Brooks, S. M.: A programmed introduction to microbiology, ed. 2, St. Louis, 1973, The C. V. Mosby Co.

Brooks, S. M.: Basic biology: a first course, St. Louis, 1972, The C. V. Mosby Co.

Brooks, S. M.: Basic chemistry: a programmed presentation, ed. 2, St. Louis, 1971, The C. V. Mosby Co.

Brooks, S. M.: Basic facts of body water and ions, ed. 3, New York, 1973, Springer Publishing Co., Inc.

Brooks, S. M.: Basic facts of medical microbiology, ed. 2, Philadelphia, 1962, W. B. Saunders Co.

Brooks, S. M.: Basic facts of pharmacology, ed. 2, Philadelphia, 1963, W. B. Saunders Co.

Brooks, S. M.: Biology simplified, Lincoln, Neb., 1968, Cliff's Notes, Inc.

Brooks, S. M.: Laboratory manual and workbook for integrated basic science, ed. 2, St. Louis, 1971, The C. V. Mosby Co.

Brooks, S. M.: McBurney's point: the story of appendicitis, Cranbury, N. J., 1969, A. S. Barnes & Co., Inc.

Brooks, S. M.: Ptomaine: the story of food poisoning, Cranbury, N. J., 1974, A. S. Barnes & Co., Inc.

Brooks, S. M.: The cancer story, Cranbury, N. J., 1973, A. S. Barnes & Co., Inc.

Brooks, S. M.: The sea inside us, New York, 1968, Hawthorn Books, Inc.

Brooks, S. M.: The V. D. story, Cranbury, N. J., 1972, A. S. Barnes & Co., Inc.

Brooks, S. M.: The world of the viruses, Cranbury, N. J., 1970, A. S. Barnes & Co., Inc.

Brown, W. V., and Bertke, E. M.: Textbook of cytology, St. Louis, 1969, The C. V. Mosby Co.

Burdon, K. L., and Williams, R.: Microbiology, ed. 6, New York, 1968, The Macmillan Co.

Burrows, W.: Textbook of microbiology, ed. 20, Philadelphia, 1970, W. B. Saunders Co.

Cannon, W. B.: The wisdom of the body, rev. ed., New York, 1963, W. W. Norton & Co., Inc.

Chen, P. S.: Chemistry: inorganic, organic, and biological, New York, 1968, Barnes & Noble Books.

Collins, R. D.: Illustrated manual of laboratory diagnosis, Philadelphia, 1968, J. B. Lippincott Co.

Davenport, H. W.: Physiology of the digestive tract, ed. 2, Chicago, 1966, Year Book Medical Publishers, Inc.

De Coursey, R. M.: The human organism, ed. 4, New York, 1974, McGraw-Hill Book Co.

Dorland's illustrated medical dictionary, ed. 25, Philadelphia, 1974, W. B. Saunders Co.

Drobner, R., and Mock, G. V.: Roe's principles of chemistry, ed. 11, St. Louis, 1972, The C. V. Mosby Co.

Dubos, R. J., and Hirsch, J. G., editors: Bacterial and mycotic infections of man, ed. 4, Philadelphia, 1965, J. B. Lippincott Co.

Escourolle, R., and Poirier, J.: Manual of basic neuropathology, Philadelphia, 1973, W. B. Saunders Co.

Eyzguirre, C.: Physiology of the nervous system, Chicago, 1969, Year Book Medical Publishers, Inc.

Faust, E., and Russel, P. F.: Craig and Faust's clinical parasitology, ed. 7, Philadelphia, 1965, Lea & Febiger.

Fawcett, D. W.: An atlas of fine structure: the cell, Philadelphia, 1966, W. B. Saunders Co.

Francis, C. C: Introduction to human anatomy, ed. 6, St. Louis, 1973, The C. V. Mosby Co.

Frankel, S., Reitman, S., and Sonnenwirth, A., editors: Gradwohl's clinical laboratory methods and diagnosis, ed. 7, St. Louis, 1970, The C. V. Mosby Co.

Frisell, W. R.: Acid-base chemistry in medicine, New York, 1968, The Macmillan Co.

Ganong, W. F.: Review of medical physiology, Los Altos, Calif., 1967, Lange Medical Publications.

Gardner, E.: Fundamentals of neurology, ed. 5, Philadelphia, 1968, W. B. Saunders Co.

Goodman, L. S., and Gilman, A.: The pharmacological basis of therapeutics, ed. 4, New York, 1970, The Macmillan Co.

Goss, C. M., editor: Gray's anatomy of the human body, ed. 29, Philadelphia, 1973, Lea & Febiger.

Goth, A.: Medical pharmacology: principles and concepts, ed. 6, St. Louis, 1972, The C. V. Mosby Co.

Grollman, S.: The human body: its structure and function, ed. 3, Riverside, N. J., 1974, Macmillan Publishing Co., Inc.

Guthrie, H. A.: Introductory nutrition, ed. 2, St. Louis, 1971, The C. V. Mosby Co.

Guyton, A. C.: Basic human physiology, Philadelphia, 1971, W. B. Saunders Co.

Guyton, A. C.: Textbook of medical physiology, ed. 4, Philadelphia, 1971, W. B. Saunders Co.

Ham, A. W., and Leeson, T. S.: Histology, ed. 6, Philadelphia, 1969, J. B. Lippincott Co.

Hardin, G.: Biology: its principles and implications, ed. 2, San Francisco, 1966, W. H. Freeman and Co. Publishers.

Harrison, T. R.: Principles of internal medicine, ed. 6, New York, 1970, McGraw-Hill Book Co.

Horsfall, F. L., and Tamm, I., editors: Viral and rickettsial infections of man, ed. 4, Philadelphia, 1965, J. B. Lippincott Co.

Joklik, W. K., and Smith, D. T.: Zinsser microbiology, ed. 15, New York, 1972, Appleton-Century-Crofts.

Kennelly, R., and Neal, E. S. B.: Chemistry: with selected principles of physics, ed. 2, New York, 1971, McGraw-Hill Book Co.

King, B. G., and Showers, M. J.: Human anatomy and physiology, ed. 6, Philadelphia, 1969, W. B. Saunders Co.

King, R. C.: A dictionary of genetics, ed. 2, New York, 1974, Oxford University Press, Inc.

Knight, C. A.: Molecular virology, New York, 1974, McGraw-Hill Book Co.

Krogman, W. M.: The human skeleton, Springfield, Ill., 1962, Charles C Thomas, Publisher.

Levine, L.: Biology of the gene, ed. 2, St. Louis, 1973, The C. V. Mosby Co.

Lisser, H., and Escamilla, R. F.: Atlas of clinical endocrinology, ed. 2, St. Louis, 1962, The C. V. Mosby Co.

Lockhart, R. D., Hamilton, G. F., and Fyfe, F. W.: Anatomy of the human body, Philadelphia, 1965, J. B. Lippincott Co.

Loewy, A. G., and Siekevitz, P.: Cell structure and function, New York, 1963, Holt, Rinehart and Winston, Inc.

Merck manual, ed. 12, Rahway, N. J., 1972, Merck, Sharp and Dohme Research Laboratories.

Miller, M. A., and Leawell, L. C.: Kimber-Gray-Stackpole's anatomy and physiology, ed. 16, New York, 1972, The Macmillan Co.

Moore, C.: Synopsis of clinical cancer, ed. 2, St. Louis, 1970, The C. V. Mosby Co.

Mountcastle, V. B., editor: Medical physiology, ed. 13, St. Louis, 1974, The C. V. Mosby Co.

Orten, J. M., and Neuhaus, O. W.: Biochemistry, ed. 8, St. Louis, 1970, The C. V. Mosby Co.

Pai, A. C.: Foundations of genetics, New York, 1974, McGraw-Hill Book Co.

Patten, B. M.: Foundations of embryology, ed. 3, 1974, New York, McGraw-Hill Book Co.

Pauling, L.: College chemistry, ed. 3, New York, 1964, W. H. Freeman and Co. Publishers.

Pauling, L.: Vitamin C and the common cold, San Francisco, 1970, W. H. Freeman and Co. Publishers.

Pitts, R. F.: Physiology of the kidney and body fluids, Chicago, 1966, Year Book Medical Publishers, Inc.

Ranson, S. W., and Clark, S. L.: The anatomy of the nervous system, ed. 10, Philadelphia, 1969, W. B. Saunders Co.

Rasch, P. J., and Roger, K. B.: Kinesiology and applied anatomy, ed. 2, Philadelphia, Lea & Febiger.

Ray, O. S.: Drugs, society, and human behavior, St. Louis, 1972, The C. V. Mosby Co.

Reisman, L. E., and Matheny, A. P., Jr.: Genetics and counseling in medical practice, St. Louis, 1969, The C. V. Mosby Co.

Rhodin, J. G.: Histology, New York, 1974, Oxford University Press, Inc.

Robbins, S., and Angell, M.: Basic pathology, Philadelphia, 1971, W. B. Saunders Co.

Rogers, T. A.: Elementary human physiology, New York, 1961, John Wiley & Sons, Inc.

Rolle, A.: Discovering the basis of life: an introduction to molecular biology, New York, 1974, McGraw-Hill Book Co.

Routh, J. I., et al.: A brief introduction to general, organic, and biochemistry, Philadelphia, 1971, W. B. Saunders Co.

Sahakian, W. S.: Psychopathology today, Itasca, Ill., 1970, F. E. Peacock Publishers, Inc.

Schottelius, B. A., and Schottelius, D. D.: Textbook of physiology, ed. 17, St. Louis, 1973, The C. V. Mosby Co.

Sleisenger, M. H., and Fordtran, J. S.: Gastrointestinal disease, Philadelphia, 1973, W. B. Saunders Co.

Smith, A. L.: Microbiology and pathology, ed. 10, St. Louis, 1972, The C. V. Mosby Co.

Smith, A. L.: Principles of microbiology, ed. 7, St. Louis, 1973, The C. V. Mosby Co.

Smith, W., editor: Mechanism of virus infections, New York, 1963, Academic Press, Inc.

Spalteholz, W.: Atlas of human anatomy, ed. 16 (revised and re-edited by R. Spanner), New York, 1967, F. A. Davis Co.

Stacey, R. W., and Santolucito, J. A.: Modern college physiology, St. Louis, 1966, The C. V. Mosby Co.

Stanbury, J. B., Wyngarden, J. B., and Fredrickson, D.: The metabolic basis of inherited diseases, ed. 3, New York, 1971, McGraw-Hill Book Co.

Stanier, R. Y., Doudoroff, M., and Adelber, E. A.: The microbial world, ed. 2, Englewood Cliffs, N. J., 1963, Prentice-Hall, Inc.

Stanley, W. M., and Valens, E. G.: Viruses and the nature of life, New York, 1961, E. P. Dutton & Co., Inc.

Stents, G. S.: Molecular biology of viruses, San Francisco, 1963, W. H. Freeman and Co. Publishers.

Swanson, H. D.: Human reproduction, New York, 1974, Oxford University Press, Inc.

Tedeschi, H.: Cell physiology, New York, 1974, Academic Press, Inc.

Tepperman, J.: Metabolic and endocrine physiology, ed. 2, Chicago, 1967, Year Book Medical Publishers, Inc.

Thompson, J. S., and Thompson, M. W.: Genetics in medicine, ed. 2, Philadelphia, 1973, W. B. Saunders Co.

Top, F. H., and Wehrle, P. F., editors: Communicable and infectious diseases, ed. 7, St. Louis, 1972, The C. V. Mosby Co.

Turner, C. D.: General endocrinology, ed. 4, Philadelphia, 1966, W. B. Saunders Co.

Ulett, G.: A synopsis of contemporary psychiatry, ed. 5, St. Louis, 1972, The C. V. Mosby Co.

Volk, W. A., and Wheeler, M. F.: Basic microbiology, ed. 3, Philadelphia, 1973, J. B. Lippincott Co.

Watson, J. D.: The molecular biology of the gene and its role in protein synthesis, New York, 1965, The Benjamin Co., Inc.

White, E. H.: Chemical background for the biological sciences (FMB series), Englewood Cliffs, N. J., 1964, Prentice-Hall, Inc.

Williams, R. H.: Textbook of endocrinology, ed. 4, Philadelphia, 1968, W. B. Saunders Co.

Williams, S. R.: Nutrition and diet therapy, ed. 2, St. Louis, 1973, The C. V. Mosby Co.

Willis, R. A.: Pathology of tumors, ed. 4, New York, 1967, Appleton-Century-Crofts.

Wilson, J. G.: Environment and birth defects, New York, 1974, Academic Press, Inc.

Wohl, M. G., and Goodhart, R. S.: Modern nutrition in health and disease, ed. 4, Philadelphia, 1968, Lea & Febiger.

Woolridge, D. E.: The machinery of the brain, New York, 1963, McGraw-Hill Book Co.

Young, J. A., and Margerison, T.: From molecule to man, New York, 1969, Crown Publishers, Inc.

INDEX

A

Abdomen, 65
Abdominal cavity, 64, 65, *66*, 98
  and serous membrane, 98
  region, *66*
  wall muscles, 158-161
Abduction, 130
Abductor, 157
Abductor pollicis, 166
ABO groups, blood, 210-211
Abortion, 404
Abscess, 116, 117
Acanthocytosis, 279
Acetabulum, 140, 141
Acetazolamide, 359
Acetoacetic acid, 271
Aceto-acetyl-coenzyme A, 271
Acetone, 271
Acetylcholine, 153, 315, 333, 334
Acetyl-coenzyme A, 82, 84, 269, 271
Acid(s), 26, 27-28, 31-32, 34-35, 37, 38,
    39, 50, 81-82, 84, 85, 86, 87
  acetoacetic, 271
  adenylic, 87
  alpha-ketoglutaric, 82, 85
  amino, 68, 74, 85, *86*, 103, 257, 258,
    272, 273, 274, 278, 282, 283,
    287, 377
  arachidonic, 257
  ascorbic, 286
  beta-hydroxybutyric, 271
  binary, 28
  carbonic, 27

Italics indicate an illustration or table.

Acid(s)—cont'd
  deoxyadenylic, 85, 86, 87
  deoxycytidylic, 85
  deoxyguanylic, 85
  deoxyribonucleic, 34, 85, 86
  deoxythymidylic, 85, 86
  fatty, 31-32, 68, 84, 257, 269, 271,
    272, 283, 377, 385
  folic, 203, 287
  folinic, 286
  glutamic, 87
  hydrobromic, 28
  hydrochloric, 27
  keto, 272, 297, 381
  lactic, 81, 82, 147, 268
  Lewis, 36-37
  linoleic, 257
  linolenic, 257
  nicotinic, 287
  nitric, 28
  nitrous, 28
  nucleic, 34-35, 50, 68, 79, 273
  oleic, 257
  palmitic, 257
  pantothenic, *38*, 288-289
  phosphatidic, 272
  pyruvic, 81, 82, 84, 147-148, 268, 269
  stearic, 39, 257
  ternary, 28
  uric, 68, 273, 277, 297
  uridylic, 86
  wastes, 297
Acid fermentation, 81, 82
Acid-base balance, 108-109, 110
  buffers, 109

Acid-base balance—cont'd
  and diarrhea, 110
  and ketone bodies, 271
  and kidneys, 296-297
  and pH, 108
  and respiration, 234-235
  and urinary system, 292
Acidemia, 108
Acid-fast bacteria, 48
Acid-fast staining, 47-48
Acidophilic cells, 369, 370
Acidosis, 377, 381
  compensated, 108
  in diabetes, 108
  and diarrhea, 110
  and ketone bodies, 271
  prevention of, 296, 297
  respiratory, 234
  uncompensated, 108
Acne dermatitis, 122
Acoustic nerve, 306, 363
Acromegaly, 370
Acromion process, 130, 139
ACTH, 270, 272, 274, 371-372, 379
Actin, 146, 147
Actinic dermatitis, 122
Actinobacillus mallei, 117
Actinomycetales, 44, 45
Actinomycetes, 45
Actinomycin D, 301
Action potential, 307, 308
Activator(s), 39
Acute anterior poliomyelitis, 341
Acute coryza, 236
Acute renal failure, 300

Emulsions, defined, 10
Enamel, tooth, 245, 247, 258, 259
Encephalitis, *49,* 57, 340-341
Encephalon, 317
Endemic typhus, cause, 51
Endocarditis, 198
Endocardium, 17
Endocrine glands, 100, *370*
Endocrine system, 69, 100, 369-385
    insulin, 380-382
    pancreas, 380-382
    pituitary gland, 369-373
    prostaglandins, 385
    and testes, 383
    thyroid gland, 373-376
Endoderm, 399
Endogenous hypertriglyceridemia, 279
Endolymph, 362, 366
Endometrium, 392, 395, 397
Endomysium, 157
Endoplasm, 72-73
Endoplasmic reticulum, 72, 73, 74, 85,
    304
Endothelium, 92
Endotoxins, 58
Energy, 2-3, 5-19, 34-35, 36, 68, 74,
    79-80, 82, 84
    and ATP, 79, 82, 268
    body, needs of, 272
    of brain, 16
    and capillarity, 7-8
    and carbohydrates, 79
    cellular, 34-35, 74, 84
    and colloids, 10-11
    and conservation, laws of, 5
    defined, 2
    and density, 7
    and dialysis, 11
    and diffusion, 8-10
    and digestive system, 68
    Einstein's equation, 5
    and electricity, 15-16
    expenditures, table of, 288
    and fats, 270
    and food, 12, 79-80
    and gas laws, 6-7
    and glucocorticoids, 377
    and glucose, 268
    and heat, 11-12
    kinetic, 5, 6, 11
    and light, 12-14

Energy—cont'd
    and liver, 16-17
    and machines, 16-17
    measurement of, 3
    and molecule, 5-6
    and osmosis, 8-10
    and oxidation, 36
    and oxygen debt, 147-148
    potential, 5
    power, defined, 3
    and radiation, 14
    and solutions, 8
    and sound, 15
    and specific gravity, 7
    and states of matter, 6
    and surface tension, 7-8
    types of, 5
    Watt, defined, 3
English system, 1
Entamoeba histolytica, 54, 262
Enterobacter aerogenes, 299
Enterobius vermicularis, 266
Enterocrinin, 252, 384, 385
Enterogastrone, 250, 385
Enterokinase, 258
Enteroviruses, *49,* 50, 264, 342
Entoderm, 399
Environment
    and attack, 93
    and heredity, 419
    and stimuli, 93
Enzymes, 33, 37-39, 67, 74, 79, 81, 82,
    87
    activators, 37, 39
    adenylcyclase, 379
    amino acid oxidases, 272
    and beta oxidation, 271
    and blood, 67
    and body heat, 274
    and carbohydrase, 39
    catechol-o-methyl transferase, 334
    cholinesterase, 334
    classification of, 39
    coenzymes, 38-39
    cytochrome, 273
    digestive, 68, 74, 256, 385
    and drugs, 39
    galactose-1-phosphate uridyltransfer-
      ase, 278
    gastric, 256
    gastric lipase, 249

Enzymes—cont'd
    glucokinase, 270
    hexokinase, 81
    hexosaminidase, 179
    intracellular, 272
    and lipid storage, 279
    and metabolic rate, 274
    and muscle contraction, 273
    pepsin, 249
    and pH, 38
    phosphorylase, 269
    protein and protoplasm, 79
    rennin, 249, 296, 377
    respiratory, 74
    salivary amylase, 247
    and temperature, 38
    in tissue, 37, 38
Eosin, 207
Eosinophil, 205, 207
Eosinophilia, 207
Epicardium, 171
Epicondyles, 139
Epidemic cerebrospinal meningitis, 340
Epidemic meningitis, 45
Epidemic typhus, 51
Epidermis, 112-113
    surface area of, *125*
Epidermophyton, 51
Epididymis, 390
Epigastric region, *66*
Epiglottis, 97, 225
Epilepsy, 327, 336, 419
Epimysium, 157
Epinephrine, 240, 270, 333, 335, 376,
    379
Epiphyseal line, 129
Epiphyses, 128, 129, 130
Epistropheus, 137
Epithelial tissue, 92-94, 95
Epithelium, 92, 98, 112
Equation(s), chemical, 36
Equator and metaphase, 76
Equatorial plane, 79
Equilibrium, 326, 361, 366
    dynamic, 37
Equine encephalitis, *49*
Erection, 390
Erepsin, 258
Erg, 3
Ergocalciferol, 285
Erysipelas, 116

Joint(s), 67
  ankle, 167
  ball-and-socket, 130
  biaxial, 130
  and bone, 129
  cavities, 100
  gliding, 130
  and gout, 277
  hinge, 130
  inflammation of, 141
  pivot, 130
  shoulder, 161
  synovial, 130
  tissue, 96
Joule, 3
Juvenile cells, 207
Juxtaglomerular apparatus, 301

K

Karyokinesis, 75
Karyotype, *416*, 417
Keratin, 112, 114
Keratitis, 359
Keratohyalin, 112
Keratomalacia, 285
Keratoses, 123
Keto acid, 272, 297, 381
Ketone bodies, 271
Ketonic acids, 108
Kidney(s), 292-297
  and acid-base balance, 296-297
  and alkalosis, 109
  artificial, 11
  and blood, 68
  and blood pressure, 184
  and dehydration, 103
  diuresis, 295-296
  and gout, 277
  and hormonal control, 296
  infection, 299
  location of, 65
  nephron, 292
  role of, 68
  stones in, 142
  tests, function of, 297
  transplantation, 300
  and urine formation, 293-295
  and water balance, 105
Kidney failure, acute, 111
Kidney stones, 376
Kilogram, 2

Kinases, 58
Kinesthetic sense, 322
Kinetic energy
  and the human body, 11
  and matter, 6
Kingdom, 43
Klebsiella pneumoniae, 236
Klebs-Loeffler bacillus, 238
Klinefelter's syndrome, 417
Knee(s), *65*, 130, 141
  cap, 127, 141
  and gout, 277
  jerk, 310
  joint, 167
Koch's postulates, 56-57
Koplik's spots, 118
Krause's corpuscle, *115*
Krebs cycle, *80, 81, 82, 84, 85*, 271, 272, 273, 381
Kupffer's cells, 192, 255
Kwashiorkor, 283
Kwell, 121
Kymograph, 180
Kyphosis, 143

L

Labia majora, 392
Labia minora, 392, 394
Labor, 373, 403
Labyrinth of inner ear, 92, 362
Lac operon, 89
Lacrimal apparatus, 351, 352-353
Lacrimal bones, 135
Lacrimal canaliculus, 352, 353
Lacrimal ducts, 348, 352
Lacrimal glands, 348, 352
Lacrimal sacs, 352, 353
Lactase, 257
Lactation, 394
Lacteals, 216, 251, 258
Lactic acid, 81, 82, 147, 268
Lactiferous ducts, 394
Lactobacillus acidophilus, 258
Lactogenic hormone, 370
Lactose, 31, 89, 256, 278
Lacunae, 97
Lamellae, 97
Lamina propria, 250
Landouzy-Déjerine muscular dystrophy, 168
Lansing enterovirus, 342

Larodopa, 337
Laryngopharynx, 224
Larynx, 68, 97, 225-226
Latent period, 151
Lateral canthus, 352
Lateral, defined, 64
Lateral horns, 310
Lateral lemniscus, 323
Lateral malleolus, 141
Lateral rectus, 351, 355
Lateral region, *66*
Lateral ventricles, 328
Latissimus dorsi, 161
Law(s)
  all-or-none, 149-150, 308
  Boyle's, 6, 228
  of capillaries, Starling's, 103, 175, 178
  Charles', 6
  of conservation, 5
  Dalton's, 6, 228
  DuBois-Reymond, 150
  gas, 6-7
  Henry's, 6-7
  Marey's, 183
  of mass action, 5, 37
  Ohm's, 16
  Poiseuille's, 184-185
  of proportions, definite, 20
  of reflection, 13
Lawn, 50
Lead compounds and cancer, 301
Leg, *65*, 141, 167
  nerves supplying, *314*
Length, 1
Lens, 92, 348, 351, 354, 355, 356, 359
Lente insulin, 382
Lenticular nucleus, 320
Lentiform nucleus, 320
Leon enterovirus, 342
Leprosy, 45, 117
Leptospira icterohaemorrhagiae, 45
Lesser omentum, 249
Lesser trochanter, 141
Lethal genes, 415
Lethargic encephalitis, *49*
Leucine, 282
Leukemia, 14, 51, 120, 207-208, 379
Leukocidins, 58
Leukocytes, 128, 205
Leukocytosis, 207
Leukopenia, 207

Protein(s)—cont'd
  and epidermis, 112
  and extracellular fluid, *107*
  and glucocorticoids, 377
  heat-labile, 58
  and hypoproteinemia, 103
  and lipids, 270
  and metabolism, *85,* 270, 272-274,
    377
  muscle, 140
  and nutrition, 282-283
  and RQ, 276
  and sodium, 103
  synthesis of, 74
Protein coat, 49
Proteoses, 257
Proteus spp, 299
Proteus vulgaris, 60
Prothrombin, 208
  formation of, 286
Prothrombin time, 209
Protista, 43, 48, 52
Protons, 21, 22, 24, 35
Protophyta, 48
Protoplasm, 46, 71, 75, 79, 95,
    272
  and chemical information, 79
  chemical processes of, 79
  constituents of, 79
  and molecular backbone, 79
  multinucleate mass of, 95
  and spores, 46
  and vitamins, 287
Protoplasmic projections, 94
Protoporphyrin IX, 273
Protozoa, 43, 52-54
Proximal, defined, 64
Proximal tubules, 295
Pseudomonadales, 44, 45
Pseudomonas aeruginosa, 45, 116-117,
    299, 368
Pseudostratified ciliated columnar epi-
    thelium, 93, 94, 223
Psilocin, 344
Psilocybin, 344
Psittacosis, 238-239
Psoas major, 167
Psoriasis, 122
PSP test, 297
Psyche, disorders of, 343
Psychedelics, 344

Psychiatry, 343
  and limbic system, 327
Psychic stimuli, 247
Psychoanaleptics, 344
Psychodysleptics, 344
Psycholeptics, 344
Psychomotor epilepsy, 336
Psychoneurotic disorders, 343
Psychoses, 343
Psychosomatic disorders, 330, 343
Psychotherapy and insulin shock, 382
Psychotic disorders, 343
Pterygoids, 157
PTT, 209
Ptyalin, 39, 256
Puberty, 122, 382, 392
Pubic arch, 141
Pubic bones, 130, 141
Pubic region, *66,* 387
Pubis, 140
Puff balls, 51
Pulex irritans, 121
Pulmonary artery, 173, 176, 192
Pulmonary circulation, 67, 192
Pulmonary stenosis, 196
Pulmonary units, 226
Pulmonary veins, 172, 176, 192
Pulp, tooth, 258
Pulse pressure, 180
Pulse wave, 180
Punnett square, 418-419
Pupil of eye, 350
  and acetylcholine, 334
  constriction of, 355
  and norepinephrine, 334
  reflex, 316
Purine, 34, 273
Purine nucleotides, 273
Purine synthesis, 277
Purkinje's fibers, 173
Purkinje system, 173-174
Pus, 69, 206, 298, 299
Putamen nucleus, 320
Pyelonephritis, 299
Pyloric sphincter, 249, 250
Pylorus, 249, 250
Pyramid(s), 323
  of kidneys, 292
Pyramidal tracts, 322, 326
Pyridoxine, 287-288, 404
Pyrimidine, 34, 35, 273

Pyrogenicity, 58
Pyrrole pigment derivates, 277
Pyruvic acid, 81, 82, 84, 147-148, 268,
    269
Pytalin, 247
Pyuria, 298

Q

QRS complex, 174, 175
Quadrate lobe, 255
Quadratus lumborum, 158, 159
Quadriceps femoris, 167
Quinacrine, 268
Quintuplets, 399

R

Rabbit fever, 117
Rabies, 57, 59, 342, 343
Racemose ducts, 100
Rad, 14
Radial muscle, 350
Radiation, 14
  and aging, *14*
  alpha rays, 14, *35*
  beta rays, 14, *35*
  and cancer, 14
  gamma rays, *14*
Radioactive iodine, 375
Radioactivity, 35-36, 419
Radioscanning, 339
Radiowaves, 14
Radium, 35
Radius, 130, 139
Ramus(i), *312,* 313, 332
Rana pipiens, 403
Rat-bite fever, 117
Rays
  alpha, 14, *35*
  beta, 14, *35*
  gamma, 14
Reabsorption, 295
  facultative, 296
  obligatory, 296
  and urine, 293
Reaction(s)
  allergen-reagin, 122
  antigen-antibody, 60
  in the cell, 84-85
  chemical, 20-39
  rate of, 37
  reversible, 37

Starch(es)—cont'd
    salivary digestion of, 256
Starling's law, 103, 175
Starvation and fat, 271
Steapsin, 39, 257
Stearic acids, 39, 257
Stearin, 39
Stenosis, 301
    aortic, 196
Stensen's duct, 247
Sterility, 387, 410
Sterilization and spores, 46
Sternocleidomastoid, 158, 161
Sternum, 127, 128, 138-139, 228
Steroids, 32, 382
Sterols, 32, 257, 270
Stethoscope, 174
Stimulants, 326
Stimulation
    and contraction, 149-153
    cortical, 330
    electrical, of vagus nerve, 332
    parasympathetic, 332-333, 334
    repetitive, 151-153
Stimulus(i), 93, 149-151, 307-309, 315
Stirrup, 135, 361, 365
St. Louis encephalitis, 49, 341
Stoma, 392
Stomach, 65, 68, 248, 249-250, 257
    cancer, 260, 261
Stones, kidney, 142, 300
Stratification, 94
Stratified squamous epithelium, 92-93,
    98
Stratum basale, 112
Stratum corneum, 112, 114
Stratum germinativum, 112
Stratum granulosum, 112
Stratum lucidum, 112, 114
Stratum spinosum, 112
Streptococcal infections, 45, 116, 142
Streptococcus, 45, 58
Streptococcus faecalis, 299
Streptococcus hemolyticus, 116
Streptococcus pyogenes, 45, 116, 142,
    300, 340
Streptococcus viridans, 198
Streptodornase, 58
Streptokinase, 58
Streptomyces, 45
Streptomyces griseus, 45

Streptomycin, 45, 47, 237
Stress
    and ACTH, 377
    and body water, 372
    and bone, 129, 333
Stress response, 377
Stretch receptors, 153, 231
Stretch reflexes, 316
Stretching and contraction, 153
Striations, 94
Striped muscle, 94
Stroke, 337-338
Stroke volume, 180
Strongyloides stercoralis, 266
Stupor, 336
Sty, 116, 359
Styloid, 132
Subacute bacterial endocarditis, 198
Subarachnoid cisterna, 328
Subarachnoid space, 311, 328
Subcutaneous nodules, 277
Subcutaneous tissue, 96, 113-114, 115
Subdural space, 327
Subliminal stimulus, 149, 151
Subliminal summation, 309
Sublingual gland, 247
Submaxillary gland, 247
Submaximal stimulus, 149
Substantia propria, 350
Succinic acid, 82, 84
Succinyl-coenzyme A, 82
Succus entericus, 252
Sucrase, 257
Sucrose, 31, 39, 256
Sudoriferous glands, 114
Sugar(s)
    blood, 327
    disaccharides, 31, 34
    and fat, 269
    and insulin, 89
    in perspiration, 106
    in urine, 69, 295
Sulci, 112, 318
Surfactants, 228
Sulfate(s)
    and ionic milliequivalent, 106
    and protoplasm, 79
Sulfuric acid, 297
Summation, 151, 152, 309
Sunburn, 122
Superficial fascia, 113

Superior, defined, 64
Superior oblique, 351
Superior rectus, 351
Superior sagittal sinus, 327
Superior vena cava, 174
Supinators, 157, 162
Supramaximal stimulus, 149
Supraspinatus, 162
Surface tension, 7-8
Suspension, 10
Suspensory ligament, 351
Sutures
    bone, 129-130
    cranium, 132
Sweat glands, 114, 352
Sympathectomy, 333
Sympathetic nervous system, 67, 330,
    333
Sympatholytics, 336
Sympathomimetics, 335
Symphysis pubis, 141, 392
    composition of, 97
Symptoms, treatment of, 62
Synapse, 315-316
Synapsis, 77
Synaptic relays, 315
Synarthroses, 129-130
Synchondrosis, 130
Syncytium, 71, 95
Syndrome
    Cushing's, 143
    low salt, 110-111
Synergist, 157
Synovial fluid, 100, 130
Synovial joint, 130
Synovial membrane, 98, 100, 130
Synthesis, 68
Synthetases, 39
Synthroid, 375
Syphilis, 45, 311, 405-410
    stages of, 406
System(s)
    autonomic nervous, 67, 69, 329-336
    binomial, 43
    body, human, 65-69
    circulatory, 60, 67, 87, 171-199, 201-
        213, 401-402
    digestive, 68, 244-279
    endocrine, 69, 369-385
    limbic, 326-327
    lymphatic, 67-68, 216-222

tRNA, 87
Trochlea, 139
Trombicula irritans, 121
Trophectoderm, 397
Trophoblast, 397
Trophozoite, 262, 263
Tropic hormone, 369
Tropomyosin and muscle, 146
Troponin and muscle, 146, 147
True bacteria, 44
True vocal cords, 225
Truncal obesity, 379
Trunk
  arteries of, *187*
  bones of, 132
  muscles of, 158
  veins of, *189*
Trypanosoma, 54
Trypsin, 38, 256, 258
Trypsinogen, 258
Tryptophan, 282, 287
Tubercle, 127
Tubercle bacillus, 237
Tuberculin test, 237
Tuberculosis, 45, 143-144, 236-238
Tuberosity, 127
Tubes, uterine, 94, 392
Tubular ducts, 100
Tubular necrosis, 300
Tubules
  distal, 296
  nephron, 295
  renal, 295
Tularemia, 45, 117
Tumor
  of acidophilic cells, 370
  of adrenal cortex, 378
  angioma, 359
  of anterior pituitary, 383
  benign, 51, 123
  of bone, 143
  brain, 338-339
  cancer, 51
  and cerebrospinal fluid, 311
  chromaffin cell, 379
  eye, 359
  intracranial, 338
  kidney, 301
  of lymphatic tissue, 221-222
  malignant, 123
  melanoma, 359

Tumor—cont'd
  myoma, 359
  neurofibroma, 359
  ovarian, 383
  pancreatic, 380, 382
  of parathyroid gland, 376
  of pituitary gland, 378
  retinoblastoma, 359
  spinal cord, 338-339
  thymic, 168
  Wilms', 301
Tunica externa, 176
Tunica intima, 176
Tunica media, 176
Turbinates, 135, 223
Turner's syndrome, 417
Tweenbrain, 317, 323
Twinning, 399
Twitch, muscle, *150,* 151
Tympanum, 361, 362, 365
Typhoid fever, 45, 60, 262
Typhoid Mary, 57
Typhus fever, 51, 120
Typing, blood, *211,* 212
Tyrosine, 278

U

UCG test, 404
Ulcer(s), 219
  peptic, 259
Ulcerative colitis, 259
Ulna, 130, 139
Ultraviolet light, 123
Ultraviolet rays, 14
  and rickets, 142
Umbilical arteries, 399
Umbilical cord, 69, 397, 403
Umbilical region, *66*
Umbilical vein, 399
Unconsciousness, 336
  and insulin overdose, 327
Unicellular glands, 93
United States customary system, 1
Universal donor, 211
Universal recipient, 211
Uracil, 34, 86, 87
  and pyrimidine bases, 34
  and RNA, 34
Uranium, 35
Urate deposits, 277
Urea, 68, 105, 114, 272, 273, 298

Urease, 38
Urecholine, 335
Uremia, 300
Ureteritis, 299
Ureters, 65, 69, 292, 297
Urethra, 297, 387, 389
Uric acid, 68, 273, 277, 297
Uridylic acid, 86
Urinary system, 45, 68-69, 93, 292-301
  acute renal failure, 300
  bladder, 297
  calculi, 300-301
  cancer in, 301
  diuresis, 295
  glomerulonephritis, 299-300
  infections, 299
  juxtaglomerular apparatus, 296, 301
  kidneys, 292-297
  micturition, 297
  nephrotic syndrome, 300
  obligatory reabsorption, 295
  Pseudomonas aeruginosa, 117
  renal hypertension, 301
  threshold value, 295
  ureters, 297
  urethra, 297
  urine, 68-69, 184, 298-299, 389
    bacteria in, 299
    blood in, 69, 298
    and dehydration, 103
    in diagnosis, 62
    and flukes, 56
    formation of, 293-295
    and sodium, 103
Urine; *see* Urinary system
Urinometer, *299*
Urochrome, 298
Uterine tubes, 94, 390, 392
Uterus, 94, 160, 373, 383, 392
  posterior view, *391*
Utricle, 362, 363, 364, 365, 366
Uvula, 244

V

Vaccine
  attenuated, 118
  BCG, 238
  measles, 118
  oral, 342
  polio, 342
  preparation of, 50, 59